Introduction to

Electromagnetic Theory

Introduction to Electromagnetic Theory

George E. Owen

Professor of Physics
The Johns Hopkins University

Dover Publications, Inc.
Mineola, New York

Illustrations by the Author

Bibliographical Note

This Dover edition, first published in 2003, is an unabridged republication of the work originally published by Allyn and Bacon, Inc., Boston, in 1963.

Library of Congress Cataloging-in-Publication Data

Owen, George E. (George Ernest), 1922-
Introduction to electromagnetic theory / George E. Owen ; [illustrations by the author].
p. cm.
Originally published: Boston : Allyn and Bacon, 1963.
Includes index.
ISBN 0-486-42830-3 (pbk.)
1. Electricity. 2. Electromagnetism. I. Title.

QC523.O9 2003
537–dc21 2003043801

Manufactured in the United States of America
Dover Publications, Inc., 31 East 2nd Street, Mineola, N.Y. 11501

To my son oğlum Stephen

Preface

Although the topics contained in the subject of electricity and magnetism are standard, this text will present some fresh viewpoints and some original material.

This manuscript was developed over a period of ten years from my course for undergraduates and first-year graduate students. The results of this course were first presented in a set of bound notes in 1956. The kind reception of these notes led me to revise and rewrite them until the text presented here was a reality.

Among the variations from the standard treatment is a development of the electric displacement field, incorporating contributions from quadrupole distributions and higher moments. Also, I have found that the use of the Dirac delta function enables the intermediate student to gain a much greater insight into the relation between point configurations and continuous distributions. In fact, by introducing the delta function with the development of the point charge potentials and fields, I have found that students are able to gain physical interpretations of both the delta function and later the Green's function. The gross features of these functions are grasped more readily in some respects than the solutions of partial differential equations which have become a standard part of a course at this level.

As far as possible, exact solutions are presented along with various far field and near field approximations. In particular, the fields of circular configurations are discussed in detail. This necessitates the introduction of the complete elliptic functions. To allow for gaps in the student's background these functions are described briefly in an appendix.

Seven appendices have been included as the final portion of this book to allow the reader a brief review of elementary differential geometry, matrices, elliptic functions, partial differential equations, Fourier series, conformal transformations, and physical constants. Perhaps the most obvious departure from the standard treatment lies in the use of matrices and matrix notation. I would compare the usual tendency to avoid matrix notation to an attitude prevalent in the thirties which neglected the more convenient vector notation. We now accept vector methods as an absolute necessity.

Matrix methods make the presentation more powerful and often simpler. Whenever the discussion could benefit, the matrix representation has been employed.

To note further some recent additions, the vector harmonics are developed and employed to describe the radiation field.

By and large the text portion alone would appear to be highly formal and to require a fair mathematical background. The book *must* be used in conjunction with the appendices. With the appendices the discussions are complete, and the reader should not find the level too advanced.

It was my intention to leave many of the applications and so-called physical discussions as problems. As a result the problems are essential to the work. For instance, applications of the formalism to such items as Brewster's law, total reflection, etc., appear in a systematic fashion in the problems.

Because the scope of this text was intended for both undergraduate and first-year graduate students, courses at the intermediate level may well omit the more advanced portions of the book. The ordering of topics is appropriate for such deletions.

I wish to thank my colleagues Professor Feldman, Professor Fulton, Professor Kerr, Professor Pevsner, Professor Madansky and Professor Franco Rasetti for their comments and suggestions.

This has been an enjoyable endeavor, and I trust that the reader will gain some new insights from this effort.

Finally, my deepest regards are extended to Mrs. Dencie Kent, who typed this manuscript so many times.

George E. Owen

Contents

Introduction to Electromagnetic Theory

Notation

Vectors which are designated by a symbol such as $\vec{A}$ in the illustrations will be represented by bold face letters such as **A** in the text.

Matrices will be represented by open face letters such as $\mathbb{S}$ in the text. The elements of a matrix are shown as symbols having two subscripts.

I. Preliminaries

Force, kinematics, and mass form the foundation of mechanics. The proper definition of the mathematical objects called force and mass is involved and in some instances is controversial. Because the theory of electricity and magnetism utilizes the concept of forces and displacements of charged bodies, it is essential that some elements of mechanics be incorporated in this development. Under the assumption that the definitions of force and mass have been established, our major interest in this section will be to review the definitions of work, potential energy, and the conservative force. Appendix A contains a brief review of vector analysis using generalized coordinates; hereafter the vector and coordinate notation can be referred to in that Appendix.

A. WORK

Assume that a point particle, m, moves along a path labeled C. If at every point on the path the particle is acted upon by a force $\mathbf{F}(x,y,z)$, then the work to move the particle from point A to point B is DEFINED as

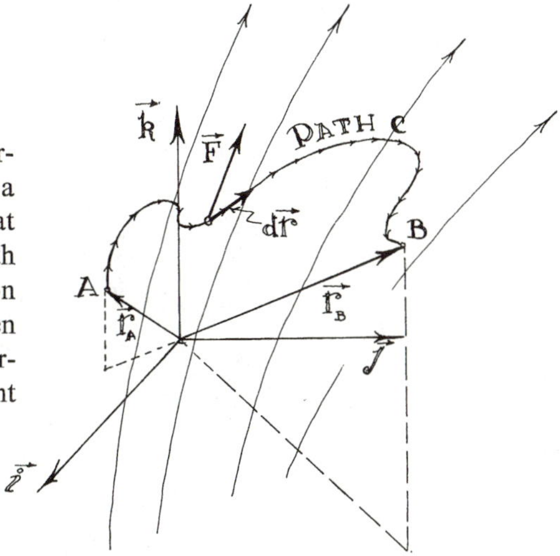

$$W_{A\to B} = \int_{A \atop \text{(path C)}}^{B} \mathbf{F}\cdot d\mathbf{r}.$$

The force $\mathbf{F}(x,y,z)$ is a vector point function,

$$\mathbf{F}(x,y,z) = F_x(x,y,z)\mathbf{i} + F_y(x,y,z)\mathbf{j} + F_z(x,y,z)\mathbf{k},$$

and

$$d\mathbf{r} = dx\,\mathbf{i} + dy\,\mathbf{j} + dz\,\mathbf{k}.$$

Considering only those vectors which are independent of time,

$$W_{A\to B} = \int_{\substack{A \\ (\text{path C})}}^{B} F_x(x,y,z)\,dx + F_y(x,y,z)\,dy + F_z(x,y,z)\,dz.$$

In generalized orthogonal coordinates ξ_m,

$$W_{A\to B} = \int_{\substack{A \\ (\text{path C})}}^{B} \sum_{j=1}^{3} F_j(\xi_1,\xi_2,\xi_3)h_j\,d\xi_j.$$

To evaluate the line integral an equation of constraint (representing the specified path) must be utilized. The space curve connecting A and B can be written

$$f(x) = g(y) = h(z).$$

Actually this equation of constraint can be considered to be three equations giving the projections of the path or space curve on the three mutually orthogonal planes xy, yz, and zx. By substituting the equations of constraint into $W_{A\to B}$, the integral is evaluated. In other words, to evaluate $W_{A\to B}$ we reduce each element of the integrand to a function of the single variable corresponding to the associated differential. If $f(x) = g(y) = h(z)$ can be solved to give

$$y = \phi(x) \quad \text{and} \quad z = \zeta(x),$$

then for instance

$$F_x(x,y,z)\,dx = F_x(x,\phi(x),\zeta(x))\,dx.$$

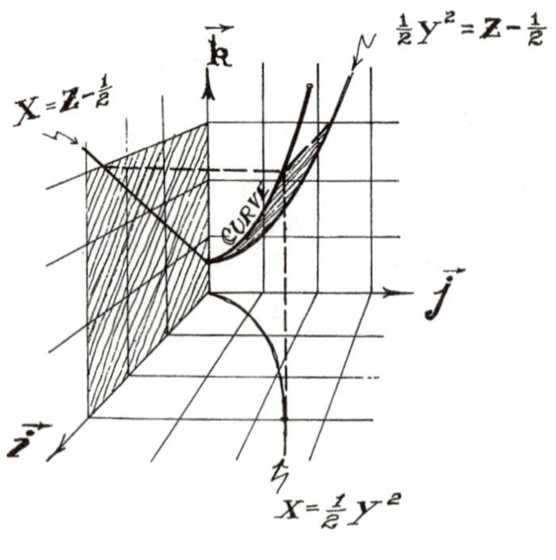

To illustrate this consider the following example. Let

$$\mathbf{F} = xy\,\mathbf{i} + z\,\mathbf{j} + \mathbf{k},$$

(where the appropriate units are implied)
and let the path be represented by

$$x = \tfrac{1}{2}y^2 = z - \tfrac{1}{2},$$

starting at $A = (0,0,\frac{1}{2})$ and terminating at $B = (3,\sqrt{6},3\frac{1}{2})$. Now evaluating,

$$W_{A\to B} = \int_{0,0,\frac{1}{2}}^{3,\sqrt{6},3\frac{1}{2}} [xy\,dx + z\,dy + dz]$$

$$= \int_0^3 \sqrt{2}(x)^{3/2}\,dx + \int_0^{\sqrt{6}} \tfrac{1}{2}(1 + y^2)\,dy + \int_{\frac{1}{2}}^{3\frac{1}{2}} dz;$$

$$W_{A\to B} = \frac{51\sqrt{6} + 35}{10}.$$

The force vector must have a constant befoıe each component carrying the appropriate units; such as newtons/m² for the first component. Then the units of W will be newton meters.

B. THE CONSERVATIVE FORCE

Under certain conditions the LINE INTEGRAL is independent of the path of integration, depending only upon the coordinates of the end points. If $\mathbf{F} \cdot \mathbf{dr}$ is an EXACT differential then it is equal to the total differential of some scalar point function V(x,y,z).

$$\mathbf{F} \cdot \mathbf{dr} = F_x(x,y,z)\,dx + F_y(x,y,z)\,dy + F_z(x,y,z)\,dz = -dV(x,y,z)$$

or in more general terms

$$\mathbf{F} \cdot \mathbf{dr} = \sum_{m=1}^{3} F_m(\xi_1,\xi_2,\xi_3)h_m\,d\xi_m.$$

The components of the vector $\mathbf{F}$ under such circumstances are directly related to the derivatives of the scalar function V, because

$$\mathbf{F} \cdot \mathbf{dr} = -dV = F_x(x,y,z)\,dx + F_y(x,y,z)\,dy + F_z(x,y,z)\,dz$$

then

$$-dV = -\frac{\partial V}{\partial x}dx - \frac{\partial V}{\partial y}dy - \frac{\partial V}{\partial z}dz,$$

and
$$F_x(x,y,z) = -\frac{\partial V}{\partial x}$$
$$F_y(x,y,z) = -\frac{\partial V}{\partial y}$$
$$F_z(x,y,z) = -\frac{\partial V}{\partial z}\,.$$

This statement of course relates the vector **F** and the gradient of V:

$$-dV = -\text{grad } V \cdot \mathbf{dr} = \mathbf{F} \cdot \mathbf{dr}$$

whence
$$\mathbf{F} = -\text{grad } V,$$

which is a statement of the exact differential relationship.

In generalized coordinates

$$F_m(\xi_1,\xi_2,\xi_3) = -\frac{1}{h_m}\frac{\partial V}{\partial \xi_m}(\xi_1,\xi_2,\xi_3).$$

If we now examine the integral under these conditions, we find

$$\int_A^B \mathbf{F} \cdot \mathbf{dr} = -\int_A^B dV = -V(x_B,y_B,z_B) + V(x_A,y_A,z_A).$$

This expansion demonstrates that the integral is independent of the particular path chosen between A and B.

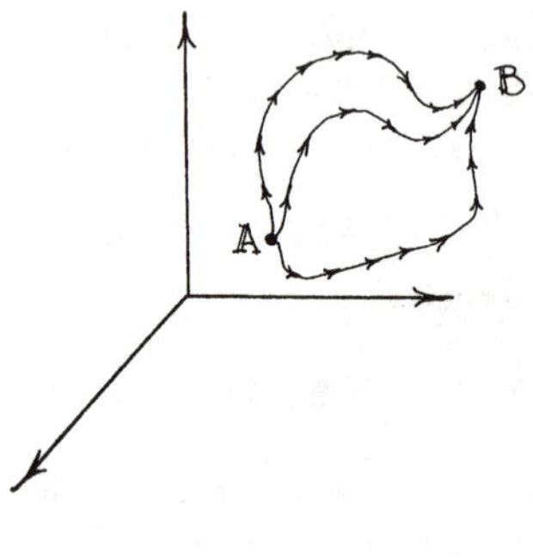

The NEGATIVE sign assigned to the relation between **F** and grad V is a convention. As we shall observe, this choice of sign defines the total energy of a system as the sum of the kinetic and potential energies. The opposite sign convention would require that the total energy to be the difference of these two constituents.

Forces which have the properties just derived are called CONSERVATIVE. A conservative force must satisfy two conditions.

(1) The vector point function representing the force must be generated from a scalar point function by

$$\mathbf{F}(\xi_1,\xi_2,\xi_3) = -\text{grad } V(\xi_1,\xi_2,\xi_3);$$

(2) The force **F** must not be an explicit function of time,

$$\mathbf{F} \neq \mathbf{F}(\xi_1,\xi_2,\xi_3;t).$$

If these two conditions are satisfied, the conservative property is concisely stated by the zero-value closed path integral

$$\oint_{\substack{\text{closed}\\\text{path}}} \mathbf{F} \cdot d\mathbf{r} = 0.$$

The time independence of **F** is implied if we assume that the integration over the closed path is performed in a nonzero time interval. Finally we should remark that the vanishing of the curl of **F** (see Appendix A) is a sufficient condition for the closed path integral to be zero (assuming that curl **F** is defined at every point on S). Stokes' theorem provides a proof of this statement. If

$$\text{curl } \mathbf{F} = 0$$

then

$$\iint_{\text{cap surface S}} (\text{curl } \mathbf{F}) \cdot \mathbf{n} \, dS = \oint_{\substack{\text{closed path}\\\text{bounding S}}} \mathbf{F} \cdot d\mathbf{r} = 0.$$

The curl of **F** is a measure of the circulation of **F**.

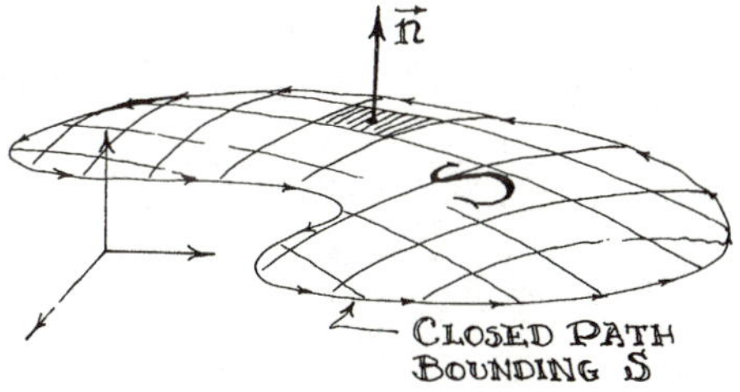

One can demonstrate further that when

$$\mathbf{F} = -\text{grad}\, V = -\sum_{m=1}^{3} \frac{1}{h_m} \frac{\partial V}{\partial \xi_m},$$

then

$$(\text{curl}\, \mathbf{F})_k = \frac{1}{h_l h_m} \left\{ \frac{\partial}{\partial \xi_l} \left(\frac{h_m}{h_m} \frac{\partial V}{\partial \xi_m} \right) - \frac{\partial}{\partial \xi_m} \left(\frac{h_l}{h_l} \frac{\partial V}{\partial \xi_l} \right) \right\} = 0.$$

C. CONSERVATION OF ENERGY

For simplicity of development consider the motion of a point mass m in a force field $\mathbf{F} = -\text{grad}\, V$. By Newton's second law,

$$\mathbf{F} = \frac{d}{dt} \mathbf{p} = \frac{d}{dt} \left\{ m \frac{d\mathbf{r}}{dt} \right\},$$

where **F** is the force acting upon a point mass moving with a velocity $\mathbf{v} = d\mathbf{r}/dt$. The mechanical momentum* is designated by

$$\mathbf{p} = m \frac{d\mathbf{r}}{dt}.$$

The first quadrature of the second law is obtained by taking the

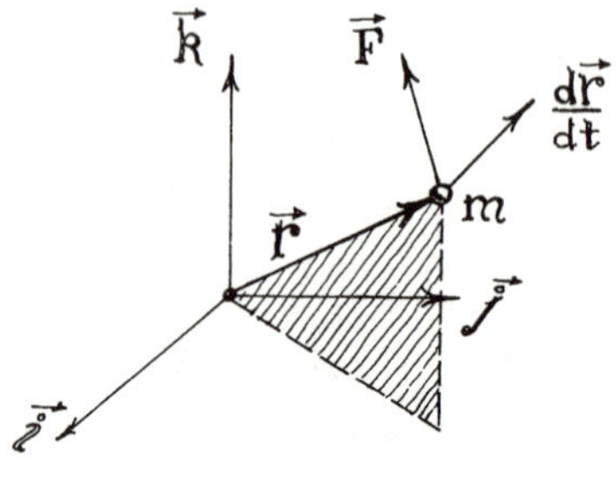

scalar product of $d\mathbf{r}/dt = \dot{\mathbf{r}}$ with the equation in question†:

$$\dot{\mathbf{r}} \cdot \mathbf{F} = \dot{\mathbf{r}} \cdot \frac{d}{dt}(m\dot{\mathbf{r}}) = \frac{d}{dt}(\tfrac{1}{2} m \dot{\mathbf{r}}^2) = \frac{d}{dt}(T).$$

* Later we shall see that a charged particle can be assigned a conjugate momentum which is a somewhat modified form of **p**.

† The symbols $d\mathbf{r}/dt$ and $\dot{\mathbf{r}}$ both represent the velocity vector and will be used interchangeably.

T by definition is the KINETIC ENERGY, $\frac{1}{2}m(\dot{\mathbf{r}})^2$.
The power dissipated is defined as

$$P = \dot{\mathbf{r}} \cdot \mathbf{F} = \frac{dT}{dt}.$$

If **F** is conservative, the power equation reduces to the equation for the conservation of energy: when

$$\mathbf{F} = -\text{grad } V,$$

then
$$\mathbf{F} \cdot \frac{d\mathbf{r}}{dt} = -(\text{grad } V) \cdot \frac{d\mathbf{r}}{dt} = -\frac{dV}{dt};$$

thus under these conditions

$$\frac{dT}{dt} + \frac{dV}{dt} = 0,$$

and
$$T + V = E = \text{a constant of the motion.}$$

$E = T + V$ is called the total energy of the system.

When the conservative force was defined we stipulated that **F** be independent of time. The consequences of having a time-dependent potential demonstrate vividly the reasons for this restriction. If we allow a force field **F** to be generated from a time-dependent scalar point function U(x,y,z;t), then

$$\mathbf{F}(x,y,z;t) = -\text{grad } U(x,y,z;t)$$

and
$$\frac{dT}{dt} = \mathbf{F} \cdot \frac{d\mathbf{r}}{dt} = -(\text{grad } U) \cdot \frac{d\mathbf{r}}{dt}.$$

When U is an explicit function of t we find an extra term in the expansion of the total derivative,

$$\frac{dU}{dt} = (\text{grad } U) \cdot \frac{d\mathbf{r}}{dt} + \frac{\partial U}{\partial t}.$$

Therefore
$$\frac{dT}{dt} = -\frac{dU}{dt} + \frac{\partial U}{\partial t},$$

and
$$\frac{dE}{dt} = \frac{d}{dt}(T + U) = \frac{\partial U}{\partial t}.$$

The total energy E under such circumstances is not constant in time.

II. The Electrostatics of Point Charges

A. DEFINITION OF CHARGE

In the natural sciences it is often difficult to prepare the definitions of the terms which are to be introduced. Strangely enough, sometimes the very basis of a science (and by this we mean the initial definitions) is a region of controversy. In the field of analytical dynamics one must proceed quite carefully in the manner of introducing the definitions of mass and force. In all cases we postulate that observers are endowed with spatial and temporal intuition. This is to say, we assume that observers can detect and measure the displacement of material bodies. In addition we assume that they can detect displacement in time; in particular, that they are endowed with the ability to discern, of two events a and b, whether event a occurs before or after event b.

Such assumptions remove the concepts of displacement, velocity, and acceleration from the realm of philosophical argument. However, even when we have established our right to space and time measurement, still greater obstacles can be encountered in attempting to define force and mass. There is a danger in this instance that two undefined quantities will be defined one in terms of the other.

The main objective is to establish a one-to-one correspondence between our observations of natural phenomena and the representation of these observations in terms of certain mathematical objects. There are several ways in which the correspondence of the objects in the mathematical representation and the observations can be presented. From a practical viewpoint any one of these approaches is sufficient for our purposes. A standard method consists of postulating the basic equations and observables of electromagnetic theory.

One then proceeds to demonstrate that the postulated quantities are consistent with all natural observations. The operational definition represents a different approach. All quantities such as mass, charge, etc., are defined in terms of a set of experimental observations. In the end quantities have reality in terms of these recipes for definition.

In order to introduce a definition of charge we will assume that displacement, time, force, and mass are quantities already defined. A body is said to be charged or to possess charge when it exhibits forces characteristic of a charging process. We now must explain what is meant by a charging process.

(1) A primitive charging process is an operation such as rubbing glass or amber with fur, silk, or other materials, which cause the bodies involved to exhibit forces attractive or repulsive. These forces cannot be attributed to gravitational forces.

(2) A more sophisticated charging operation involves the use of a battery. For instance a simple battery can be constructed by placing a Zn or Mg rod in water. A number of ions

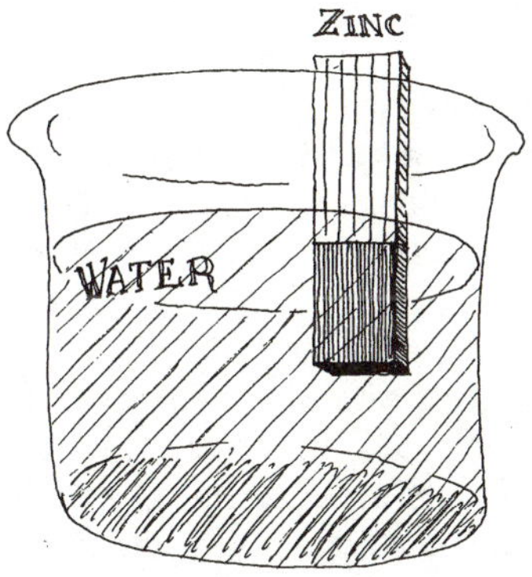

separate from the metallic terminal and leave it charged or in a state of electrification. External bodies can now be charged by bringing them into contact with the terminal, and after charging will show the forces characteristic of this process.

The so-called forces characteristic of the charging process can be detected in a number of ways. If several bodies such as pithballs, suspended by insulating strings from a common point, are charged, when brought together they will assume positions of equilibrium consistent with the electrical forces. It is possible to take one charged body and use it to place (induce) charge on an electroscope. The

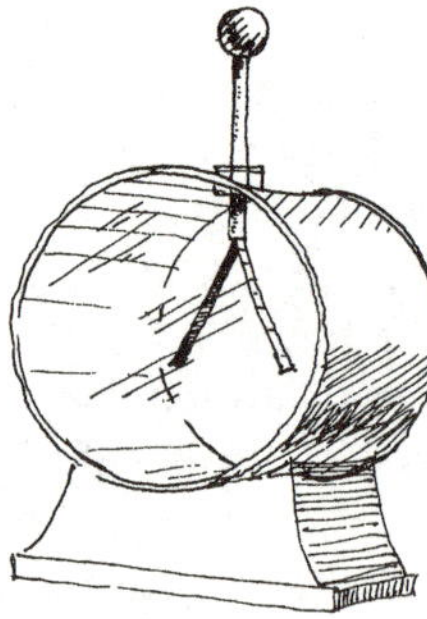

displacement of the leaves of the instrument then indicates the presence of charge.

There are but two kinds of electrical charge, positive and negative. Verification of this statement lies in experimental evidence, that

the mutual attraction of three or more
charged bodies has never been observed.

Experimental evidence also indicates that a lower limit of electrical charge exists. This discrete lower limit is the magnitude of the electron or proton charge.

The unit of charge which will be used in this work is the coulomb. The standard coulomb can be specified from Coulomb's law, from the faraday, or indirectly from the standard ampere.

Present day practice utilizes the standard ampere as a means of measuring the standard coulomb. This technique is by far the most accurate, uniform, and convenient. The unit ampere corresponds to the rate of flow of one coulomb per second across a specified boundary. Because the ampere and coulomb are used as standards, the units to be employed throughout the discussions to follow will be the rationalized MKS units. This procedure also has become standard. Since lengthy discussions of units tend to distract rather than enlighten, the comparison of units will be confined to a table in Appendix G at the termination of this study.

B. COULOMB'S LAW

The forces characteristic of charged bodies have been alluded to previously. By and large the forces set up between extended charged bodies can have a relatively complicated dependence upon the laboratory coordinates. Fortunately the force fields set up by extended charged bodies can be understood by a linear superposition of the fields of all of the infinitesimal volumes of charge which make up the body in question. There is no a priori reason why the total field should be obtainable by linear superposition; however, experimental observation demonstrates the reliability of this principle.

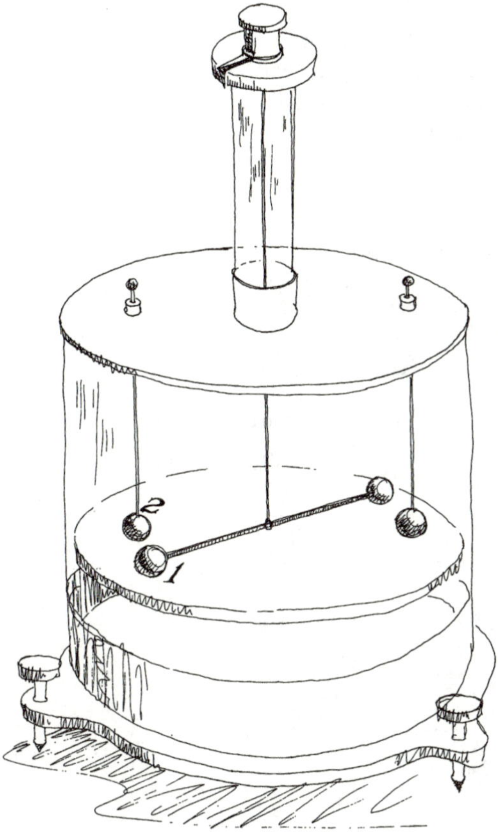

Because the fields can be constructed essentially by superposing fields of point charges, we shall begin with a study of the forces between point charges.

A true point charge of course may not be realized in an actual experiment. The forces between point charges, however, can be indicated in the limit, if a series of force experiments are performed with a set of charged bodies, a smaller body being used in each succeeding experiment. Alternatively, the forces can be determined by use of a torsion balance (page 11). By charging the two bodies employed to different degrees one can establish quite quickly that the magnitude of the force $\mathbf{F}_{12}$ between two very small charged bodies 1 and 2 is proportional to the PRODUCT of the charges,

$$|\mathbf{F}_{12}| \propto q_1 q_2.$$

Next one can demonstrate that the force between charged bodies is a function of the distance between them. In fact, in the limit as the dimensions of the bodies become very much smaller than their separation $\mathbf{r}_{12}$, the force depends upon the inverse square of the separation:

$$|\mathbf{F}_{12}| \propto \frac{1}{|\mathbf{r}_{12}|^2}.$$

Further experiments would indicate that the force between two point charges q_1 and q_2 has the direction of the line of centers connecting the two points:

$$\mathbf{F}_{12} \propto kq_1q_2 \frac{\mathbf{r}_{12}}{|\mathbf{r}_{12}|^3} = kq_1q_2 \frac{(\mathbf{r}_1 - \mathbf{r}_2)}{|\mathbf{r}_1 - \mathbf{r}_2|^3};$$

where $\mathbf{r}_1$ = the position vector of q_1

$\mathbf{r}_2$ = the position vector of q_2.

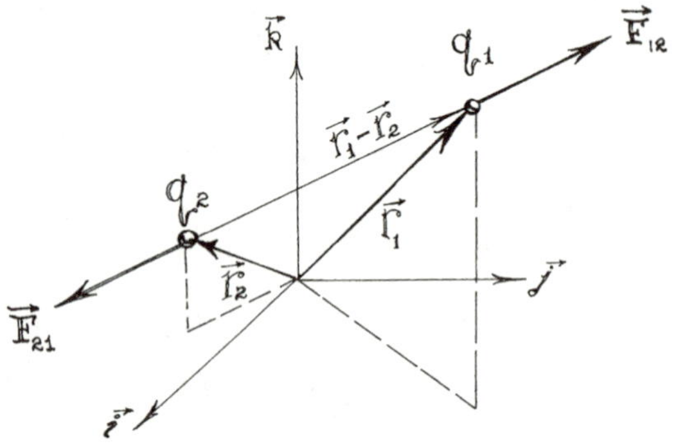

The constant of proportionality in rationalized mks units is determined from the velocity of light and the magnetic permeability of free space. The reasons for these dependences will be seen in later sections (see Chapters X and XI). The constant of proportionality is

$$k = 10^9 \text{ meters/farad,}$$

where k, written in terms of the PERMITTIVITY of free space ϵ_0, is

$$k = \frac{1}{4\pi\epsilon_0}.$$

ϵ_0, the permittivity of free space, is $8.85(10)^{-12}$ farad/meter. Thus

$$\mathbf{F}_{12} = \frac{q_1 q_2}{4\pi\epsilon_0} \frac{(\mathbf{r}_1 - \mathbf{r}_2)}{|\mathbf{r}_1 - \mathbf{r}_2|^3},$$

or

$$\mathbf{F}_{12} = \frac{q_1 q_2}{4\pi\epsilon_0} \frac{(x_1 - x_2)\mathbf{i} + (y_1 - y_2)\mathbf{j} + (z_1 - z_2)\mathbf{k}}{\{(x_1 - x_2)^2 + (y_1 - y_2)^2 + (z_1 - z_2)^2\}^{3/2}}.$$

In these units we should find a force of $9(10)^9$ newtons between two point charges one meter apart, each having a charge of one coulomb.

In electrostatic units (esu), the unit of charge is defined as the charge on each of two point bodies one centimeter apart which produces a mutual force of interaction of one dyne. Thus the constant of proportionality in the esu system is dimensionless and equal to unity.

The relation between ϵ_0, the permittivity of free space, μ_0, the magnetic permeability of free space, and c, the velocity of light in a vacuum, is

$$\epsilon_0 \mu_0 = \frac{1}{c^2}.$$

c is a measured quantity equal to $2.998(10)^8$ meters/sec while μ_0 is taken arbitrarily as

$$\mu_0 = 4\pi(10)^{-7} \text{ henry/meter.}$$

The relation connecting ϵ_0, μ_0, and c comes from the equations governing the propagation of electromagnetic waves in free space (see Chapters X and XI). This relation fixes ϵ_0, once c and μ_0 are determined: $\epsilon_0 = 1/c^2\mu_0$.

Under the conditions considered for the two point charges, the force exerted by q_2 on q_1, $\mathbf{F}_{12}$, and that which q_1 exerts upon q_2, $\mathbf{F}_{21}$, are equal in magnitude and opposite in direction:

$$\mathbf{F}_{21} = -\mathbf{F}_{12}.$$

Thus Newton's third law is satisfied. This relation has been implied already in the vector form of Coulomb's law; by interchanging the indices in the equation for $\mathbf{F}_{12}$, we find

$$\mathbf{F}_{21} = \frac{q_1 q_2}{4\pi\epsilon_0} \frac{(\mathbf{r}_2 - \mathbf{r}_1)}{|\mathbf{r}_2 - \mathbf{r}_1|^3} = -\mathbf{F}_{12}.$$

C. THE ELECTRIC FIELD INTENSITY $\mathscr{E}$

1. THE PRINCIPLE OF LINEAR SUPERPOSITION

Experimentally it has been verified that the total force $\mathbf{F}_t$ on a test charge Q_t in the presence of N charges q_j is given by the vector sum

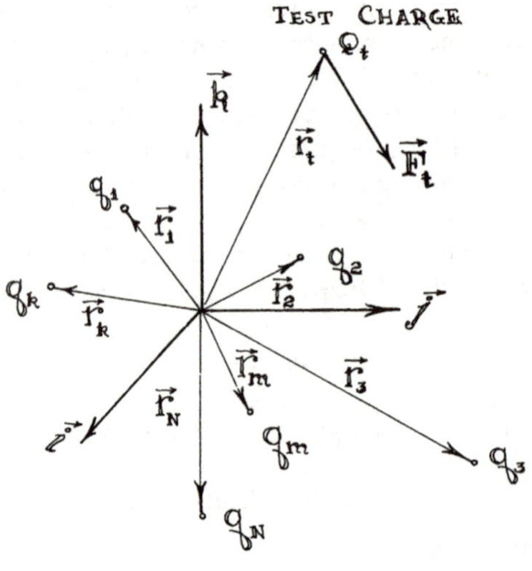

of all of the separate Coulomb interactions, $\mathbf{F}_{tj}$:

$$\mathbf{F}_t = \sum_{j=1}^{N} \frac{Q_t q_j}{4\pi\epsilon_0} \frac{(\mathbf{r}_t - \mathbf{r}_j)}{|\mathbf{r}_t - \mathbf{r}_j|^3},$$

or

$$\mathbf{F}_t = Q_t \sum_{j=1}^{N} \frac{q_j}{4\pi\epsilon_0} \frac{(\mathbf{r}_t - \mathbf{r}_j)}{|\mathbf{r}_t - \mathbf{r}_j|^3}.$$

2. THE ELECTRIC FIELD INTENSITY

Consider a region of space which contains N point source charges q_j. In general we must consider that the N source charges occupy certain positions of equilibrium. This equilibrium is the final result of the electrostatic forces between the source charges coupled with certain mechanical constraints which must be specified for each particular situation. For instance, in Chapter IV we will find that when charges are placed on the surface of a conductor they are constrained to remain on the surface. This situation can be represented as a geometric constraint.

Our purpose is now to obtain a measurement of the net force field set up by the N point charges, at any point in space. To accomplish

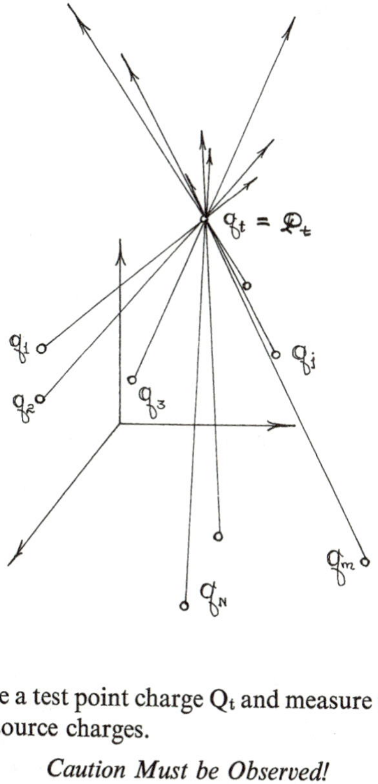

this we introduce a test point charge Q_t and measure the force exerted on it by the N source charges.

Caution Must be Observed!

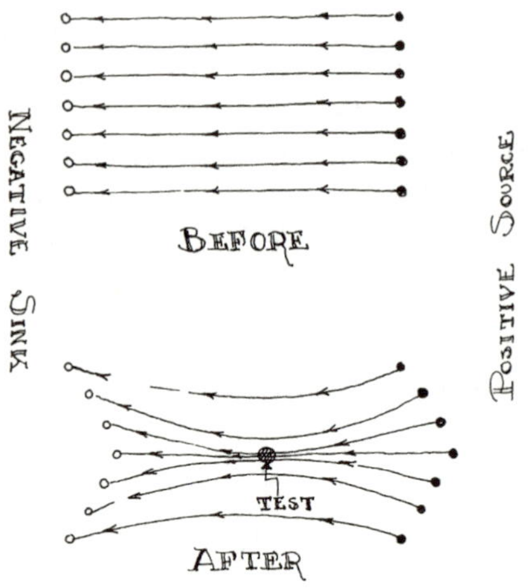

If the magnitude of the test charge Q_t is too large we may obtain a measurable rearrangement of the source charges. In other words, the act of making the measurement tends to alter the initial situation we wish to measure. Our desire therefore is to introduce the test charge without influencing the positions of the source charges. This of course is impossible.

There is a technique, however, which will allow us to gauge the original force field in the limit as the magnitude of the test charge approaches zero. A series of experiments can be performed with a continual decrease in magnitude of Q_t. In the limit as Q_t goes to zero we will obtain the magnitude and direction of the original force field of the N source charges.

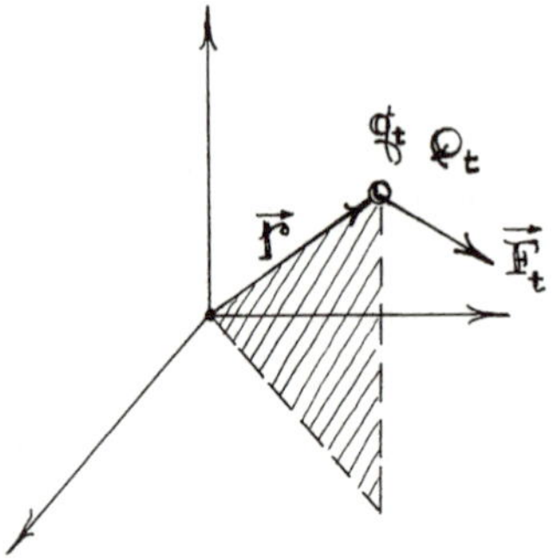

To illustrate this we show an example plot of the magnitude of the net force $\mathbf{F}_t$ as a function of Q_t. It is quite clear that when Q_t approaches zero the force on Q_t goes to zero. On the other hand, the ratio of $\mathbf{F}_t$ to Q_t has a nonzero value in the limit.

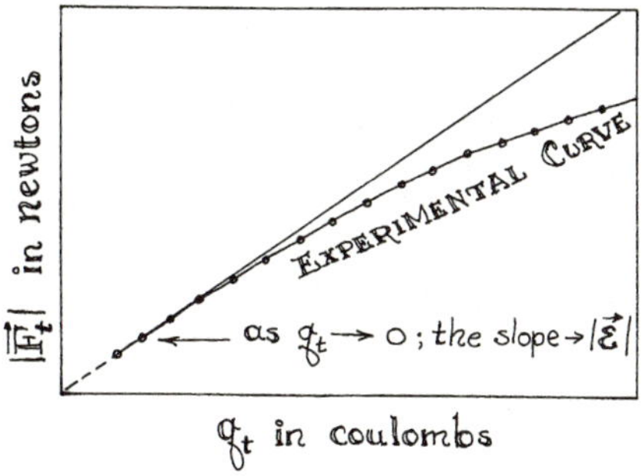

To represent the field of the source charges we define the ELECTRIC FIELD INTENSITY $\mathscr{E}$ as the net force per unit charge exerted on the test charge Q_t as the magnitude of Q_t approaches zero:

$$\mathscr{E} = \lim_{Q_t \to 0} \frac{\mathbf{F}_t}{Q_t}.$$

This vector field is characteristic of the N source charges only. In the diagram we see that the magnitude of $\mathscr{E}$ is measured as the slope of the $|\mathbf{F}_t|$ vs. Q_t curve in the linear region near $Q_t = 0$.

We are now in a position to compute the electric field intensity for our N source charges. If Q_t is at a point P, defined by the position vector $\mathbf{r}_p$, then

$$\mathbf{F}_t = Q_t \sum_{j=1}^{N} \frac{q_j}{4\pi\epsilon_0} \frac{(\mathbf{r}_t - \mathbf{r}_j)}{|\mathbf{r}_t - \mathbf{r}_j|^3},$$

and

$$\mathscr{E}_p = \lim_{Q_t \to 0} \frac{\mathbf{F}_t}{Q_t} = \sum_{j=1}^{N} \frac{q_j}{4\pi\epsilon_0} \frac{(\mathbf{r}_p - \mathbf{r}_j)}{|\mathbf{r}_p - \mathbf{r}_j|^3}.$$

The field of a single point charge q situated at a point designated by $\mathbf{r}_p$ is therefore

$$\mathscr{E}_p = \frac{q}{4\pi\epsilon_0} \frac{(\mathbf{r}_p - \mathbf{r})}{|\mathbf{r}_p - \mathbf{r}|^3}.$$

Once this result has been obtained, the fundamental aspects of electrostatics are at our disposal. Using point charges only for the remaining development of this chapter we can now study various special properties of this fundamental equation. After demonstrating that $\mathscr{E}$ is conservative, we will develop the energy stored in a charge ensemble and then investigate the electrical properties of special configurations of point charges.

D. THE ELECTROSTATIC POTENTIAL

The electrostatic field intensity $\mathscr{E}$ is a conservative vector field. This can be demonstrated in various ways. A series of path integrations

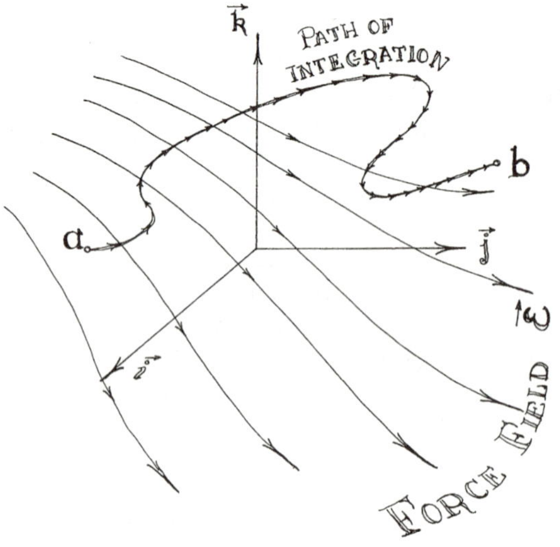

would show that the work done in moving a test charge from a point a to a point b is independent of the path taken between a and b:

$$W_{ab} = \int_a^b \mathbf{F}_t \cdot d\mathbf{r} = Q_t \int_a^b \mathscr{E}_p \cdot d\mathbf{r}.$$

The simplest method to show that $\mathscr{E}$ is conservative is to demonstrate that the vector point function $\mathscr{E}(x,y,z)$ can be generated from a scalar point function V(x,y,z) by

$$\mathscr{E}_p = -\operatorname{grad}_p V_p.$$

Examine the field of a single point charge q:

$$\mathscr{E}_p = \frac{q}{4\pi\epsilon_0}\frac{(\mathbf{r}_p - \mathbf{r})}{|\mathbf{r}_p - \mathbf{r}|^3}.$$

We assert that

$$-\operatorname{grad}_p\left\{\frac{q}{4\pi\epsilon_0}\frac{1}{|\mathbf{r}_p - \mathbf{r}|} + \text{const.}\right\} = \frac{q}{4\pi\epsilon_0}\frac{(\mathbf{r}_p - \mathbf{r})}{|\mathbf{r}_p - \mathbf{r}|^3}.$$

If the reader is in doubt about the details of this operation he should expand $|\mathbf{r}_p - \mathbf{r}|^{-1}$ in terms of its components and take the gradient term by term.

Notice that the gradient represents variations in the coordinates of the point P, at which the field is measured, wherefore the derivatives are taken with respect to the variables x_p, y_p, and z_p.

$$-\operatorname{grad}\frac{1}{|\mathbf{r}_p - \mathbf{r}|} = -\sum_{m=1}^{3}\boldsymbol{\epsilon}_m \frac{1}{h_m^{(p)}}\frac{\partial}{\partial \xi_m^{(p)}}\left\{\sum_{j=1}^{3}(h_j^{(p)}\xi_j^{(p)} - h_j\xi_j)^2\right\}^{-1/2},$$

$$= \sum_{m=1}^{3}\frac{(x_m^{(p)} - x_m)\boldsymbol{\epsilon}_m}{\left\{\sum_{j=1}^{3}(x_j^{(p)} - x_j)^2\right\}^{3/2}}.$$

Once we establish that there is a V such that $\mathscr{E} = -\operatorname{grad} V$, the field can be recognized as a *conservative field.*

Going back to the more general problem of N source charges, the principle of superposition implies that

$$\mathscr{E}_p = \sum_{j=1}^{N}\frac{q_j}{4\pi\epsilon_0}\frac{(\mathbf{r}_p - \mathbf{r}_j)}{|\mathbf{r}_p - \mathbf{r}_j|^3}.$$

Using the result for the single charge, we find that

$$-\operatorname{grad}_p\left\{\sum_{j=1}^{N}\frac{q_j}{4\pi\epsilon_0}\frac{1}{|\mathbf{r}_p - \mathbf{r}_j|} + \text{const.}\right\} = \sum_{j=1}^{N}\frac{q_j}{4\pi\epsilon_0}\frac{(\mathbf{r}_p - \mathbf{r}_j)}{|\mathbf{r}_p - \mathbf{r}_j|^3}.$$

In other words, the linear superposition in $\mathscr{E}$ corresponds to a sum of the individual point potentials of each source charge q_j. If we add N potentials corresponding to each value of q we obtain the desired result. The total potential function for $\mathscr{E}_p$ is

$$V_p(x_p,y_p,z_p,) = \sum_{j=1}^{N}\frac{q_j}{4\pi\epsilon_0}\frac{1}{|\mathbf{r}_p - \mathbf{r}_j|} + \text{const.},$$

and

$$\mathscr{E}_p = -\operatorname{grad}_p V_p.$$

Once again we consider the work done moving a point charge Q_t (of negligible magnitude) from point a to point b, in the presence of N source charges q_j. Because $\mathscr{E}$ is conservative, the integration is independent of the path connecting a and b:

$$\begin{aligned} W_{ab} &= Q_t \int_a^b \mathscr{E}_p \cdot d\mathbf{r} = -Q_t \int_a^b (\text{grad } V) \cdot d\mathbf{r} \\ &= -Q_t \int_a^{a_6} dV = -Q_t(V_b - V_a). \end{aligned}$$

The difference $(V_b - V_a)$ is called the potential difference, and we have shown that

The negative of the potential difference between two points is equal to the positive work done per unit test charge.

The potential difference is *positive* when the work is contributed by the external mechanism which moves the test charge. The electrical energy increases.

The potential difference is *negative* if the electrical system of N charges dissipates energy in moving Q_t. This energy is absorbed by the external mechanism controlling Q_t.

In the case of N source charges the potential difference is

$$V_b - V_a = \sum_{j=1}^{N} \frac{q_j}{4\pi\epsilon_0} \left\{ \frac{1}{|\mathbf{r}_b - \mathbf{r}_j|} - \frac{1}{|\mathbf{r}_a - \mathbf{r}_j|} \right\},$$

assuming of course that the constant (which sets the zero of potential) is the same in both V_b and V_a and cancels in the difference.

When the source consists of a single point charge q at the origin $(r = 0)$, then

$$V_b - V_a = \frac{q}{4\pi\epsilon_0} \left\{ \frac{1}{r_b} - \frac{1}{r_a} \right\}.$$

Because the functional form of V goes as $1/r$, the work required to bring a test charge from infinity to a point $\mathbf{r}_p$ depends upon the coordinates specified by $\mathbf{r}_p$. Thus it is convenient to take the zero of potential at infinity. Then the constant appearing in $V(x_p,y_p,z_p)$ is ZERO, and

$$V_p = -\int_{\infty}^{\mathbf{r}_p} \mathscr{E}(\mathbf{r}') \cdot d\mathbf{r}'.$$

For N source charges, the potential at P is then

$$V_p = \sum_{j=1}^{N} \frac{q_j}{4\pi\epsilon_0} \frac{1}{|\mathbf{r}_p - \mathbf{r}_j|}.$$

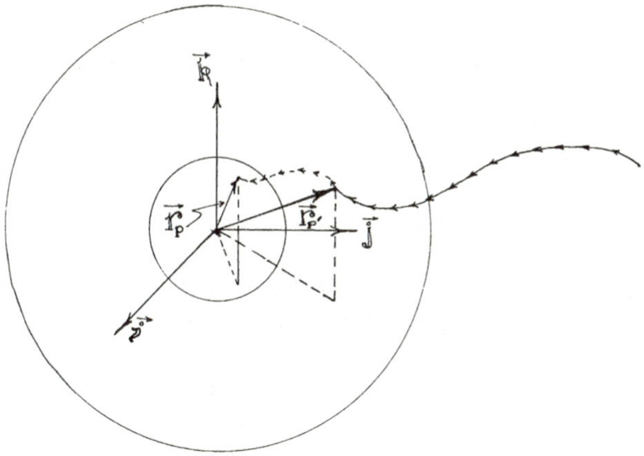

In Chapter III these sums over discrete point charges will be transformed to integrals over a continuous distribution of charge. In many cases the development of the vector field $\mathscr{E}_p$ from the integral form can become quite complex and tedious, and the development of the scalar potential function is usually much easier. Therefore the usual procedure to obtain the vector field will consist of first solving for the potential function, after which the vector field can be obtained quite readily by taking the gradient of the scalar function.

E. THE TOTAL ENERGY STORED IN A SYSTEM OF POINT CHARGES

The work required to bring a point charge q_1 from ∞ to a point designated by $\mathbf{r}_1$ in a field-free region is ZERO. However, if a second charge q_2 is brought from ∞ to a point $\mathbf{r}_2$, the region is no longer field-free, and WORK is done by the agent bringing up the second charge. As a result, electrostatic energy (positive or negative depending upon the relative signs of q_1 and q_2) is stored in the system of two charges.

The symbol for the total energy stored in the electrostatic system will be U. U is equal to the negative of the work done by the field, or to the positive work done by an external agency:

$$U = -W_{q_2} = -q_2 \int_{\infty}^{\mathbf{r}_2} \mathscr{E}_{21} \cdot d\mathbf{r}_2 = +q_2 \int_{\infty}^{\mathbf{r}_2} (\text{grad}_2 V_{21}) \cdot d\mathbf{r}_2,$$

$$U = q_2 \left\{ \frac{q_1}{4\pi\epsilon_0} \frac{1}{|\mathbf{r}_2 - \mathbf{r}_1|} \right\}.$$

The term grad_2 implies a variation in the coordinates of charge q_2.

This equation has been written in such a manner that the storage is considered to be concentrated in q_2. Actually in a system composed of two charges the energy can be considered to be shared between the two.

Let us call V_{21} the potential at q_2 arising from the presence of q_1,

$$V_{21} = \frac{q_1}{4\pi\epsilon_0} \frac{1}{|\mathbf{r}_2 - \mathbf{r}_1|}.$$

In the same manner, q_2 sets up a potential V_{12} at the position of q_1:

$$V_{12} = \frac{q_2}{4\pi\epsilon_0} \frac{1}{|\mathbf{r}_1 - \mathbf{r}_2|}.$$

Because $|\mathbf{r}_2 - \mathbf{r}_1| = |\mathbf{r}_1 - \mathbf{r}_2|$, the ratio of V_{21} to V_{12} is equal to the ratio of q_1 to q_2. For the same reason, the energies stored by each charge in a doublet are equal. Define

$$U_{21} = q_2 V_{21} = \frac{q_2 q_1}{4\pi\epsilon_0} \frac{1}{|\mathbf{r}_2 - \mathbf{r}_1|},$$

and

$$U_{12} = q_1 V_{12} = \frac{q_1 q_2}{4\pi\epsilon_0} \frac{1}{|\mathbf{r}_2 - \mathbf{r}_1|}.$$

Then

$$U = \tfrac{1}{2}(U_{12} + U_{21}) = \tfrac{1}{2}(q_1 V_{12} + q_2 V_{21}),$$

or

$$U = \tfrac{1}{2} \sum_{m=1}^{2} \sum_{\substack{n=1 \\ n \neq m}}^{2} q_m V_{mn}.$$

The factor $\frac{1}{2}$ appears because the pair energy is shared equally by each member of a pair.

Thus in general if there are N charges q_j in a region, every pair of charges $q_m q_n$ shares the corresponding interaction energy.

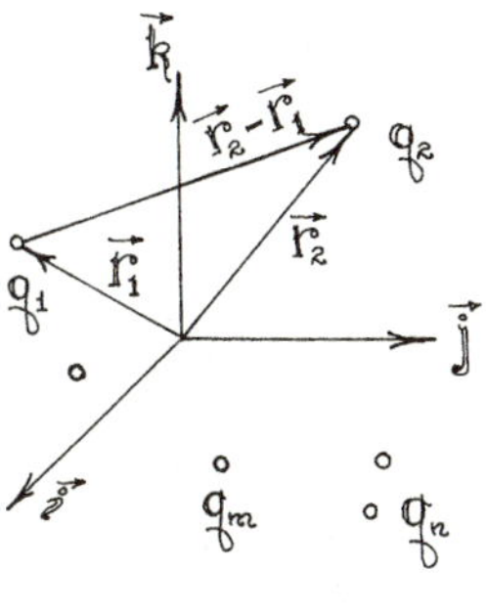

The total potential at the position of the mth charge in the system can be derived from the other N − 1 charges:

$$V_m = \sum_{\substack{n=1 \\ n \neq m}}^{N} \frac{q_n}{4\pi\epsilon_0} \frac{1}{r_{mn}} = \sum_{\substack{n=1 \\ n \neq m}}^{N} V_{mn},$$

where $$r_{mn} = |\mathbf{r}_m - \mathbf{r}_n|.$$

The energy stored in the mth charge is

$$U_m = \tfrac{1}{2} \sum_{\substack{n=1 \\ n \neq m}}^{N} U_{mn}$$

or $$U_m = \frac{q_m}{2} \sum_{\substack{n=1 \\ n \neq m}}^{N} V_{mn} = \tfrac{1}{2} q_m V_m.$$

The factor of $\frac{1}{2}$ appears for the reason that every U_{mn} is associated with an equal U_{nm} from the sharing of the doublet energy between q_m and q_n. Thus the total energy stored in this system of N charges is

$$U_{total} = \sum_{m=1}^{N} U_m = \tfrac{1}{2} \sum_{m=1}^{N} \sum_{\substack{n=1 \\ n \neq m}}^{N} U_{mn},$$

or $$U_{total} = \tfrac{1}{2} \sum_{m=1}^{N} q_m \sum_{\substack{n \neq m \\ n=1}}^{N} V_{mn} = \tfrac{1}{2} \sum_{m=1}^{N} q_m V_m.$$

In closing we should comment on the quadratic properties of the function U. The doublet potential V_{mn} is proportional to q_n and is written

$$V_{mn} = \frac{q_n}{4\pi\epsilon_0} \frac{1}{|\mathbf{r}_m - \mathbf{r}_n|}.$$

Designating the geometric portion of this equation by the symbols P_{mn}, where

$$P_{mn} = \frac{1}{4\pi\epsilon_0} \frac{1}{|\mathbf{r}_m - \mathbf{r}_n|}, \quad m \neq n,$$

we note the absence of the diagonal elements P_{mm}; this represents a neglect of the self-energy of a point charge. This is reasonable when one considers that in a given equilibrium problem the amount of charge assigned to each point charge is assumed fixed. Therefore, the work necessary to set up an individual point charge is of no concern here.*

Returning to our discussion of V_{mn}, the geometric terms P_{mm} (which are symmetric, $P_{mn} = P_{nm}$) are incorporated in the following manner:

$$V_{mn} = P_{mn} q_n \quad (P_{nn} = 0).$$

Thus

$$U = \tfrac{1}{2} \sum_{m=1}^{N} q_m \sum_{n=1}^{N} P_{mn} q_n,$$

and we can write U in a form quadratic in the q's,

$$U = \tfrac{1}{2} \sum_{m=1}^{N} \sum_{n=1}^{N} q_m P_{mn} q_n.$$

Here we have specified that the diagonal elements of the $\mathbb{P}$ matrix, P_{kk}, are zero,

$$P_{kk} = 0.$$

This is to be contrasted to an analogous equation which will be set up in Section IV-D concerning extended charged conductors. In the case of charged conductors there will be diagonal elements p_{kk} which will definitely not be zero. This is apparent because a given charged conductor can itself be represented as an ensemble of point charges. Again the energy stored is derived from the external work required to construct the ensemble.

By employing the matrix notation outlined in Appendix B we can write the ensemble of point charges as a column vector **q** (of N

* The self-energy of a single electron, however, is a problem of primary interest in modern electrodynamics.

dimensions):

$$\mathbf{q} = \begin{bmatrix} q_1 \\ q_2 \\ q_3 \\ \cdot \\ \cdot \\ \cdot \\ q_N \end{bmatrix}.$$

The geometric factors P_{mm} form a square $N \times N$ matrix $\mathbf{P}$, where

$$\mathbf{P} = \begin{pmatrix} 0 & P_{12} & P_{13} & \dots & P_{1N} \\ P_{21} & 0 & P_{23} & \dots & P_{2N} \\ P_{31} & P_{32} & 0 & \dots & P_{3N} \\ \cdot & & & & \cdot \\ \cdot & & & & \cdot \\ \cdot & & & & \cdot \\ P_{N1} & & & & 0 \end{pmatrix}$$

Then the potential at the position of each charge can be written as a vector $\mathbf{V}$ with elements V_m. V_m is the potential set up by the remaining $N - 1$ charges. Each V_m is derived from only $N - 1$ charges since the self energy has been neglected.

$$\mathbf{V} = \begin{bmatrix} V_1 \\ V_2 \\ V_3 \\ \cdot \\ \cdot \\ \cdot \\ V_N \end{bmatrix}.$$

Our development of V_m becomes a LINEAR TRANSFORMATION of the vector **q**,

$$V_m = \sum_{n=1}^{N} P_{mn} q_n,$$

or

$$\mathbf{V} = \mathbf{P} \cdot \mathbf{q}.$$

By constructing $\mathbb{P}^{-1}$ we can write $\mathbf{q} = \mathbb{P}^{-1} \cdot \mathbf{V}$. Finally, the total energy stored in the system of N charges is

$$U = \tfrac{1}{2} \sum_{m=1}^{N} q_m V_m,$$

or

$$U = \tfrac{1}{2}\tilde{\mathbf{q}} \cdot \mathbf{V}$$

where

$$\tfrac{1}{2}\tilde{\mathbf{q}} \cdot \mathbf{V} = \tfrac{1}{2}[q_1, q_2, q_3, \ldots, q_N] \begin{bmatrix} V_1 \\ V_2 \\ \cdot \\ \cdot \\ \cdot \\ V_N \end{bmatrix}.$$

Substituting for $\mathbf{V}$ we obtain the compact form

$$U = \tfrac{1}{2}\tilde{\mathbf{q}} \cdot \mathbb{P} \cdot \mathbf{q},$$

or

$$U = \tfrac{1}{2}\tilde{\mathbf{V}} \cdot (\mathbb{P}^{-1} \cdot \mathbb{P} \cdot \mathbb{P}^{-1}) \cdot \mathbf{V};$$

$$U = \tfrac{1}{2}\tilde{\mathbf{V}} \cdot \mathbb{P}^{-1} \cdot \mathbf{V}.$$

Above $\tilde{\mathbf{q}}$ and $\tilde{\mathbf{V}}$ are the required *row* matrix representations of transposed vectors. This form is used in matrix multiplication when multiplying from the left by a vector (Appendix B).

It is perhaps interesting to observe the effect of a similarity transformation $\mathbb{S}$ upon $\mathbb{P}$ (see Appendix B). $\mathbb{P}$ is symmetric and can be diagonalized. The eigenvalues of $\mathbb{P}$ represent a set of diagonal elements λ which are associated with unique charge distributions characteristic of the particular geometry $\mathbb{P}$. This is analogous to reducing the equation to normal coordinates, in this case the q_m'.

The eigenvalue problem is developed if we ask for an orthogonal transformation $\mathbb{S}$ establishing a very special system of charges (retaining their original geometric positions) which produce a set of potential vectors $\mathbf{V}'$ parallel to $\mathbf{q}'$ where $\mathbf{q}'$ is the eigenvector of $\mathbb{P}$. In other words, we intend that the diagonalization will make the ratio of any two charges q_m and q_n equal to the ratio of the corresponding potentials at those points. Since

$$\mathbf{V} = \mathbb{P} \cdot \mathbf{q},$$

if we vary all of the magnitudes of the charges while keeping their original N positions in space fixed, it is then possible to obtain N solutions of the type

$$\mathbf{V}' = \lambda \mathbf{q}'$$

such that

$$\mathbf{P} \cdot \mathbf{q}' = \lambda \mathbf{q}'.$$

Consequently

$$|\mathbf{P} - \lambda \mathbf{I}| = 0.$$

As an example of this transformation, the system of two charges q_1 and q_2 can be diagonalized. For this special case

$$P_{11} = 0 \qquad P_{12} = \frac{1}{4\pi\epsilon_0} \frac{1}{|\mathbf{r}_1 - \mathbf{r}_2|}$$

$$P_{21} = \frac{1}{4\pi\epsilon_0} \frac{1}{|\mathbf{r}_2 - \mathbf{r}_1|} \qquad P_{22} = 0.$$

If $|\mathbf{P} - \lambda \mathbf{I}| = 0$, the two roots λ_1 and λ_2 are

$$\lambda_1 = +P_{12} \quad \text{and} \quad \lambda_2 = -P_{12}.$$

For $\lambda = \lambda_1 = +P_{12}$ the components of the charge vector are obtained from

$$\begin{pmatrix} 0 & P_{12} \\ P_{21} & 0 \end{pmatrix} \begin{bmatrix} q_{11}' \\ q_{21}' \end{bmatrix} = +P_{12} \begin{bmatrix} q_{11}' \\ q_{21}' \end{bmatrix}$$

giving

$$\mathbf{q}_1' = q \begin{bmatrix} 1 \\ 1 \end{bmatrix}.$$

The other eigenvalue $\lambda = \lambda_2 = -P_{12}$ gives

$$\mathbf{q}_2' = q \begin{bmatrix} 1 \\ -1 \end{bmatrix}.$$

As one would expect, if the two charges are equal in magnitude the total energy stored is

$$U = \frac{\pm 2}{2} P_{12} q^2 = P_{12} q^2,$$

or

$$U = \frac{\pm q^2}{4\pi\epsilon_0} \frac{1}{|\mathbf{r}_1 - \mathbf{r}_2|}.$$

F. GAUSS'S LAW FOR POINT CHARGES

The experimental fact that the electric field of a point charge depends upon the inverse square of the distance from the charge leads to several theorems concerning the relationships between the charges in a closed region and the electric field lines crossing the boundaries of that region.

1. THE ELECTRIC DISPLACEMENT VECTOR

To obtain a consistent notation in later chapters we shall introduce the vacuum *electric displacement vector* **D**. Our remarks to follow will be correct although quite limited in that we speak now only of point charges in a vacuum. Later it will be demonstrated that the electric displacement vector **D** will always be a measure of the singlet free charge present in a system of free and bound charges.

For a vacuum containing a set of N point charges associated with a net electric field intensity $\mathscr{E}$, the electric displacement vector **D** is defined as

$$\mathbf{D} = \epsilon_0 \mathscr{E}.$$

The directions of **D** and $\mathscr{E}$ *in a vacuum* are the same. **D**, however, has the dimensions of charge per unit area.

The electric displacement field at a point P of a point charge q located at a point **r** is then

$$\mathbf{D}_p = \frac{q}{4\pi} \frac{(\mathbf{r}_p - \mathbf{r})}{|\mathbf{r}_p - \mathbf{r}|^3}.$$

The **D** field at P when there are N charges is

$$\mathbf{D}_p = \sum_{j=1}^{N} \frac{q_j}{4\pi} \frac{(\mathbf{r}_p - \mathbf{r}_j)}{|\mathbf{r}_p - \mathbf{r}_j|^3}.$$

2. THE SOLID ANGLE

Consider an arbitrary point O and a surface element ΔS. The orientation of the surface element is specified by a unit surface normal **n**.

If the equation of a surface containing ΔS is specified by a scalar point function $\Phi(x,y,z) = 0$, then the unit normal at any point **r** on the surface is given by

$$\frac{\text{grad } \Phi}{|\text{grad } \Phi|} = \mathbf{n}.$$

The element ΔS is located relative to O by the position vector **r**. If a spherical surface of radius **r** is constructed with O as its center, the projection of ΔS on the spherical surface provides us with a measure of the solid angle subtended by ΔS at O.

Designate the projection of ΔS on the sphere of radius r as $\Delta\Lambda$. The area $\Delta\Lambda$ is the area encompassed by the elements common to O and the periphery of ΔS.

The solid angle, $\Delta\Omega$, is defined as the ratio of $\Delta\Lambda$ to r^2, the area of the sphere divided by 4π. The units of this quantity $\Delta\Omega$ are steradians, and the solid angle subtended by a closed surface about a point is 4π steradians.

Alternatively, the solid angle is measured taking the area cut out by the elements drawn between the circumference of ΔS and O on

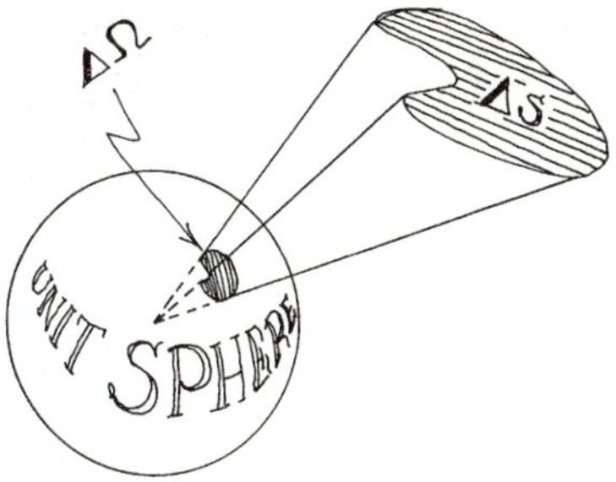

the surface of a sphere having a unit radius. If the elements describe an area $\Delta\Lambda'$ on the surface of the unit sphere, then $\Delta\Omega = \Delta\Lambda'$. It follows that the solid angle encompassed by the unit sphere is equal to its surface area (4π). Thus

$$\Delta\Omega = \frac{\Delta\Lambda}{r^2}.$$

If the dimensions of ΔS are small compared to r, then we use the projection of ΔS on a plane perpendicular to r to give ΔA ($\Delta A \simeq \Delta\Lambda$ if $\Delta A \ll r^2$),

$$\Delta A \simeq \Delta S \cos \sphericalangle_{\mathbf{n}}^{\mathbf{r}} = \Delta S\, \mathbf{n} \cdot \frac{\mathbf{r}}{r}.$$

Finally*

$$\Delta\Omega \simeq \frac{\Delta S \mathbf{n} \cdot \mathbf{r}}{r^3} = \frac{\mathbf{r}}{r^3} \cdot \mathbf{n}\, \Delta S.$$

* When r^2 is of the same order as ΔA the inverse square definition of Ω_{approx} breaks down.

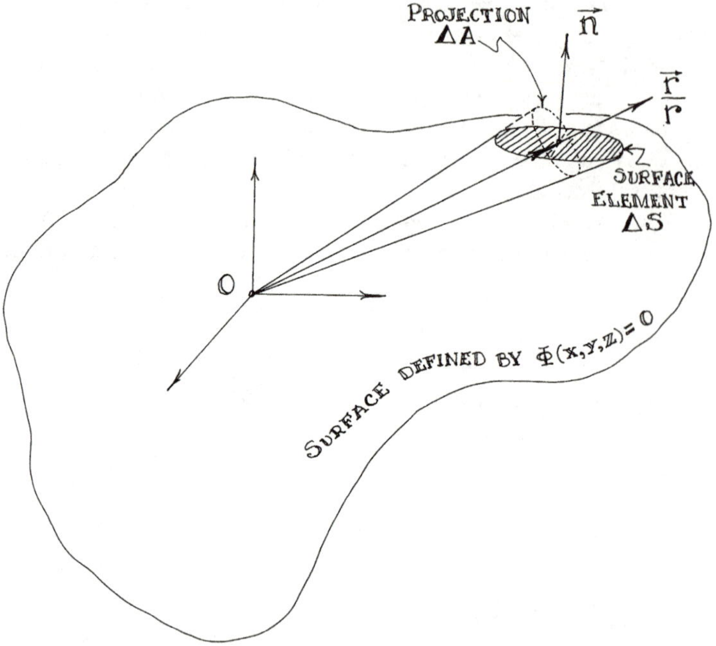

The total solid angle subtended at a point O by a surface S can be developed by integration,

$$\Omega_S = \iint_S \frac{\mathbf{r}}{r^3} \cdot \mathbf{n}\, dS.$$

If the point O from which the total solid angle is being measured lies

INSIDE

of a *closed* surface S, then Ω is the total solid angle about the point, i.e., 4π. On the other hand, if the point O lies

OUTSIDE

of a closed surface S, then for every positive $(\mathbf{r}/\mathbf{r}^3) \cdot \mathbf{n}\, dS$ there is an equal and opposite $(\mathbf{r}'/\mathbf{r}'^3) \cdot \mathbf{n}'\, dS'$ which exactly cancels it. Thus the total solid angle subtended by a closed surface S when the point lies outside of S is ZERO:

$$\iint_{S(\text{closed})} \frac{\mathbf{r}}{r^3} \cdot \mathbf{n}\, dS \begin{array}{ll} = 4\pi & \text{for O inside S,} \\ = \ \ 0 & \text{for O outside S.} \end{array}$$

As an exercise the reader can demonstrate that the point O is inside S if every straight line element L drawn from and originated at O intersects the surface an *odd* number of times. If the element intersects an *even* number of times, the point O is outside S.

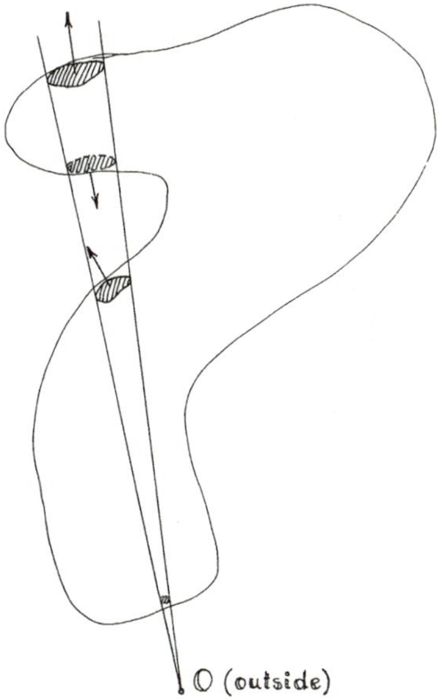

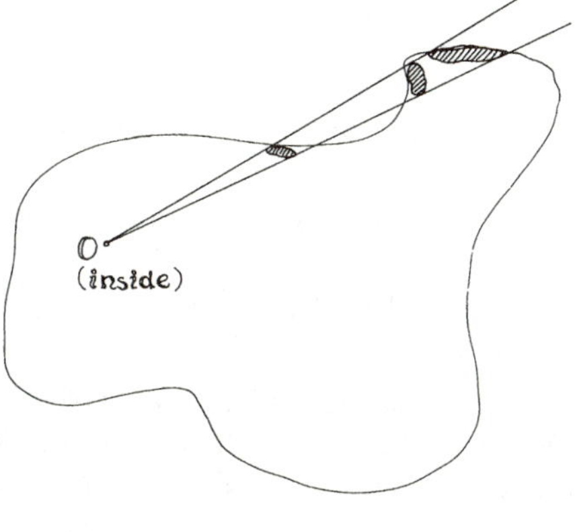

3. GAUSS'S LAW

Consider a point charge q at the origin. The integral of the normal component of **D** over a closed surface S assumes a particularly simple form, the result being a further expression of the inverse square law. Take

$$\iint_{S(\text{closed})} \mathbf{D}\cdot\mathbf{n}\,dS = \frac{q}{4\pi}\iint_{S(\text{closed})} \frac{\mathbf{r}}{r^3}\cdot\mathbf{n}\,dS$$

$$= q \quad (\text{for q inside S})$$
$$= 0 \quad (\text{for q outside S}).$$

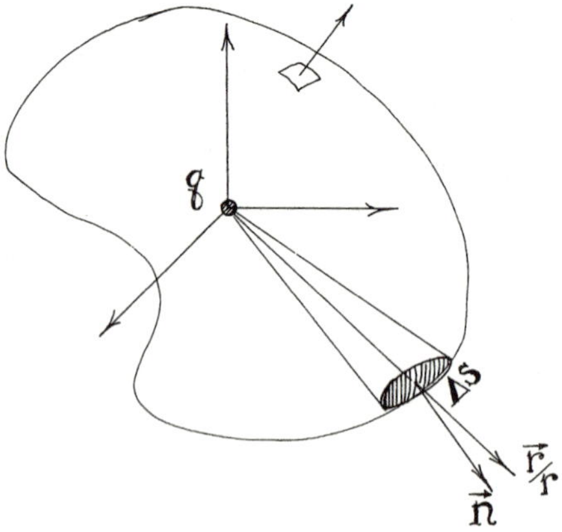

We write this as

$$\iint_{S(\text{closed})} \mathbf{D}\cdot\mathbf{n}\,dS = q(\text{enclosed}).$$

The result is of the same form when N charges q_m are considered:

$$\iint_{S(\text{closed})} \mathbf{D}\cdot\mathbf{n}\,dS = \sum_{m=1}^{N}\frac{q_m}{4\pi}\iint_{S}\frac{(\mathbf{r}-\mathbf{r}_m)}{|\mathbf{r}-\mathbf{r}_m|^3}\cdot\mathbf{n}\,dS.$$

Assume that the first k charges are inside of S, i.e., $q_1, q_2, \ldots, q_k$, and that the remaining N − k charges are outside. The surface integral of **D** now resolves into the sum of N separate integrals.

In each case the origin for the integration can be shifted to the particular position of the charge in question. Define $\mathbf{r}'_m = \mathbf{r} - \mathbf{r}_m$; then

$$\iint_{S(\text{closed})} \mathbf{D}\cdot\mathbf{n}\,dS = \sum_{m=1}^{k}\frac{q_m}{4\pi}\iint_{S(\text{closed})}\frac{\mathbf{r}'_m}{r'^3_m}\cdot\mathbf{n}\,dS + \sum_{m=k+1}^{N}\frac{q_m}{4\pi}\iint_{S(\text{closed})}\frac{\mathbf{r}'_m}{r'^3_m}\cdot\mathbf{n}\,dS$$

$$= \sum_{m=1}^{k} q_m + 0 = \text{total charge enclosed}$$

$$= q(\text{enclosed}).$$

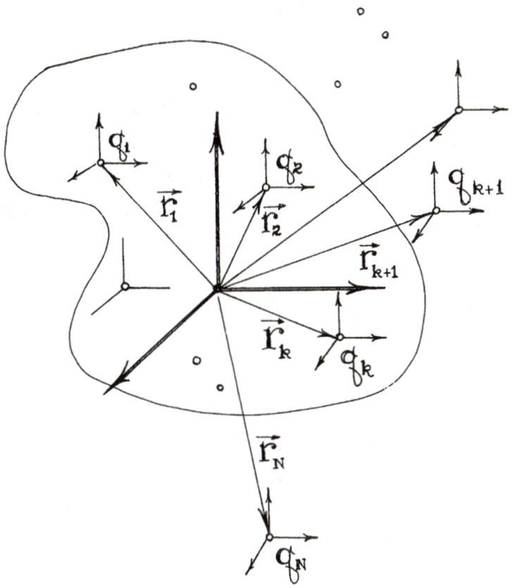

The application of Gauss's law to point charges seems to lead to a null quantity when the divergence theorem is employed. This is not the case, however, and there are several methods by which the divergence theorem can be shown to hold for point charge problems.

According to the divergence theorem,

$$\iint_{S(\text{closed})} \mathbf{D}\cdot\mathbf{n}\,dS = \iiint_{\tau_s \text{ enclosed by } S} \operatorname{div}\mathbf{D}\,d\tau$$

where $d\tau$ = the volume element

$= h_1h_2h_3\,d\xi_1\,d\xi_2\,d\xi_3$

or $d\tau = dx\,dy\,dz.$

Regard a point charge at the origin O. When O lies inside of S (S will be understood to be closed),

$$\iint_S \mathbf{D}\cdot\mathbf{n}\,dS = \iiint_{\tau_S} \operatorname{div}\left(\frac{q}{4\pi}\frac{\mathbf{r}}{r^3}\right) d\tau.$$

Now apparently

$$\operatorname{div}\left(\frac{\mathbf{r}}{r^3}\right) = \sum_{n=1}^{3} \frac{\partial}{\partial x_n}\left(\frac{x_n}{r^3}\right) = 0.$$

If we are not very careful this volume integral might be taken as zero everywhere, indicating some inconsistency in our procedure. Notice that the term $\mathbf{r}/r^3$ is singular at $r = 0$, and the divergence is not defined at this point.

To avoid the singular point a small sphere of radius R is constructed about the charge q, and this spherical surface S'_R is connected to the outer surface S by a hollow tube of negligible volume and surface. The closed surface composed of S, S'_R and the negligible surface of the connecting tube now encloses zero charge (the sphere is being utilized to exclude the singular point).

Thus taking the normal surface integration in the tube as zero and letting the radius of the sphere approach zero,

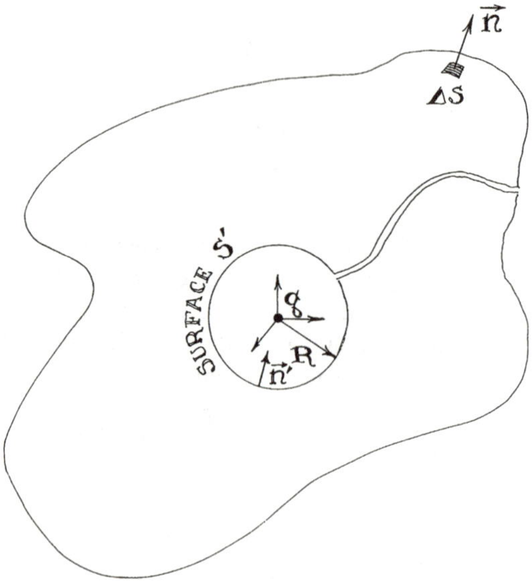

$$\iint_S \mathbf{D}\cdot\mathbf{n}\,dS + 0 + \lim_{R\to 0}\iint_{S_R'} \mathbf{D}\cdot\mathbf{n}'\,dS' = 0,$$

and $$\iint_S \mathbf{D}\cdot\mathbf{n}\,dS = -\lim_{R\to 0}\left\{\frac{-q}{4\pi}\int_0^\pi \frac{2\pi}{R^2} R^2 \sin\theta\, d\theta\right\} = q.$$

In Chapter III in connection with Poisson's equation we shall find that

$$\mathrm{div}_r\left\{\frac{(\mathbf{r}-\mathbf{r}')}{|\mathbf{r}-\mathbf{r}'|^3}\right\} = 4\pi\delta(\mathbf{r}'-\mathbf{r}),$$

where $\delta(\mathbf{r}'-\mathbf{r})$ is the Dirac delta function, a function of zero width and arbitrarily large amplitude.

The delta function is defined only under the integral sign, having the value 1 when the volume considered encloses the point defined by **r** and the value zero when the volume *excludes* the point defined by **r**:

$$\iiint_{\tau'} \delta(\mathbf{r}'-\mathbf{r})\,d\tau' = \begin{matrix} 1 & \text{if } \tau' \text{ contains } \mathbf{r} \\ 0 & \text{if } \tau' \text{ excludes } \mathbf{r}; \end{matrix}$$

also* $$\iiint_{\text{All Space}} \delta(\mathbf{r}'-\mathbf{r})f(\mathbf{r}')\,d\tau' = f(\mathbf{r}).$$

The three-dimensional delta function for cartesian coordinates is the product of three one-dimensional delta functions:

$$\delta(\mathbf{r}'-\mathbf{r}) = \delta(x'-x)\delta(y'-y)\delta(z'-z).$$

The function $\delta(x_n'-x_n)$ has the property

$$\int_{-\infty}^{\infty}\delta(x_n'-x_n)\,dx_n' = 1.$$

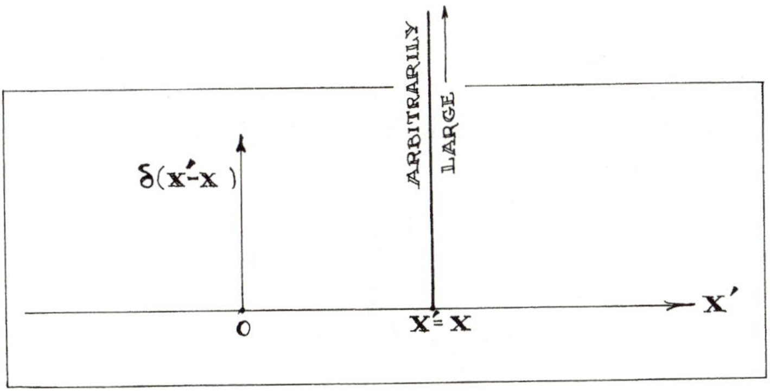

* An all-space integration implies an integration from minus infinity to plus infinity for all three cartesian coordinates.

Therefore, the application of the divergence theorem to Gauss's law leads to the correct result when we consider these properties of div $(\mathbf{r}/r^3)$ (the proof of this statement lies in the preceding two-surface integration).

Returning to the problem for a point charge at the origin,

$$\operatorname{div}(\mathbf{r}/r^3) = 4\pi\delta(\mathbf{r} - 0).$$

For the point charge at $r = 0$,

$$\iint_S \mathbf{D} \cdot \mathbf{n}\, dS = \iiint_\tau \frac{q}{4\pi} \operatorname{div}\left(\frac{\mathbf{r}}{r^3}\right) d\tau$$

$$= \frac{q}{4\pi} \iiint 4\pi\delta(\mathbf{r} - 0)\, d\tau = \begin{matrix} q & \text{if 0 is inside S} \\ 0 & \text{if 0 is outside S.} \end{matrix}$$

Later we shall find that this result accounts for the volume charge density ρ of the point charge: $\rho_{\text{point}} = q\delta(\mathbf{r} - 0)$. This is an example of a case in which the charge is at $r = 0$.

G. POINT CHARGE MULTIPOLES: THE SINGLET

One point charge is designated as a singlet charge. Our discussions thus far have been concerned mainly with the superposition of the fields and potentials of singlet charges.

If in a given charge distribution of N charges q_m the net charge is not zero, the ensemble is characterized by a singlet potential and the singlet electric field in the FAR FIELD REGION, i.e. $|\mathbf{r}_m| \ll |\mathbf{r}_p|$.

Let the position vectors of the N charges be designated by $\mathbf{r}_m$. If

$$\sum_{m=1}^{N} q_m = q_1 + q_2 + \ldots + q_N \neq 0,$$

then the electric potential at a given point P (specified by $\mathbf{r}_p$) goes as $1/r_p$ for $|\mathbf{r}_m| \ll |\mathbf{r}_p|$. This, of course, means that the distance r_p is much greater than any of the distances r_m of the charges from the origin. This will be demonstrated in the multipole expansion in Section J of this chapter.

To restate: if

$$\sum_{m=1}^{N} q_m \neq 0$$

then*

$$V_p \xrightarrow[r_p \gg r_m]{} \frac{\left\{\sum_{m=1}^{N} q_m\right\}}{4\pi\epsilon_0} \frac{1}{r_p}.$$

* The long arrow means "behaves like the function to follow," under the conditions stated below the arrow.

H. POINT CHARGE MULTIPOLES: THE ELECTRIC DIPOLE

If the net charge of an ensemble of charges is zero, then the far field potential to first order in $1/r^n$ will exhibit the characteristic shape of a higher electric multipole. The second multipole is the dipole.

The classical electric dipole is formed from two charges equal in magnitude but of opposite sign. These charges are separated by a distance $\boldsymbol{l}$ where the vector $\boldsymbol{l}$ initiates on the negative charge and terminates on the positive charge. For simplicity of development we place the charges on the z axis equidistant from the origin.

The dipole moment $\mathbf{p}$ is defined as

$$\mathbf{p} = q\boldsymbol{l}.$$

1. THE POTENTIAL OF THE ELECTRIC DIPOLE

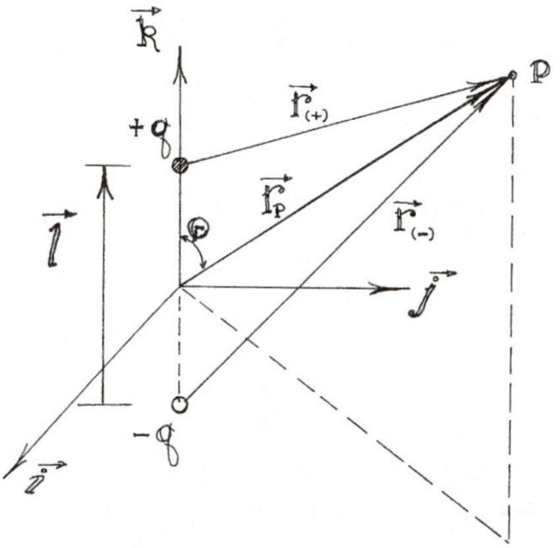

The potential at a point P (designated by $\mathbf{r}_p$) set up by $\mathbf{p}$ is

$$V = \frac{q}{4\pi\epsilon_0}\left\{\frac{1}{r_+} - \frac{1}{r_-}\right\} = \frac{q}{4\pi\epsilon_0}\left\{\frac{1}{|\mathbf{r}_p - (\boldsymbol{l}/2)|} - \frac{1}{|\mathbf{r}_p + (\boldsymbol{l}/2)|}\right\},$$

where r_+ and r_- are the vectors connecting $+q$ and $-q$ to P. This expression is exact and provides V for any $\mathbf{r}$ except for the undefined values at $\pm\boldsymbol{l}/2$.

The electric field is given by

$$\mathscr{E} = -\text{grad } V.$$

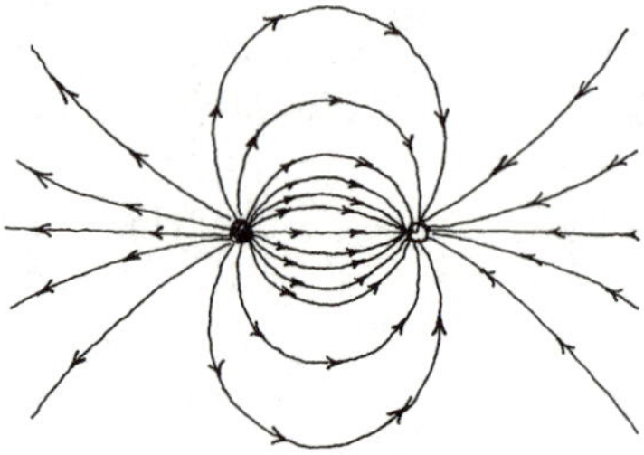

NEAR FIELD OF A DIPOLE

2. THE FAR FIELD POTENTIAL OF THE ELECTRIC DIPOLE

$\mathbf{r}$ makes an angle θ with the z axis, and the vectors $\mathbf{r}_+$ and $\mathbf{r}_-$ are

$$\mathbf{r}_+ = \mathbf{r} - (\mathbf{l}/2)$$
$$\mathbf{r}_- = \mathbf{r} + (\mathbf{l}/2),$$

while the magnitudes of $\mathbf{r}_+$ and $\mathbf{r}_-$ are

$$r_+ = r_p \left\{1 + (l/2r)^2 - \frac{\mathbf{l}\cdot\mathbf{r}}{r^2}\right\}^{1/2}$$
$$r_- = r_p \left\{1 + (l/2r)^2 + \frac{\mathbf{l}\cdot\mathbf{r}}{r^2}\right\}^{1/2}.$$

If we now make the far field assumption that $|\mathbf{l}| \ll r$ and expand the square roots dropping terms which go as l/r_p^n,

then
$$r_+ \simeq r - \frac{l}{2}\cos\theta$$
$$r_- \simeq r + \frac{l}{2}\cos\theta.$$

Substituting this result into the equation for the electric potential of the dipole,

$$V_{\text{dipole}} \simeq \frac{q}{4\pi\epsilon_0}\left\{\frac{l\cos\theta}{[r^2 - (l^2/4)\cos^2\theta]}\right\} \simeq \frac{ql}{4\pi\epsilon_0}\frac{\cos\theta}{r^2}.$$

Now $$ql \cos\theta = \mathbf{p} \cdot \mathbf{r}/r;$$

therefore $$V_{\text{dipole}} \simeq \frac{\mathbf{p} \cdot \mathbf{r}}{4\pi\epsilon_0 r^3} = -\frac{\mathbf{p} \cdot \nabla(1/r)}{4\pi\epsilon_0}.$$

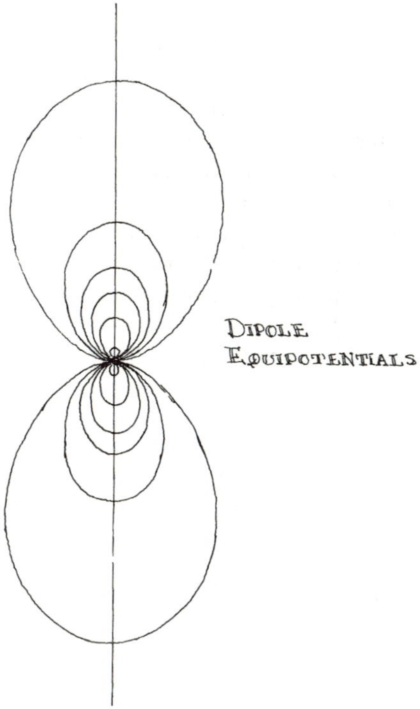

This potential is cylindrically symmetric, depending on θ but not on ϕ.

3. THE FAR FIELD OF THE ELECTRIC DIPOLE

A direct attempt to approximate $\mathscr{E}$ in the far field would have proved to be quite tedious. We observe the advantage of computing the potential first and then finding $\mathscr{E}$. Starting with

$$V \simeq \frac{\mathbf{p} \cdot \mathbf{r}}{4\pi\epsilon_0 r^3}$$

(where $\mathbf{r}$ = vector from origin to the point of examination), we find

$$\mathscr{E} = -\text{grad } V = -\frac{1}{4\pi\epsilon_0}\nabla\left(\frac{\mathbf{p}\cdot\mathbf{r}}{r^3}\right) = -\frac{1}{4\pi\epsilon_0 r^3}\nabla(\mathbf{p}\cdot\mathbf{r}) - \frac{(\mathbf{p}\cdot\mathbf{r})}{4\pi\epsilon_0}\nabla\left(\frac{1}{r^3}\right);$$

now
$$\nabla(\mathbf{p}\cdot\mathbf{r}) = \mathbf{p};$$

thus
$$\mathscr{E} = \left\{\frac{3(\mathbf{p}\cdot\mathbf{r})\mathbf{r} - r^2\mathbf{p}}{4\pi\epsilon_0 r^5}\right\}.$$

The shape of the field lines can be obtained by examining the lines in the yz plane ($x = 0$). These are the lines which are everywhere perpendicular to the equipotential surfaces. The variation in field strength is shown by variations in the density of the field lines. The field lines also provide the direction of the electric field at every point, since the direction is always tangent to the field line.

Because the direction of $\mathscr{E}$ is tangent to the field lines, the slope of the field line at a given point is equal to the ratio of the field components. In other words

$$\frac{dz}{dy} = \frac{\mathscr{E}_z}{\mathscr{E}_y} = -\frac{(y^2 - 2z^2)}{3yz},$$

which is the differential equation for the field lines. The solution, employing an integrating factor $y^{-7/3}$, is

$$f(y,z) = \tfrac{3}{2}y^{2/3} + \tfrac{3}{2}z^2/y^{4/3} = \text{const.}$$

A single line is given by

$$y^2 + z^2 = cy^{4/3}.$$

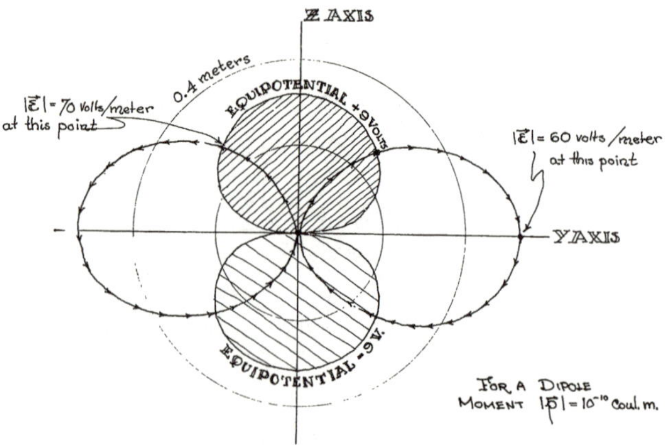

In the same plane (x = 0) the intersection of the surfaces of constant equipotential and the plane is

$$V = \frac{pz}{4\pi\epsilon_0(y^2 + z^2)^{3/2}},$$

or

$$y^2 + z^2 = \left\{\frac{pz}{4\pi\epsilon_0 V}\right\}^{2/3}.$$

An example set of field lines and equipotential curves are shown in the accompanying diagram.

4. THE POINT DIPOLE

Often it is advantageous to consider the dipole as a point quantity. In such considerations the far field and far field potentials become the exact expressions for the electric field and potential of this point dipole.

The point electrostatic dipole is constructed by a limiting process. The extended dipole consists of two charges equal in magnitude but of opposite sign connected by a vector $\boldsymbol{l}$ directed from $-q$ to $+q$. The charge q characterizing the dipole moment is allowed to become arbitrarily large while $|\boldsymbol{l}|$ approaches zero in such a manner that $\mathbf{p}$ remains finite and nonzero.

$$\mathbf{p}_{\text{point}} = \lim_{\substack{q\to\infty \\ l\to 0}} q\boldsymbol{l}.$$

The self-energy U_{self} of a dipole is derived from the interaction of the two charges. From our previous discussion the total energy stored in the doublet system is

$$U_{\text{self}} = \tfrac{1}{2}\sum_{m=1}^{2} q_m V_m = \frac{-q^2}{4\pi\epsilon_0 l},$$

or

$$U_{\text{self}} = \frac{-p^2}{4\pi\epsilon_0 l^3}.$$

The self-energy of a point singlet charge is undefined, and in a similar manner the self-energy of a point dipole is also undefined. In other words, $\lim_{l\to 0} (1/l^3)$ is an undefined quantity. The point dipole can be used in many problems, but not in those involving the self-energy of the dipole.

5. THE INTERACTION OF THE ELECTRIC DIPOLE WITH AN EXTERNAL FIELD

Assume that an external field $\mathscr{E}_0$ is established by an ensemble of source charges. Throughout this book we shall use the name "external field" to designate a field $\mathscr{E}_0$ which is not altered by the introduction of a small set of additional test charges. In the problem at hand we shall therefore employ a field which to all intents and purposes is unaltered by the introduction of a small electrostatic dipole **p**.

Let $\mathbf{r}_+$ and $\mathbf{r}_-$ represent the position vectors of $+q$ and $-q$ (the dipole charges). The energy of interaction of these two charges in the external field is

$$U = +qV(\mathbf{r}_+) - qV(\mathbf{r}_-).$$

Notice that here the factor of $\frac{1}{2}$ is not present, since this energy term does not in any manner account for the effect of the dipole charges upon the sources of the field.

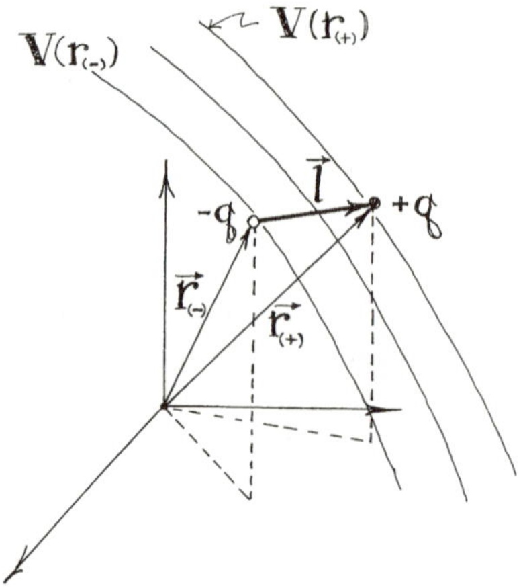

We can expand $V(\mathbf{r}_+)$ about the point $\mathbf{r}_-$ keeping only the first two terms in the Taylor's expansion,

$$V(\mathbf{r}_+) = V(\mathbf{r}_-) + (\boldsymbol{l} \cdot \nabla)V + \text{h.t.}$$

where $\boldsymbol{l}$ is $(\mathbf{r}_+ - \mathbf{r}_-)$ and h.t. means higher order terms. Then substituting for $V(\mathbf{r}_+)$,

$$U = q\boldsymbol{l} \cdot \text{grad}\, V = -\mathbf{p} \cdot \mathscr{E}.$$

The position of stable equilibrium, occurring when $\mathbf{p}$ is parallel to $\mathscr{E}$, corresponds to the minimum of the potential energy. When $\mathbf{p}$ and $\mathscr{E}$ are antiparallel the system is in unstable equilibrium.

The translational force on $\mathbf{p}$ is obtained by taking the gradient of U,

$$\mathbf{F} = -\text{grad}\, U = +\text{grad}\, (\mathbf{p} \cdot \mathscr{E}).$$

If $\mathbf{p}$ is fixed such that $\partial \mathbf{p}/\partial x_m = 0$ then $\mathbf{F}$ has the form

$$\mathbf{F} = \nabla(\mathbf{p} \cdot \mathscr{E}) = (\mathbf{p} \cdot \nabla)\mathscr{E} + (\mathscr{E} \cdot \nabla)\mathbf{p} + \mathbf{p} \times (\nabla \times \mathscr{E}) + \mathscr{E} \times (\nabla \times \mathbf{p})$$
$$= (\mathbf{p} \cdot \nabla)\mathscr{E}.$$

This latter form is exhibited when one computes the torque about a given point in space.

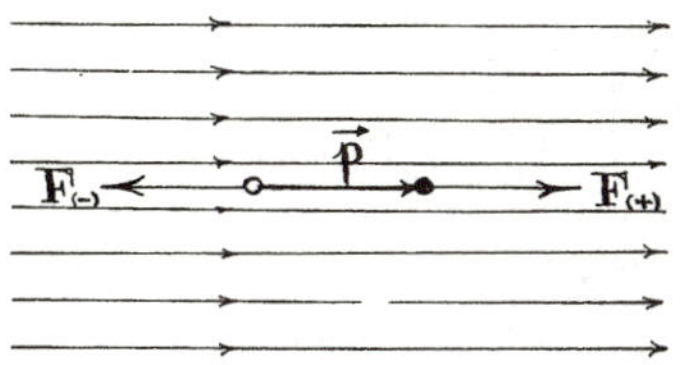

HOMOGENEOUS FIELD

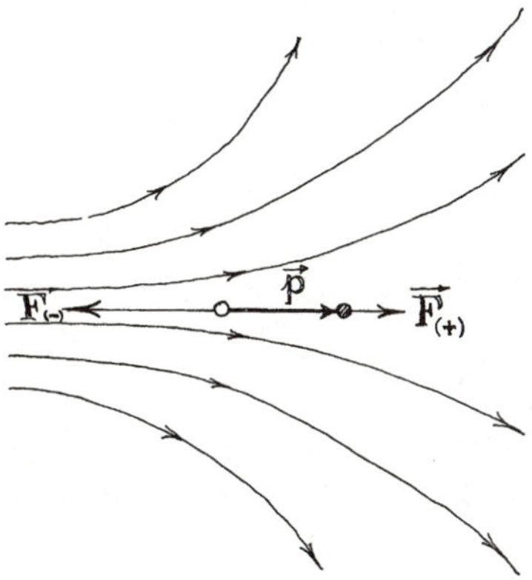

INHOMOGENEOUS FIELD

If on the other hand **p** is not a function of the coordinates, the interpretation of F is quite simple. Because $\mathbf{F} = (\mathbf{p} \cdot \nabla)\mathscr{E}$, the dipole will not experience a translational force unless the electric field is *inhomogeneous*. (In a homogeneous field the net force on the center of mass of the dipole is zero, although there can be a net torque about the center of mass.) In an inhomogeneous field the variation of $\mathscr{E}$ over the distance $\boldsymbol{l}$ produces unequal forces on the two charges of the dipole, so that the sum of the forces can be nonzero.

As an example consider the force exerted upon a dipole $\mathbf{p} = p\mathbf{k}$ situated at a point $\mathbf{r} = y\mathbf{j}$ (on the y axis), by a point charge Q at the origin. The field of the point charge is inhomogeneous and produces an initial translational force on **p** in the positive z direction. In addition to the translational force there is a torque both about the center of the dipole and about the origin O.

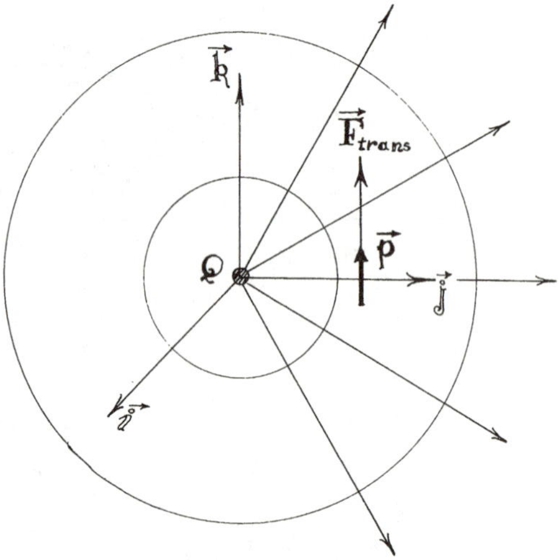

The torque $\mathscr{T}$ exerted upon a dipole $\mathbf{p} = q\boldsymbol{l}$ in an external field can be computed utilizing the individual forces exerted upon the charges. The electric field at $\mathbf{r}_+$ is

$$\mathscr{E}(\mathbf{r}_+) = \mathscr{E}(\mathbf{r}_-) + [(\boldsymbol{l} \cdot \nabla)\mathscr{E}]_{\mathbf{r}_-} + \text{h.t.}$$

At $\mathbf{r}_-$ the field is $\mathscr{E}(\mathbf{r}_-)$. The forces on the charges are

$$\mathbf{F}_+ = q\mathscr{E}(\mathbf{r}_+)$$
$$\simeq q\mathscr{E}(\mathbf{r}_-) + q[(\boldsymbol{l}\cdot\nabla)\mathscr{E}]_{\mathbf{r}-},$$

while
$$\mathbf{F}_- = q\mathscr{E}(\mathbf{r}_-).$$

The torque on $\mathbf{p}$ is

$$\mathscr{T} = \sum_{m=1}^{2} \mathbf{r}_m \times \mathbf{F}_m = \mathbf{r}_+ \times \mathbf{F}_+ + \mathbf{r}_- \times \mathbf{F}_-.$$

Substituting for $\mathbf{F}_+$ and collecting terms,

$$\mathscr{T} = q(\mathbf{r}_+ - \mathbf{r}_-) \times \mathscr{E}(\mathbf{r}_-) + (\mathbf{r}_+) \times [(\mathbf{p}\cdot\nabla)\mathscr{E}].$$

In the limit as $|\boldsymbol{l}| \rightarrow 0$,

$$\mathscr{T} = \mathbf{p} \times \mathscr{E} + \mathbf{r} \times [(\mathbf{p}\cdot\nabla)\mathscr{E}].$$

The term $(\mathbf{p} \times \mathscr{E})$ represents the torque exerted about the center of charge of the dipole. This torque tends to orient $\mathbf{p}$ in the direction corresponding to stable equilibrium. The second term, $\mathbf{r} \times [(\mathbf{p}\cdot\nabla)\mathscr{E}]$, represents the torque arising from the presence of a translational force $(\mathbf{p}\cdot\nabla)\mathscr{E}$ whose value depends on the geometry, i.e., on the location of the origin O.

6. THE DIPOLE-DIPOLE INTERACTION

When a dipole $\mathbf{p}_2$ is brought into the vicinity of a dipole $\mathbf{p}_1$ there is a dipole pair interaction. This can be readily computed by taking the

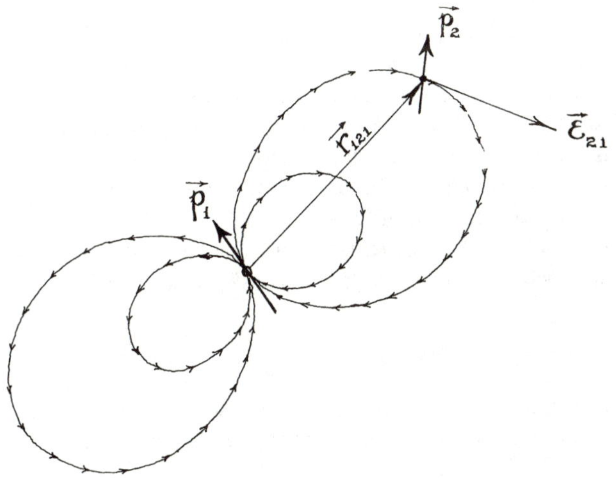

field interactions of $\mathbf{p}_2$ with the field $\mathbf{p}_1$, which we designate as $\mathscr{E}_{21}$.

$$\mathscr{E}_{21} = \frac{3(\mathbf{p}_1 \cdot \mathbf{r}_{21})\mathbf{r}_{21} - \mathbf{r}_{21}{}^2\mathbf{p}_1}{4\pi\epsilon_0 \mathbf{r}_{21}{}^5},$$

where

$$\mathbf{r}_{21} = \mathbf{r}_2 - \mathbf{r}_1.$$

$$U = -\mathbf{p}_2 \cdot \mathscr{E}_{21} = \frac{-3(\mathbf{p}_1 \cdot \mathbf{r}_{21})(\mathbf{p}_2 \cdot \mathbf{p}_{21}) + \mathbf{r}_{21}{}^2(\mathbf{p}_1 \cdot \mathbf{p}_2)}{4\pi\epsilon_0 \mathbf{r}_{21}{}^5},$$

or by assigning the energy symmetrically and considering N point dipoles,

$$U = \frac{1}{2} \sum_{m=1}^{N} \sum_{\substack{n=1 \\ m \neq n}}^{N} \left\{ \frac{3(\mathbf{p}_m \cdot \mathbf{r}_{mn})(\mathbf{p}_n \cdot \mathbf{r}_{nm}) + \mathbf{r}_{mn}{}^2(\mathbf{p}_m \cdot \mathbf{p}_n)}{4\pi\epsilon_0 \mathbf{r}_{mn}{}^5} \right\}.$$

This last relation holds for an ensemble of N point dipoles. For a single dipole-dipole interaction we merely set N = 2.

Notice the change in sign on the first term which took place when the indices were interchanged on one of the connecting vectors $\mathbf{r}_{mn}$.

The dipole-dipole potential terms are encountered quite often in atomic and molecular problems.

I. POINT CHARGE MULTIPOLES: THE RECTANGULAR QUADRUPOLE

The singlet is a pole of order zero, and the dipole is a pole of order one. The pole of order two is the quadrupole. In contrast to the singlet and doublet which have unique charge distributions, the quadrupole may have a number of geometric arrangements all of which exhibit characteristic quadrupole fields.

The reasons for this lack of unique geometry are apparent when we consider that on the one hand the dipole or doublet has a geometry of two points which can be associated with only one connecting vector, while the quadrupole can be thought of as being composed of two dipoles directed oppositely (to cancel the over-all dipole moment) with their geometric centers separated by a distance d.

1. THE EXACT POTENTIAL

The simplest of all quadrupoles is the square quadrupole. The charges of this configuration lie upon the corners of a square and

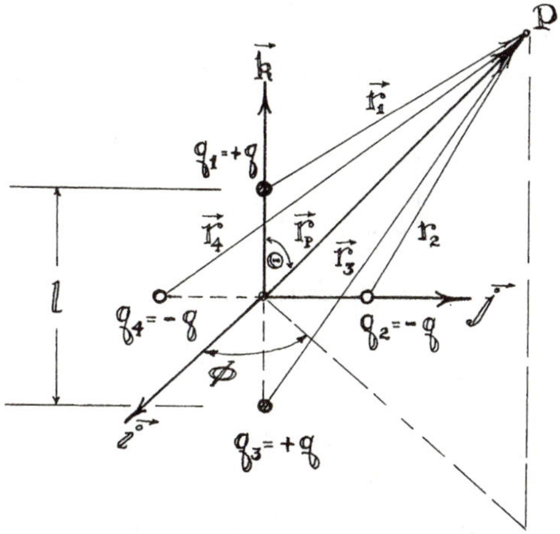

alternate in sign. From the diagram we see that in our particular example the square quadrupole is made up of two dipoles $\pm q(l/\sqrt{2})\,(\mathbf{j} + \mathbf{k})$ with separation of centers $l/\sqrt{2}$.

Because the charge distribution is no longer axially symmetric the field and potential will *not* have cylindrical symmetry relative to the z axis. This, of course, implies that in addition to an r,θ dependence we can expect a ϕ dependence in the potential.

If we label the charges clockwise as shown in the diagram, q_1 and q_3 are equal to $+q$ while q_2 and q_4 are $-q$. The vectors connecting q_j and the point of examination P are labeled $\mathbf{r}_j$.

The exact (near field and far field) potential of the square quadrupole is

$$V = \sum_{j=1}^{4} V_j = \sum_{j=1}^{4} \frac{q_j}{4\pi\epsilon_0} \frac{1}{r_j} = \frac{q}{4\pi\epsilon_0} \left\{ \frac{1}{r_1} - \frac{1}{r_2} + \frac{1}{r_3} - \frac{1}{r_4} \right\}.$$

NOW!

$$\mathbf{r}_1 = \mathbf{r}_p - \frac{l}{2}\mathbf{k}, \qquad r_1 = \left\{ r_p^2 + \frac{l^2}{4} - r_p l \cos\theta \right\}^{1/2};$$

$$\mathbf{r}_2 = \mathbf{r}_p - \frac{l}{2}\mathbf{j}, \qquad r_2 = \left\{ r_p^2 + \frac{l^2}{4} - r_p l \sin\theta \sin\phi \right\}^{1/2};$$

$$\mathbf{r}_3 = \mathbf{r}_p + \frac{l}{2}\mathbf{k}, \qquad r_3 = \left\{ r_p^2 + \frac{l^2}{4} + r_p l \cos\theta \right\}^{1/2};$$

$$\mathbf{r}_4 = \mathbf{r}_p + \frac{l}{2}\mathbf{j}, \qquad r_4 = \left\{ r_p^2 + \frac{l^2}{4} + r_p l \sin\theta \sin\phi \right\}^{1/2}.$$

Here θ and ϕ are the spherical coordinates of r_p.

2. THE FAR FIELD APPROXIMATION

When $l \ll r_p$, then (expanding the square root expressions)

$$r_1 \simeq r_p - \frac{l}{2}\cos\theta; \qquad r_2 \simeq r_p - \frac{l}{2}\sin\theta\sin\phi;$$

$$r_3 \simeq r_p + \frac{l}{2}\cos\theta; \qquad r_4 \simeq r_p + \frac{l}{2}\sin\theta\sin\phi.$$

Substituting into the expression for the quadrupole potential,

$$V_p = \frac{q}{4\pi\epsilon_0}\left\{\frac{r_1 + r_3}{r_1 r_3} - \frac{r_2 + r_4}{r_2 r_4}\right\}$$

and $V_p \simeq \dfrac{ql^2}{8\pi\epsilon_0}\dfrac{(\cos^2\theta - \sin^2\theta\sin^2\phi)}{r_p^3} = \dfrac{ql^2}{8\pi\epsilon_0}\dfrac{(z_p^2 - y_p^2)}{r_p^5}$

The electric field intensity in this approximation is

$$\mathscr{E}_p = -\text{grad } V_p$$

$$\simeq \frac{ql^2}{8\pi\epsilon_0}\left\{\frac{5x_p(y_p^2 - z_p^2)\mathbf{i} + y_p(-2x_p^2 + 3y_p^2 - 7z_p^2)\mathbf{j} + z_p(2x_p^2 + 7y_p^2 - 3z_p^2)\mathbf{k}}{r_p^7}\right\}$$

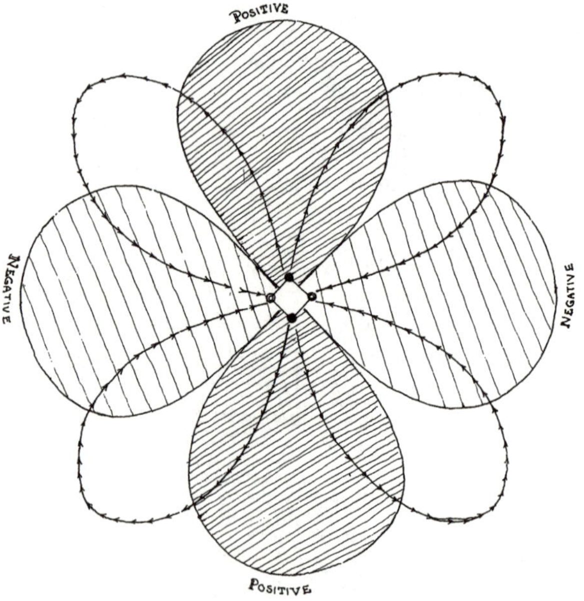

In polar coordinates in the yz plane,

$$\mathscr{E}_p(x_p = 0) \simeq \frac{ql^2}{8\pi\epsilon_0}\left\{\frac{3\cos 2\theta\,\boldsymbol{\epsilon}_r + 2\sin 2\theta\,\boldsymbol{\epsilon}_\theta}{r_p^{\,4}}\right\}.$$

3. THE INTERACTION OF THE RECTANGULAR QUADRUPOLE WITH AN EXTERNAL FIELD

Suppose a rectangular quadrupole is constructed from four charges equal in magnitude but of alternating sign. The charges lie at the corners of a rectangle having edges $|\boldsymbol{l}|$ and $|\mathbf{d}|$.

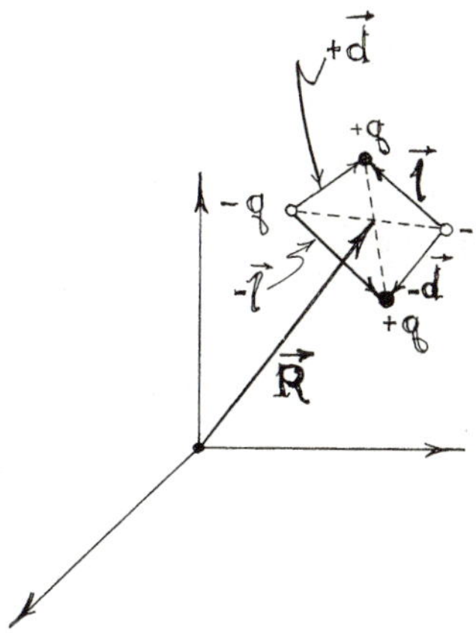

With the labeling shown in the diagram, the center of symmetry of the rectangle is given by

$$\mathbf{R} = \frac{\sum_{j=1}^{4} |q_j|\,\mathbf{r}_j}{\sum_{j=1}^{4} |q_j|}$$

and the location vectors of the charges are the following:

$$\begin{aligned}
\text{for}\quad q_1 &= +q, & \mathbf{r}_1 &= \mathbf{R} + \tfrac{1}{2}(\boldsymbol{l} - \mathbf{d});\\
q_2 &= -q, & \mathbf{r}_2 &= \mathbf{R} + \tfrac{1}{2}(\boldsymbol{l} + \mathbf{d});\\
q_3 &= +q, & \mathbf{r}_3 &= \mathbf{R} + \tfrac{1}{2}(-\,\boldsymbol{l} + \mathbf{d});\\
q_4 &= -q, & \mathbf{r}_4 &= \mathbf{R} - \tfrac{1}{2}(\boldsymbol{l} + \mathbf{d}).
\end{aligned}$$

The sign convention is set by requiring that $\pm \boldsymbol{l}$ and $\pm \mathbf{d}$ terminate on positive charges. The potential energy U of this system in the presence of an external field $\mathscr{E}$ with an associated potential V is

$$U = \sum_{j=1}^{4} q_j V(\mathbf{r}_j).$$

The potential functions $V(\mathbf{r}_j)$ can be expanded about the symmetry point designated by **R**. For instance,

$$V(\mathbf{r}_1) = V(\mathbf{R}) + \left[\left\{\frac{1}{2}(\boldsymbol{l} - \mathbf{d}) \cdot \nabla\right\} V\right]_{\mathbf{r}=\mathbf{R}}$$
$$+ \frac{1}{2!}\left[\left\{\frac{1}{2}(\boldsymbol{l} - \mathbf{d}) \cdot \nabla\right\}\left\{\frac{1}{2}(\boldsymbol{l} - \mathbf{d}) \cdot \nabla\right\} V\right]_{\mathbf{r}=\mathbf{R}} + \text{h.t.}$$

The last term can be written

$$-\left[\left\{\frac{1}{2}(\boldsymbol{l} - \mathbf{d}) \cdot \nabla\right\}\left\{\frac{1}{2}(\boldsymbol{l} - \mathbf{d}) \cdot \mathscr{E}\right\}\right]_{\mathbf{r}=\mathbf{R}}.$$

By expanding each $V(\mathbf{r}_j)$ in this manner and carrying out the sum over j we find that the first two terms in the expansion cancel. As a result the interaction is specified by the third (quadratic) term. After summing over j we obtain

$$U = +\tfrac{1}{4}q\{\boldsymbol{l} \cdot \nabla(\mathbf{d} \cdot \mathscr{E}) + \mathbf{d} \cdot \nabla(\boldsymbol{l} \cdot \mathscr{E})\}.$$

If we think of the rectangular quadrupole as two dipoles $\pm \mathbf{p} = \pm q\boldsymbol{l}$ separated by a distance $|\mathbf{d}|$,

$$U = +\tfrac{1}{4}\{\mathbf{p} \cdot \nabla(\mathbf{d} \cdot \mathscr{E}) + \mathbf{d} \cdot \nabla(\mathbf{p} \cdot \mathscr{E})\}.$$

This expression is symmetric in $\boldsymbol{l}$ and **d**. If we had characterized the quadrupole by two dipoles $\pm \mathbf{p}' = \pm q\mathbf{d}$, the equation for U would have had the same form.

The translational force on the quadrupole in the external field is

$$\mathbf{F} = -\text{grad}\, U = -\tfrac{1}{4}\,\text{grad}\,\{\mathbf{p} \cdot \nabla(\mathbf{d} \cdot \mathscr{E}) + \mathbf{d} \cdot \nabla(\mathbf{p} \cdot \mathscr{E})\}$$
$$= -\frac{q}{2}\,\text{grad div}\,\{\boldsymbol{l}(\mathbf{d} \cdot \mathscr{E}) + \mathbf{d}(\boldsymbol{l} \cdot \mathscr{E})\}.$$

The interaction of the quadrupole with an external field is more complex than that of a singlet or dipole. To illustrate the behavior of a special quadrupole we construct an inhomogeneous external field specified by

$$V = -z\mathscr{E}_0 e^{-y^2/2\alpha^2}$$

where α has the dimensions of a length. The associated field is

$$\mathcal{E} = +\mathcal{E}_0 e^{-y^2/2\alpha^2} \left\{ \frac{-zy}{\alpha^2} \mathbf{j} + \mathbf{k} \right\}.$$

According to our formalism the potential energy of a quadrupole specified by q and

$$\boldsymbol{l} = l\mathbf{k}, \quad \mathbf{d} = d\mathbf{j}$$

is

$$U = -\frac{qld}{\alpha^2} y\mathcal{E}_0 \, e^{-y^2/2\alpha^2},$$

with the minimum at $y = +\alpha$. The translational force is then

$$\mathbf{F} = \frac{qld}{\alpha^2} \mathcal{E}_0(1 - y^2/\alpha^2) \, e^{-y^2/2\alpha^2}.$$

It is perhaps interesting to see that the position of stable equilibrium does not lie at the maximum of the field but rather to one side or the other of it, depending on the orientation of the quadrupole.

J. THE MULTIPOLE EXPANSION

Systematic behavior can be attributed to the multipoles discussed. The singlet potential varies as $1/r$ and is isotropically distributed. The dipole potential varies as $1/r^2$, and also varies as $\cos\theta$ when $\mathbf{p}$ is aligned along the z axis. In addition we have noticed that a rectangular quadrupole can be represented as a superposition of two of the next lowest multipoles (dipoles). In the same manner the dipole is a superposition of two singlets. In each case the two lower multipoles are arranged to cancel, thus producing the next higher multipole. Without further analysis we might guess that the octupole (the multipole following the quadrupole) could be formed from two quadrupoles arranged in such a manner that the quadrupole characteristics cancel. Further, we might conclude that the octupole potential varies as $1/r^4$.

Given a set of N charges q_m we can expand the far field potential ($r_p \gg r_j$) of this set in a Taylor's series which finally will represent a series of multipoles of ascending order.

The potential of N point charges q_m at a point P is

$$V_p = \sum_{m=1}^{N} \frac{q_m}{4\pi\epsilon_0} \frac{1}{|\mathbf{r}_p - \mathbf{r}_m|}.$$

If $r_m \ll r_p$ for all m we can expand V_p about the origin in a Taylor's series. This expansion is actually an expansion of each of the terms $|\mathbf{r}_p - \mathbf{r}_m|^{-1}$. Expanding* about $r = 0$ (the origin),

$$\frac{1}{|\mathbf{r}_p - \mathbf{r}_j|} = \left(\frac{1}{|\mathbf{r}_p - \mathbf{r}|}\right)_{r=0} + \frac{1}{1!}\sum_{n=1}^{3}\left\{\frac{\partial}{\partial x_n}\left(\frac{1}{|\mathbf{r}_p - \mathbf{r}|}\right)\right\}_{r=0}(x_n^{(j)} - 0)$$

$$+ \frac{1}{2!}\sum_{m=1}^{3}\sum_{n=1}^{3}\left\{\frac{\partial^2}{\partial x_m \partial x_n}\left(\frac{1}{|\mathbf{r}_p - \mathbf{r}|}\right)\right\}_{r=0}(x_m^{(j)} - 0)(x_n^{(j)} - 0) + \text{h.t.}$$

Notice that by $x_3^{(j)}$ we mean z_j.

Term by term this series is

$$\frac{1}{|\mathbf{r}_p - \mathbf{r}_j|} = \frac{1}{r_p} + \left\{\frac{x_p x_j}{r_p^3} + \frac{y_p y_j}{r_p^3} + \frac{z_p z_j}{r_p^3}\right\}$$
$$+ 2^{\text{nd}} \text{ order terms, etc.}$$

For convenience the notation can be altered.

CHANGE IN NOTATION!

The arrows indicate that the first symbol will now be written in the form which follows it. Let

$$x_m \to x_1^{(m)} \quad \text{and} \quad x_p \to x_1^{(p)}$$
$$y_m \to x_2^{(m)} \quad \text{and} \quad y_p \to x_2^{(p)}$$
$$z_m \to x_3^{(m)} \quad \text{and} \quad z_p \to x_3^{(p)}.$$

Writing a Taylor's expansion for each particle and using the relation

$$\sum_{n=1}^{3}\left(\frac{1}{r_p^3}\right)x_n^{(p)}x_n^{(j)} = -\left\{\nabla_p\left(\frac{1}{r_p}\right)\right\}\cdot \mathbf{r}_j,$$

we find

$$V_p = \frac{1}{4\pi\epsilon_0}\left(\frac{1}{r_p}\right)\sum_{m=1}^{N} q_m - \frac{1}{4\pi\epsilon_0}\left\{\nabla_p\left(\frac{1}{r_p}\right)\right\}\cdot\sum_{m=1}^{N}(q_m\mathbf{r}_m)$$

$$+ \frac{1}{2!}\frac{1}{4\pi\epsilon_0}\sum_{m=1}^{N} q_m\left\{\sum_{n=1}^{3}\sum_{k=1}^{3}(3x_n^{(m)}x_k^{(m)} - r_m^2\,\delta_{nk})x_n^{(p)}x_k^{(p)}\right\} + \text{h.t.}$$

The form of the third term will be discussed in detail.

The first term in the series represents the effective singlet contribution. If $\sum_{m=1}^{N} q_m$ does not vanish, then this term predominates in the far field since it varies as $1/r_p$ while all of the higher terms vary as

* $x_n^{(j)}$ is the nth coordinate of the jth charge; e.g., $x_2^{(10)}$ is the y coordinate of the charge $q_{(10)}$.

$1/r_p^n$ where $1 < n$:

$$V_{\text{singlet}} \rightarrow \frac{1}{4\pi\epsilon_0} \sum_{m=1}^{N} q_m \left(\frac{1}{r_p}\right).$$

The second term in the series is the effective dipole contribution. This term is nonzero if $\sum_{m=1}^{N} q_m \mathbf{r}_m$ does not vanish:

$$V_{\text{dipole}} \rightarrow -\frac{1}{4\pi\epsilon_0} \nabla_p \left(\frac{1}{r_p}\right) \cdot \sum_{m=1}^{N} q_m \mathbf{r}_m.$$

The third term in the series represents the quadrupole contribution to the potential. Because the quadrupole potential varies as $1/r_p^3$ it is not significant at large r_p unless the singlet and dipole terms vanish. The quadrupole potential term is quadratic in the coordinates of the charges, and as a result it can be represented in terms of a second-order tensor having three symmetry axes or eigenvectors.

To develop the quadrupole tensor let us manipulate the third term in V_p in some detail. Rearranging the form shown previously,

$$\begin{aligned} V_{\text{quad}} \rightarrow \frac{1}{2!\,4\pi\epsilon_0} \sum_{m=1}^{N} q_m \left(\frac{1}{r_p^5}\right) \{ & (2x_m^2 - y_m^2 - z_m^2) x_p^2 \\ & + 2(3x_m y_m) x_p y_p + (-x_m^2 + 2y_m^2 - z_m^2) y_p^2 \\ & + 2(3y_m z_m) y_p z_p + (-x_m^2 - y_m^2 + 2z_m^2) z_p^2 \\ & + 2(3z_m x_m) z_p x_p \}. \end{aligned}$$

We can then write (using the more compact notation)

$$V_{\text{quad}} = \frac{1}{2!} \sum_{m=1}^{N} \frac{q_m}{4\pi\epsilon_0 r_p^5} \sum_{j=1}^{3} \sum_{k=1}^{3} \{ x_j^{(p)} (3x_j^{(m)} x_k^{(m)} - r_m^2 \delta_{jk}) x_k^{(p)} \},$$

or by interchanging the order of summing

$$V_{\text{quad}} = \frac{1}{8\pi\epsilon_0 r_p^5} \sum_{j=1}^{3} \sum_{k=1}^{3} x_j^{(p)} Q_{jk} x_k^{(p)};$$

where the *components of the quadrupole tensor* Q_{jk} are

$$Q_{jk} = \sum_{m=1}^{N} q_m (3x_j^{(m)} x_k^{(m)} - r_m^2 \delta_{jk}).$$

If we write out a few of these terms we see that

$$Q_{11} = \sum_{m=1}^{N} q_m \{ (2x_1^{(m)})^2 - (x_2^{(m)})^2 - (x_3^{(m)})^2 \},$$

$$Q_{12} = \sum_{m=1}^{N} q_m (3x_1^{(m)} x_2^{(m)}), \quad \text{etc.}$$

Now we write the position vector $\mathbf{r}_p$ as a column matrix,

$$\mathbf{r}_p = \begin{bmatrix} x_1^{(p)} \\ x_2^{(p)} \\ x_3^{(p)} \end{bmatrix},$$

and the matrix Q as

$$Q = \begin{pmatrix} Q_{11} & Q_{12} & Q_{13} \\ Q_{21} & Q_{22} & Q_{23} \\ Q_{31} & Q_{32} & Q_{33} \end{pmatrix}.$$

With this representation

$$V_{quad} = \frac{1}{8\pi\epsilon_0 r_p^5} \{\tilde{\mathbf{r}}_p \cdot Q \cdot \mathbf{r}_p\},$$

where $\tilde{\mathbf{r}}_p$ is the transpose of $\mathbf{r}_p$ (i.e., a row matrix).

For a system of N point charges the Q matrix will contain "off diagonal" terms. However, the Q matrix is symmetric and therefore can be diagonalized by a spatial rotation.

The elements of the rotation matrix are found from the eigenvector problem for Q:

$$Q \cdot \mathbf{R}_j = \lambda_j \mathbf{R}_j,$$

where the λ_j's are the roots of

$$|Q - \lambda\, I| = 0.$$

The elements of the normalized $\mathbf{R}_j$'s, i.e., the X_{mj}, make up the components S'_{mj} of S^{-1},

$$X_{mj} = S'_{mj};$$

where

$$\mathbf{R}_j = \begin{bmatrix} X_{1j} \\ X_{2j} \\ X_{3j} \end{bmatrix}$$

and

$$\tilde{\mathbf{R}}_j \cdot \mathbf{R}_j = \sum_{k=1}^{3} X_{kj}^2 = 1.$$

The eigenvalues λ_j make up the elements of the diagonal quadrupole tensor Q' obtained by a rotation of the coordinates S,

$$Q' = \begin{pmatrix} \lambda_1 & 0 & 0 \\ 0 & \lambda_2 & 0 \\ 0 & 0 & \lambda_3 \end{pmatrix} = S \cdot Q \cdot S^{-1}.$$

In the transformed axes the charge distribution behaves like an equivalent rectangular quadrupole.

To further understand this matrix it is instructive to compute the quadrupole matrix elements of a square quadrupole in which the charges are *not* aligned along the coordinate axes. Assume charges +q at

$$\left(0, \frac{l}{2\sqrt{2}}, \frac{l}{2\sqrt{2}}\right) \quad \text{and} \quad \left(0, -\frac{l}{2\sqrt{2}}, -\frac{l}{2\sqrt{2}}\right);$$

place charges −q at

$$\left(0, \frac{l}{2\sqrt{2}}, -\frac{l}{2\sqrt{2}}\right) \quad \text{and} \quad \left(0, -\frac{l}{2\sqrt{2}}, \frac{l}{2\sqrt{2}}\right).$$

The charge placement is shown in the figure.

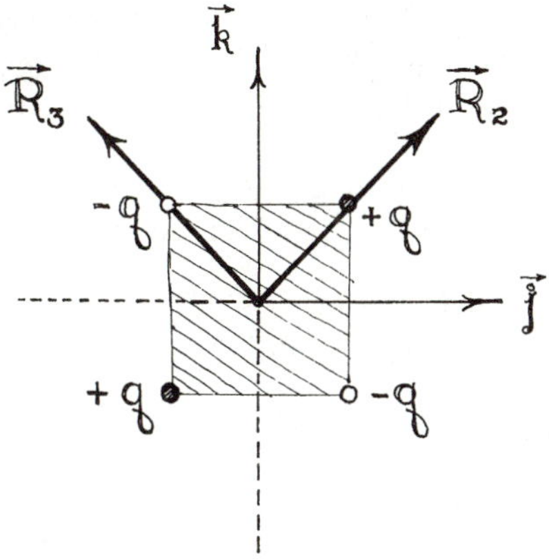

According to our formula for the elements of Q we find that all are zero except $Q_{23} = Q_{32} = \frac{3}{2}ql^2$, or

$$Q = \tfrac{3}{2}ql^2 \begin{pmatrix} 0 & 0 & 0 \\ 0 & 0 & 1 \\ 0 & 1 & 0 \end{pmatrix}.$$

Solving for the eigenvalues,

$$|Q - \lambda I| = -\lambda^3 + \lambda = 0,$$

or

$$\lambda_1 = 0, \quad \lambda_2 = -\tfrac{3}{2}ql^2,$$
$$\lambda_3 = +\tfrac{3}{2}ql^2.$$

The corresponding eigenvectors obtained are

$$\mathbf{R}_1 = \begin{bmatrix} 1 \\ 0 \\ 0 \end{bmatrix}, \quad \mathbf{R}_2 = \frac{1}{\sqrt{2}} \begin{bmatrix} 0 \\ 1 \\ 1 \end{bmatrix}, \quad \mathbf{R}_3 = \frac{1}{\sqrt{2}} \begin{bmatrix} 0 \\ -1 \\ 1 \end{bmatrix}.$$

The diagram shows the $\mathbf{R}_j$'s and their relation to the q's. Using the components of the eigenvectors we find that

$$S^{-1} = \begin{pmatrix} 1 & 0 & 0 \\ 0 & \frac{1}{\sqrt{2}} & \frac{-1}{\sqrt{2}} \\ 0 & \frac{1}{\sqrt{2}} & \frac{1}{\sqrt{2}} \end{pmatrix}.$$

Some care must be exercised concerning the phase of $\mathbf{R}_3$. If we had taken $-\mathbf{R}_3$ (an equally valid choice), then $|S^{-1}| = -1$ corresponding to an improper rotation (a rotation and a reflection through the origin).

Constructing S, knowing that $S = S^{-1}$ for an orthogonal transformation, we find as a check on our method that

$$S \cdot \mathbf{Q} \cdot S^{-1} = \tfrac{3}{2} q l^2 \begin{pmatrix} 0 & 0 & 0 \\ 0 & 1 & 0 \\ 0 & 0 & -1 \end{pmatrix}.$$

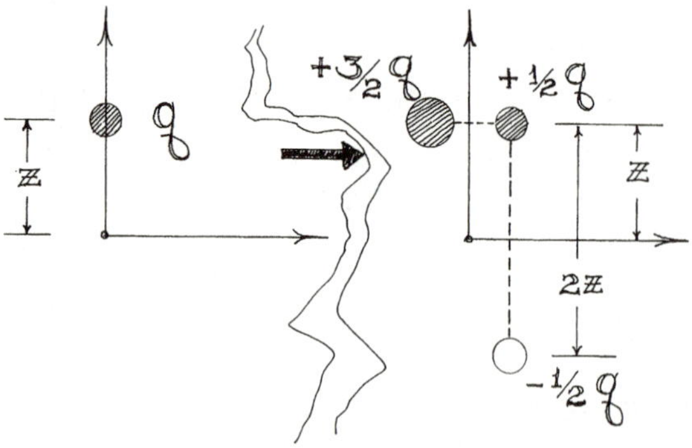

A GEOMETRIC MULTIPOLE

A clear distinction should be made betweeen GEOMETRIC and INTRINSIC multipoles. For instance, a single point charge $+q$ at a point on the z axis $\mathbf{r} = z\mathbf{k}$ has a *geometric dipole moment*. In other words

$$\sum_{m=1}^{1} |q_m|\, \mathbf{r}_m \neq 0;$$

in fact,

$$\sum_{m=1}^{1} |q_m|\, \mathbf{r}_m = qz\mathbf{k}.$$

To avoid geometric effects, the origin can always be shifted to the center of charge. This point is analogous to the center of mass in mechanics except that in electrostatics both positive and negative quantities are involved. Location of the center of charge is provided by a vector $\mathbf{R}$,*

$$\mathbf{R} = \frac{\sum_{m=1}^{N} |q_m|\, \mathbf{r}_m}{\sum_{m} |q_m|}$$

Thus in our singlet charge problem above, $\mathbf{R} = z\mathbf{k}$. Then

$$\sum_{m=1}^{1} |q_m|\, \mathbf{r}_m' = 0$$

where

$$\mathbf{r}_m' = \mathbf{r}_m - \mathbf{R}.$$

In the examples concerning the single dipole and the square quadrupole the origin was placed at the center of charge.

An INTRINSIC MULTIPOLE is one which is nonzero when expanded about the center of charge.

In conclusion we can rewrite our Taylor's expansion in terms of our abbreviated notation, with the expansion performed about the center of charge. Given N charges q_m positioned by vectors $\mathbf{r}_m$, we define

$$\mathbf{R} = \frac{\sum_{m=1}^{N} |q_m|\, \mathbf{r}_m}{\sum_{m=1}^{N} |q_m|}$$

$$\mathbf{p} = \sum_{m=1}^{N} q_m \mathbf{r}_m'$$

where $\quad \mathbf{r}_m' = \mathbf{r}_m - \mathbf{R} \quad$ and $\quad \mathbf{r}_p' = \mathbf{r}_p - \mathbf{R}.$

* We use only the magnitude of the charges to locate the center of charge.

Also,
$$Q_{ij} = \sum_{m=1}^{N} q_m(3x_i^{(m)\prime}x_j^{(m)\prime} - r_m'^2\delta_{ij}).$$
Then
$$V_p = \frac{1}{4\pi\epsilon_0}\left\{\sum_{m=1}^{N} q_m\right\}\frac{1}{r_p'} - \frac{1}{4\pi\epsilon_0}\nabla_p\left(\frac{1}{r_p'}\right)\cdot \mathbf{p}$$
$$+ \frac{1}{8\pi\epsilon_0}\left(\frac{1}{r_p'^5}\right)\{\tilde{\mathbf{r}}_p'\cdot \mathrm{Q}\cdot \mathbf{r}_p'\} + \text{h.t.}$$

To analyze, first shift the origin to the center of charge and then expand about that point.

Finally it should be kept in mind that a charge distribution is characterized by its lowest nonvanishing term in the multipole expansion.

A few features of symmetric multipoles are listed in the table.

Pole	Order of pole	r_p dependence	θ dependence	Characteristic sum
Singlet	0	$1/r$	1	$\sum_m q_m$
Dipole	1	$1/r^2$	$\cos\theta$	$\sum_m q_m\mathbf{r}_m$
Quadrupole	2	$1/r^3$	$\cos^2\theta$	$\sum_m q_m(3x_j^{(m)}x_k^{(m)} - r_m{}^2\,\delta_{jk})$
Octupole	3	$1/r^4$	$\cos^3\theta$	etc.

III. Distributions of Charge

A. SINGLET DISTRIBUTIONS

1. INTEGRAL FORMS OF $\mathscr{E}$ AND V

Assume that singlet charges are distributed continuously throughout a volume confined within a closed surface S, the volume charge density being ρ. $\rho(x_1,x_2,x_3)$ is the quantity of charge per unit volume. If the charge contained in a volume element $\Delta\tau$ is Δq, then

$$\rho = \lim_{\Delta\tau\to 0} \frac{\Delta q}{\Delta\tau}.$$

In order to avoid unnecessary separations of the charge distribution into point charges, surface distributions, and volume distributions, we can assume that the charge density function ρ contains all of these. Point charges will appear as delta functions, while surface

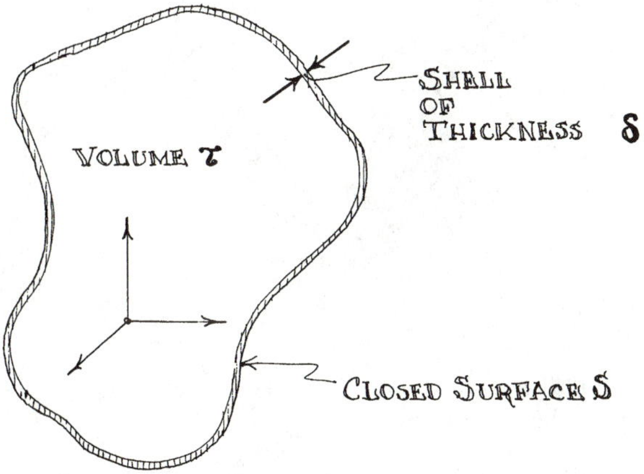

charge densities σ can be thought of as a volume distribution in a shell of uniform, small thickness δ. The conventional surface density σ is then

$$\sigma = \lim_{\Delta S \to 0} \frac{\Delta q}{\Delta S}.$$

Set
$$\Delta q = \rho \Delta S \delta;$$
then
$$\sigma = \rho\delta, \quad \text{with } \delta \text{ arbitrarily small.}$$

The superposition principle has been demonstrated for point charges. In dealing with the electric field and the potential of a charged volume element, we shall treat the element as a point charge. The total field and potential at any point of examination P is merely the sum of the contributions from all charged volume elements $\rho\Delta\tau$.

If the field produced at P by $\rho\Delta\tau$ is $\Delta\mathscr{E}_p$, then

$$\Delta\mathscr{E}_p = \frac{\rho\Delta\tau}{4\pi\epsilon_0} \frac{(\mathbf{r}_p - \mathbf{r})}{|\mathbf{r}_p - \mathbf{r}|^3},$$

with
$$\Delta V_p = \frac{\rho\Delta\tau}{4\pi\epsilon_0} \frac{1}{|\mathbf{r}_p - \mathbf{r}|}.$$

$\mathbf{r}$ is the vector locating the volume element $\Delta\tau$.

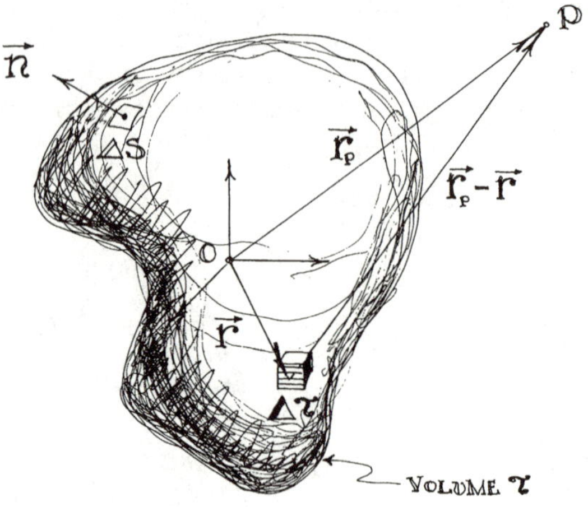

Summing over the volume τ which encloses the charge distribution and taking the limit as $\Delta\tau \to 0$,

$$\mathscr{E}_p = \lim_{\Delta\tau \to 0} \sum_{\text{all}\,\Delta\tau} \Delta\mathscr{E}_p,$$

giving $$\mathscr{E}_p = \frac{1}{4\pi\epsilon_0}\iiint_\tau \frac{\rho(x_1,x_2,x_3)(\mathbf{r}_p - \mathbf{r})\,d\tau}{|\mathbf{r}_p - \mathbf{r}|^3}.$$

Expanded in terms of cartesian components this integral is

$$\mathscr{E}_p = \frac{1}{4\pi\epsilon_0}\iiint_\tau \frac{\rho(x,y,z)\{(x_p - x)\mathbf{i} + (y_p - y)\mathbf{j} + (z_p - z)\mathbf{k}\}\,dx\,dy\,dz}{\{(x_p - x)^2 + (y_p - y)^2 + (z_p - z)^2\}^{3/2}}.$$

The potential at P is obtained in the same manner,

$$V_p = \frac{1}{4\pi\epsilon_0}\iiint_\tau \frac{\rho(x,y,z)\,dx\,dy\,dz}{\{(x_p - x)^2 + (y_p - y)^2 + (z_p - z)^2\}^{1/2}}.$$

During this development the integrations have been limited to a specified volume of interest, τ. In general if ρ is specified for every point in space the integrations can be extended to all space. In other words, the integration over each coordinate x_j then extends from $-\infty$ to $+\infty$. Consider the case implied in the previous integrals: there ρ is considered to be nonzero inside τ and zero for all points exterior to τ. With this qualification we can define the integrals as extending over all space,

$$V_p = \frac{1}{4\pi\epsilon_0}\iiint_{\text{All Space}} \frac{\rho\,d\tau}{|\mathbf{r}_p - \mathbf{r}|}.$$

In principle these integral expressions represent the solutions for most of the problems in electrostatics. However, except in a few cases they cannot be solved in a convenient form. For instance, in many problems the charge density may not be explicitly prescribed at all points in space; instead certain constraints or boundary conditions may be placed directly upon the fields or potentials. As an example, for a grounded conductor the condition is that the potential on the surface of the conductor is constant and equal to zero. The surface charge is then that unique charge distribution consistent with this boundary condition.

2. APPLICATION OF THE INTEGRAL SOLUTIONS

a. THE ELECTRIC FIELD OF A STRAIGHT LINE OF CHARGE. Consider a line of charge extending along the z axis from $-L/2$ to $+L/2$. The charge per unit length is λ in units of coulombs per meter.

The contribution to the field at P by a length element dz at z is

$$d\mathscr{E}_p = \frac{(\lambda\,dz)}{4\pi\epsilon_0}\frac{(\mathbf{r}_p - z\mathbf{k})}{|\mathbf{r}_p - z\mathbf{k}|^3},$$

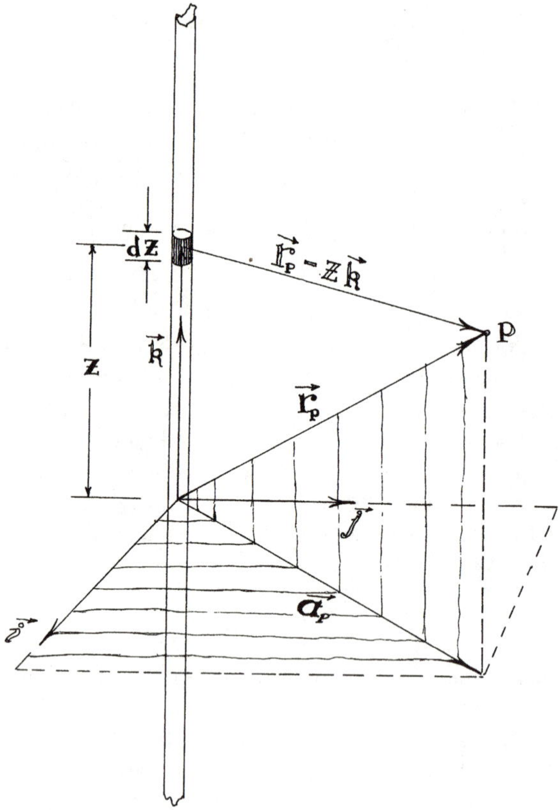

and $$\mathscr{E}_p = \frac{\lambda}{4\pi\epsilon_0}\int_{-L/2}^{L/2} \frac{x_p\mathbf{i} + y_p\mathbf{j} + (z_p - z)\mathbf{k}}{\{x_p^2 + y_p^2 + (z_p - z)^2\}^{3/2}}\,dz.$$

Integrating, we find

$$\mathscr{E}_p = \frac{\lambda}{4\pi\epsilon_0}\left\{\frac{\mathbf{a}_p}{a_p^2}\left(\frac{z_p + \frac{L}{2}}{\left|\mathbf{r}_p + \frac{L}{2}\mathbf{k}\right|} - \frac{z_p - \frac{L}{2}}{\left|\mathbf{r}_p - \frac{L}{2}\mathbf{k}\right|}\right) + \mathbf{k}\left(\frac{1}{\left|\mathbf{r}_p - \frac{L}{2}\mathbf{k}\right|} - \frac{1}{\left|\mathbf{r}_p + \frac{L}{2}\mathbf{k}\right|}\right)\right\}$$

where $$\mathbf{a}_p = x_p\mathbf{i} + y_p\mathbf{j}.$$

One fundamental property of the charge distribution that should be noticed in a development of this type is the behavior of the far field ($L \ll |\mathbf{r}_p|$). Some care must be exercised for such approximations. For instance, if we assume that the two terms in the $\mathbf{k}$ component each go as $1/r_p$, then for large r_p the $\mathbf{k}$ component seems to vanish. This, however, is not correct, and if the terms in the denominator are expanded carefully the results are quite different. Expanding the term $|\mathbf{r}_p \pm (L/2)\mathbf{k}|^{-1}$,

$$\frac{1}{\left|\mathbf{r}_p \pm \frac{L}{2}\mathbf{k}\right|} \simeq \frac{1}{r_p} \frac{1}{\left(1 \pm \frac{Lz}{r_p^2}\right)^{1/2}}$$

$$\simeq \frac{1}{r_p}\left(1 \mp \frac{Lz}{2r_p^2}\right).$$

Using this expansion in the expression for $\mathscr{E}_p$ we find

$$\mathscr{E}_p \xrightarrow[L \ll r_p]{} \frac{L\lambda}{4\pi\epsilon_0} \frac{\mathbf{r}_p}{r_p^3}.$$

This result is typical of all far field calculations for nonvanishing distributions of charge near the origin. When the net charge of the distribution is *not zero*, the distribution gives rise to a field which to first order behaves as that of a point charge at the origin. Notice in the expression above that $L\lambda$ is the total charge in the line distribution.

The calculation usually presented in connection with this problem is the near field problem. Here we let $r_p \ll L$. Then

$$\mathscr{E}_p \xrightarrow[r_p \ll L]{} \frac{\lambda}{2\pi\epsilon_0} \frac{\mathbf{a}_p}{a_p^2} = \frac{\lambda}{2\pi\epsilon_0} \frac{(x_p\mathbf{i} + y_p\mathbf{j})}{(x_p^2 + y_p^2)}.$$

This is the result ordinarily obtained for a line of charge of infinite length. In such a case we let $L \to \infty$, thus imposing a nonphysical condition on the problem; namely, that the amount of charge be infinite. Thus as the point of examination is removed to large distances from the origin an infinite amount of charge acts to produce the observed field. For this reason the potential function for the infinite wire has no natural zeros. By inspection, the potential for the field

$$\mathscr{E}_p = \frac{\lambda}{2\pi\epsilon_0} \frac{\mathbf{a}_p}{a_p^2}$$

is

$$V_p = -\frac{\lambda}{2\pi\epsilon_0} \log \frac{a_p}{C},$$

where C is a constant having the dimensions of length.

It is apparent that the logarithmic potential does not have a natural zero at either $r_p \to \infty$ or $r_p \to 0$. If we examine the fields which vary as r^n where n is a positive or negative integer, we see that the corresponding potentials vary as r^{n+1} *except* when $n = -1$; under this condition the potential varies as log r. This latter potential is of interest in that it separates those potentials which have zeros at $r = 0$ from those which have zeros at $r \to \infty$. As remarked, the logarithmic potential does not have a zero at infinity or at $r = 0$. Ordinarily, however, we are interested in the difference of potential, and therefore the zero of the field does not enter.

b. THE ELECTRIC FIELD OF A CHARGED RING. A uniformly charged circular ring is located in the xy plane with the center of the circle at the origin. The radius is R. To solve this problem conveniently we convert to cylindrical coordinates.

The charge per unit length on the ring is λ, and an element of

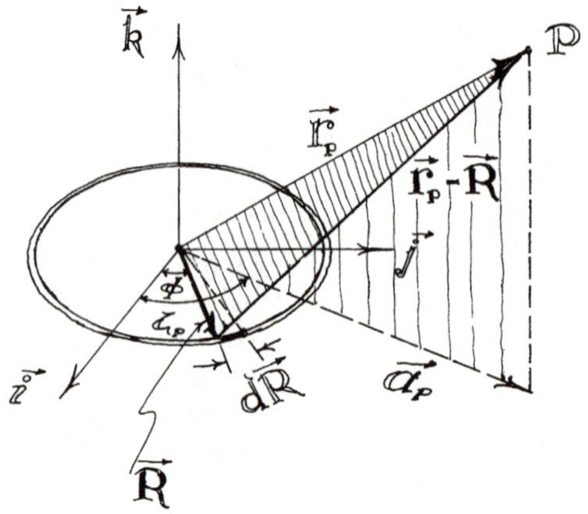

length on the circle is dR where the position of the length element is $\mathbf{R}$:

$$\mathbf{R} = R(\cos\phi\,\mathbf{i} + \sin\phi\,\mathbf{j})$$

and

$$dR = R\,d\phi.$$

The vector connecting the length element and P is

$$\mathbf{r}_p - \mathbf{R} = (x_p - R\cos\phi)\mathbf{i} + (y_p - R\sin\phi)\mathbf{j} + z_p\mathbf{k}.$$

Computing the contribution of λ dR to the potential and summing over the circle,

$$V_p = \frac{\lambda R}{4\pi\epsilon_0}\int_0^{2\pi}\frac{d\phi}{\{R^2 + r_p^2 - 2Ra_p\cos(\phi - \zeta_p)\}^{1/2}},$$

where $$\mathbf{a}_p = x_p\mathbf{i} + y_p\mathbf{j},$$

and $$\zeta_p = \tan^{-1}(y_p/x_p).$$

Once this problem is set up the only remaining operation is the conversion of this integral to the form of a complete elliptic function. Extract $\{(R + a_p)^2 + z_p^2\}^{1/2}$ from the denominator. By adding and subtracting $(2a_pR)$ inside the denominator, we obtain

$$V_p = \frac{Rk}{2\pi\epsilon_0(Ra_p)^{1/2}}\int_0^{\pi/2}\frac{d\alpha}{\sqrt{1 - k^2\sin^2\alpha}}$$

where $$k^2 = \frac{4a_pR}{(R + a_p)^2 + z_p^2}$$

and $$\sin^2\alpha = \tfrac{1}{2}\{1 + \cos(\phi - \zeta_p)\}.$$

Finally, referring to Appendix C, for the complete elliptic functions of the first kind, K(k),

$$V_p = \frac{\lambda k}{2\pi\epsilon_0}\left(\frac{R}{a_p}\right)^{1/2}K(k).$$

Thus for any point in space we can compute k and look up K(k) to evaluate V_p.

In order to obtain the electrostatic field it is necessary to take the gradient of V_p. Because of the symmetry, this is most readily done in cylindrical coordinates. First, we notice that k is *not* a function of ϕ, and therefore $\mathscr{E}_p$ will have only a and z components*:

$$\mathscr{E}_p = -\text{grad}\, V_p = -\frac{\partial V_p}{\partial a_p}\boldsymbol{\epsilon}_a - \frac{\partial V_p}{\partial z_p}\boldsymbol{\epsilon}_3.$$

Referring again to Appendix C we remind the reader that because

$$V_p = f(a_p,z_p)K(k)$$

where $k = k(a_p,z_p)$,

$$\frac{\partial V_p}{\partial a_p} = \frac{\partial f}{\partial a_p}K(k) + f\frac{\partial k}{\partial a_p}\frac{\partial K}{\partial k}$$

* Because of the possible confusion between the function k(a,z) and the z base vector **k**, we use $\boldsymbol{\epsilon}_3$ as the z base vector.

and $$\frac{\partial V_p}{\partial z_p} = \frac{\partial f}{\partial z_p} K(k) + f \frac{\partial k}{\partial z_p} \frac{\partial K}{\partial k} .$$

Carrying out the appropriate differentiations,

$$\mathscr{E}_p = \frac{\lambda}{8\pi\epsilon_0} \left(\frac{R}{a_p}\right)^{1/2} \frac{k}{a_p} \left\{ 2\left(K(k) - \frac{R^2 - a_p^2 + z_p^2}{(R - a_p)^2 + z_p^2} E(k) \right) \boldsymbol{\epsilon}_a + \frac{k^2 z_p}{R} \frac{E(k)}{(1 - k^2)} \boldsymbol{\epsilon}_3 \right\}$$

The function E(k) is the complete elliptic function of the second kind.

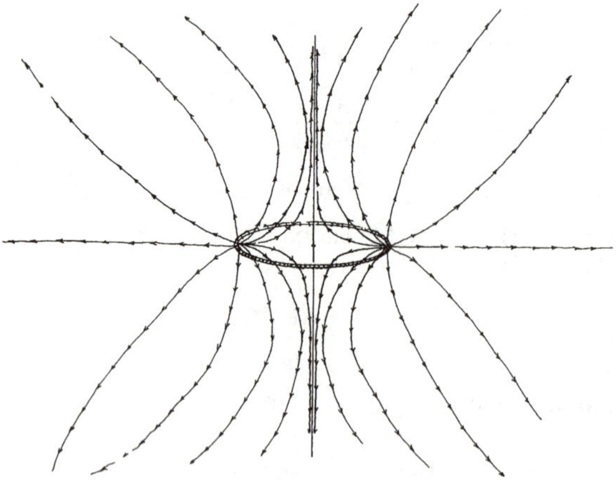

SKETCH: FIELD LINES OF THE CHARGED RING

The field lines evaluated at the plane containing the loop ($z_p = 0$) all lie in the plane. As an exercise the reader can demonstrate that

$$\mathscr{E}_p(z = 0) = \mathscr{E}(a_p)\boldsymbol{\epsilon}_a.$$

The most familiar calculation in connection with this problem is the integration which provides the *axial field and potential* ($a_p = 0$). The result is obtained from our general expression using the condition $a_p = 0$, which in turn gives $k = 0$. Because $K(0) = \pi/2$ and $E(0) = \pi/2$,

$$\mathscr{E}_{\text{axial}} = \frac{\lambda R z_p \boldsymbol{\epsilon}_3}{2\epsilon_0(R^2 + z_p^2)^{3/2}},$$

whence
$$V_{\text{axial}} = \frac{\lambda R}{2\epsilon_0(R^2 + z_p^2)^{1/2}}.$$

These latter results can be achieved by taking $a_p = 0$ in the initial integrals for $\mathscr{E}_p$ and V_p. A word of caution is called for in this connection, however. If V_{axial} is computed and then one derives $\mathscr{E}_{\text{axial}}$ by taking the gradient, the electric field is being generated from a constrained potential. It turns out in this particular example that the correct result is obtained; however, in most instances fields should be derived by taking the gradient of the general potential function. If then a particular configuration is desired, the constraint can be imposed after the differentiation has been made.

Finally, we should examine our general form of V_p or $\mathscr{E}_p$ in the far field. Once again one would guess that to first order the potential should go as $1/r_p$ and the field as $\mathbf{r}_p/r_p^3$. When
$$R \ll r_p, \quad k^2 \to \frac{4Ra_p}{r_p^2};$$
then
$$V_p \xrightarrow[R \ll r_p]{} \frac{\lambda R}{2\epsilon_0 r_p} = \frac{2\pi R\lambda}{4\pi\epsilon_0}\frac{1}{r_p}$$
which is the potential of a point charge of magnitude $2\pi R\lambda$.

3. THE INTEGRAL MULTIPOLE EXPANSION

In Section II-J the multipole expansion was carried out for a system of N point charges. Because the continuum is an equivalent problem we can employ the expansion techniques of that section and let the sums go to integrals.

Starting with the potential function
$$V_p = \frac{1}{4\pi\epsilon_0}\int\!\!\!\int_\tau\!\!\!\int \frac{\rho(x,y,z)\,dx\,dy\,dz}{|\mathbf{r}_p - \mathbf{r}|}$$
we expand $|\mathbf{r}_p - \mathbf{r}|^{-1}$ about $r = 0$ and integrate the series term by term. The reader is referred to Section II-J for the details of this expansion.

As before, the first three terms in the expansion are the singlet, dipole, and quadrupole contributions:
$$V_p = \frac{1}{4\pi\epsilon_0}\left\{\int\!\!\!\int_\tau\!\!\!\int \rho\,d\tau\right\}\frac{1}{r_p} - \frac{1}{4\pi\epsilon_0}\nabla_p\left(\frac{1}{r_p}\right)\cdot\int\!\!\!\int_\tau\!\!\!\int \rho\mathbf{r}\,d\tau$$
$$+ \frac{1}{8\pi\epsilon_0}\left(\frac{1}{r_p^5}\right)\{\tilde{\mathbf{r}}_p\cdot\mathbf{Q}\cdot\mathbf{r}_p\} + \text{h.t.}$$

The quadrupole tensor is developed in the same fashion as before, having nine elements each defined by the relation

$$Q_{jk} = \iiint_\tau \rho(3x_j x_k - r^2 \delta_{jk})\, d\tau.$$

Here x_j is the jth component of $\mathbf{r}$ which locates the position of the charge volume $d\tau$ relative to the origin.

Given a charge distribution $\rho(x_1,x_2,x_3)$ enclosed in volume τ, the far field (r_p much larger than the dimensions of τ) is characterized by the lowest nonvanishing term in the multipole expansion.

B. DISTRIBUTIONS OF DOUBLETS

1. POTENTIALS AND FIELDS

The methods of Section III-A can be applied quite well for all distributions of charge as long as distributions of *point* dipoles and other *point* multipoles are excluded. Our knowledge of atomic and molecular structure leads us to believe that the concept of a point dipole is a mathematical device and that point dipoles do not exist at a microscopic level. In the macroscopic theory of dielectrics, however, the concept of the point dipole is quite useful, and in many instances the properties of dielectrics can be conveniently described by the integral representations of the fields and potentials of a volume distribution of dipoles.

During the development to follow the reader should keep in mind the similarity of the steps in this section to those in the section covering the fields of singlet charges.

Assume that there exists a volume distribution of dipole moments $\mathbf{p}_v(x_1,x_2,x_3)$. If the total dipole moment contained in a volume element $\Delta\tau$ is $\Delta\mathbf{P}$, then

$$\mathbf{p}_v = \lim_{\Delta\tau \to 0} \frac{\Delta \mathbf{P}}{\Delta\tau}.$$

This definition of a volume distribution can be extended to include single point dipoles (using the derivative of the delta function) and surface distributions of dipoles.

The units of $\mathbf{p}_v$ are coulombs/m^2.

Previously we computed the potential and field of a point dipole. The fields and potentials of a volume distribution of doublets can be computed utilizing the superposition principle.

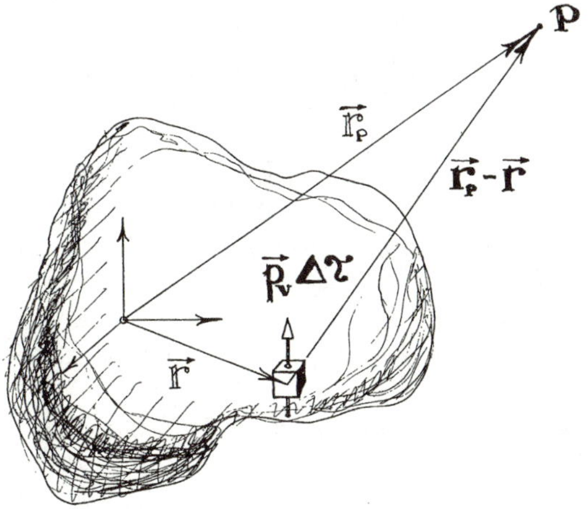

The dipole moment of a volume element $\Delta\tau$ located at $\mathbf{r}$ is $\mathbf{p}_v\Delta\tau$. At a point P the contribution to the potential and field by $\mathbf{p}_v\Delta\tau$ is

$$\Delta V_p = \frac{(\mathbf{p}_v\Delta\tau)\cdot(\mathbf{r}_p - \mathbf{r})}{4\pi\epsilon_0\,|\mathbf{r}_p - \mathbf{r}|^3},$$

and

$$\Delta\mathscr{E}_p = \frac{3\{(\mathbf{p}_v\Delta\tau)\cdot(\mathbf{r}_p - \mathbf{r})\}(\mathbf{r}_p - \mathbf{r}) - (\mathbf{r}_p - \mathbf{r})^2\mathbf{p}_v\Delta\tau}{4\pi\epsilon_0\,|\mathbf{r}_p - \mathbf{r}|^5}.$$

The potential and field at any point P produced by this volume distribution of dipoles is obtained by integrating the above expressions over τ (the volume containing the dipoles),

$$V_p = \frac{1}{4\pi\epsilon_0}\iiint_\tau \frac{\mathbf{p}_v(\mathbf{r})\cdot(\mathbf{r}_p - \mathbf{r})}{|\mathbf{r}_p - \mathbf{r}|^3}\,d\tau,$$

and

$$\mathscr{E}_p = \frac{1}{4\pi\epsilon_0}\iiint_\tau \frac{\{3\mathbf{p}_v\cdot(\mathbf{r}_p - \mathbf{r})\}(\mathbf{r}_p - \mathbf{r}) - (\mathbf{r}_p - \mathbf{r})^2\mathbf{p}_v}{|\mathbf{r}_p - \mathbf{r}|^5}\,d\tau.$$

These integrals then provide a complete description of the field produced by the doublet distribution.

The integral for the potential function V_p can be converted to a form in which the *dipole distribution* is described in terms of an *equivalent induced singlet charge distribution.* We first notice that

$$-\nabla_p \left\{\frac{1}{|\mathbf{r}_p - \mathbf{r}|}\right\} = +\nabla_r \left\{\frac{1}{|\mathbf{r}_p - \mathbf{r}|}\right\};$$

therefore
$$V_p = \frac{1}{4\pi\epsilon_0} \iiint \mathbf{p}_v(\mathbf{r}) \cdot \nabla_r \left\{\frac{1}{|\mathbf{r}_p - \mathbf{r}|}\right\} d\tau.$$

Now

$$\mathbf{p}_v \cdot \nabla_r \left\{\frac{1}{|\mathbf{r}_p - \mathbf{r}|}\right\} = \mathrm{div}_r \left\{\frac{\mathbf{p}_v}{|\mathbf{r}_p - \mathbf{r}|}\right\} - \frac{1}{|\mathbf{r}_p - \mathbf{r}|} \mathrm{div}_r\, \mathbf{p}_v.$$

Employing the divergence theorem for the first term in the expansion,

$$V_p = \frac{1}{4\pi\epsilon_0} \iint_{S_\tau} \frac{\mathbf{p}_v \cdot \mathbf{n}}{|\mathbf{r}_p - \mathbf{r}|} dS + \frac{1}{4\pi\epsilon_0} \iiint_\tau \frac{-\mathrm{div}\, \mathbf{p}_v}{|\mathbf{r}_p - \mathbf{r}|} d\tau.$$

This potential expansion has the same form as that for a singlet surface and volume distribution σ_I and ρ_I,

$$V_p = \frac{1}{4\pi\epsilon_0} \iint_{S_\tau} \frac{\sigma_I}{|\mathbf{r}_p - \mathbf{r}|} dS + \frac{1}{4\pi\epsilon_0} \iiint_\tau \frac{\rho_I}{|\mathbf{r}_p - \mathbf{r}|} d\tau.$$

Thus $\sigma_I = \mathbf{p}_v \cdot \mathbf{n} =$ the normal component of the dipole density at the surface S;

and $\rho_I = -\mathrm{div}\, \mathbf{p}_v =$ divergence of $\mathbf{p}_v$ in the volume element $\Delta\tau$.

The physical interpretation of these equations can be demonstrated graphically. Regard the simplest geometry for $\mathbf{p}_v$, that in

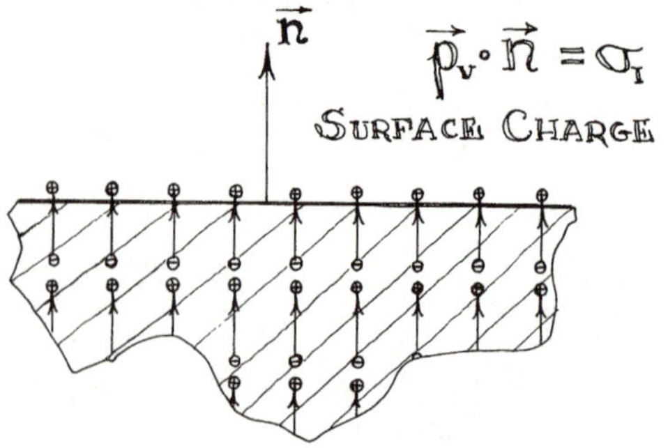

which $\mathbf{p}_V$ is parallel to $\mathbf{n}$ at the surface. If we now consider $\mathbf{p}_V$ as a group of extended dipoles of charges of magnitude q separated by a distance l, we perceive that the positive charges all lie on the surface, and thus the net surface charge is positive and is given by $\mathbf{p}_V \cdot \mathbf{n}$.

In the case of $-\text{div}\, \mathbf{p}_V = \rho_I$ consider an inhomogeneous array of extended dipoles (not point dipoles) in a cubical volume element. Here, as illustrated, the divergence is negative because the density decreases as we move in a positive direction. If we measure the net charge in the cube, there is more positive charge than negative because of the inhomogeneity. The induced volume density is positive because div $\mathbf{p}_V$ is negative, and $\rho_I = -\text{div}\, \mathbf{p}_V$.

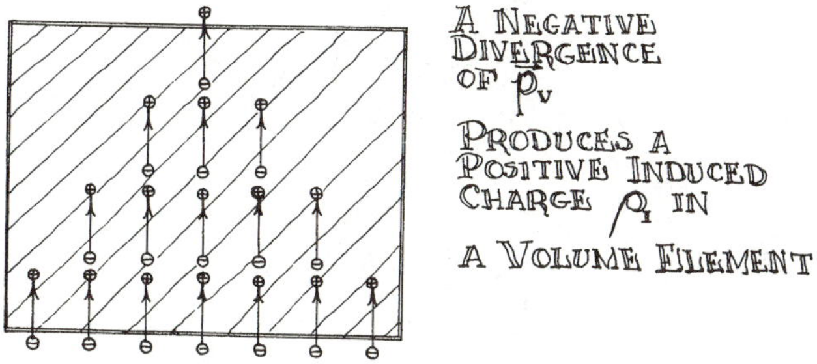
A NEGATIVE
DIVERGENCE
OF p_V
PRODUCES A
POSITIVE INDUCED
CHARGE ρ_I IN
A VOLUME ELEMENT

From the picture of the cube we notice that the measure of ρ_I depends on the number of dipoles crossing the closed surface of the cube. Thus we can expect that ρ_I and σ_I are mutually dependent. This dependence is formally introduced by remembering that the total charge in a doublet distribution is zero:

$$\iint_{S_\tau} \sigma_I \, dS + \iiint_{\tau} \rho_I \, d\tau = 0,$$

or

$$\iint_{S_\tau} \mathbf{p}_V \cdot \mathbf{n} \, dS = -\iiint_{\tau} (-\text{div}\, \mathbf{p}_V) \, d\tau.$$

This last equation is simply the divergence theorem, which demonstrates the consistency of the argument.

2. APPLICATION OF THE INTEGRAL FORMS

a. AXIAL ELECTRIC FIELD AND POTENTIAL OF A HOMOGENEOUS DISTRIBUTION OF DIPOLES IN A CYLINDER. Consider a cylindrical volume of radius R and height 2L containing a homogeneous volume

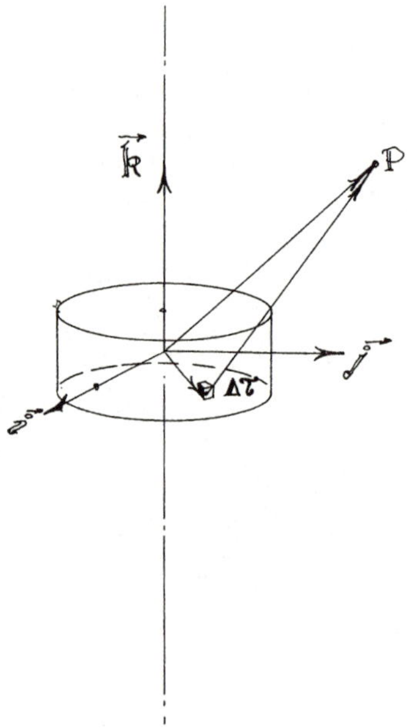

distribution of dipoles $\mathbf{p}_V = p_V\mathbf{k}$ where the scalar p_V is a constant. The cylinder is oriented with its symmetry axis along the z axis and the center at the origin.

The potential at any point P has been given previously and is

$$V_P = \frac{1}{4\pi\epsilon_0}\iiint_\tau \frac{\mathbf{p}_V \cdot (\mathbf{r}_P - \mathbf{r})}{|\mathbf{r}_P - \mathbf{r}|^3}\, d\tau.$$

As an example, take $a_P = 0$; in other words compute V_P at a point on the z axis. Converting to cylindrical coordinates and integrating over ϕ,

$$V_{axis} = \frac{p_V}{2\epsilon_0}\int_0^R \int_{-L}^{+L} \frac{(z_P - z)a}{\{a^2 + (z_P - z)^2\}^{3/2}}\, da\, dz.$$

At first sight this integral appears to be integrable in a straightforward manner. On inspection, however, it is found that care must be exercised in passing through the zeros of $(z_p \pm L)$. The integrated result is

$$V_{axis} = \frac{p_V}{2\epsilon_0}\{[R^2 + (z_p - L)^2]^{1/2} - [R^2 + (z_p + L)^2]^{1/2} - [(z_p - L)^2]^{1/2} + [(z_p + L)^2]^{1/2}\}.$$

Considering the last two square roots, for

$$L < z_p, \quad \sqrt{(z_p - L)^2} \to +(z_p - L),$$
$$z_p < L, \quad \sqrt{(z_p - L)^2} \to -(z_p - L);$$

also when

$$-L < z_p, \quad \sqrt{(z_p + L)^2} \to +(z_p + L),$$
$$z_p < -L, \quad \sqrt{(z_p + L)^2} \to -(z_p + L).$$

Using these relations we obtain the axial potentials for the three regions. For $L < z_p$,

$$V_{axis} = \frac{p_V}{2\epsilon_0}\{[R^2 + (z_p - L)^2]^{1/2} - [R^2 + (z_p + L)^2]^{1/2} + 2L\},$$

$$V_{axis} \xrightarrow[L \ll R]{} \frac{p_V L}{\epsilon_0}\left\{1 - \frac{z_p}{R}\right\}.$$

For $-L \leqslant z_p \leqslant L$,

$$V_{axis} = \frac{p_V}{2\epsilon_0}\{[R^2 + (z_p - L)^2]^{1/2} - [R^2 + (z_p + L)^2]^{1/2} + 2z_p\},$$

$$V_{axis} \xrightarrow[L \ll R]{} \frac{p_V z_p}{\epsilon_0}\left\{1 - \frac{L}{R}\right\}.$$

In the region $z_p \leqslant -L$,

$$V_{axis} = \frac{p_V}{2\epsilon_0}\{[R^2 + (z_p - L)^2]^{1/2} - [R^2 + (z_p + L)^2]^{1/2} - 2L\},$$

$$V_{axis} \xrightarrow[L \ll R]{} \frac{-p_V L}{\epsilon_0}\left\{1 + \frac{z_p}{R}\right\}.$$

These expressions are quite useful in giving us an idea of the effect of a homogeneous distribution of dipole moments all having the same orientation. From the expression for the potential inside the dielectric we can compute the axial electric field,

$$\mathscr{E}_{axis} = -\operatorname{grad}_p V_{axis}$$
$$= \frac{p_V}{2\epsilon_0}\left\{-\frac{(z_p - L)}{[R^2 + (z_p - L)^2]^{1/2}} + \frac{(z_p + L)}{[R^2 + (z_p + L)^2]^{1/2}} - 2\right\}\mathbf{k}.$$

If $R \gg L$ then we perceive that $\mathscr{E}$ is almost constant along the axis. Expanding the expression to first order in terms (L/R),

$$\mathscr{E}_{\text{axis}} \xrightarrow[L \ll R]{} \frac{-\mathbf{p}_V}{\epsilon_0}\left\{1 - \frac{L}{R}\right\}$$

where

$$\mathbf{p}_V = p_V \mathbf{k}.$$

The z_p dependence enters as a term $\frac{3}{2}\left(\frac{L}{R}\right) z_p^2$, which can be neglected.

On examining the three equations for the axial potential it is observed that $\mathscr{E}_p$ reverses direction at the faces of the cylindrical volume, i.e., at $z_p = \pm L$.

Because of the induced surface charge density $\sigma_I = \mathbf{p}_V \cdot \mathbf{n}$, the potential function has cusps at $z_p = \pm L$.

Although this problem was performed as an exercise we shall find that in dielectrics the induced dipole moment per unit volume in a dielectric medium has the effect illustrated here. In the dielectric, the dipole moment density $\mathbf{p}_V$ establishes an electric field opposing the external field. The net effect is to reduce the fields in a dielectric medium.

b. THE AXIAL FIELD OF A HOMOGENEOUSLY POLARIZED SPHERE. To further illustrate technique, let us set up the problem on the z axis, i.e., again compute the axial field for $\mathbf{p}_V = p_V \mathbf{k}$.

$$V_{\text{axis}} = \frac{1}{4\pi\epsilon_0} \iiint_{\text{Sphere}} \frac{\mathbf{p}_V \cdot (\mathbf{r}_p - \mathbf{r})}{|\mathbf{r}_p - \mathbf{r}|^3}\, d\tau.$$

This integral can be solved conveniently by converting to spherical coordinates. The location of the volume element

$$d\tau = r^2\, dr \sin\theta\, d\theta\, d\phi$$

is

$$\mathbf{r} = r\boldsymbol{\epsilon}_r.$$

A point P on the axis is

$$z_p \mathbf{k} = z_p \cos\theta\, \boldsymbol{\epsilon}_r - z_p \sin\theta\, \boldsymbol{\epsilon}_\theta.$$

Then

$$z_p \mathbf{k} - \mathbf{r} = (z_p \cos\theta - r)\boldsymbol{\epsilon}_r - z_p \sin\theta\, \boldsymbol{\epsilon}_\theta.$$

The dipole moment per unit volume is oriented along the z axis; thus

$$\mathbf{p}_V = p_V \cos\theta\, \boldsymbol{\epsilon}_r - p_V \sin\theta\, \boldsymbol{\epsilon}_\theta.$$

Finally,

$$V_p = \frac{p_v}{4\pi\epsilon_0}\int_0^R\int_0^\pi\int_0^{2\pi}\frac{(z_p - r\cos\theta)r^2\,dr\,\sin\theta\,d\theta\,d\phi}{\{z_p{}^2 + r^2 - 2rz_p\cos\theta\}^{3/2}}.$$

This integral can be integrated directly with the restriction that when $z_p < R$, the integration over r must be separated into an integral over $0 \to z_p$ plus an integral over $z_p \to R$.

Inside the sphere, $z_p < R$,

$$V_{axis} = \frac{p_v}{3\epsilon_0} z_p$$

and

$$\mathscr{E}_{axis} = \frac{-p_v}{3\epsilon_0}.$$

Outside,

$$R < z_p,$$

$$V_{axis} = \frac{p_v}{3\epsilon_0}\frac{R^3}{z_p{}^2},$$

$$\mathscr{E}_{axis} = \frac{2p_v}{3\epsilon_0}\frac{R^3}{z_p{}^3},$$

and the negatives of these two values for $z_p < -R$.

C. DISTRIBUTIONS OF QUADRUPOLES

1. THE POTENTIALS AND FIELDS

Very briefly, since the method of development is analogous to that of Section III-B, we can construct the potentials and fields in the presence of a distribution of point quadrupoles.

This particular aspect of continua is ordinarily omitted from texts. One finds, however, that in special circumstances the presence of an inhomogeneous distribution of quadrupoles can appreciably alter the electric fields at a point.

We construct a volume distribution of the quadrupole tensor such that the quadrupole tensor in $\Delta\tau$ is ΔQ, and

$$Q_v = \lim_{\Delta\tau\to 0}\frac{\Delta Q}{\Delta\tau},$$

where

$$\Delta Q_{jk} = \sum_{n=1}^{N} q_n(3x_j^{(n)}x_k^{(n)} - r_n{}^2\delta_{jk}).$$

We have assumed N charges in $\Delta\tau$.

The potential at a point P set up by ΔQ is ΔV_P. The total potential from a quadrupole distribution in a volume τ is*

$$V_P = \frac{1}{8\pi\epsilon_0} \iiint_\tau (\tilde{\mathbf{r}}_P - \tilde{\mathbf{r}}) \cdot \mathbf{Q}_v(\mathbf{r}) \cdot \frac{(\mathbf{r}_P - \mathbf{r})}{|\mathbf{r}_P - \mathbf{r}|^5}\, d\tau.$$

Before applying this integral to specific geometries it can be demonstrated that, like the dipole distribution $\mathbf{p}_v$, $\mathbf{Q}_v$ can produce an induced volume charge density ρ_Q and an induced surface charge density σ_Q. In addition the quadrupole distribution produces an induced surface distribution of dipole moments.

The integrand can be written

$$\frac{1}{3}\tilde{\mathbf{r}}' \cdot \mathbf{Q}_v \cdot \nabla_r\left(\frac{1}{r'^3}\right),$$

where

$$\mathbf{r}' = \mathbf{r}_P - \mathbf{r}.$$

Using the expansion of div $(a\mathbf{A})$ we find that

$$V_P = \frac{1}{24\pi\epsilon_0} \iiint_\tau \left\{ \widetilde{\text{div}} \left[\frac{1}{r'^3}(\mathbf{Q}_v \cdot \mathbf{r}')\right] - \frac{1}{r'^3}\, \widetilde{\text{div}}(\mathbf{Q}_v \cdot \mathbf{r}')\right\} d\tau.$$

We have employed the relation

$$\widetilde{\text{div}}\,(\mathbf{Q}_v \cdot \mathbf{r}') = \text{div}\,(\widetilde{\tilde{\mathbf{r}} \cdot \mathbf{Q}_v}).$$

The first term in V_P can be converted to a surface integral over the closed surface S,

$$V_P = \frac{1}{4\pi\epsilon_0}\left\{\frac{1}{6}\iint_S \frac{\mathbf{r}'}{r'^3} \cdot (\mathbf{Q}_v \cdot \mathbf{r}')\, dS - \frac{1}{6}\iiint_\tau \frac{\text{Tr}\,\mathbf{Q}_v}{r'^3}\, d\tau \right.$$

$$\left. - \frac{1}{6}\iiint_\tau (\widetilde{\text{div}}\,\mathbf{Q}_v) \cdot \frac{\mathbf{r}'}{r'^3}\, d\tau. \right.$$

As an exercise the reader should demonstrate that

$$\widetilde{\text{div}}\,(\mathbf{Q}_v \cdot \mathbf{r}') = \text{Tr}\,\mathbf{Q}_v + (\widetilde{\text{div}}\,\mathbf{Q}_v) \cdot \mathbf{r}'.$$

Tr $\mathbf{Q}_v$ is the trace of $\mathbf{Q}_v = \sum_{j=1}^{3} Q_{jj}$. We can immediately see that this term is zero:

$$\text{Tr}\,\mathbf{Q}_v = \sum_{j=1}^{3}\sum_{n=1}^{N} q_n(3x_j^{(n)2} - r_n^2\delta_{jj}) = 0,$$

* The term $(\tilde{\mathbf{r}}_P - \tilde{\mathbf{r}})$ is the transpose of the column vector $(\mathbf{r}_P - \mathbf{r})$:

$$(\tilde{\mathbf{r}}_P - \tilde{\mathbf{r}}) = [(x_P - x), (y_P - y), (z_P - z)].$$

while

$$\widetilde{\text{div}}\, Q_V = \left(\frac{\partial}{\partial x_1}, \frac{\partial}{\partial x_2}, \frac{\partial}{\partial x_3}\right)\begin{pmatrix} Q_{11} & Q_{12} & Q_{13} \\ Q_{21} & Q_{22} & Q_{23} \\ Q_{31} & Q_{32} & Q_{33} \end{pmatrix}$$

$$= \sum_{k=1}^{3} \sum_{j=1}^{3} \frac{\partial Q_{kj}}{\partial x_k} \boldsymbol{\epsilon}_j.$$

These terms in V_p can be associated with an induced dipole surface density,

$$\mathbf{p}_S = \tfrac{1}{6}\{Q_V \cdot \mathbf{n}\}$$

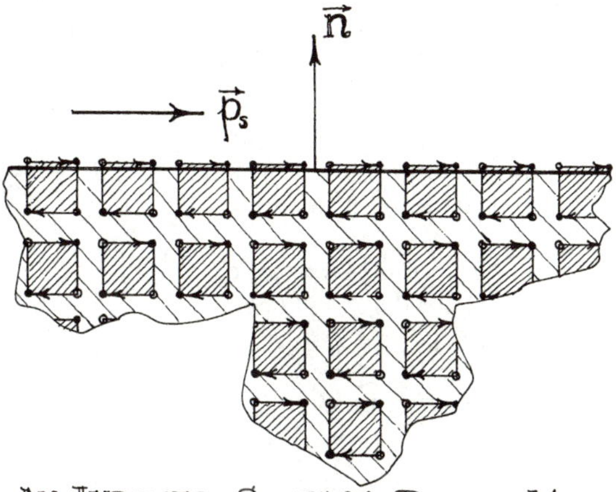

AN INDUCED SURFACE DIPOLE MOMENT $\vec{p}_s$ PRODUCED BY

$$Q_V \cdot \vec{n} = Q_0 \begin{pmatrix} 0 & 0 & 0 \\ 0 & 0 & 1 \\ 0 & 1 & 0 \end{pmatrix} \begin{bmatrix} 0 \\ 0 \\ 1 \end{bmatrix} = Q_0 \begin{bmatrix} 0 \\ 1 \\ 0 \end{bmatrix}$$

in units of dipole moment per unit area, and an induced dipole moment volume density of

$$\mathbf{p}_Q = -\tfrac{1}{6}\widetilde{\{\widetilde{\text{div}}\, Q_V\}}.$$

Because we are interested in the induced *singlet* charge densities, the $\mathbf{p}_Q$ term can be reduced, by the methods used in Section III-B,

INDUCED VOLUME DIPOLE MOMENT

PRODUCED BY

$$\{\widetilde{\text{div } \underline{Q}}\} = \left\{ [\tfrac{\partial}{\partial x_1}, \tfrac{\partial}{\partial x_2}, \tfrac{\partial}{\partial x_3}] \begin{pmatrix} 0 & 0 & 0 \\ 0 & 0 & \tfrac{Q}{4}x_3 \\ 0 & \tfrac{Q}{4}x_3 & 0 \end{pmatrix} \right\} = \begin{bmatrix} 0 \\ Q/4 \\ 0 \end{bmatrix}$$

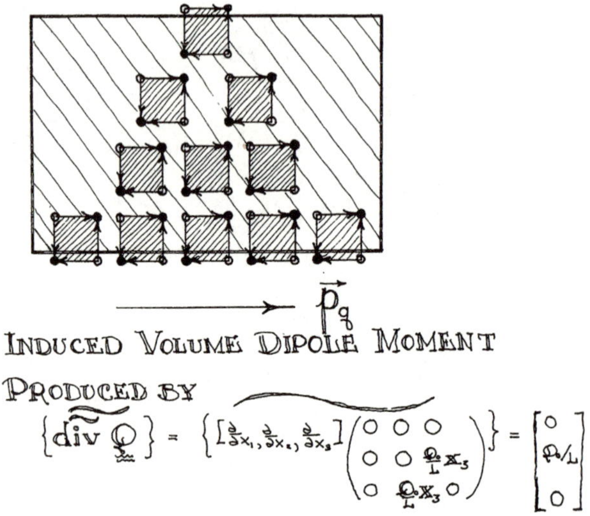

to an induced surface charge density σ_Q and an induced volume charge density ρ_Q. Write the last integral in V_p as*

$$-\frac{1}{6}\int_\tau \!\!\int\!\!\int \mathbf{p}_Q \cdot \nabla_r \left(\frac{1}{r'}\right) d\tau = -\frac{1}{6}\int\!\!\int_S \frac{1}{r'} (\mathbf{p}_Q \cdot \mathbf{n}) \, dS + \frac{1}{6}\int_\tau\!\!\int\!\!\int \frac{1}{r'} (\text{div } \mathbf{p}_Q) \, d\tau.$$

Thus

$$\sigma_Q = -\tfrac{1}{6}\{\widetilde{\text{div } Q}_v\} \cdot \mathbf{n} = \mathbf{p}_Q \cdot \mathbf{n},$$

and†

$$\rho_Q = \tfrac{1}{6}\,\widetilde{\text{div}}\,\{\widetilde{\text{div } Q}_v\}$$

where

$$\widetilde{\text{div}}\,\{\widetilde{\text{div } Q}_v\} = \sum_{j=1}^{3} \sum_{k=1}^{3} \frac{\partial^2 Q_{jk}}{\partial x_j \, \partial x_k}.$$

This analysis is no more difficult than that of the dipole distribution. In Section B we required a knowledge of the spatial dependence of the three components of $\mathbf{p}_v(\mathbf{r})$ while here we require the spatial dependence of the nine components of Q_v, i.e., $Q_{jk}(x_1,x_2,x_3)$.

2. AN EXAMPLE

The previous discussion can be further illuminated by computing the potentials and fields along the z axis of a homogeneous spherical

---

* Once the components of $\mathbf{p}_Q$ have been computed in the matrix representation, the vector can be written in terms of its base vector expansion and utilized in this form.
† The use of the transpose in the equation for ρ_Q is necessary if these symbolic expressions are to be meaningful in the matrix notation.

distribution of quadrupoles having a fixed orientation in space. Let q be the charge magnitude characterizing a rectangular quadrupole, and A the area of the rectangle. Then

$$Q_v = Q_0 \begin{pmatrix} 0 & 0 & 0 \\ 0 & 0 & 1 \\ 0 & 1 & 0 \end{pmatrix},$$

where

$$Q_0 = \frac{3}{2} \frac{qA}{\tau_Q}.$$

In spherical coordinates,

$$V_p = \frac{Q_0}{8\pi\epsilon_0} \int_0^R \int_0^\pi \int_0^{2\pi} \frac{\{(z_p - r\cos\theta)^2 - r^2 \sin^2\theta \sin^2\phi\} r^2\, dr \sin\theta\, d\theta\, d\phi}{\{r^2 + z_p^2 - 2z_p r\cos\theta\}^{5/2}}.$$

Remembering that the radial integral must be separated into two parts, $0 \to z_p$ and $z_p \to R$, we get

inside, $-R < z_p < R$,

$$V_{axis} = \frac{Q_0}{6\epsilon_0}$$

$$\mathscr{E}_{axis} = 0;$$

outside, $R < z_p$ (use plus sign) and $z_p < R$ (use minus sign),

$$V_{axis} = \pm \frac{Q_0}{6\epsilon_0} \frac{R^3}{z_p^3}$$

$$\mathscr{E}_{axis} = \pm \frac{Q_0}{2\epsilon_0} \frac{R^3}{z_p^4} \mathbf{k}.$$

This result for the interior solution is not surprising in view of our general development in the first article of Section III-C. A homogeneous Q_v will not produce an induced surface or volume charge density.

D. GAUSS'S LAW AND POISSON'S EQUATION

1. GAUSS'S LAW

In Chapter II an important theorem was derived, namely, that the surface integral of the normal component of $\epsilon_0\mathscr{E}$ over a closed surface S is equal to the total charge enclosed in S. This charge is either free charge (singlet) or bound charge. The bound charges usually refer to the volume distributions of dipoles. Distributions of higher moments can be incorporated as equivalent induced dipole moments.

Because we have computed the induced charge densities for quadrupole distributions, and because it is now apparent that all of the higher multipole distributions can produce induced charge, these distributions will be included in the final statement of Gauss's law.

In most treatments these higher order contributions are neglected. In many cases this omission results from the realization that in the most familiar experimental problems the higher order contributions are small. In the discussion here the higher order terms will be indicated, and the quadrupole terms inserted in detail. The reason for doing this lies in the fact that soon we may be able to detect these effects experimentally, and it certainly is possible, as in the case of the dipole dielectric, that under certain favorable experimental conditions the effects may be appreciable.

By analogy with the dipole and quadrupole developments we can see that the octupole contribution must also be capable of accounting for an induced volume density of charge

$$\rho_{oct} \rightarrow \sum_{i=1}^{3} \sum_{j=1}^{3} \sum_{k=1}^{3} \frac{\partial^3 O_{ijk}}{\partial x_i\, \partial x_j\, \partial x_k},$$

where

$$V_p \rightarrow \frac{1}{4\pi\epsilon_0} \iiint_\tau \frac{\rho_{oct}}{r'}\, d\tau.$$

We will assume that this term is very small since it goes as ql^3 where l is of atomic dimensions.

Regard a volume τ containing a distribution of multipoles: singlet, dipole, quadrupole, etc. Construct a closed surface S which contains part of the distribution. Because the volume densities can be defined to include the surface charge, the closed surface integral*

* When surface charges densities σ are separated out as distinct from strictly volume densities ρ, and only surface densities are present, the relation between $\mathbf{D}$ and σ is elementary. For a single closed surface S carrying *only* a surface charge σ,

$$\iint_S \mathbf{D} \cdot \mathbf{n}\, dS = \iint_S \sigma\, ds,$$

$$\mathbf{D} \cdot \mathbf{n} = \sigma \quad \text{at every point on the surface.}$$

This relation is important in the problem of the charged conducting surface.

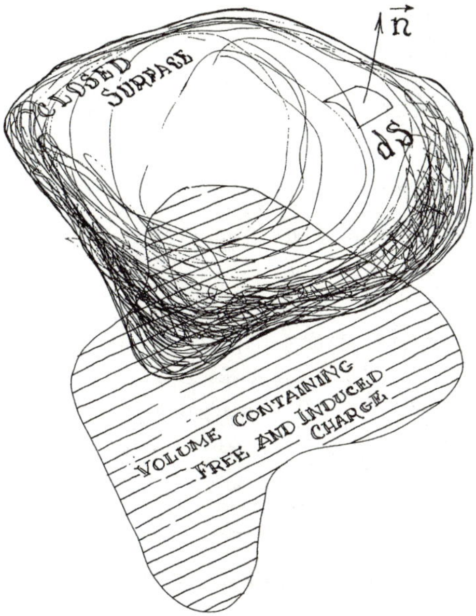

of the normal component of $\epsilon_0\mathscr{E}$ is equal to the total charge *enclosed* in S, i.e., the free plus induced charge.

$$\iint_S \epsilon_0\mathscr{E}\cdot\mathbf{n}\,dS = q_{\text{total}} = \iiint_\tau \{\rho_{\text{singlet}}(\mathbf{r}) + \rho_{\text{dipole}}(\mathbf{r}) + \rho_Q(\mathbf{r}) + \text{h.t.}\}\,d\tau.$$

By the divergence theorem and using the previously developed forms for ρ_{dipole} and ρ_Q,

$$\iiint_\tau \text{div}\,(\epsilon_0\mathscr{E})\,d\tau = \iiint_\tau \{\rho_{\text{singlet}} - \text{div}\,\mathbf{p}_V + \tfrac{1}{6}\,\widetilde{\text{div}}\,\widetilde{(\widetilde{\text{div}}\cdot\mathbf{Q}_V)} + \text{h.t.}\}\,d\tau,$$

$$\text{or}\qquad \iiint_\tau \widetilde{\text{div}}\,\{\epsilon_0\mathscr{E} + \mathbf{p}_V + \mathbf{p}_Q + \text{h.t.}\}\,d\tau = \iiint_\tau \rho_{\text{singlet}}\,d\tau.$$

The *electric displacement vector* is defined by the equation*

$$\iiint_\tau \text{div}\,\mathbf{D}\,d\tau = \iiint_\tau \rho_{\text{free}}\,d\tau.$$

* ρ_{singlet} is customarily referred to as the free or mobile charge. One assumes that volumes of bound charge are neutral and can produce only moments higher than that of the singlet.

This equation is the integral form of Gauss's law. Thus

$$\operatorname{div} \mathbf{D} = \rho_{\text{free}},$$

which is Gauss's law in the form of a differential equation. By definition then

$$\mathbf{D} = \epsilon_0 \mathscr{E} + \mathbf{p}_V - \tfrac{1}{6}\widetilde{(\widetilde{\operatorname{div}} \mathrm{Q}_V)} + \text{h.t.}$$

Previously it was stated that ordinarily **D** is defined by the equation

$$\mathbf{D} = \epsilon_0 \mathscr{E} + \mathbf{p}_V.$$

As we see, this is an exact equality if the definition of $\mathbf{p}_V$ is extended to include the induced volume density of dipole moments produced by volume distributions of the higher moments.

Gauss's law is usually presented in the form

$$\iint_{\text{S(closed)}} \mathbf{D} \cdot \mathbf{n}\, dS = q \text{ (enclosed)}$$

where the normal component of **D** over S is related to the charge enclosed in S.

When a problem involves a surface with symmetry such that the direction of **D** can be observed to be normal to the surface, this form of Gauss's law presents a particularly convenient method for computing the magnitude of **D**. Spherical and cylindrical symmetries are the most common examples xf this.

2. POISSON'S EQUATION

Rewriting the differential form of Gauss's law which relates $\mathscr{E}$, $\mathbf{p}_V$, Q_V, and ρ_{free} at every point in space, we find that

$$\begin{aligned} \operatorname{div} \epsilon_0 \mathscr{E} &= -\epsilon_0 \nabla^2 V \\ &= \rho_{\text{free}} - \operatorname{div} \mathbf{p}_V + \tfrac{1}{6}\, \widetilde{\operatorname{div}}\, \widetilde{(\widetilde{\operatorname{div}} \mathrm{Q}_V)} + \text{h.t.} \end{aligned}$$

whence $$\nabla^2 V = -\frac{\rho_{\text{free}}}{\epsilon_0} + \frac{\operatorname{div} \mathbf{p}_V}{\epsilon_0} + \frac{\operatorname{div} \mathbf{p}_Q}{\epsilon_0} + \text{h.t.}$$

where ∇^2 is the Laplacian operator in appropriate coordinates.

In the absence of bound charge we obtain Poisson's equation:

$$\nabla^2 V = -\frac{\rho_{\text{free}}}{\epsilon_0}.$$

3. APPLICATIONS OF GAUSS'S LAW

Gauss's law is useful in problems which involve a charge symmetry, or in general developments (such as the discussion of the field inside of a charged conductor at equilibrium).

a. FIELDS AND POTENTIALS OF A HOMOGENEOUSLY CHARGED SPHERE. We take as the radius R, $\mathbf{p}_v = 0$, h.t. $= 0$, and

$$\rho = \rho_0, \quad r < R$$
$$\rho = 0, \quad R < r.$$

Outside the sphere ($R \leqslant r$): Construct a concentric spherical surface of radius $r \geqslant R$. From the symmetry of the problem the field on the surface of the sphere is of constant magnitude and parallel to the surface normal at every point. Thus our surface

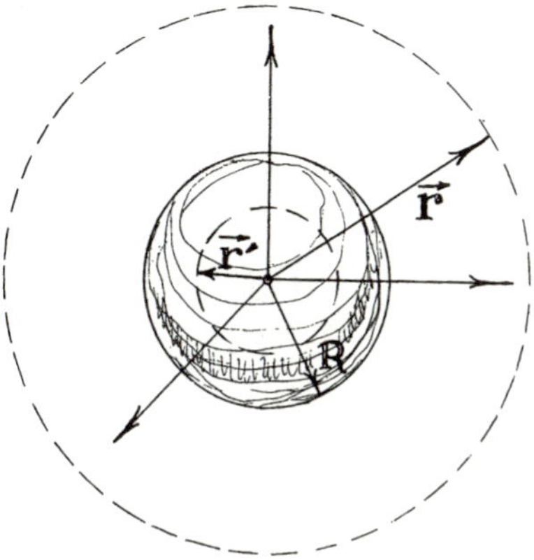

integral reduces to a product of $|\mathbf{D}|$ times the area of the sphere.

The total charge enclosed in $S = 4\pi r^2$ is the total charge enclosed in the spherical volume $\frac{4}{3}\pi R^3$. Therefore by Gauss's law

$$|\mathbf{D}|\, 4\pi r^2 = \tfrac{4}{3}\pi R^3 \rho_0$$

and

$$|\mathbf{D}| = \frac{R^3\rho_0}{3\epsilon_0}\frac{1}{r^2}.$$

Because $\mathbf{D}$ is parallel to the surface normal $\mathbf{r}/r$ at every point,

$$\mathscr{E} = \frac{\mathbf{D}}{\epsilon_0} = \frac{R^3\rho_0}{3\epsilon_0}\frac{\mathbf{r}}{r^3}.$$

This is a familiar result which indicates that the external field is equivalent to that of a point charge at the origin. We also see that

$$V_p = \frac{R^3 \rho_0}{3\epsilon_0} \frac{1}{r}.$$

Inside the sphere ($r \leqslant R$): Again construct a concentric spherical surface S, this time of radius $r' \leqslant R$. By the same methods,

$$|\mathbf{D}'| \, 4\pi r'^2 = \tfrac{4}{3}\pi r'^3 \rho_0.$$

The total charge enclosed in S' is less than the total charge of the sphere. Then

$$\mathscr{E}' = \frac{\mathbf{D}'}{\epsilon_0} = \frac{\rho_0}{3\epsilon_0} \mathbf{r}',$$

while

$$V' = -\frac{\rho_0}{6\epsilon_0} r'^2 + C.$$

The constant C is evaluated by remembering that V must be continuous across the boundary at R. In other words

$$V'(r' = R) = V(r = R),$$

giving

$$\frac{R^2 \rho_0}{3\epsilon_0} = -\frac{R^2 \rho_0}{6\epsilon_0} + C,$$

$$C = \frac{R^2 \rho_0}{2\epsilon_0}.$$

Because there is no surface charge ($\sigma = \lim_{\delta \to 0} \rho_0 \delta = 0$), the electric field is also continuous across the boundary.

b. INFINITELY LONG CYLINDRICAL DISTRIBUTIONS. Fields and potentials of distributions having cylindrical symmetry about the z axis and extending from $-\infty$ to $+\infty$ along this axis can be readily obtained using Gauss's law.

The employment of concentric cylindrical surfaces allows one to compute $\mathscr{E}$ and then V by methods similar to those in the preceding example. Examples of this geometry will be taken up as problems.

c. CAVITIES. Cavities in singlet charge distributions $\rho(\mathbf{r})$ can be taken into account by a method of superposition. A cavity in a region defined by a surface S' which encloses a volume τ' can be considered as a negative charge distribution $-\rho(\mathbf{r})$ which cancels identically the positive volume density $\rho(\mathbf{r})$ in the region bounded by S′.

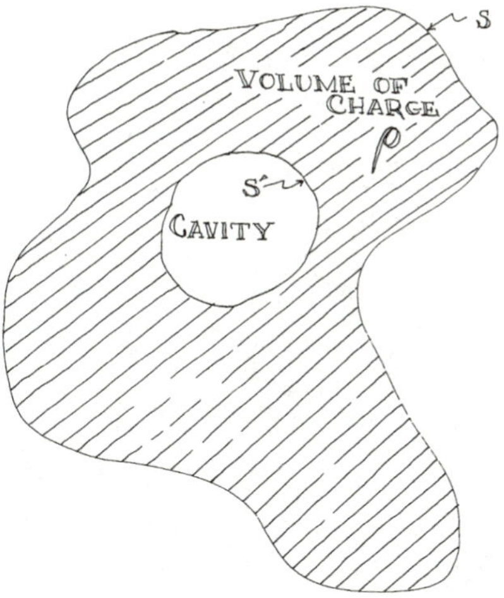

For instance, a spherical cavity in an infinite sea of positive charge $+\rho_0$ where the volume density is constant throughout will act as a negative charge distribution. Because the infinite sea of uniformly distributed positive charge has a zero field at every point, the insertion of a spherical cavity is equivalent to the problem of a negatively charged spherical volume in a vacuum.

That the infinite sea of charge of constant volume density has a zero field at every point can be understood by a relatively simple calculation:

$$\mathscr{E}_p = \frac{\rho_0}{4\pi\epsilon_0}\int_{-\infty}^{\infty}\int_{-\infty}^{\infty}\int_{-\infty}^{\infty}(\mathbf{r}_p - \mathbf{r})\frac{dx\,dy\,dz}{|\mathbf{r}_p - \mathbf{r}|^3} = 0.$$

If we now form a spherical cavity of radius R by superposing a charge density $-\rho_0$ in the volume of the sphere, the resultant field at any point (taking the sphere center at the origin) is

$$\mathscr{E} = -\frac{\rho_0}{3\epsilon_0}R^2\frac{\mathbf{r}}{r^3} \quad \text{for} \quad R \leqslant r,$$

and

$$\mathscr{E} = -\frac{\rho_0}{3\epsilon_0}\mathbf{r} \quad \text{for} \quad r \leqslant R.$$

We also find that two cavities imbedded in a uniform sea of charge repel one another. The original representation of the positron was formulated upon this basis. The positron was viewed as a *hole*

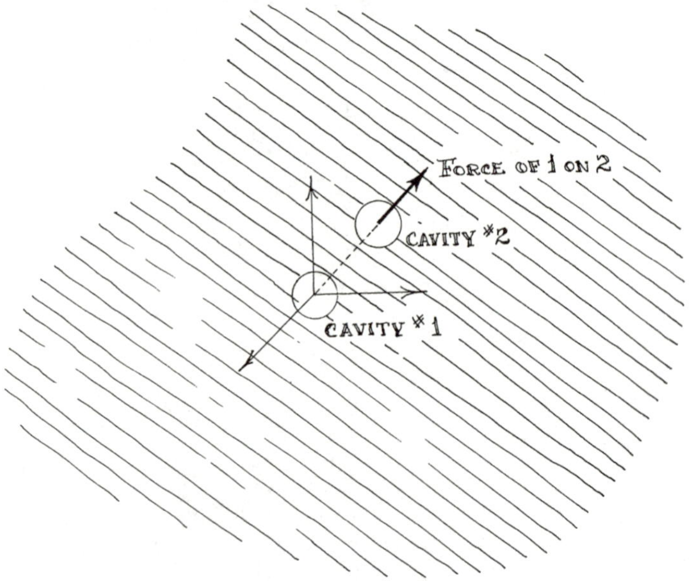

in an infinite sea of negative charge, and as such, the hole exhibited the properties of a positively charged particle. The annihilation of the positron was expressed as the superposition of a negative electron and the hole. This superposition produces a uniform sea of charge having a zero electric field everywhere. This picture of the positron, although of historic interest, has now been superseded by a somewhat different representation.

4. APPLICATIONS OF POISSON'S EQUATION

In Appendix D the discussion of the method of separation of variables, and later the development of the solutions of linear differential equations, indicate the types of solutions we might expect for equations involving the Laplacian operator.

To illustrate the use of the differential equation in solving problems in electrostatics let us first investigate the problem discussed in the previous section.

a. THE POTENTIAL OF A HOMOGENEOUSLY CHARGED SPHERE OF RADIUS R. Poisson's equation takes two forms, one for the exterior solution (excluded origin), and one for the interior solution.

Exterior solution, $R \leqslant r$:

$$\nabla^2 V = 0;$$

interior solution, $r \leqslant R$:

$$\nabla^2 V = -\rho_0/\epsilon_0.$$

There is no ϕ dependence; therefore, the most general solution for the exterior region is

$$V = \sum_{L=0}^{\infty} a_L R_L(r) P_L(\cos\theta).$$

The solutions R(r) are r^L or $r^{-(L+1)}$.

As has been stated r^L is well behaved near the origin but diverges for large r (except for fields derived from the first two terms), while $r^{-(L+1)}$ is divergent at $r = 0$ but is well behaved in regions in which r is arbitrarily large. Thus for $R \leqslant r$ we choose the series

$$V(r) = \sum_{L=0}^{\infty} a_L r^{-(L+1)} P_L$$

with the boundary conditions that

$$V(r \to \infty) = 0$$

and

$$V(r = R) \neq V(\theta).$$

The two terms involving r^0 and r do not contribute and have been omitted.

Only the first term involving $P_0(\cos\theta)$ is independent of θ. Therefore

$$V(r) = a_0 \frac{1}{r},$$

where

$$a_0 = \frac{\rho_0 R^3}{3\epsilon_0}.$$

The solution of the inhomogeneous equation in the region $r \leqslant R$ is somewhat more complicated. Appendix D indicates that the general inhomogeneous equation can be solved using a Green's function $G(\mathbf{r},\mathbf{r}')$. Then*

$$V(\mathbf{r}) = \frac{1}{\epsilon_0} \iiint_{\tau} G(\mathbf{r},\mathbf{r}')\rho_0 \, d\tau.$$

For the electrostatic problem it is demonstrated in Appendix D that

$$G(\mathbf{r},\mathbf{r}') = \frac{1}{4\pi\,|\mathbf{r} - \mathbf{r}'|},$$

which leads us back to our original integral form

$$V(\mathbf{r}) = \frac{\rho_0}{4\pi\epsilon_0} \iiint_{\text{Sphere}} \frac{1}{|\mathbf{r} - \mathbf{r}'|} \, d\tau.$$

As an exercise the reader should show that the solution V(r) in the region $r \leqslant R$ is

$$V(r) = \frac{-\rho_0 r^2}{3\epsilon_0} + \frac{\rho_0 R^2}{2\epsilon_0}.$$

E. THE ENERGY STORED IN THE ELECTROSTATIC FIELD

In Chapter II an expression was derived for the energy stored in a system containing N point charges,

$$U = \tfrac{1}{2}\sum_{j=1}^{N} q_j V_j,$$

where V_j was the potential at the position of the jth charge. V_j is established by all of the charges in the system with the exception of the jth charge. Thus self-energies of points are specifically excluded. The q_j's are associated with the free charge ρ_{free} in the continuum. Thus the energy stored in a volume element $\Delta\tau$ carrying a volume charge density $\rho_f(\mathbf{r})$ is

$$\Delta U = [\rho_f(\mathbf{r})\,\Delta\tau]V(\mathbf{r}).$$

The energy stored in a finite charge system bounded by the surface S is

* The Greens function is a solution of Poisson's Equation when the inhomogeniety is the Dirac delta function.

$$\nabla^2 G(\mathbf{r},\mathbf{r}') = -\delta(\mathbf{r}' - \mathbf{r}),$$ see Appendix E.

$$U = \tfrac{1}{2} \iiint_{\tau} \rho_f(\mathbf{r}) V(\mathbf{r})\, d\tau = \tfrac{1}{2} \iiint_{\text{All Space}} \rho_f(\mathbf{r}) V(\mathbf{r})\, d\tau.$$

We have specifically included surface charges σ_f in our definition of ρ_f. From Gauss's law,

$$\rho_f = \operatorname{div} \mathbf{D}$$

therefore

$$U = \tfrac{1}{2} \iiint_{\text{All Space}} V \operatorname{div} \mathbf{D}\, d\tau.$$

Because

$$V \operatorname{div} \mathbf{D} = \operatorname{div} (V\mathbf{D}) - \mathbf{D} \cdot \operatorname{grad} V,$$

and

$$-\operatorname{grad} V = \mathscr{E},$$

$$U = \tfrac{1}{2} \iiint_{\text{All Space}} \operatorname{div} (V\mathbf{D})\, d\tau + \tfrac{1}{2} \iiint_{\text{All Space}} \mathbf{D} \cdot \mathbf{E}\, d\tau.$$

The first integral vanishes. This is demonstrated by changing the volume integral to a surface integral. Our original assumption was that the charge system (and as a result the volume containing ρ) was finite. If such be the case then the potential function V(r) and **D** vanish at infinity.

When we convert the "all space" volume integral to a surface integral, at infinity the V(r) and **D** functions vanish everywhere on this *particular* surface. Thus this surface integral is identically zero, and

$$U = \tfrac{1}{2} \iiint_{\text{All Space}} \mathbf{D} \cdot \mathscr{E}\, d\tau.$$

In using the volume integral of $\mathbf{D} \cdot \mathscr{E}$ we have merely changed the representation. The quantity

$$u = \tfrac{1}{2}\mathbf{D} \cdot \mathscr{E}$$

is called the energy density of the field and has units of joules per cubic meter. Later during the discussion of the radiation field we shall find that the interpretation of u is much more straightforward. During the process of forming the charge system, the motion of the charges can be thought of as undergoing a quasistatic energy transport. The energy transported across a closed surface surrounding the charges can be equated to the energy density of the final static field.

U can be derived directly without resort to considering the point charge problem. Regard a finite volume τ containing $\rho(\mathbf{r})$. The potential at every point in space is V.

Bring up an infinitesimal element of charge δq. The work required to bring δq into the system is

$$\delta W_1 = \iiint_\tau V\,\delta\rho\,d\tau + \iint_{\substack{\text{All Charged}\\ \text{Surfaces}}} V\,\delta\sigma\,dS.$$

The addition of δq increases V(r) everywhere by δV. The increased energy to first order is

$$\delta W_2 = \iiint_\tau \rho\,\delta V\,d\tau + \iint_{\text{All S}} \sigma\,\delta V\,dS.$$

If the system is closed (all field producing charges are incorporated in the integration), the work δW_1 must equal δW_2. Therefore

$$\delta U = \tfrac{1}{2}(\delta W_1 + \delta W_2) = \tfrac{1}{2}\iiint_\tau (\rho\,\delta V + V\,\delta\rho)\,d\tau + \tfrac{1}{2}\iint_{\text{All S}} (\sigma\,\delta V + V\,\delta\sigma)\,dS.$$

This is a complete differential, leading to

$$U = \tfrac{1}{2}\iiint_{\text{All Space}} \rho V\,d\tau + \tfrac{1}{2}\iint_{\text{All S}} \sigma V\,dS.$$

The inclusion of the charged surfaces of conductors into the volume integral does not alter the result.

In performing a derivation starting with

$$\delta U = \iiint_{\text{All Space}} V(r)\,\delta(\operatorname{div}\mathbf{D})\,d\tau = \iint_{S\to\infty} V\,\delta\mathbf{D}\cdot\mathbf{n}\,dS + \iiint_{\text{All Space}} \mathscr{E}\cdot\delta\mathbf{D}\,d\tau,$$

we have a lack of uniqueness.

Even though the surface integral can be brought to zero, the fact that we are dealing with a differential $\delta\mathbf{D}$ allows the addition of *any* vector $\mathbf{G}$ which has the property that

$$\int_{S\to\infty} \delta\mathbf{G}\cdot\mathbf{n}\,dS = 0.$$

Because of this our solution in terms of the field quantities is not unique.

Note again that the interpretation of $\frac{1}{2}\mathbf{D} \cdot \mathscr{E}$ is different. The formula

$$U = \tfrac{1}{2} \iiint_{\text{All Space}} \rho V \, d\tau + \tfrac{1}{2} \iint_{\substack{\text{All Charged} \\ \text{Surfaces}}} \sigma V \, dS$$

implies that the energy belongs to the charges and not to the field. The difference between the two points of view can be traced back to different concepts of electrostatics, i.e., action at a distance exemplified by charges and distances, vs. field theory which is stated in terms of local interaction of charges with fields.

There are further limitations of these derivations. $\delta W_1 = \delta W_2$ assumes no loss of energy such as by radiation or by heating material substances. It should have been stated that the system remains at a constant temperature. Except in free space, δW_1 is not the total increase in energy; some energy will always go into dielectric heating. δW_2 is the maximum energy that can be regained from the field at constant temperature. Thermodynamically this is the Helmholtz free energy A. If Λ is the internal energy, S the entropy and T the temperature, then

$$A = \Lambda - TS,$$

$$dA = d\Lambda - T\,dS - S\,dT;$$

$$đW \leqslant T\,dS - d\Lambda$$

or

$$đW \leqslant -dA - S\,dT.$$

If the reaction is reversible (if the equality signs above hold), then for T = const. (dT = 0),

$$(dW)_T = -(dA)_T = \iiint_{\text{All Space}} \mathscr{E} \cdot \delta\mathbf{D} \, d\tau.$$

Reactions always proceed until the free energy is a minimum. In the electrical case the mutual forces between charges (consistent with whatever constraints are imposed) cause them to move until the free energy is a minimum. This is Thomson's theorem.

IV. Conductors and Dielectrics

A. ELECTRICAL BEHAVIOR OF MATERIALS

Physical matter exhibits electrical behavior by the extent of its electrical conductivity and its ability to polarize. In addition, the magnetic properties characterize the matter under consideration. In this section we will be interested only in the limits of conductivity and electrical polarizability.

Real materials form a continuous range from nearly ideal conductors to nearly ideal insulators.

In an ideal conductor electrical charges move freely under the influence of electrical fields (within the constraints imposed by the geometry of the conductor). Metals are examples of conductors.

At the other end of the range of material media lie the ideal insulators which do not conduct free charges in the presence of electric fields. In practice no material is an ideal insulator; all insulators have some conductive properties no matter how small. In addition, at a certain maximum field strength the insulation characteristics of an insulator decrease in what is known as dielectric breakdown.

By and large insulators behave as dielectrics. In a dielectric the charged particles are bound and the net charge of the substance is zero. Under the influence of an external field the atomic or molecular constituents of the dielectric polarize, producing a net volume distribution of dipole moments. This is not the complete description. Some substances are composed of molecules which possess a permanent dipole moment as contrasted to the induced moment which is actuated by an external field. These permanent dipole or polar molecules, as they are called, tend to assume preferred orientations in the direction of an external field. However, thermal motion coupled with collisions produces a statistical distribution of the alignments. This influence of thermal motions produces an over-all temperature dependence.

The charge carriers in a real conductor such as a metal are the electrons. On the other hand, semiconductors possess both positive and negative charge carriers. Holes in the electron sea provide the positive charge carriers. Electrolytes have both positive and negative ion charge carriers.

B. CONDUCTORS

1. THE STATIC EQUILIBRIUM OF A CHARGED CONDUCTOR

a. THE STATIC ELECTRIC FIELD. The charged conductor in static equilibrium by definition is in a state in which all of the charges are *at rest*.

If the charges are at rest on the surface there is *no* component of an electric field *tangent* to the surface. Thus the electric fields must be NORMAL to the surfaces of all conductors in static equilibrium.

Because the static electric field is normal to the corresponding equipotential surfaces, the surface of a conductor in static equilibrium is an EQUIPOTENTIAL SURFACE.

b. THE SURFACE CHARGE. In stable equilibrium, all the charge resides on the surface of a charged conductor. We demonstrate this in the following fashion.

The field inside of a charged conductor is zero unless there is free charge in the interior region. This is proven quite readily by Gauss's law. Let the surface of the conductor be S. Construct a second closed surface S′ just inside S. If the net charge on the interior is zero, then the normal component of **D** (or $\epsilon_0\mathscr{E}$ in this case) integrated over S′ is zero. Because $\iint_{S'} \mathbf{D}\cdot\mathbf{n}\,dS'$ is zero for *any* S′, D must be zero, and therefore $\mathscr{E}$ must be zero.

As a result, the only way in which we can create an electric field inside of S is to place free charges inside of S. This, however, violates our condition of equilibrium since the free charge in the presence of a field will move until the net force on the charges is zero. This, of course, takes place only when the charge is on the surface.

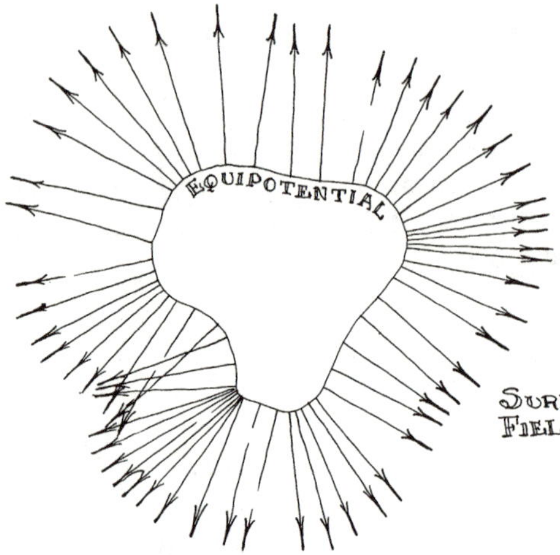

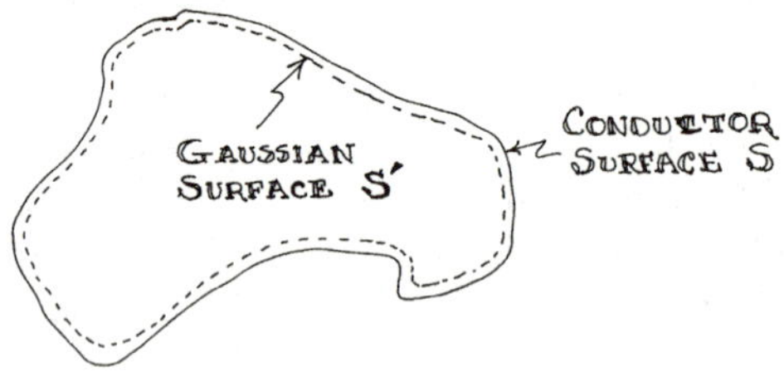

At the beginning care was taken to stipulate *stable* equilibrium. This was done because in certain conductors possessing symmetry it is possible in principle to place one point charge at the center of symmetry in a state of unstable equilibrium. The conducting sphere is an example of this type. A point charge at the exact center is in unstable equilibrium, i.e., the net force is zero. Any small displacement from the symmetry point, however, will set the charge in motion and it will move to the surface.

C. TESTS OF THE INVERSE SQUARE LAW. From the previous discussion it becomes apparent that because the derivation of Gauss's law depends explicitly upon the inverse square law, an experimental test of Gauss's law should be at the same time a test of the inverse square law.

One can show by direct integration that a necessary and sufficient condition that the field be zero inside of a charged spherical conducting shell is that the field of an element of charge on the surface vary as $1/r^2$, where r is the distance from the surface element to a point inside the surface.

Experimentally the tests can be made in several ways; the majority of these are variations of the experiments employing "Cavendish spheres." The set of Cavendish spheres consists of two concentric conducting spherical shells insulated from one another. The outer sphere has a small aperture cut in it. Attached is a port which covers and uncovers the aperture. The experiment consists of two basic operations which are diagrammed in the accompanying figures. Let us call the two operations A and B.

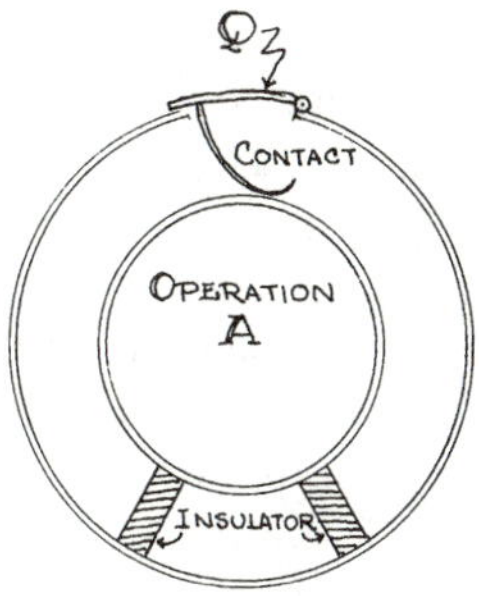

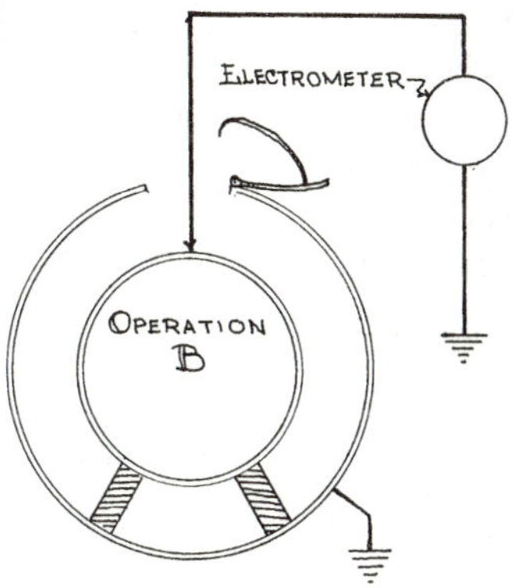

Operation A:

(1) The two spheres are initially uncharged. The port is closed, and thereby electrical contact is made between the inner and outer spheres.

(2) A charge Q is placed upon the surface of the outer sphere.

(3) If a field exists between the inner and outer spheres, charge will flow between the two.

Operation B:

(1) The port is opened, and electrical contact is broken between the two spheres.

(2) An electroscope is connected to the inner sphere. The electroscope then will provide an indication which is a measure of the charge deposited upon the inner sphere.
(3) Within experimental measurement the inner sphere is found to have zero charge on it. Therefore the inverse square law can be verified to the limits of accuracy of this experiment.

The present accuracy of experimental testing of the inverse square law shows that $|\mathscr{E}|$ varies as $1/r^{(2\pm\delta)}$ where $\delta < 2(10)^{-9}$ [Plimton and Lawton, *Phys. Rev.*, **50**, 1066 (1936)].

2. FIELDS NEAR THE SURFACES OF CHARGED CONDUCTORS

Problems involving charged conductors in static equilibrium require that the surfaces of the conductors be equipotentials of the problem. This can be stated in the form of a requirement that the field be normal to the conducting surface (using Gauss's law) or as a direct boundary condition on the solution for the potential function.

The choice of technique for solving problems depends in large measure upon the insight and experience of the person involved. In the paragraphs to follow methods will be described for the solution of special problems. Once the reader is familiar with the various techniques available he should be in a position to apply the most useful technique in a given exercise.

a. THE FIELD AT THE SURFACE OF A CHARGED CONDUCTOR. Consider a closed conducting surface S with a surface charge density σ at every point on the surface. Then

$$\iint_S (\mathbf{D}\cdot\mathbf{n} - \sigma)\, dS = 0$$

A sufficient condition that this integral be zero is that

$$\sigma = \mathbf{D}\cdot\mathbf{n}.$$

Since **D** is parallel to **n** at every point *on* the surface S,

$$\mathbf{D} = \sigma\mathbf{n}.$$

This result can also be obtained by a limiting process. If we construct a small cylindrical disk-shaped surface with faces A_1 and A_2 equal to dS parallel to the intercept on S, then **n** the normal at

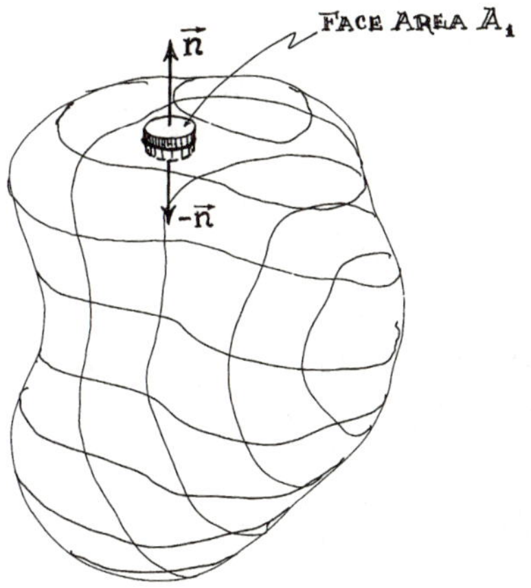

dS is perpendicular to the faces of the disk. The field inside the surface S is zero. Thus by Gauss's law

$$\mathbf{D} \cdot \mathbf{n}\, A_1 - 0 \cdot \mathbf{n}\, A_2 = \sigma A,$$

or

$$\sigma = \mathbf{D} \cdot \mathbf{n}.$$

We have assumed here that dS is sufficiently small that **D** is everywhere parallel to **n** to first order. In the limit, second-order infinitesimals will drop out.

With this result we see that a semi-infinite uniformly charged conducting plane will produce a uniform field perpendicular to the plane. Assume that the xy plane is a conducting plane and that the volume in region $z \leqslant 0$ is conducting. If this plane carries a uniform surface charge σ_0, the electric field in the region $0 < z$ is given by

$$\mathscr{E} = \frac{\mathbf{D}}{\epsilon_0} = \frac{\sigma \mathbf{n}}{\epsilon_0}.$$

The fact that **D** on the surface is equal to $\sigma\mathbf{n}$ also allows the simple qualitative description of the manner in which the surface charge density varies as the radius of curvature of the surface.

Consider the charged conducting surface shown in the figure. The potential is constant over the surface. To first order the potential on the surface of radius of curvature R at the extremity

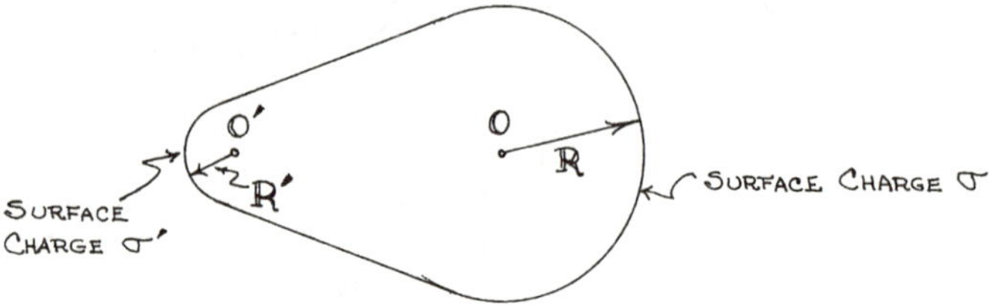

behaves like a charged conducting sphere of radius R,

$$V \simeq \frac{\sigma R \, d\Omega}{4\pi\epsilon_0}.$$

The potential on a surface of smaller radius of curvature R′ behaves at the extremity as

$$V \simeq \frac{\sigma' R' \, d\Omega}{4\pi\epsilon_0}.$$

Because the V's are the same,

$$\frac{\sigma}{\sigma'} \to \frac{R'}{R}.$$

The surface having the smallest radius of curvature has the highest surface charge density and thereby the highest electric field.

If the radius of curvature is made very small, large fields can be developed. This is a method used to produce field emission when extremely high fields are required.

b. FIELDS OF CYLINDERS AND SPHERES. The fields produced by single charged conducting cylinders of infinite length or by single charged conducting spheres can be readily calculated using Gauss's law. Examples of these will be left as exercises.

3. FORCES ON CHARGED CONDUCTORS

A charged conductor experiences pressure at every point on its surface. These pressures arise from two sources. The mutual interaction of the surface charges produces a surface pressure. For example, a uniformly charged sphere is in a state in which all of the surface charges mutually repel. This repulsive surface interaction produces a net uniform pressure outward along the surface normal.

In addition to the surface interaction of the charges, an external field will exert a pressure. For instance, an uncharged conducting sphere in an external field will polarize, the positive charge lying mostly in one hemisphere and the negative charge in the other. The integrated pressure can give rise to a net translational force if the external field is inhomogeneous.

In both cases the pressure depends upon the energy density of the total field at the surface of the conductor, where the energy density u has been defined as

$$u = \tfrac{1}{2}\mathbf{D} \cdot \mathscr{E}.$$

Consider a conductor having a closed surface S and fields **D** and $\mathscr{E}$ defined at every point on the surface. Displace a small surface

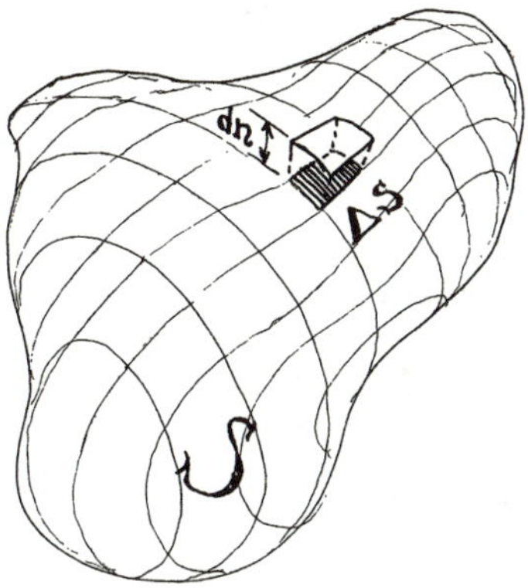

element ΔS in the direction of the surface normal **n** by a distance Δn. The volume swept out is ΔS Δn, and the change in energy associated with this displacement is

$$\Delta U = -\tfrac{1}{2}\mathbf{D} \cdot \mathscr{E}\, \Delta S\, \Delta n.$$

The minus sign occurs because U decreases as ΔS is displaced outward. This is seen from a limiting argument. Take a charged spherical conductor of radius R. Let $R \rightarrow \infty$ (with the total charge fixed); in this limit the total energy of the system goes to zero.

The total derivative of U is

$$\Delta U = \frac{dU}{dn}\Delta n = \text{grad } U \cdot \Delta \mathbf{r}$$

where $\quad \Delta n = \mathbf{n} \cdot \Delta \mathbf{r}.$

Then $\quad \Delta U = \text{grad } U \cdot \Delta \mathbf{r} = \{-\tfrac{1}{2}(\mathbf{D} \cdot \mathscr{E})\, \Delta S \mathbf{n}\} \cdot \Delta \mathbf{r},$
giving

$$-\text{grad } U_{\Delta S} = \Delta \mathbf{F} = \tfrac{1}{2}(\mathbf{D} \cdot \mathscr{E})\, \Delta S \mathbf{n}.$$

The pressure on the surface element ΔS is

$$\mathbf{P} = \frac{d\mathbf{F}}{dS} = \lim_{\Delta S \to 0} \frac{\Delta \mathbf{F}}{\Delta S} = \tfrac{1}{2}(\mathbf{D} \cdot \mathscr{E})\mathbf{n}.$$

Restating this result, the pressure $\mathbf{P}$ on the surface of a conductor is

$$\mathbf{P} = \tfrac{1}{2}(\mathbf{D} \cdot \mathscr{E})\mathbf{n}.$$

Since $\mathbf{D} = \sigma \mathbf{n}$, in the absence of dielectric media at the surface,

$$\mathscr{E} = \frac{\mathbf{D}}{\epsilon_0} = \frac{\sigma}{\epsilon_0}\mathbf{n};$$

thus $\quad \mathbf{P} = \dfrac{\sigma^2}{2\epsilon_0}\mathbf{n}$ (conductor in a vacuum).

The net force on a charged conductor is obtained by integrating the pressure over the surface S:

$$\mathbf{F} = \iint_S \mathbf{P}\, dS = \tfrac{1}{2}\iint_S (\mathbf{D} \cdot \mathscr{E})\mathbf{n}\, dS.$$

If the conductor is in a vacuum,

$$\mathbf{F} = \frac{1}{2\epsilon_0}\iint \sigma^2 \mathbf{n}\, dS.$$

4. INDUCED CHARGE

a. INTRODUCTION. When electrical matter is placed in an external electric field $\mathscr{E}_0$, the field exerts forces upon the intrinsic charge carried by the matter in question. The charges will assume equilibrium configurations consistent with the shape and intensity of the field and with the electrical properties of the medium.

In the case of a conductor, the free charge will flow upon the surface until the conductor surface is an *equipotential of the final field.*

The charges of a dielectric, although not free, can separate to form a volume distribution of dipoles. The final field is then a superposition of the original field plus the field set up by the dipoles.

In all cases the final electric field differs from the original external field according to the constraints imposed by the presence of the electrical matter.

There are several basic types of problems involving induced charge on conductors.

In the problem of charges induced on an isolated uncharged conductor placed in an external field, the following conditions hold. The surface of the conductor becomes an equipotential of the final field, and therefore the final field is perpendicular to the surface. The *total* charge on the conductor is unchanged, and in this instance it is zero. Thus the induced charge distribution consists of variations between positive and negative charge densities. At great distances the field is not altered from the original form (in the absence of the conductor).

For the problem involving a charged conductor in an external field the statements are similar, except that the *total* charge on the conductor is not zero but maintains the initial value given. In addition, the far field now consists of a superposition of the initial external field plus the field of a point charge having a charge magnitude equal to the net charge on the conductor.

A second class of problems arises when the potential of the conductor is specified. For instance a grounded conductor can be placed in an external field. Then not only must the surface be an equipotential of the final field, but in addition this surface must be at zero potential. The total charge on the conductors in these problems will not be zero.

Problems involving conductors can be solved by several techniques; the method of analysis employed is a matter of convenience. The results must be the same, and so the methods are in some sense equivalent. When point charges supply the external field one can often find a solution by the method of images. The method of images requires some experience, and it is equivalent to an expansion in multipoles (similar to the expansion in terms of Legendre polynomials in the case of Laplace's equation in spherical coordinates).

b. THE METHOD OF IMAGES. The application of the method of images to problems requires some insight. This insight can be advanced to a great extent by demonstration of the solutions for several standard cases.

The utility of this method lies in the fact that it is possible in many examples, concerning the fields and potentials of *continuous charge* distributions, to establish a *point charge configuration* which produces fields and equipotentials in a *restricted region*, which are completely equivalent to the fields of the distributions.

For the simpler problems involving image charges, the point charge multipoles can be examined to find the charged conductor problems to which they are equivalent.

(1) *The Charged Conducting Sphere in a Vacuum.* Previously it was demonstrated that a conducting sphere of radius R carrying a homogeneous surface density σ_0 is equivalent in the region $R \leqslant r$ to a point charge of magnitude $4\pi R^2\sigma_0$ located at the point corresponding to the center of the sphere.

Notice very carefully that the two problems produce the same potential and field configurations in the RESTRICTED REGION $R \leqslant r$. The equivalence is *not* present when the point of examination is taken inside the surface of the sphere. The existence of equivalence in restricted regions is characteristic of the image method.

(2) *The Point Charge above a Semi-Infinite Grounded Conducting Plane.* Consider the near field of a dipole lying along the z axis.

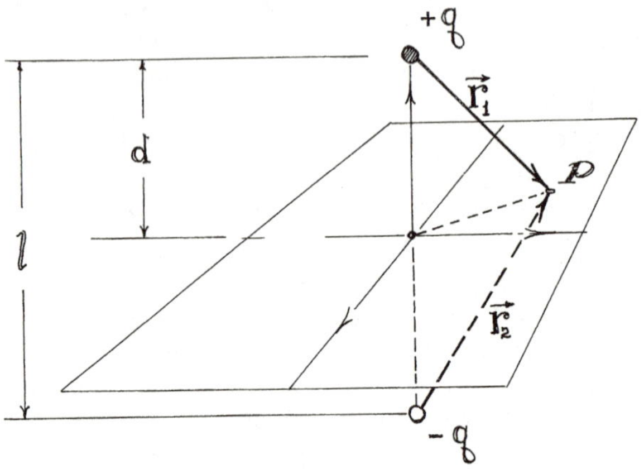

The terminal charges are $+q$ and $-q$. The separation is 2d, and the charges are centered about the origin. The potential is

$$V_p = \frac{q}{4\pi\epsilon_0}\left(\frac{1}{r_1} - \frac{1}{r_2}\right).$$

At every point in the xy plane the potential is zero. Therefore in the region $z \geqslant O$ (or for a mirror image problem $z \leqslant O$) this problem is completely equivalent to the problem of a point charge above a semi-infinite grounded conducting plane (the xy plane in this case).

Assume that $+q$ is at $d\mathbf{k}$ in this problem. The charge $-q$ at $-d\mathbf{k}$ is called the IMAGE CHARGE for this particular problem.

In conclusion we see that if a charge $+q$ is placed a distance d above a grounded conducting plane, the fields in the region *above* the

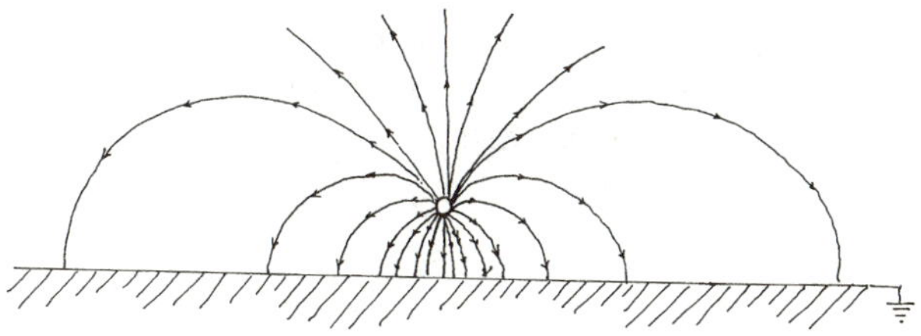

NEAR FIELD: POINT CHARGE OUTSIDE A GROUNDED CONDUCTING PLANE

plane can be developed by replacing the plane by an *image* charge $-q$.

At a point P above the plane ($0 \leqslant z_p$)

$$V_p = \frac{q}{4\pi\epsilon_0}\left\{\frac{1}{[x_p^2 + y_p^2 + (z_p - d)^2]^{1/2}} - \frac{1}{[x_p^2 + y_p^2 + (z_p + d)^2]^{1/2}}\right\},$$

while

$$\mathscr{E}_p = -\text{grad}_p V_p$$

$$= \frac{q}{4\pi\epsilon_0}\left\{\frac{(x_p\mathbf{i} + y_p\mathbf{j} + (z_p - d)\mathbf{k})}{[x_p^2 + y_p^2 + (z_p - d)^2]^{3/2}} - \frac{(x_p\mathbf{i} + y_p\mathbf{j} + (z_p + d)\mathbf{k})}{[x_p^2 + y_p^2 + (z_p + d)^2]^{3/2}}\right\}.$$

Since $$|\mathbf{D}_p(z_p = 0)| = \sigma(x,y),$$

and because $$\mathbf{D}_p = \epsilon_0\mathscr{E}_p,$$

$$\sigma = |\mathbf{D}(z_p = 0)| = -\frac{qd}{2\pi[x_p^2 + y_p^2 + d^2]^{3/2}}.$$

Finally the total induced charge on the plane is

$$q_{\text{induced}} = \iint_{x,y\text{ plane}} \sigma\, dx\, dy = -q;$$

$-q$ being the value of the image charge. This implies that all of the lines originating on the charge $+q$ above the plane terminate on

FIELD

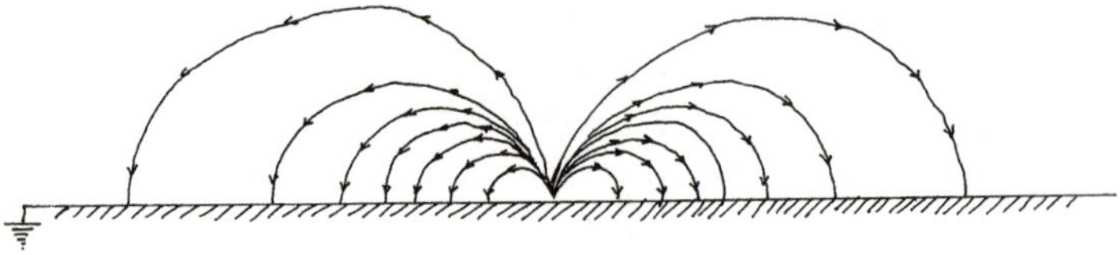

POINT CHARGE ABOVE A GROUNDED SEMI-INFINITE PLANE

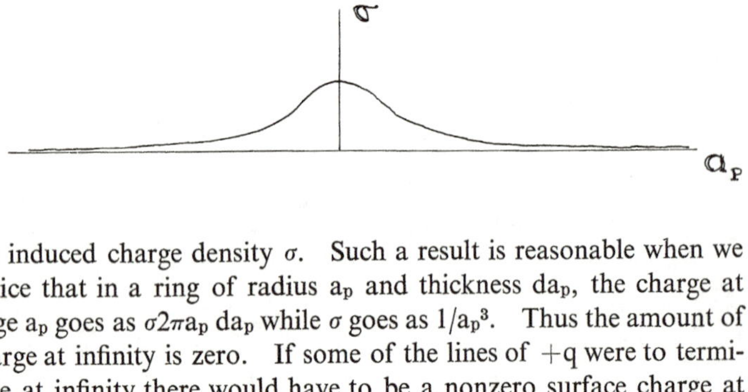

the induced charge density σ. Such a result is reasonable when we notice that in a ring of radius a_p and thickness da_p, the charge at large a_p goes as $\sigma 2\pi a_p\, da_p$ while σ goes as $1/a_p^3$. Thus the amount of charge at infinity is zero. If some of the lines of $+q$ were to terminate at infinity there would have to be a nonzero surface charge at $z_p \to \infty$, but such is not the case.

The far field ($d \ll r_p$) of a point charge above a grounded conducting plane is identical with the far field of a dipole of length 2d.

(3) *The Field of a Point Charge Near a Grounded Conducting Right Angle.* If we examine the planes of zero potential for a rectangular quadrupole, we find that the planes parallel to the sides of the rectangle and passing through the center of symmetry are planes of zero potential.

$$V_p = \frac{q}{4\pi\epsilon_0}\left(\frac{1}{r_1} - \frac{1}{r_2} + \frac{1}{r_3} - \frac{1}{r_4}\right).$$

In the region $0 \leqslant y$ and $0 \leqslant z$ this problem is equivalent to that of a point charge $+q$ outside and a distance a from the xz plane and a

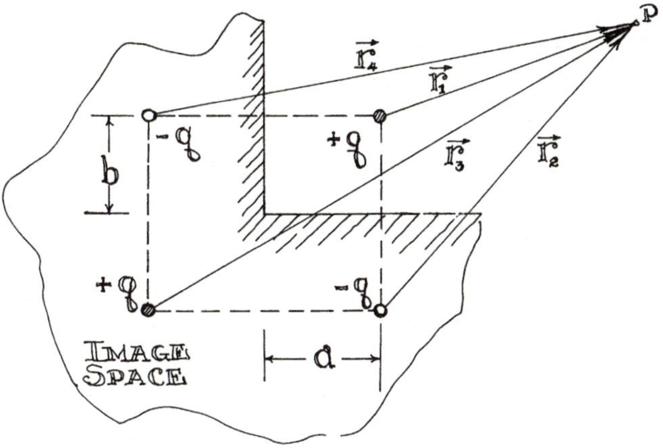

GROUNDED CONDUCTING RIGHT ANGLE

distance b from the xy plane. In this case there are THREE IMAGES, two negative and one positive.

To find $\mathscr{E}_p$ in the region of interest, compute $-\text{grad}\, V_p$. Then the induced charge distribution on the two planes is given by

$$\sigma(y_p = 0) = |\mathbf{D}(y_p = 0, z_p \geqslant 0)|$$
$$\sigma(z_p = 0) = |\mathbf{D}(z_p = 0, y_p \geqslant 0)|.$$

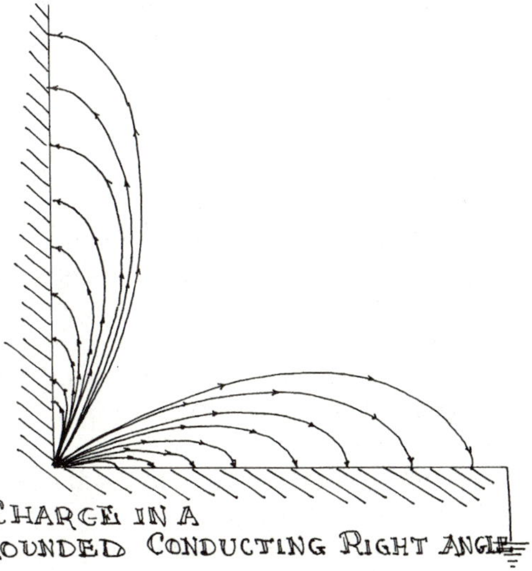

POINT CHARGE IN A
GROUNDED CONDUCTING RIGHT ANGLE

Again the far field ($a \ll r_p$, $b \ll r_p$) is made up of one potential lobe of the rectangular quadrupole.

(4) *A Point Charge in a Conducting Corner.* From the symmetry, the charge $+q$ together with seven image charges forms part of an octupole field in the exterior region.

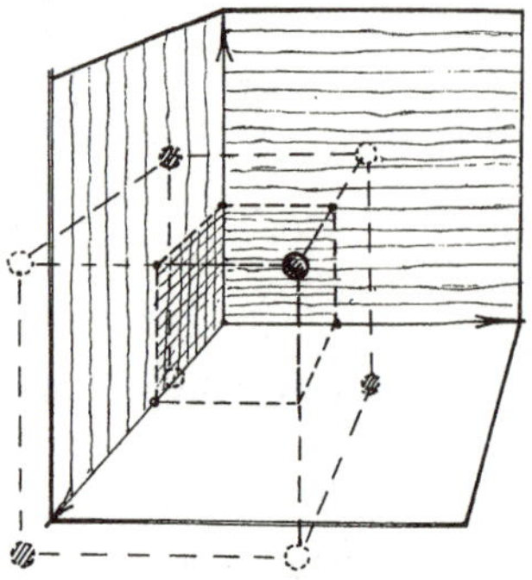

GROUNDED CONDUCTING CORNER

(5) *A Point Charge Outside of a Grounded Spherical Conductor.* A number of point charge-plus-conductor problems have been discussed. In each a solution was achieved by employing a series of images. Most problems of this type can be solved by expanding in a series of images; however, the series may be infinite. Our hope in a given problem is that the series converges rapidly.

There are several unique problems which can be solved not by a simple multipole but by some linear combination of multipoles. The example of a point charge outside of a spherical conductor can be represented by assuming a point charge plus a dipole configuration. In the particular example to follow, the sphere is grounded.

In many cases the qualitative characteristics of the images can be found by examining the obvious physical conditions of the problem. Let us state below what we know immediately about the physical configuration of the point charge outside of a grounded conducting sphere.

(a) The induced charge on the sphere will be *negative*; therefore the image should be negative.

(b) Not all of the lines of q need terminate on the sphere.

(c) The induced charge will have its highest concentration in the region of the surface of the sphere nearest the external point charge.

Let +q lie on the axis a distance d from the center of a grounded conducting sphere of radius R. Assume an image charge −mq a

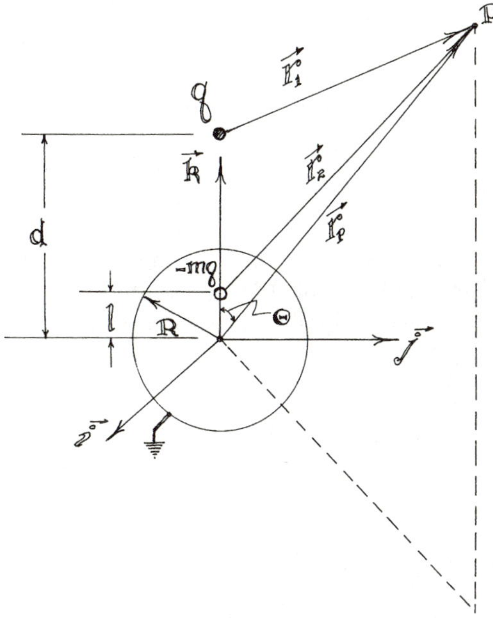

distance $l\mathbf{k}$ from the center, where $l < R$ and m is a number less than 1.

The potential of +q plus that of the image −mq is

$$V_p = \frac{q}{4\pi\epsilon_0}\left(\frac{1}{r_1} - \frac{m}{r_2}\right),$$

where
$$r_1 = (r_p^2 + d^2 - 2r_p d\cos\theta)^{1/2}$$
$$r_2 = (r_p^2 + l^2 - 2r_p l\cos\theta)^{1/2}.$$

The boundary condition of the problem requires that V_p vanish when $r_p = R$,

$$V_p(r_p = R) = 0.$$

or

$$\left(\frac{1}{r_1} = \frac{m}{r_2}\right)_{r_p=R}.$$

Squaring both sides of this last equation and equating the coefficients of 1 and of cos θ respectively,

$$R^2 + d^2 = \frac{1}{m^2}(R^2 + l^2)$$

$$2Rd = 2Rl/m^2.$$

Thus

$$m^2 = l/d,$$

$$l = R^2/d,$$

$$m = R/d.$$

The potential for points $R \leqslant r_p$ is then

$$V_p = \frac{q}{4\pi\epsilon_0}\left\{\frac{1}{[r_p{}^2 + d^2 - 2r_pd\cos\theta]^{1/2}} - \frac{1}{\left[\left(\frac{r_p{}^2d^2}{R^2}\right) + R^2 - 2rd\cos\theta\right]^{1/2}}\right\}.$$

The far field produced is that of a point charge $+[1 - (R/d)]q$ plus that of an *intrinsic* dipole of moment $\mathbf{p} = (R/d)(d - l)q\mathbf{k}$. In view of the fact that the origin has been taken at the center of the

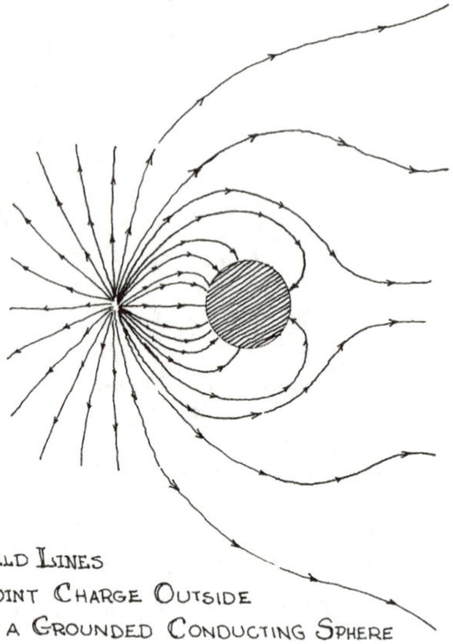

FIELD LINES
POINT CHARGE OUTSIDE
OF A GROUNDED CONDUCTING SPHERE

sphere, a geometric dipole field arises from the displacement of the point charge, and a geometric quadrupole effect is incorporated because of the displacement from the origin of the center of the dipole.

The induced surface charge distribution can be computed from $-$ grad V_p in spherical coordinates;

$$-\operatorname{grad} V_p = -\left\{\frac{\partial V_p}{\partial r_p}\boldsymbol{\epsilon}_r + \frac{1}{r_p}\frac{\partial V_p}{\partial \theta}\boldsymbol{\epsilon}_\theta + \frac{1}{r_p \sin\theta}\frac{\partial V_p}{\partial \phi}\boldsymbol{\epsilon}_\phi\right\}.$$

At the surface of the sphere the field must be radial since this surface is an equipotential. Then

$$\sigma(\theta) = |\mathbf{D}(r_p = R)| = \left| -\epsilon_0\left(\frac{\partial V}{\partial r}\right)_{r_p=R}\right|$$

$$= -\frac{q}{4\pi}\left\{\frac{(d^2 - R^2)}{R(d^2 + R^2 - 2Rd\cos\theta)^{3/2}}\right\},$$

Another advantage of the image technique lies in the ease with which the forces can be computed. The force on $+q$ is merely the attractive force exerted upon it by the image.

When the surface of the sphere is at a known potential or when the sphere is uncharged, a second image can be employed at the center of the sphere. This, of course, will maintain the surface at $r_p = R$ as an equipotential.

A straightforward, two-dimensional analogue of this problem is that of a line of charge outside of, and axially parallel to, a grounded cylindrical conductor. The analysis of this problem will be left as an exercise.

(6) *The Uncharged Conducting Sphere in an External Uniform Electric Field.* Customarily this problem is solved as a boundary value problem satisfying Laplace's equation. However, it is instructive to observe that an examination of the physical situation allows one to guess a set of image charges which also provide an exact solution.

Orient the **k** axis in the direction of the constant field; the problem has cylindrical symmetry.

The induced charge is positive on the half of the sphere in the direction of the field and negative in the opposite direction.

Because of the symmetry and the fact that the total charge is zero, the induced charge polarizes the sphere.

We assume that in the exterior region ($R < r_p$) the sphere produces a dipole field. This satisfies the condition that the external

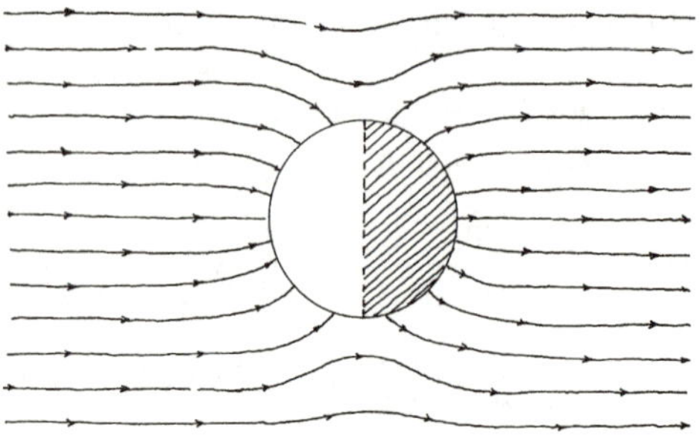

CONDUCTING SPHERE IN A CONSTANT EXTERNAL FIELD

constant field $\mathscr{E}_0$ be relatively unchanged at very large distances from the sphere.

If the effect of the uncharged sphere is to produce the field of a dipole in the exterior region, then the total potential is the superposition of the dipole potential and the original potential of the constant field $\mathscr{E}_0$;

$$V_{\text{total}} = \frac{p \cos\theta}{4\pi\epsilon_0 r_p^{\,2}} - \mathscr{E}_0 r_p \cos\theta + V_0.$$

When $r_p = R$ the surface potential of the sphere is a constant. In other words the surface potential cannot be a function of θ. Therefore

$$V_{\text{total}}\,(r_p = R) = V_0,$$

and
$$\frac{p \cos\theta}{4\pi\epsilon_0 R^2} = \mathscr{E}_0 R \cos\theta.$$

The image dipole moment is

$$p = 4\pi\epsilon_0 R^3 \mathscr{E}_0.$$

Finally, for $R \leqslant r_p$,

$$V_{\text{total}} = \mathscr{E}_0 \cos\theta \left(\frac{R^3}{r_p^{\,2}} - r_p\right) + V_0.$$

The electric field is merely $-$ grad V_T, giving

$$\mathscr{E}_p = \mathscr{E}_0 \cos\theta \left(\frac{2R^3}{r_p^{\,3}} + 1\right)\boldsymbol{\epsilon}_r + \mathscr{E}_0 \sin\theta \left(\frac{R^3}{r_p^{\,3}} - 1\right)\boldsymbol{\epsilon}_\theta.$$

At $r_p = R$, $\sigma = |\mathbf{D}_{r=R}|$, and the total induced charge on one half the surface is

$$q_+ = \int_0^{\pi/2} \sigma 2\pi R^2 \sin\theta \, d\theta = 3\pi\epsilon_0 \mathscr{E}_0 R^2.$$

The effective charge separation of the image dipole is

$$\frac{|\mathbf{p}|}{q_+} = \frac{4}{3} R.$$

C. SOLUTIONS OF LAPLACE'S EQUATION

(1) *The Uncharged Conducting Sphere in an External Uniform Electric Field.* As indicated in the previous section, this problem has cylindrical symmetry. In spherical coordinates there is no dependence upon the azimuthal angle ϕ.

In the region outside the sphere the charge is zero and the potential must obey Laplace's equation. Appendix D shows that two possible sets of solutions are available for the equation

$$\nabla^2 V = 0;$$

in spherical coordinates the series solutions are

$$V(\mathbf{r}) = \sum_{L=0}^{\infty} a_L r^L P_L(\theta)$$

or

$$V(\mathbf{r}) = \sum_{L=0}^{\infty} b_L r^{-(L+1)} P_L(\theta).$$

The first solution is well behaved at the origin; however, all terms above the first two produce electric fields which diverge for large r. The first two terms account for a zero field and a constant field independent of r.

The second series is divergent at $r = 0$ but is well behaved at large r.

Because our problem has an excluded origin and must be defined at large r, the most general solution consists of the two applicable terms from the first series and all of the terms from the second series:

$$\begin{aligned} V(\mathbf{r}_p) &= \sum_{L=0}^{1} a_L r_p{}^L P_L(\theta) + \sum_{L=0}^{\infty} b_L r_p^{-(L+1)} P_L(\theta) \\ &= a_0 + a_1 r_p \cos\theta + \frac{b_0}{r_p} + \frac{b_1}{r_p{}^2}\cos\theta \\ &\quad + \frac{b_2}{2r_p{}^3}(3\cos^2\theta - 1) + \dots . \end{aligned}$$

V_P must go as $a_0 + a_1 r_P \cos\theta$ at large r_P since the initial external field is constant. This condition has been utilized in omitting all a_L for $1 < L$. Then

$$\mathscr{E}_P \xrightarrow[\text{large } r_P]{} -\operatorname{grad}(a_0 + a_1 r_P \cos\theta) = \mathscr{E}_0 \mathbf{k},$$

giving

$$a_1 = -\mathscr{E}_0.$$

At the surface of the sphere ($r_P = R$), V_P is independent of θ, i.e., is constant. Equating coefficients of $P_L(\theta)$, this constraint requires that

$$a_1 R = -\frac{b_1}{R^2}$$

and that

$$b_L = 0 \quad \text{for } 1 < L.$$

The third condition is that the total charge on the sphere be zero. The relation between $|\mathbf{D}|$ and σ on the surface must be employed.

$$\mathscr{E}_r(r_P = R) = -\left[\frac{\partial V_P}{\partial r_P}\right]_R = -a_1 \cos\theta + \frac{b_0}{R^2} + \frac{2b_1}{R^3}\cos\theta.$$

The total charge is

$$q_{\text{total}} = \epsilon_0 \int_0^\pi \left\{a_1 \cos\theta + \frac{b_0}{R^2} + \frac{2b_1}{R^3}\cos\theta\right\} 2\pi R^2 \sin\theta \, d\theta,$$

$$= 4\pi\epsilon_0 b_0 = 0.$$

Therefore

$$b_0 = 0,$$

and

$$V_P = a_0 - \mathscr{E}_0 r_P \cos\theta + \frac{\mathscr{E}_0 R^3}{r_P^{\,2}}\cos\theta.$$

This problem illustrates quite well the use of the boundary value method.

If the fields had been exterior to a surface S_1 and interior to a surface S_2 ($R_1 < R_2$), the entire solution for V_P would have been required,

$$V_P = \sum_{L=0}^{\infty} a_L r^L P_L + \sum_{L=0}^{\infty} b_L r^{-(L+1)} P_L.$$

The values of the a_L's and b_L's could then be evaluated by similar methods. The total charges on the surfaces plus a knowledge of the values of the surface potentials would provide a solution (see figure).

d. POISSON'S EQUATION. The differential equation

$$\nabla^2 V = -\frac{\rho(r,\theta,\phi)}{\epsilon_0}$$

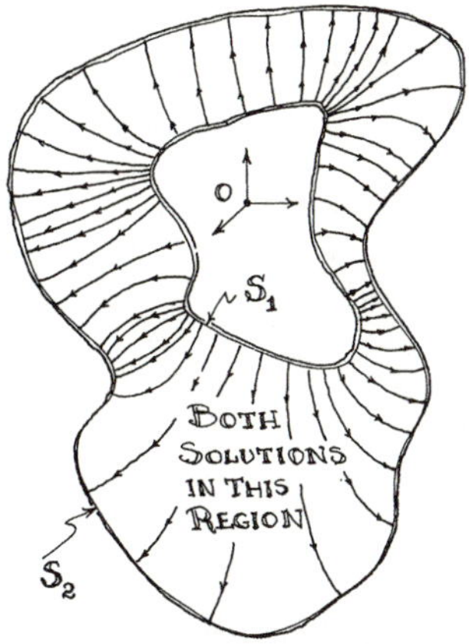

is inhomogeneous. The general solution, when portions of ρ are induced, is

$$V_p = \frac{1}{4\pi\epsilon_0}\iiint \frac{\rho\, d\tau}{|\mathbf{r}_p - \mathbf{r}|}$$

plus the general solution to Laplace's equation (i.e., the homogeneous equation). If ρ is completely specified the integral solution will suffice.

If both V and ρ are functions of one variable, say r, then one merely looks for a solution to the resulting inhomogeneous linear differential equation.

C. DIELECTRICS

1. INTRODUCTION

At the beginning of this chapter it was indicated that the two types of electrical matter of primary interest to us in our study of electrostatics were conductors and dielectrics. An ideal dielectric has no mobile singlet charge, and under the influence of an external electric field the constituent atoms and or molecules will be polarized either by induction or orientation to produce a net volume dipole moment.

Microscopically dielectric media are subclassified according to the action of the atoms or molecules. When no external field is present *nonpolar molecules* (or atoms) exhibit no electric dipole moment.

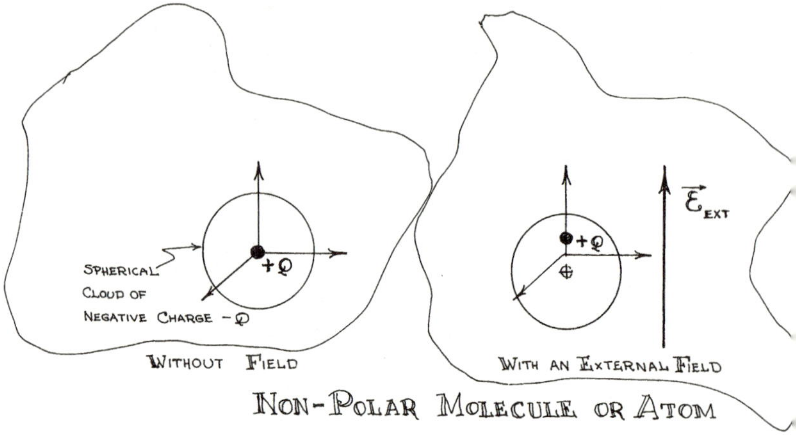

In the presence of a field there is a net separation of the centers of negative and positive charge.

On the other hand, *polar molecules* possess a permanent dipole moment. When there is no external field the orientation is generally random. As mentioned previously the imposition of an external

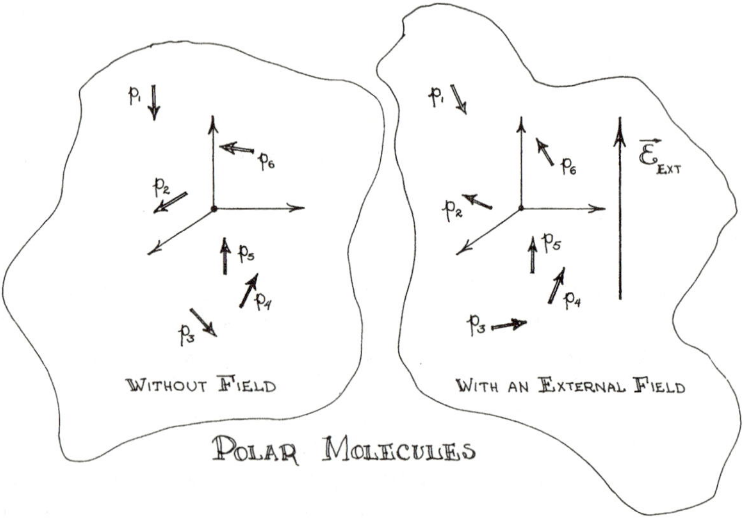

field produces a net orientation in the direction of the field. Interaction with neighboring particles causes the alignment to be statistically distributed. The greater number are aligned with the field while fewer have orientations antiparallel to the field. This distribution of alignments in many materials obeys Maxwell-Boltzmann statistics, and as a result there is a temperature dependence of the alignment and the net dipole moment per unit volume.

2. THE RELATION BETWEEN $\mathscr{E}$ AND D IN A DIELECTRIC

The discussion of fields inside and outside a volume distribution of dipoles in Section III-B presented the general microscopic properties with which we shall be concerned. In that discussion a quantity **D** was defined. The divergence of **D** provided a measure of the *mobile* or free charge density. Another statement of this relation was that the integral of the normal component of **D** over a closed surface equals the net free or mobile charge enclosed in that surface:

$$\iint_{S(\text{closed})} \mathbf{D} \cdot \mathbf{n}\, dS = q_{\text{free}},$$

and

$$\operatorname{div} \mathbf{D} = \rho_{\text{free}}.$$

To obtain these equations we defined **D** as

$$\mathbf{D} = \epsilon_0 \mathscr{E} + \mathbf{p}_V - \tfrac{1}{6}\widetilde{(\operatorname{div} \tilde{\mathbf{Q}}_V)} + \text{h.t.}$$

For the present we shall content ourselves with problems which *neglect* all multipole contributions above $\mathbf{p}_V$. The higher order contributions may prove interesting in time. This is an open question, however, since the present description of electrical matter, which does not include the quadrupole induced components of $\mathbf{p}_V$, has proven to be quite adequate in most cases.

The discussion will not be limited to isotropic dielectrics. We will consider in some detail the behavior of **D** and $\mathscr{E}$ at a boundary and in certain anisotropic media. The boundary surface of a dielectric represents an inhomogeneity in the medium. A surface discontinuity will produce cusps in the electric field lines. These cusps are representative of the refraction of the field at the boundary.

The equations describing the refraction can be developed readily using the two general characteristics of **D** and $\mathscr{E}$: that the normal component of **D** is conserved across a boundary (Gauss's law), and that the tangential component of $\mathscr{E}$ is conserved (conservative field).

Consider a portion ΔS of the boundary surface between a medium labeled (1) and a medium labeled (2). The electric field in (1) at ΔS we will designate as $\mathscr{E}^{(1)}$, and that in (2) at ΔS as $\mathscr{E}^{(2)}$. Now construct a rectangular mathematical pillbox enclosing ΔS and having faces parallel to ΔS. The thickness of the box, t, is perpendicular to the boundary and of negligible magnitude.

Assuming that there are no free charges on the surface, according

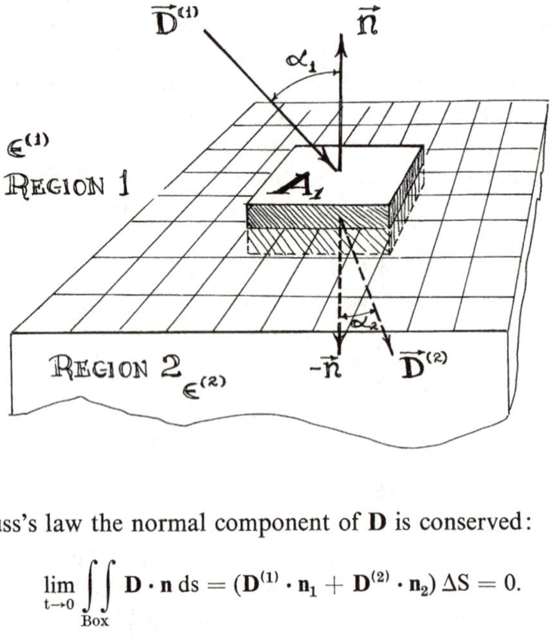

to Gauss's law the normal component of **D** is conserved:

$$\lim_{t\to 0} \iint_{\text{Box}} \mathbf{D}\cdot\mathbf{n}\,ds = (\mathbf{D}^{(1)}\cdot\mathbf{n}_1 + \mathbf{D}^{(2)}\cdot\mathbf{n}_2)\,\Delta S = 0.$$

Because $$\mathbf{n}_1 = -\mathbf{n}_2 = \mathbf{n},$$

therefore $$\mathbf{D}^{(1)}\cdot\mathbf{n} = \mathbf{D}^{(2)}\cdot\mathbf{n}.$$

Together with this result we use the condition that $\mathscr{E}$ is conservative. Taking a closed line integral which contains the two segments $\boldsymbol{l}$ and $-\boldsymbol{l}$ parallel to the surface and arbitrarily close (one on each side) plus two infinitesimal segments $\pm$t parallel to **n**,

$$\lim_{t\to 0} \oint \mathscr{E}\cdot d\mathbf{r} = \lim_{t\to 0} \{\mathscr{E}^{(1)}\cdot\boldsymbol{l}_1 - \mathscr{E}^{(1)}\cdot\mathbf{n}t + \mathscr{E}^{(2)}\cdot\boldsymbol{l}_2 + \mathscr{E}^{(2)}\cdot\mathbf{n}t\},$$

where $$l_1 = -l_2 = l.$$

Then $$\mathscr{E}^{(1)} \cdot l = \mathscr{E}^{(2)} \cdot l.$$

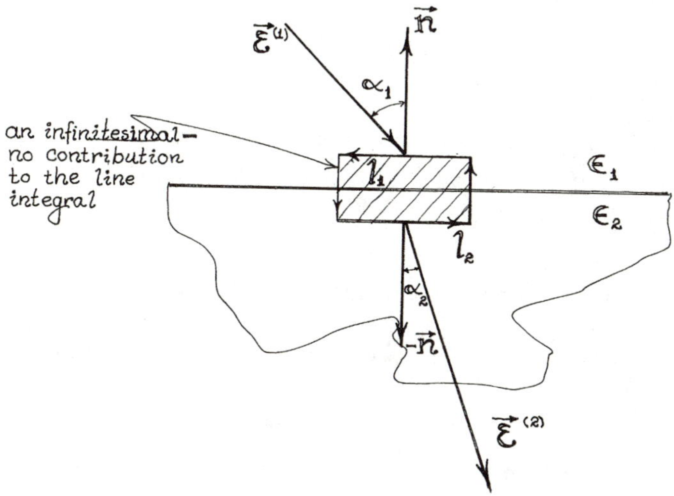

Call the angle of incidence α_1 (the angle between $\mathscr{E}^{(1)}$ and $-\mathbf{n}$);

$$\alpha_1 = \measuredangle\, {}^{-\mathbf{n}}_{\mathscr{E}^{(1)}}$$
$$\alpha_2 = \measuredangle\, {}^{\mathbf{n}}_{\mathscr{E}^{(2)}}.$$

Then $$\mathbf{D}^{(1)} \cdot \mathbf{n} = |\mathbf{D}^{(1)}| \cos \alpha_1$$
$$\mathbf{D}^{(2)} \cdot \mathbf{n} = |\mathbf{D}^{(2)}| \cos \alpha_2.$$

Also $$\mathscr{E}^{(1)} \cdot l = -|\mathscr{E}^{(1)}| \sin \alpha_1 = -|\mathscr{E}^{(2)}| \sin \alpha_2.$$

Combining these results,

$$\frac{|\mathscr{E}^{(1)}|}{|\mathbf{D}^{(1)}|} \tan \alpha_1 = \frac{|\mathscr{E}^{(2)}|}{|\mathbf{D}^{(2)}|} \tan \alpha_2.$$

3. THE DIELECTRIC TENSOR AND THE ELECTRIC SUSCEPTIBILITY

a. GENERAL INTRODUCTION. The polarization vector $\mathbf{p}_V$ is, in most instances, a function of the electric field $\mathscr{E}$ *in* the medium of $\mathbf{p}_V$:

$$\mathbf{p}_V = \mathbf{p}_V(\mathscr{E}).$$

In the simplest situation the polarization vector $\mathbf{p}_v$ can be represented by a linear transformation of the electric field vector.

A linear relation between $\mathbf{p}_v$ and $\mathscr{E}$ implies that *each* component of the vector $\mathbf{p}_v$ is given by a linear combination of the components of $\mathscr{E}$:

$$\frac{1}{\epsilon_0}(\mathbf{p}_v)_j = X_{j1}\mathscr{E}_1 + X_{j2}\mathscr{E}_2 + X_{j3}\mathscr{E}_3,$$

where $\mathscr{E}_n$ is the nth component of the field vector. Further the effects of saturation will cause the coefficients X_{mn} to be field dependent.

Referring to the notation outlined in Appendix B we can write the matrix transformation of $\mathscr{E}$ as

$$\mathbf{p}_v = \epsilon_0 \mathbf{X} \cdot \mathscr{E}.$$

Here $\mathbf{p}_v$ and $\mathscr{E}$ are column vectors, each having three components. $\mathbf{X}$ is a 3×3 matrix of nine elements.

The tensor (or matrix) $\mathbf{X}$ is called the *electric susceptibility tensor.*

According to Gauss's law,

$$\mathbf{D} = \epsilon_0 \mathscr{E} + \epsilon_0 \mathbf{X} \cdot \mathscr{E} + \text{h.t.}$$

Having neglected the higher order terms we write $\mathscr{E} = \mathbf{I} \cdot \mathscr{E}$, where $\mathbf{I}$ is the identity or unit matrix. Thus

$$\mathbf{D} = \epsilon_0(\mathbf{I} + \mathbf{X}) \cdot \mathscr{E}.$$

The quantity $(\mathbf{I} + \mathbf{X})$ is known as the dielectric tensor $\boldsymbol{\epsilon}$; then

$$\mathbf{D} = \epsilon_0 \boldsymbol{\epsilon} \cdot \mathscr{E}.$$

This representation is concise and represents $\mathbf{D}$ as a linear transformation of the electric field vector $\mathscr{E}$,

$$D_j = \sum_{n=1}^{3} \epsilon_0 \epsilon_{jn} \mathscr{E}_n.$$

In the inverse representation,

$$\mathscr{E} = \frac{1}{\epsilon_0} \boldsymbol{\epsilon}^{-1} \cdot \mathbf{D}:$$

that is, $\mathscr{E}$ is a transformation of $\mathbf{D}$. $\boldsymbol{\epsilon}$ is a nonzero matrix, and thus it is always possible to find its inverse.

By the rules for the addition of matrices, the elements of $\boldsymbol{\epsilon}$ are

$$\epsilon_{mn} = \delta_{mn} + X_{mn}.$$

δ_{mn} is the Kronecker delta, and X_{mn} is the mn^{th} element of the electric susceptibility tensor. Since $\mathbf{X}$ is a function of $\mathscr{E}$, $\boldsymbol{\epsilon}$ is also.

b. ISOTROPIC DIELECTRICS. Most of our interest will center about homogeneous isotropic dielectrics in which $\mathbf{p}_V$ is parallel to $\mathscr{E}$. Then

$$\mathbf{p}_V = \epsilon_0 \chi \mathbf{I} \cdot \mathscr{E} = \epsilon_0 \chi \mathscr{E},$$

and

$$\mathbf{D} = \epsilon_0 (1 + \chi) \mathscr{E} = \epsilon_0 \epsilon \mathscr{E},$$

with

$$\epsilon = 1 + \chi.$$

Here χ and ϵ are scalars. This implies that in the case of isotropic media $\mathbf{X}$ is a diagonal matrix with *all three elements the same.* Later we shall find that the tensor $\mathbf{X}$ can always be diagonalized; however, for the anisotropic problem the three diagonal elements are not equal.

The polarization vector was said to be usually $\mathscr{E}$ dependent. When such is the case the normal polarization is expected to go to zero with $\mathscr{E}$. However it does not always behave in this manner. A ferroelectric medium might show a hysteresis effect. When the external exciting field is reduced to zero the internal electric field of the dielectric may not go to zero. A similar behavior is observed in the magnetic polarization of a ferromagnet when the exciting currents are reduced to zero.

For the particular case of the isotropic dielectric, $\mathbf{D}$ is parallel to $\mathscr{E}$. Thus for a scalar susceptibility the equation representing the refraction at a boundary takes on a particularly simple form. Let the susceptibility and dielectric constants be $\chi^{(1)}$ and $\epsilon^{(1)}$ in the first medium and $\chi^{(2)}$ and $\epsilon^{(2)}$ in the second. Since

$$\frac{|\mathscr{E}^{(1)}|}{|\mathbf{D}^{(1)}|} \tan \alpha_1 = \frac{|\mathscr{E}^{(2)}|}{|\mathbf{D}^{(2)}|} \tan \alpha_2,$$

and

$$\mathbf{D}^{(m)} = \epsilon_0 \epsilon^{(m)} \mathscr{E}^{(m)},$$

thus

$$\frac{\epsilon^{(1)}}{\epsilon^{(2)}} = \frac{\tan \alpha_1}{\tan \alpha_2}.$$

If the first medium is a vacuum ($\chi^{(1)} = 0$) then $\epsilon^{(1)} = 1$.

The standard operational definitions of $\mathscr{E}$ and D inside an isotropic dielectric utilize two cavities, one a needleshaped cavity with the long axis parallel to the lines of $\mathscr{E}$, and the other a thin disk-shaped cavity with faces perpendicular to the $\mathscr{E}$ lines. In the needle-shaped cavity the cross-sectional area is assumed sufficiently small that the induced charge $\sigma_I = \mathbf{p}_V \cdot \mathbf{n}$ on the end surfaces does not contribute to the cavity field. Then the field in the needle-shaped cavity is the electric field of the dielectric. In the disk-shaped cavity, however, the induced surface charge on the faces is maximized and the cavity field is a measure of **D**,

$$\mathbf{D}_{\text{diel}} = \epsilon_0 \mathscr{E}^{\text{disk}}_{\text{cavity}} .$$

These definitions can be seen to be highly dependent upon end effects and edge effects produced by the induced charge. Thus our original definition of $\mathscr{E}$ in a medium as the force per unit charge exerted upon a test charge of a vanishingly small magnitude is by and large the best definition.

Although our original definition is microscopic, the present level of science enables such definitions to be used in a practical way. The cavity definitions are a holdover from the early development of electromagnetic theory and are macroscopic definitions. Furthermore, they are not sufficient when one encounters anisotropic dielectrics.

C. ISOTROPIC DIELECTRICS; INDUCED DIPOLES. A simple idealized model of an atom of a monatomic gas can be constructed to provide a useful illustration of the induced dipole moment. We assume that a gas atom in a zero external field consists of a positively charged point nucleus of charge $+Ze$ and a homogeneous spherical volume distribution $-\rho_0$ of negative charge surrounding the positive nucleus. Let the radius of the spherical distribution be R.

We view the negative spherical distribution as the electronic cloud about the nucleus; therefore the total charge in the negative distribution must be $-Ze$.

$$-\tfrac{4}{3}\pi R^3 \rho_0 = -Ze,$$

$$\rho_0 = \frac{3Ze}{4\pi R^3} .$$

The separation between atoms is assumed large in order that their mutual interaction be essentially zero. This assumption corresponds to a low gas pressure.

In the presence of an external field $\mathscr{E}_0$ the center of negative charge shifts antiparallel to the field. We assume that to first order the volume remains spherical. The positive nucleus shifts in the direction of the field.

Two forces act upon each charge center: The nucleus experiences the force of the external field F_{ext} tending to separate the charges, and when the positive and negative charge centers are separated there is a restoring force F_{res} exerted between the negative charge and the nucleus.

At equilibrium the vector sum of the two forces is zero. The

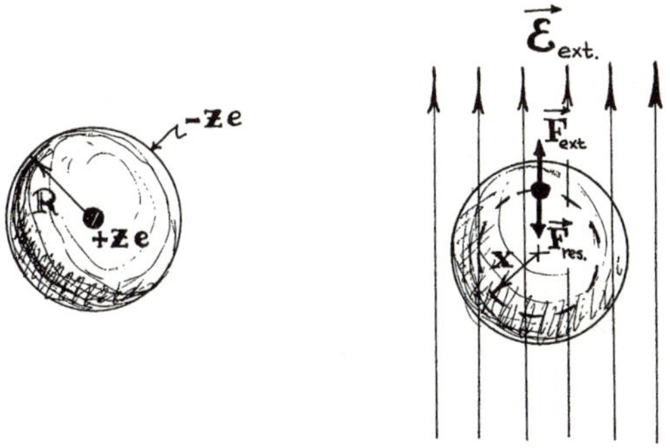

forces on +Ze are then

$$\mathbf{F}_{ext} + \mathbf{F}_{res} = 0.$$

The force exerted by the external field $\mathcal{E}_0$ is

$$\mathbf{F}_{ext} = Ze\mathcal{E}_0.$$

The internal restoring force can be computed utilizing Gauss's law to provide the force field of the negative charge. At a distance x from the center of the sphere,

$$\mathcal{E}_{intern} = \frac{-Ze}{\epsilon_0 4\pi R^3} \, x \, \frac{\mathcal{E}_0}{|\mathcal{E}_0|};$$

then $$\mathbf{F}_{res} = +Ze\mathcal{E}_{intern}.$$

Rewriting the equation for equilibrium,

$$Ze\mathcal{E}_0 - \frac{(Ze)^2}{4\pi\epsilon_0 R^3} x = 0,$$

giving $$(Ze)x = 4\pi\epsilon_0 R^3 \mathcal{E}_0.$$

This result provides the dipole moment of an atom. If there are n_v atoms per unit volume, then because the displacement $\mathbf{p}_v$ is oriented in the direction of $\mathcal{E}_0$,

$$\mathbf{p}_v = 4\pi\epsilon_0 n_v R^3 \mathcal{E}_0.$$

From this the electric susceptibility is

$$\chi = 4\pi R^3 n_v.$$

Interesting is the fact that this is merely three times the volume excluded by the atoms. The same result for χ is obtained if one computes the electric susceptibility of a uniform volume distribution of isolated conducting spheres of radius R.

This idealized result gives a constant susceptibility. It seems clear that our model, poor as it might be, becomes worse as x increases. We can assume that the cloud remains undistorted only so long as $x \ll R$. Distortion of the negative charge volume will make χ a function of $\mathscr{E}$.

If we take R to be about 10^{-10} meter and $n_v \simeq 3(10)^{25}$ atoms/m^3, we find that

$$\chi \simeq 0.004,$$

a value which compares well with measured susceptibilities of many gases at temperatures and pressures appropriate to n_v.

d. ISOTROPIC DIELECTRICS: POLAR MOLECULES. The polar molecule is classified as a microscopic system which possesses a permanent electric dipole moment. This moment, which will be designated as $\mathbf{p}_m$, is always present in the molecule and tends to align in the direction of an external field $\mathscr{E}_0$. In other words the molecule tends to orient in a position of stable equilibrium in which $\mathscr{E}_0$ and $\mathbf{p}_m$ are parallel.

The mutual interaction of the molecules plays an important role. The collisions occurring because of the kinetic motion of the molecules produces a statistical alignment. That is to say, the alignments of the individual moments are statistically distributed from a parallel orientation to an antiparallel orientation.

A net dipole moment is produced because the statistical distribution requires that the largest number of molecules lie in the states of lowest energy (states near stable equilibrium).

To compute the distribution of orientations, the Maxwell-Boltzmann statistics are used. If n_1 molecules are in an energy state U_1 and if n_2 molecules are in a state U_2, where

$$K + U_1 = -p_m\mathscr{E}_0 \cos\theta_1 + K$$

(in other words the energy is described completely by the orientation angle θ_1 and the kinetic energy K), then

$$n_2 e^{(U_2 + K)/kT} = n_1 e^{(U_1 + K)/kT}$$

or

$$n_2 = n_1 e^{-(U_2 - U_1)/kT}$$

where T is the absolute temperature and k is the Boltzmann constant.

If the number of molecules between n and n + dn is dn(θ), we can write

$$dn = e^{-U(\theta)/kT}\, d\Omega_\theta = Ce^{-U(\theta)/kT}\, 2\pi \sin\theta\, d\theta.$$

The constant C can be evaluated by requiring that the total number of vectors $\mathbf{p}_m$ per unit volume (with orientations between $\theta = 0$ and $\theta = \pi$) be n_v:

$$n_v = 2\pi C \int_0^\pi e^{+(p_m \mathscr{E}_0 \cos\theta/kT)} \sin\theta\, d\theta,$$

and

$$C = \frac{n_v p_m \mathscr{E}_0/kT}{4\pi \sinh(p_m \mathscr{E}_0/kT)}.$$

The over-all polarization in direction of $\mathscr{E}_0$ is designated as dp_v:

$$\begin{aligned} dp_v &= p_m \cos\theta\, dn(\theta) \\ &= p_m \cos\theta\, Ce^{(p_m \mathscr{E}_0 \cos\theta/kT)}\, 2\pi \sin\theta\, d\theta. \end{aligned}$$

After integration over θ we obtain the Langevin equation:

$$p_v = p_m n_v \left\{\coth\left(\frac{p_m \mathscr{E}_0}{kT}\right) - \left(\frac{kT}{p_m \mathscr{E}_0}\right)\right\}.$$

When $p_m \mathscr{E}_0/kT$ is small this expression can be expanded, and

$$p_v \simeq \frac{p_m^2 \mathscr{E}_0 n_v}{3kT} = \epsilon_0 \chi \mathscr{E}_0,$$

which illustrates our original comment that the susceptibility would be temperature dependent.

e. ANISOTROPIC DIELECTRICS. To illustrate the anisotropic dielectric we shall employ a rectangular lattice of particles. We shall further assume that the restoring forces developed in the polarization are different for the three directions defined by the edges of the rectangular lattice. The edges of the lattice are aligned along the x_1, x_2, and x_3 axes (the x,y,z axes). When an electric field is set up parallel to the jth axis we obtain polarization only in the jth direction. The electric displacement vector **D** for this *particular symmetry* (i.e., $\mathscr{E}$ parallel to a symmetry axis) is parallel to $\mathscr{E}$:

$$D_j = \epsilon_0 \mathscr{E}_j + p_j,$$

where p_j is the polarization along the jth axis when $\mathscr{E}$ is aligned along the jth axis. Remember that we are *beginning* by examining only those

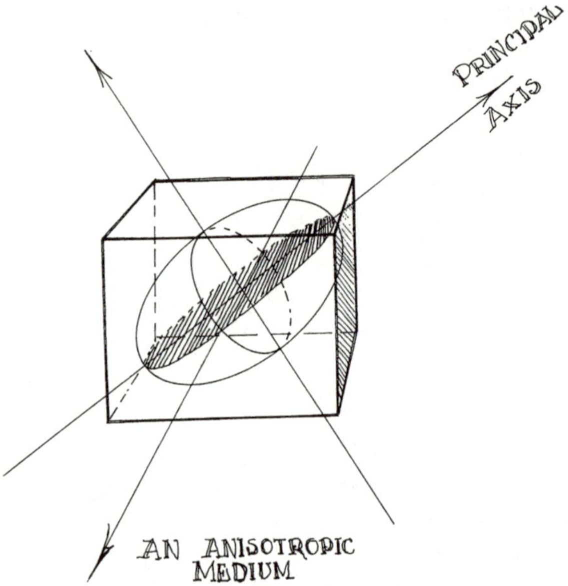

configurations which give $\mathbf{p}_v$ parallel to $\mathscr{E}$; when $\mathscr{E}$ does not lie parallel to an axis of symmetry, $\mathbf{p}_v$ will not necessarily be parallel.

This can be seen quite well if one thinks of the case in which $\mathscr{E}$ lies somewhere between the symmetry axes x_1 and x_2. If we assume that the restoring forces are different along these axes, the electric field component $\mathscr{E}_1$ will produce a polarization component p_1 while $\mathscr{E}_2$ will set up p_2. Although p_1 may be proportional to $\mathscr{E}_1$ and p_2 proportional to $\mathscr{E}_2$, *the constants of proportionality can be different.* Therefore the resultant $\mathbf{p}_v$ vector will not be parallel to the resultant $\mathscr{E}$ vector.

This argument will indicate our reasons for setting up the susceptibilities in terms of the symmetry axes.

In spite of the fact that the p_j would correspond to a different susceptibility for each j (for a given field magnitude), therefore we can assume that the polarization is given by

$$p_j = \epsilon_0 \chi_{jj} \mathscr{E}_j$$

when $\mathscr{E}$ is aligned along the jth symmetry axis. Then

$$D_j = \epsilon_0(1 + \chi_{jj})\mathscr{E}_j$$

with the same restrictions upon the alignment of $\mathscr{E}$.

We argue then that **D** and $\mathscr{E}$ are parallel in this anisotropic dielectric if $\mathscr{E}$ and **D** lie along an axis of symmetry. This is a situation analogous to that in classical mechanics in which the angular momentum $\mathscr{L}$ and the angular velocity $\boldsymbol{\omega}$ of a rigid body are parallel when the rotation of the body is about a principal axis.

For symmetry alignments $\mathscr{E}'$ and $\mathbf{D}'$,

$$\mathbf{D}' = \epsilon_0(\mathbb{I} + \boldsymbol{\chi})\mathscr{E}';$$

or
$$\mathbf{D}' = \epsilon_0 \begin{pmatrix} (1+\chi_{11}) & 0 & 0 \\ 0 & (1+\chi_{22}) & 0 \\ 0 & 0 & (1+\chi_{33}) \end{pmatrix} \cdot \mathscr{E}'$$

where $\mathscr{E}$ can be only of the form

$$\begin{bmatrix} \mathscr{E}_0 \\ 0 \\ 0 \end{bmatrix}, \quad \begin{bmatrix} 0 \\ \mathscr{E}_0 \\ 0 \end{bmatrix}, \quad \text{or} \quad \begin{bmatrix} 0 \\ 0 \\ \mathscr{E}_0 \end{bmatrix}.$$

In order to convert this restricted form to the form of a general dielectric tensor $\boldsymbol{\epsilon}$ we can employ an orthogonal transformation (or rotation). Before transforming we can observe some general properties of $\boldsymbol{\epsilon}$. The energy density is

$$u = \tfrac{1}{2}\mathbf{D} \cdot \mathscr{E} = \tfrac{1}{2}\mathscr{E} \cdot \mathbf{D}.$$

The reader should notice that $\mathscr{E}$ and **D** are real vectors and that their inner product can be commuted. This commutation demonstrates that $\boldsymbol{\epsilon}$ is a symmetric tensor. If

$$\mathbf{D} = \epsilon_0 \boldsymbol{\epsilon} \cdot \mathscr{E},$$

then in matrix notation

$$u = \tfrac{1}{2}\tilde{\mathbf{D}} \cdot \mathscr{E} = \tfrac{1}{2}(\widetilde{\epsilon_0 \boldsymbol{\epsilon} \cdot \mathscr{E}}) \cdot \mathscr{E} = \tfrac{1}{2}\epsilon_0 \tilde{\mathscr{E}} \cdot \tilde{\boldsymbol{\epsilon}} \cdot \mathscr{E}.$$

Also,
$$u = \tfrac{1}{2}\tilde{\mathscr{E}} \cdot \mathbf{D} = \frac{\epsilon_0}{2}(\tilde{\mathscr{E}} \cdot \boldsymbol{\epsilon} \cdot \mathscr{E}).$$

Thus
$$\tilde{\boldsymbol{\epsilon}} = \boldsymbol{\epsilon}$$

or
$$\epsilon_{mn} = \tilde{\epsilon}_{mn} = \epsilon_{nm};$$

and the matrix is symmetric.

In Appendix B it is demonstrated that a symmetric matrix can be diagonalized by an orthogonal transformation and that such a matrix will have orthogonal eigenvectors or principal axes. Although our original example had a specific rectangular geometry, we see from the preceding proof that because $\boldsymbol{\epsilon}$ is a symmetric matrix, anisotropic dielectrics have orthogonal principal axes. In other words if $\mathscr{E}$ is parallel to one of these axes, **D** and $\mathbf{p}_v$ will be parallel to $\mathscr{E}$.

Going back to our diagonal susceptibility $\boldsymbol{\chi}$ where

$$\boldsymbol{\chi} = \begin{pmatrix} \chi_{11} & 0 & 0 \\ 0 & \chi_{22} & 0 \\ 0 & 0 & \chi_{33} \end{pmatrix},$$

we write $\mathscr{E}'$ and $\mathbf{D}'$ in terms of another set of bases by a rotation S:

$$\mathscr{E}' = \mathrm{S} \cdot \mathscr{E}$$

and

$$\mathbf{D}' = \mathrm{S} \cdot \mathbf{D}.$$

The unprimed $\mathscr{E}$ and **D** are no longer constrained to lie along a principal axis.

The energy density in terms of $\mathbf{D}'$ and $\mathscr{E}'$ is

$$\mathrm{u} = \tfrac{1}{2}\tilde{\mathscr{E}}' \cdot \mathbf{D}' = \frac{\epsilon_0}{2}\tilde{\mathscr{E}}' \cdot (\mathrm{I} + \boldsymbol{\chi}) \cdot \mathscr{E}'.$$

As stated before, this result is set up in a preferred orientation. After rotating to any arbitrary set of orthogonal bases,

$$\mathscr{E}' = \mathrm{S} \cdot \mathscr{E} \quad \text{and} \quad \tilde{\mathscr{E}}' = \tilde{\mathscr{E}} \cdot \mathrm{S}^{-1},$$

(where $\tilde{\mathrm{S}} = \mathrm{S}^{-1}$) and substituting into the expression for u, we find

$$\mathrm{u} = \frac{\epsilon_0}{2}\tilde{\mathscr{E}} \cdot \mathrm{S}^{-1} \cdot (\mathrm{I} + \boldsymbol{\chi}) \cdot \mathrm{S} \cdot \mathscr{E} = \frac{\epsilon_0}{2}\tilde{\mathscr{E}} \cdot \boldsymbol{\epsilon} \cdot \mathscr{E}$$

giving $\quad \boldsymbol{\epsilon} = \mathrm{S}^{-1} \cdot (\mathrm{I} + \boldsymbol{\chi}) \cdot \mathrm{S} = \mathrm{I} + \mathrm{S}^{-1} \cdot \boldsymbol{\chi} \cdot \mathrm{S}.$

This equation defines the general form of the dielectric tensor and the general form of the susceptibility tensor X. $\boldsymbol{\epsilon}$ and X are both 3×3 symmetric matrices,

$$\mathrm{X} = \mathrm{S}^{-1} \cdot \boldsymbol{\chi} \cdot \mathrm{S}$$

and

$$\boldsymbol{\epsilon} = \mathrm{I} + \mathrm{X},$$

remembering that the $\boldsymbol{\chi}$ matrix is diagonal.

Usually X and $\boldsymbol{\epsilon}$ are specified initially; for instance

$$\boldsymbol{\epsilon} = \begin{pmatrix} \epsilon_{11} & \epsilon_{12} & \epsilon_{13} \\ \epsilon_{21} & \epsilon_{22} & \epsilon_{23} \\ \epsilon_{31} & \epsilon_{32} & \epsilon_{33} \end{pmatrix},$$

or
$$\mathbb{X} = \begin{pmatrix} X_{11} & X_{12} & X_{13} \\ X_{21} & X_{22} & X_{23} \\ X_{31} & X_{32} & X_{33} \end{pmatrix}$$

with
$$\epsilon_{jk} = \delta_{jk} + X_{jk},$$
and
$$X_{jk} = X_{kj}.$$

When $\mathbb{\epsilon}$ and $\mathbb{X}$ are specified in terms of these components it is often of interest to find the location of the principal axes and the magnitude of the eigenvalues. This requires that the eigenvalue problem be solved,
$$\mathbb{\epsilon} \cdot \mathbf{R} = \lambda \mathbf{R}.$$

The three eigenvalues λ_j are roots of
$$|\mathbb{\epsilon} - \lambda \mathbb{I}| = 0,$$
and the jth eigenvector $\mathbf{R}_j$ can then be found using $\lambda_j = (1 + \chi_{jj})$,
$$(\mathbb{\epsilon} - \lambda_j \mathbb{I}) \cdot \mathbf{R}_j = 0,$$
or
$$\sum_{n=1}^{3} \{\epsilon_{mn} - \delta_{mn}\lambda_j\} R_{nj} = 0.$$

The eigenvalues and eigenvectors can also be obtained from the $\mathbb{X}$ matrix,
$$\mathbb{X} \cdot \mathbf{R} = \chi \mathbf{R}.$$

Here the eigenvalues are $\chi_{jj} = \lambda_j - 1$.

The diagonal forms $(\mathbb{I} - \boldsymbol{\chi})$ and $\boldsymbol{\chi}$ can be developed from
$$(\mathbb{I} + \boldsymbol{\chi}) = \mathbb{S} \cdot \mathbb{\epsilon} \cdot \mathbb{S}^{-1}$$
where the elements of $\mathbb{S}^{-1}$ are formed from the components of the normalized eigenvectors.

Just as in the case of the rigid body problem in analytic dynamics, it is simpler in many instances to work problems in those coordinates aligned along the principal axes.

f. MATRIX APPROACH TO THE REFRACTION AT THE SURFACE OF AN ANISOTROPIC DIELECTRIC. The conservation of the normal component of **D** at the boundary of a dielectric can be represented in terms of matrices in a particularly powerful way.

Establish a set of cartesian coordinates at a point on the surface in question with the $\boldsymbol{\epsilon}_3$ base vector* pointing along the outward

* The reader must always distinguish between the general base vector $\boldsymbol{\epsilon}_m$ and the dielectric tensor $\mathbb{\epsilon}$. These quantities are both symbolized by epsilon; however one appears in bold face (the vector) and the other in open face (the matrix).

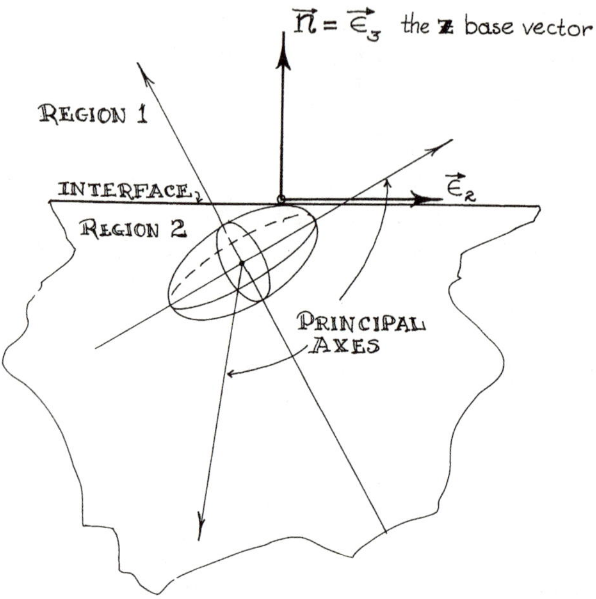

surface normal **n**. Our equation

$$\mathbf{D}^{(0)} \cdot \mathbf{n} = \mathbf{D}^{(d)} \cdot \mathbf{n}$$

now becomes

$$D_3^{(0)} = D_3^{(d)}.$$

Here the superscript (0) implies a quantity evaluated at the surface in the vacuum; the superscript (d) indicates that the quantity is evaluated in the dielectric.

If we now write **D** as a column vector

$$\mathbf{D}^{(m)} = \begin{bmatrix} D_1^{(m)} \\ D_2^{(m)} \\ D_3^{(m)} \end{bmatrix},$$

we can then construct a matrix $\mathbb{N}$ which conserves the third component (the normal component) across the boundary:

$$\mathbb{N} = \begin{pmatrix} 0 & 0 & 0 \\ 0 & 0 & 0 \\ 0 & 0 & 1 \end{pmatrix}; \qquad N_{jk} = \delta_{3k}.$$

The conservation of the normal component of **D** is now expressed as

$$\mathbb{N} \cdot \mathbf{D}^{(0)} = \mathbb{N} \cdot \mathbf{D}^{(d)}.$$

The other boundary condition, developed in Part 2 of the present section, is that the tangential component of $\mathscr{E}$ is conserved. Using the same set of base vectors, this implies that

$$\mathscr{E}_1^{(0)} = \mathscr{E}_1^{(d)}$$

and

$$\mathscr{E}_2^{(0)} = \mathscr{E}_2^{(d)}.$$

This condition can be represented with a matrix $\mathbb{T}$ where

$$\mathbb{T} = \begin{pmatrix} 1 & 0 & 0 \\ 0 & 1 & 0 \\ 0 & 0 & 0 \end{pmatrix}.$$

Then the equation for the conservation of the tangential component becomes

$$\mathbb{T} \cdot \mathscr{E}^{(0)} = \mathbb{T} \cdot \mathscr{E}^{(d)}.$$

These matrices prove very useful in solving the matrix problems involving **D** and $\mathscr{E}$ at a boundary. One important property of $\mathbb{T}$ and $\mathbb{N}$ is that their sum equals the unit matrix $\mathbb{I}$,

$$\mathbb{T} + \mathbb{N} = \mathbb{I}.$$

To obtain the conditions of refraction at the surface of the anisotropic dielectric, the equations connecting $\mathscr{E}$ and **D** for a given medium are

$$\mathbf{D}^{(m)} = \epsilon_0 \boldsymbol{\epsilon}^{(m)} \cdot \mathscr{E}^{(m)}$$

where $\boldsymbol{\epsilon}^{(0)} = \mathbb{I}$ for this case in which there is a vacuum in the first region. Substituting into the normal component equation,

$$\mathbb{N} \cdot (\epsilon_0 \boldsymbol{\epsilon}^{(0)} \cdot \mathscr{E}^{(0)}) = \mathbb{N} \cdot (\epsilon_0 \boldsymbol{\epsilon}^{(d)} \cdot \mathscr{E}^{(d)}),$$

or

$$\mathbb{N} \cdot \mathscr{E}^{(0)} = \mathbb{N} \cdot \boldsymbol{\epsilon}^{(d)} \cdot \mathscr{E}^{(d)};$$

also,

$$\mathbb{T} \cdot \mathscr{E}^{(0)} = \mathbb{T} \cdot \mathscr{E}^{(d)}.$$

Adding these last two equations we get

$$(\mathbb{N} + \mathbb{T}) \cdot \mathscr{E}^{(0)} = \mathscr{E}^{(0)} = (\mathbb{T} + \mathbb{N} \cdot \boldsymbol{\epsilon}^{(d)}) \cdot \mathscr{E}^{(d)}.$$

The matrix $\boldsymbol{\Gamma} = (\mathbb{T} + \mathbb{N} \cdot \boldsymbol{\epsilon}^{(d)})$ has a particularly simple form,

$$\boldsymbol{\Gamma} = \mathbb{T} + \mathbb{N} \cdot \boldsymbol{\epsilon}^{(d)} = \begin{pmatrix} 1 & 0 & 0 \\ 0 & 1 & 0 \\ \epsilon_{31} & \epsilon_{32} & \epsilon_{33} \end{pmatrix}.$$

Finally our problem is reduced to finding the inverse of Γ. Since

$$\mathscr{E}^{(0)} = \Gamma \cdot \mathscr{E}^{(d)},$$

it follows that

$$\mathscr{E}^{(d)} = \Gamma^{-1} \cdot \mathscr{E}^{(0)}.$$

Using the methods outlined in Appendix B the reader should demonstrate that

$$\Gamma^{-1} = \frac{1}{\epsilon_{33}} \begin{pmatrix} \epsilon_{33} & 0 & 0 \\ 0 & \epsilon_{33} & 0 \\ -\epsilon_{31} & -\epsilon_{32} & 1 \end{pmatrix}$$

and

$$\mathscr{E}^{(d)} = \begin{bmatrix} \mathscr{E}_1^{(0)} & 0 & 0 \\ \mathscr{E}_2^{(0)} & 0 & 0 \\ -\frac{\epsilon_{31}}{\epsilon_{33}} \mathscr{E}_1^{(0)} & -\frac{\epsilon_{32}}{\epsilon_{33}} \mathscr{E}_2^{(0)} & +\frac{1}{\epsilon_{33}} \mathscr{E}_3^{(0)} \end{bmatrix}.$$

Thus the equation for refraction is

$$\frac{\tan \alpha^{(0)}}{\tan \alpha^{(d)}} = \frac{\mathscr{E}_3^{(d)}}{\mathscr{E}_3^{(0)}} = -\frac{\epsilon_{31}\mathscr{E}_1^{(0)}}{\epsilon_{33}\mathscr{E}_3^{(0)}} - \frac{\epsilon_{32}\mathscr{E}_2^{(0)}}{\epsilon_{33}\mathscr{E}_3^{(0)}} + \frac{1}{\epsilon_{33}}$$

If

$$\mathscr{E}_1^{(0)} = \mathscr{E}_2^{(0)} = 0,$$

then

$$\alpha^{(d)} = \alpha^{(0)} = 0.$$

If $\mathscr{E}_3^{(0)} = 0$, there is no problem.

A special case occurs when

$$\epsilon_{31} \frac{\mathscr{E}_1^{(0)}}{\mathscr{E}_3^{(0)}} + \epsilon_{32} \frac{\mathscr{E}_2^{(0)}}{\mathscr{E}_3^{(0)}} = 1;$$

for such a condition there is a critical reflection ($\alpha^{(d)} \to \pi/2$). Suppose the vector $\mathscr{E}^{(0)}$ lies in the (1,3) plane (i.e., $\mathscr{E}_2^{(0)} = 0$); then critical reflection occurs when

$$\tan \alpha^{(0)} = \frac{1}{\epsilon_{31}}.$$

If the first medium is also an anisotropic dielectric, then

$$(\mathrm{T} + \mathrm{N} \cdot \boldsymbol{\epsilon}^{(1)}) \cdot \mathscr{E}^{(1)} = (\mathrm{T} + \mathrm{N} \cdot \boldsymbol{\epsilon}^{(2)}) \cdot \mathscr{E}^{(2)}.$$

The components of one vector can be obtained in terms of the second by taking an inverse (defining $\Gamma^{\mathrm{m}} = \mathrm{T} + \mathrm{N} \cdot \boldsymbol{\epsilon}^{\mathrm{m}}$):

$$\mathscr{E}^{(1)} = \Gamma^{(1)-1} \cdot \Gamma^{(2)} \cdot \mathscr{E}^{(2)},$$

or

$$\mathscr{E}^{(1)} = \frac{1}{\epsilon_{33}^{(1)}} \begin{pmatrix} \epsilon_{33}^{(1)} & 0 & 0 \\ 0 & \epsilon_{33}^{(1)} & 0 \\ -\epsilon_{31}^{(1)} & -\epsilon_{32}^{(1)} & 1 \end{pmatrix} \begin{pmatrix} 1 & 0 & 0 \\ 0 & 1 & 0 \\ \epsilon_{31}^{(2)} & \epsilon_{32}^{(2)} & \epsilon_{33}^{(2)} \end{pmatrix} \begin{bmatrix} \mathscr{E}_1^{(2)} \\ \mathscr{E}_2^{(2)} \\ \mathscr{E}_3^{(2)} \end{bmatrix};$$

$$\mathscr{E}^{(1)} = \frac{1}{\epsilon_{33}^{(1)}} \begin{pmatrix} \epsilon_{33}^{(1)} & 0 & 0 \\ 0 & \epsilon_{33}^{(1)} & 0 \\ (-\epsilon_{31}^{(1)} + \epsilon_{31}^{(2)}) & (-\epsilon_{32}^{(1)} + \epsilon_{32}^{(2)}) & \epsilon_{33}^{(2)} \end{pmatrix} \begin{bmatrix} \mathscr{E}_1^{(2)} \\ \mathscr{E}_2^{(2)} \\ \mathscr{E}_3^{(2)} \end{bmatrix};$$

giving*

$$\begin{bmatrix} \mathscr{E}_1^{(1)} \\ \mathscr{E}_2^{(1)} \\ \mathscr{E}_3^{(1)} \end{bmatrix} = \frac{1}{\epsilon_{33}^{(1)}} \begin{bmatrix} \epsilon_{33}^{(1)} \mathscr{E}_1^{(2)} \\ \epsilon_{33}^{(1)} \mathscr{E}_2^{(2)} \\ (-\epsilon_{31}^{(1)} + \epsilon_{31}^{(2)})\mathscr{E}_1^{(2)} + (-\epsilon_{32}^{(1)} + \epsilon_{32}^{(2)})\mathscr{E}_2^{(2)} + \epsilon_{33}^{(2)} \mathscr{E}_3^{(2)} \end{bmatrix}.$$

Then

$$\frac{\tan \alpha_1}{\tan \alpha_2} = \frac{\epsilon_{33}^{(1)} \mathscr{E}_3^{(2)}}{[(-\epsilon_{31}^{(1)} + \epsilon_{31}^{(2)})\mathscr{E}_1^{(2)} + (-\epsilon_{32}^{(1)} + \epsilon_{32}^{(2)})\mathscr{E}_2^{(2)} + \epsilon_{33}^{(2)} \mathscr{E}_3^{(2)}]}.$$

Let us now consider the *anisotropic wedge* as a linear transformation.

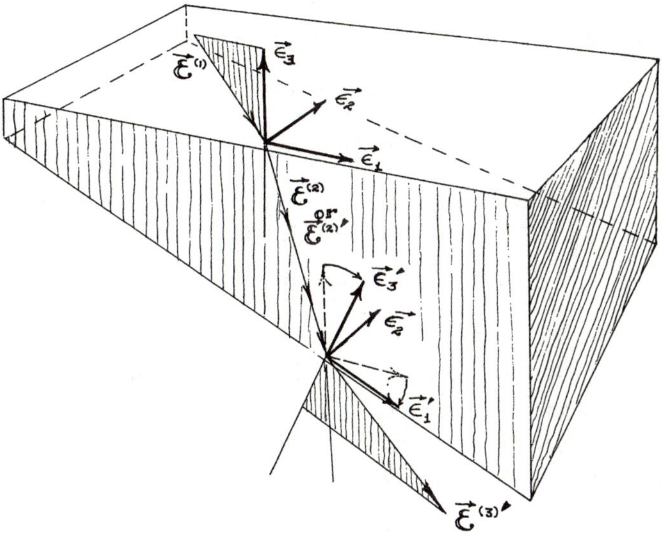

* The reader should keep in mind the symbols differentiating a column vector from a square matrix. The column vector and row vector are enclosed in *brackets*, while the square matrix is enclosed in *parentheses*.

Assume that the problem consists of an external field $\mathscr{E}^{(1)}$ in a vacuum incident upon a dielectric wedge for which the dielectric tensor is $\boldsymbol{\epsilon}$.

Orienting the axes perpendicular and parallel to the first (or incoming) face of the wedge,

$$\mathscr{E}^{(1)} = (\mathbf{T} + \mathbf{N}\cdot\boldsymbol{\epsilon})\cdot\mathscr{E}^{(2)}$$
$$= \boldsymbol{\Gamma}\cdot\mathscr{E}^{(2)},$$

or

$$\mathscr{E}^{(2)} = \boldsymbol{\Gamma}^{-1}\cdot\mathscr{E}^{(1)}.$$

At the emergent surface we rotate coordinates to a new set of axes parallel to and perpendicular to the emergent face. This corresponds to a rotation of the bases to give $\mathscr{E}^{(2)}$ as $\mathscr{E}^{(2)\prime}$. In terms of these bases

$$\mathscr{E}^{(2)\prime} = \mathbf{S}\cdot\mathscr{E}^{(2)},$$

or

$$\mathscr{E}^{(2)\prime} = \mathbf{S}\cdot\boldsymbol{\Gamma}^{-1}\cdot\mathscr{E}^{(1)}.$$

The emerging field vector $\mathscr{E}^{(3)}$ is given by a rotated tensor $\boldsymbol{\epsilon}' = \mathbf{S}\cdot\boldsymbol{\epsilon}\cdot\mathbf{S}^{-1}$. Then

$$(\mathbf{T} + \mathbf{N}\cdot\boldsymbol{\epsilon}')\cdot\mathscr{E}^{(2)\prime} = \mathscr{E}^{(3)\prime}.$$

Now we rotate back to the original bases,

$$\mathscr{E}^{(3)} = \mathbf{S}^{-1}\cdot\mathscr{E}^{(3)\prime}.$$

Thus

$$\mathscr{E}^{(3)} = \mathbf{S}^{-1}\cdot(\mathbf{T} + \mathbf{N}\cdot\boldsymbol{\epsilon}')\cdot\mathbf{S}\cdot\boldsymbol{\Gamma}^{-1}\cdot\mathscr{E}^{(1)}.$$

The entire transformation from region (1) to region (3) can therefore be written in terms of a single transformation $\mathbf{M}$,

$$\mathscr{E}^{(3)} = \mathbf{M}\cdot\mathscr{E}^{(1)},$$

where

$$\mathbf{M} = \mathbf{S}^{-1}\cdot(\mathbf{T} + \mathbf{N}\cdot\boldsymbol{\epsilon}')\cdot\mathbf{S}\cdot\boldsymbol{\Gamma}^{-1}.$$

To illustrate the use of the formalism we shall compute the potential drop across a rectangular slab of anisotropic dielectric. In this example the dielectric tensor in the base system parallel to the sides of the slab will be taken as

$$\boldsymbol{\epsilon} = \begin{pmatrix} 6 & 0 & 0 \\ 0 & \frac{62}{5} & \frac{24}{5} \\ 0 & \frac{24}{5} & \frac{48}{5} \end{pmatrix}.$$

The electric susceptibility tensor $\mathbb{X}$ is $\boldsymbol{\epsilon} - \mathbb{I}$;

$$\mathbb{X} = \begin{pmatrix} 5 & 0 & 0 \\ 0 & \frac{57}{5} & \frac{24}{5} \\ 0 & \frac{24}{5} & \frac{43}{5} \end{pmatrix}.$$

The eigenvalues of the susceptibility tensor are obtained from

$$|\mathbb{X} - \chi\mathbb{I}| = \begin{vmatrix} (5-\chi) & 0 & 0 \\ 0 & (\frac{57}{5}-\chi) & \frac{24}{5} \\ 0 & \frac{24}{5} & (\frac{43}{5}-\chi) \end{vmatrix} = 0,$$

or $$(5-\chi)(\tfrac{57}{5}-\chi)(\tfrac{43}{5}-\chi) - \frac{(24)^2}{25}(5-\chi) = 0.$$

One eigenvalue is $$\chi_{11} = 5.$$

Then $$(\tfrac{57}{5}-\chi)(\tfrac{43}{5}-\chi) - \frac{(24)^2}{25} = 0;$$

giving $$\chi_{22} = 5,$$

and $$\chi_{33} = 15.$$

The first two eigenvalues are degenerate, indicating that the dielectric ellipsoid is an elliposid of revolution, the symmetry axis being that axis corresponding to the eigenvalue $\chi_{33} = 15$. To obtain this axis we solve for the corresponding eigenvector.

$$\{\mathbb{X} - \chi_{33}\mathbb{I}\} \cdot \mathbf{R}_3 = 0,$$

or $$\sum_{n=1}^{3} \{X_{mn} - \delta_{mn}\chi_{33}\} R_{n3} = 0.$$

We find that

$$\frac{1}{5}\begin{pmatrix} -50 & 0 & 0 \\ 0 & -18 & 24 \\ 0 & 24 & -32 \end{pmatrix} \begin{bmatrix} R_{13} \\ R_{23} \\ R_{33} \end{bmatrix} = 0.$$

Solving for the R_{m3} and normalizing,

$$R_{13} = 0$$
$$R_{23} = \tfrac{4}{5}$$
$$R_{33} = \tfrac{3}{5}.$$

Thus the symmetry axis of the ellipsoid lies in the x_2,x_3 plane and makes an angle of 53° relative to the x_3 axis.

These preliminaries are not essential to our problem. However, they do give some insight into the elementary properties of the dielectric.

According to our formalism, if the external field is incident perpendicular to the x_1,x_2 face, i.e.,

$$\mathscr{E}^{(0)} = \begin{bmatrix} 0 \\ 0 \\ -\mathscr{E}_0 \end{bmatrix},$$

then in the dielectric

$$\mathscr{E}^{(d)} = (\mathbf{T} + \mathbf{N} \cdot \boldsymbol{\epsilon})^{-1} \cdot \mathscr{E}^{(0)}.$$

Here $$(\mathbf{T} + \mathbf{N} \cdot \boldsymbol{\epsilon})^{-1} = \Gamma^{-1} = \tfrac{5}{45} \begin{pmatrix} \frac{48}{5} & 0 & 0 \\ 0 & \frac{48}{5} & 0 \\ 0 & -\frac{24}{5} & 1 \end{pmatrix};$$

therefore $$\mathscr{E}^{(d)} = \begin{bmatrix} 0 \\ 0 \\ -\frac{1}{9}\mathscr{E}_0 \end{bmatrix}$$

If the thickness of the slab is t, the potential drop across the slab is

$$V_{\text{drop}} = \tfrac{1}{9}\mathscr{E}_0 t.$$

It is interesting in this example to observe that $\mathbf{p}_v$ and $\mathbf{D}$ are indeed not aligned with $\mathscr{E}^{(d)}$:

$$\mathbf{p}_v = \epsilon_0 \mathbf{X} \cdot \mathscr{E}^{(d)} = \frac{\epsilon_0}{5} \begin{pmatrix} 25 & 0 & 0 \\ 0 & 57 & 24 \\ 0 & 24 & 43 \end{pmatrix} \begin{bmatrix} 0 \\ 0 \\ -\frac{1}{9}\mathscr{E}_0 \end{bmatrix}$$

or $$\mathbf{p}_v = \frac{\epsilon_0}{5} \begin{bmatrix} 0 \\ -\frac{8}{3}\mathscr{E}_0 \\ -\frac{43}{9}\mathscr{E}_0 \end{bmatrix},$$

and $$\mathbf{D} = \epsilon_0 \begin{bmatrix} 0 \\ -\frac{8}{15}\mathscr{E}_0 \\ -\frac{16}{15}\mathscr{E}_0 \end{bmatrix}.$$

4. THE CLAUSIUS-MOSSOTTI EQUATION

In the preceding section the microscopic behavior of some idealized systems was reviewed. In light of these developments we must be concerned with the effective field at the position of a polarized atom or molecule in a dielectric. This concern is an admission that the system of dipoles is not in fact a distribution of point dipoles but rather a distribution of systems having an extension of the order of the Bohr radius or greater.

For this reason the effective field at the position of a polarized system is a superposition of the ideal internal field of a dielectric $\mathscr{E}^{(d)}$ plus the field produced by an interaction with the nearest neighbors, $\mathscr{E}^{(n)}$:

$$\mathscr{E}_{\text{eff}} = \mathscr{E}^{(d)} + \mathscr{E}^{(n)}.$$

The standard method for computing $\mathscr{E}^{(n)}$, the nearest neighbor contribution, is to cut a spherical cavity about the dipole in question

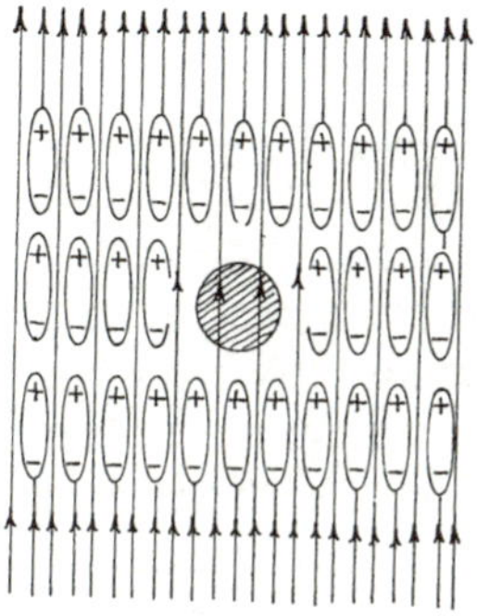

with the center of the cavity at the position of the dipole. As a first-order approximation one can then assume that the surrounding medium is made up of a homogeneous distribution of point dipoles. The contribution $\mathscr{E}^{(n)}$ is then computed from the surface charge induced upon the interior surface of the cavity. This computation is similar to that for a homogeneously polarized dielectric sphere except that the field at the center now has the opposite sign. Referring to Section III-B-4 where this computation was made, the field at the center of the cavity should be

$$\mathscr{E}^{(n)} = +\frac{\mathbf{p}_V}{3\epsilon_0},$$

and
$$\mathscr{E}_{\text{eff}} = \mathscr{E}^{(d)} + \frac{\mathbf{p}_V}{3\epsilon_0}.$$

The net polarization at the center of the cavity is assumed to be induced by the effective field:

$$\mathbf{p}_V = \alpha\epsilon_0\mathscr{E}_{\text{eff}},$$

where α is called the *polarizability* of the medium. Then

$$\mathbf{p}_V = \alpha\epsilon_0\left\{\mathscr{E}^{(d)} + \frac{\mathbf{p}_V}{3\epsilon_0}\right\}.$$

Collecting like terms,

$$\mathbf{p}_V = \frac{3\epsilon_0\alpha}{(3-\alpha)}\mathscr{E}^{(d)}.$$

Because the electric susceptibility χ is defined for an isotropic dielectric as

$$\mathbf{p}_V = \epsilon_0\chi\mathscr{E}^{(d)},$$

the polarizability α in terms of χ is

$$\alpha = \frac{3\chi}{3+\chi}.$$

This is the Clausius-Mossotti equation.

Our α is dimensionless; the conventional definition of polarizability, α', is α divided by the number of particles per unit volume, n_V, or

$$\alpha' = \alpha/n_V.$$

As stated, this development is restricted to an isotropic dielectric.

In the case of an anisotropic medium the nearest neighbor contribution will be a linear transformation of $\mathbf{p}_V$; however, it will not necessarily be directly proportional to $\mathbf{p}_V$. The $\mathscr{E}^{(n)}$ for each anisotropy is a special case, and in general

$$\mathscr{E}^{(n)} = \mathbf{G}\cdot\frac{\mathbf{p}_V}{\epsilon_0},$$

where $\mathbf{G}$ is a constant tensor transformation. (In the isotropic dielectric $\mathbf{G} = \frac{1}{3}\mathbf{I}$.)

Using this more general formalism,

$$\mathbf{p}_V = \epsilon_0 \mathbf{A} \cdot \mathscr{E}_{\text{eff}} = \epsilon_0 \mathbf{A} \cdot \left\{ \mathscr{E}^{(d)} + \mathbf{G} \cdot \frac{\mathbf{p}_V}{\epsilon_0} \right\}.$$

The polarizability $\mathbf{A}$ is also a 3×3 matrix. Collecting terms,

$$\mathbf{p}_V = \epsilon_0 (\mathbf{I} - \mathbf{A} \cdot \mathbf{G})^{-1} \cdot \mathbf{A} \cdot \mathscr{E}^{(d)} = \epsilon_0 \mathbf{X} \cdot \mathscr{E}^{(d)}.$$

The polarizability matrix is then given in terms of the electric susceptibility matrix by

$$\mathbf{A} = (\mathbf{I} + \mathbf{G} \cdot \mathbf{X})^{-1} \cdot \mathbf{X}.$$

Ferroelectric effects or permanent polarization effects set in when the effective field at the molecule is composed of the nearest neighbor field only. Then for an isotropic medium

$$\mathscr{E}_{\text{eff}} = \mathbf{p}_V / 3\epsilon_0.$$

Since

$$\mathbf{p}_V = \alpha \epsilon_0 \mathscr{E}_{\text{eff}},$$

the condition which must be met is that

$$\alpha = 3.$$

Nonferroelectric dielectrics have $\alpha < 3$. The field inside of an isolated ferroelectric material is given by the equation for $\mathbf{D}$ in the special case when $\mathbf{D} = 0$,

$$\mathscr{E}_{\text{ferro}} = -\mathbf{p}_V / \epsilon_0.$$

In terms of the orienting field of nearest neighbors,

$$\mathscr{E}_{\text{ferro}} = -3\mathscr{E}^{(n)}.$$

The anisotropic dielectric, if ferroelectric, has an internal field of

$$\mathscr{E}_{\text{ferro}} = -\mathbf{G}^{-1} \cdot \mathscr{E}^{(n)}.$$

Both $\mathbf{G}^{-1}$ and $\mathscr{E}^{(n)}$ can be computed from the specified properties of the dielectric, i.e., $\boldsymbol{\epsilon}$. The condition for ferroelectricity in an anisotropic medium is that $\mathbf{A} = \mathbf{G}^{-1}$.

5. SPECIAL PROBLEMS

a. THE FIELD OF A HOMOGENEOUSLY POLARIZED DIELECTRIC SPHERE. The axial field was computed in Section III-B-4 . In this problem there are no free charges, therefore

$$\operatorname{div} \mathbf{D} = 0$$

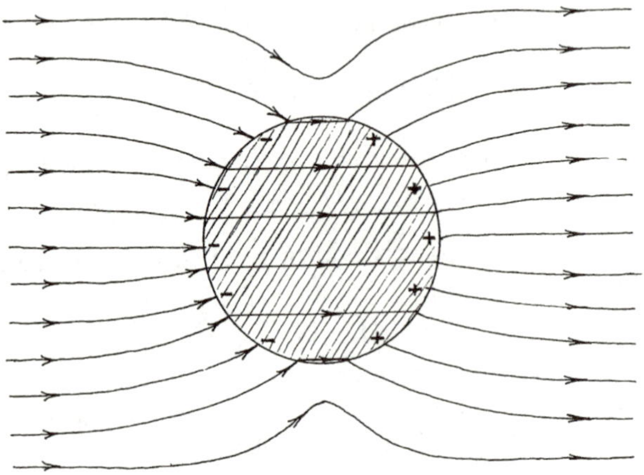

DIELECTRIC SPHERE IN A
CONSTANT EXTERNAL FIELD

and $$\operatorname{div}\left(\mathscr{E} + \frac{\mathbf{p}_V}{\epsilon_0}\right) = 0$$

In the body of the sphere $\mathbf{p}_V$ is a constant. As a result

$$\operatorname{div} \mathscr{E} = 0,$$

and $$\nabla^2 V = 0.$$

The appropriate solution is $\mathscr{E} = -\mathbf{p}_V/\epsilon_0$ = constant for $r \leqslant R$. Because $\mathbf{p}_V = 0$ for $R < r$ (where R is the radius of the sphere) the exterior potential also obeys Laplace's equation. The interior solution for V (call it V_1) is

$$V_1 = \sum_{L=0}^{\infty} a_L r^L P_L(\theta).$$

Because of the constraint $\mathbf{p}_V$ = const. and $\mathscr{E} = -\mathbf{p}_V/\epsilon_0$, if $\mathbf{p}_V$ is in the z direction then

$$V_1 = V_0 + \frac{p_V}{\epsilon_0} r \cos \theta.$$

Thus $a_0 = V_0$, $a_1 = p_V/\epsilon_0$, and all of the remaining coefficients are zero.

The potential exterior to the sphere must go as a point dipole in the far field. The general* exterior solution is V_2,

$$V_2 = \sum_{L=0}^{\infty} b_L r^{-(L+1)} P_L(\theta).$$

If

$$V_2 \xrightarrow[r\ \text{large}]{} \frac{b_1 \cos\theta}{r^2},$$

then b_0 must be zero, for otherwise the r^{-1} term would control the far field.

The final boundary condition is that V_1 must match V_2 at every point on the surface of the sphere:

$$V_1(r = R) = V_0 + \frac{p_V}{\epsilon_0} R \cos\theta$$

$$V_2(r = R) = b_1 \frac{\cos\theta}{R^2}.$$

Equating coefficients of like powers of $\cos\theta$ we find

$$V_0 = 0$$

$$b_1 = \frac{p_V R^3}{\epsilon_0}.$$

In conclusion

$$V_1 = \frac{p_V}{\epsilon_0} r \cos\theta, \qquad r \leqslant R,$$

and

$$V_2 = \frac{p_V R^3}{\epsilon_0} \frac{\cos\theta}{r^2}, \quad R \leqslant r.$$

b. A POINT CHARGE OUTSIDE OF A SEMI-INFINITE DIELECTRIC MEDIUM. Assume that the volume below the x_1,x_2 plane is filled with a homogeneous isotropic dielectric medium having a susceptibility χ. A point charge q is placed a distance d above the x_1,x_2 plane on the x_3 axis. The field lines are refracted at the surface of the dielectric and a nonuniform surface charge density is induced. This induced surface charge density has circular symmetry about the x_3 axis.

The fields in regions (1) and (2) will certainly not appear as those of a single point charge q. This problem, because it deals with a point charge, can be solved by a method of images.

Consider first region (1). The induced surface charge is negative

* The first two terms of V_1 can be included here because they produce zero and constant external fields.

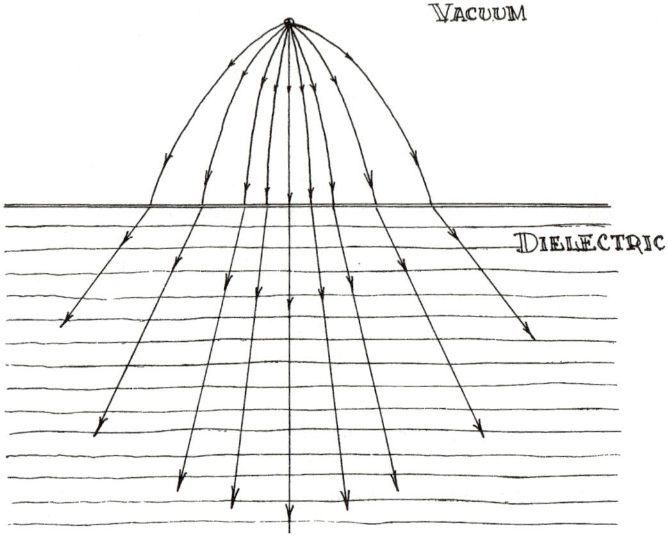

and therefore a first attempt at a solution can be made by assuming a negative image $-q'$ a distance d' below the surface. Then

$$V_1 = \frac{q}{4\pi\epsilon_0}\left\{\frac{1}{[a^2 + (x_3 - d)^2]^{1/2}} - \frac{q'}{4\pi\epsilon_0}\frac{1}{[a^2 + (x_3 + d')^2]^{1/2}}\right\}$$

where $$a^2 = x_1^2 + x_2^2.$$

In region (2) the field should be radial with an image q'' at $x_3 = d''$ as a source. However, the field intensity is necessarily reduced by the induced surface charge.

$$V_2 = \frac{q''}{4\pi\epsilon_0\epsilon_2}\frac{1}{[a^2 + (x_3 - d'')^2]^{1/2}}.$$

On the surface ($x_3 = 0$), V_1 must equal V_2 at every point. The reader should demonstrate that consistent solutions exist only for $d = d' = d''$. Then at $x_3 = 0$, $V_1(x_3 = 0) = V_2(x_3 = 0)$, giving

$$q - q' = q''/\epsilon_2.$$

We have two other boundary conditions to meet, one of which is that $\mathbf{D}^{(1)} \cdot \mathbf{n} = \mathbf{D}^{(2)} \cdot \mathbf{n}$ at the surface. Thus

$$-\epsilon_0\left[\frac{\partial V_1}{\partial x_3}\right]_{x_3=0} = -\epsilon_0\epsilon_2\left[\frac{\partial V_2}{\partial x_3}\right]_{x_3=0},$$

giving $$-q - q' = -q''.$$

The two equations for q′ and q″ yield definitions in terms of q:

$$q' = \left\{\frac{\chi_2}{1+\epsilon_2}\right\}q$$

and

$$q'' = \left\{\frac{2\epsilon_2}{1+\epsilon_2}\right\}q.$$

Physically the net effect of the dielectric is to cause the lines of $\mathscr{E}$ to assume a higher concentration about the x_3 axis. Then the lines tend to an angle of incidence which is less than that for a single charge. The refraction causes the lines in the dielectric to appear again as if they diverge radially from a point at $x_3 = d$.

D. ELECTROSTATIC CAPACITY AND CONDENSERS

1. GREEN'S RECIPROCITY THEOREM

In the case of N point charges, we demonstrated in Section II-E that the potential at the position of the jth charge was

$$V_j = \sum_{m=1}^{N} P_{jm}q_m$$

with $P_{mm} = 0$. This was written in a more compact form as

$$\mathbf{V} = \mathbf{P}\cdot\mathbf{q}.$$

The reciprocity theorem is concerned with variations of the charges and potentials V_j and q_m in a fixed geometry. If all charge magnitudes q_j are changed to q_j', the potentials change to V_m' according to

$$\mathbf{V}' = \mathbf{P}\cdot\mathbf{q}'.$$

Because the geometry is fixed, **P** remains the same. If we multiply **V** from the left by $\tilde{\mathbf{q}}'$ and $\mathbf{V}'$ from the left by $\tilde{\mathbf{q}}$ we find that

$$\tilde{\mathbf{q}}'\cdot\mathbf{V} = \tilde{\mathbf{q}}'\cdot\mathbf{P}\cdot\mathbf{q} = \tilde{\mathbf{q}}\cdot\mathbf{P}\cdot\mathbf{q}' = \tilde{\mathbf{q}}\cdot\mathbf{V}'.$$

Writing this in terms of the appropriate sums,

$$\tilde{\mathbf{q}}'\cdot\mathbf{V} = \tilde{\mathbf{q}}\cdot\mathbf{V}'$$

implies that

$$\sum_{m=1}^{N}\sum_{n=1}^{N} q_m'P_{mn}q_n = \sum_{n=1}^{N}\sum_{m=1}^{N} q_mP_{mn}q_n'.$$

Previously we demonstrated that P is symmetric; therefore, because the q's are real the equality has been proven:

$$\sum_{j=1}^{N} q_j' V_j = \sum_{j=1}^{N} q_j V_j'.$$

This relation can be extended to a series of conductors if we combine all points of equal V_j into a single term. Then our relation applies to a system of conductors.

Thus if q_m and q_m' represent the total charge on the mth conductor while V_m and V_m' are the respective potentials, the relation still holds.

2. COEFFICIENTS OF POTENTIAL AND CAPACITY

Consider M conductors each carrying a total charge Q_m at a potential V_m. The potential on the mth conducting surface is given as a linear combination of the total charges Q_j on each conductor. This theorem can be proved using the linear relationship developed for N point charges q_j. It has been shown that

$$V_j = \sum_{n=1}^{N} P_{jn} q_n,$$

where V_j is the potential at the position of the jth charge,

$$\mathbf{V} = \mathbf{P} \cdot \mathbf{q}.$$

To prove the theorem for a system of M conductors we utilize the inverse relation

$$\mathbf{q} = \mathbf{P}^{-1} \cdot \mathbf{V},$$

or

$$\begin{bmatrix} q_1 \\ q_2 \\ \cdot \\ \cdot \\ \cdot \\ q_N \end{bmatrix} = \begin{pmatrix} P'_{11} & P'_{12} & \cdots & P'_{1N} \\ P'_{21} & & & \cdot \\ \cdot & & & \cdot \\ \cdot & & & \cdot \\ \cdot & & & \cdot \\ P'_{N1} & \cdots & \cdots & P'_{NN} \end{pmatrix} \begin{bmatrix} V_1 \\ V_2 \\ \cdot \\ \cdot \\ \cdot \\ V_N \end{bmatrix}.$$

Assume now that we group the charges corresponding to a single conductor. If charges q_r to q_{s+r} lie upon the rth conductor we

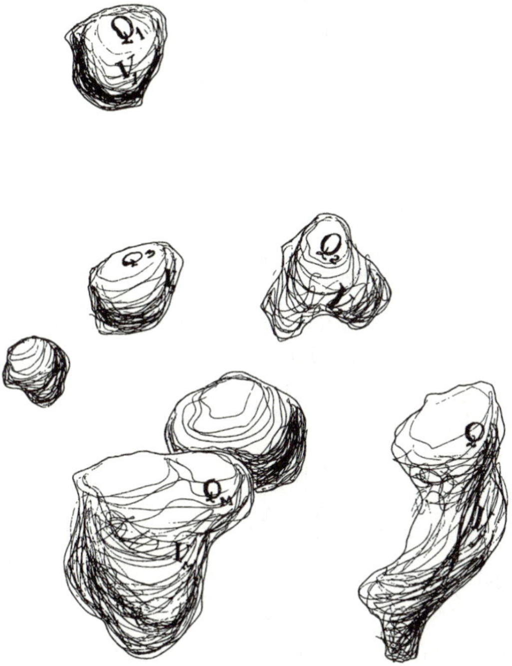

can designate the total charge on conductor r as Q_r:

$$Q_r = q_r + q_{r+1} + \dots + q_{r+s} = \sum_{n=0}^{s} q_{r+n}.$$

Each q_{r+n} is given as

$$q_{r+n} = \sum_{j=1}^{N} P'_{r+n,\,j} V_j, \quad n \leqslant s.$$

Therefore

$$Q_r = \sum_{n=0}^{s} \sum_{j=1}^{N} P'_{r+n,\,j} V_j.$$

The form of this equation is clear when a few terms are written out:

$$q_r = P'_{r1}V_1 + P'_{r2}V_2 + \dots + (P'_{rr} + P'_{r,\,r+1} + \dots + P'_{r.\,r+s})V_r + \dots$$
$$q_{r+1} = P'_{r+1,\,1}V_1 + \dots + (P'_{r+1,\,r} + \dots + P'_{r+1,\,r+s})V_r + \dots$$
$$q_{r+s} = P'_{r+s,\,1}V_1 + \dots + (P'_{r+s,\,r} + \dots + P'_{r+s,\,r+s})V_r + \dots.$$

When the charges are added to form Q_m and all on a given conductor are grouped as shown above, then

$$Q_r = \sum_{n=0}^{s} q_{r+n} = \sum_{k=1}^{M} p'_{rk} V_k.$$

The grouping of the V_j's must give M terms (i.e., the number of conductors).

It is clear that the elements p'_{mk} are made up of the grouping of the $P'_{r+n,r+k}$:

$$p'_{rr} = \sum_{j=0}^{s} \sum_{k=0}^{s} P'_{r+j,\, r+k}.$$

Now consider the kth conductor. Assume that the sequence q_k to q_{k+i} lie upon the kth conductor. Then

$$p'_{rk} = \sum_{j=0}^{s} \sum_{m=0}^{i} P'_{r+j,\, k+m}.$$

Because $$P'_{mp} = P'_{pm},$$

$$p'_{rk} = \sum_{m=0}^{i} \sum_{j=0}^{s} P'_{k+m,\, r+j} = p'_{kr}.$$

This demonstrates that

$$\mathbf{Q} = \mathbf{p}^{-1} \cdot \mathbf{V}$$

where the components of **Q** are the total charges on each conductor, while the jth component of **V** corresponds to the potential of the jth conductor.

$\mathbf{p}^{-1}$ is an M × M symmetric matrix.

The reader can demonstrate that $\mathbf{p}^{-1}$ is a nonzero matrix.

Multiplying the equation for **Q** by $\mathbf{p}$, then

$$\mathbf{V} = \mathbf{p} \cdot \mathbf{Q},$$

or $$V_j = \sum_{n=1}^{M} p_{jn} Q_n.$$

The components p_{jn} are known as the coefficients of potential. The inverse matrix $\mathbf{p}^{-1}$ is ordinarily designated as $\mathbf{C}$ where

$$p'_{mn} = C_{mn} = \frac{(-1)^{m+n} \text{ minor } p_{nm}}{|\mathbf{p}|}.$$

When reviewing this development one should observe that the grouping of the q_m's and the V_j's is done symmetrically and represents a contraction from an N-dimensional space to an M-dimensional space.

As shown, $\mathbb{C}$ and $\mathbb{p}$ are symmetric matrices with $\mathbb{C} = \mathbb{p}^{-1}$. The diagonal elements of $\mathbb{C}$, C_{mm}, are known as the *coefficients of capacity* and are positive. The off-diagonal elements C_{jk} $(j \neq k)$ are called the *coefficients of induction.* The latter coefficients are negative numbers.

$$\mathbf{Q} = \mathbb{C} \cdot \mathbf{V}.$$

In the same manner one can demonstrate that the energy stored in a system of M charged conductors is

$$U = \tfrac{1}{2}\tilde{\mathbf{Q}} \cdot \mathbf{V} = \tfrac{1}{2}\tilde{\mathbf{Q}} \cdot \mathbb{p} \cdot \mathbf{Q},$$

or

$$U = \tfrac{1}{2}\sum_{i=1}^{M} Q_i V_i = \tfrac{1}{2}\sum_{j=1}^{M}\sum_{i=1}^{M} Q_j p_{ji} Q_i.$$

Also

$$U = \tfrac{1}{2}\tilde{\mathbf{V}} \cdot \mathbf{Q} = \tfrac{1}{2}\tilde{\mathbf{V}} \cdot \mathbb{C} \cdot \mathbf{V},$$

or

$$U = \tfrac{1}{2}\sum_{r=1}^{M}\sum_{s=1}^{M} V_r C_{rs} V_s.$$

3. THE CONDENSER

a. DEFINITION. A condenser (or capacitor) is an isolated system of two and only two charged conductors. The total charge carried by each conductor must be equal in magnitude, and the two charges must have opposite signs. The condition of isolation implies that the field between the two conductors is shielded from any other conducting bodies.

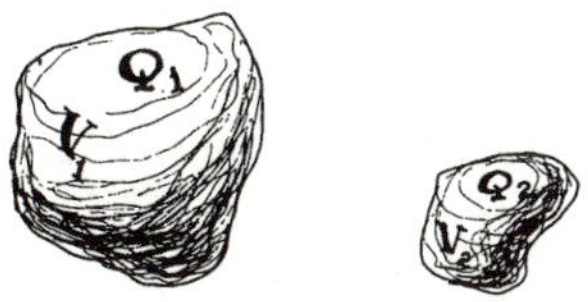

We shall derive relatively simple rules for the combination of condensers. One must be careful, for many seemingly simple condenser problems may not be subject to analysis as a single condenser because of the presence of a third body. A few examples will be shown later in which grounds at infinity make up the third body.

b. CHARGE AND POTENTIAL DIFFERENCE IN A CONDENSER SYSTEM. As stated, this is a system of two conducting bodies carrying charges $Q_1 = Q$ and $Q_2 = -Q$.

The *capacitance* C of a condenser is a scalar quantity relating the potential difference and the *positive* charge $+Q$.

From the preceding section, the potential distribution in a two-body system is

$$\mathbf{V} = \mathbf{p} \cdot \mathbf{Q},$$

or

$$V_1 = p_{11}Q_1 + p_{12}Q_2 = p_{11}Q - p_{12}Q$$
$$V_2 = p_{21}Q_1 + p_{22}Q_2 = p_{21}Q - p_{22}Q.$$

The potential difference is $\Delta V = V_1 - V_2$.

$$V_1 - V_2 = \Delta V = \{p_{11} - p_{12} - p_{21} + p_{22}\}Q$$
$$= \{p_{11} + p_{22} - 2p_{12}\}Q.$$

Then Q, the magnitude of the charge, is directly proportional to ΔV, the potential difference between the two bodies:

$$Q = C\,\Delta V$$

where

$$C = \left(\frac{1}{p_{11} + p_{22} - 2p_{12}}\right).$$

It is possible to relate the capacitance C to the elements of $\mathbf{C}$ by the inverse relation $\mathbf{p} = \mathbf{C}^{-1}$.

The stored energy can be computed as before,

$$U = \tfrac{1}{2}\widetilde{\mathbf{Q}} \cdot \mathbf{p} \cdot \mathbf{Q}$$
$$= \tfrac{1}{2}\{Q_1p_{11}Q_1 + Q_1p_{12}Q_2 + Q_2p_{21}Q_1 + Q_2p_{22}Q_2\}$$
$$= \tfrac{1}{2}\{p_{11} + p_{22} - 2p_{12}\}Q^2$$

or

$$U = \frac{Q^2}{2C}.$$

Together with this result we can write an equivalent equation,

$$U = \tfrac{1}{2}C(\Delta V)^2 = \tfrac{1}{2}Q\,\Delta V.$$

c. SERIES AND PARALLEL COMBINATIONS. The combinations are developed quite readily by noticing the common electrostatic quantity in each configuration.

(1) Consider N condensers of capacitance C_k in parallel. All C_j's have the same potential difference (let ΔV be written as V):

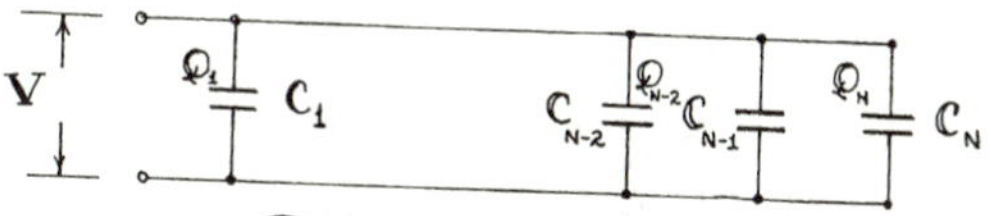

PARALLEL COMBINATION

$$V = \frac{Q_1}{C_1} = \frac{Q_2}{C_2} = \ldots = \frac{Q_N}{C_N}.$$

The total charge on the parallel combination is Q_T,

$$Q_T = Q_1 + Q_2 + \ldots + Q_N = \sum_{i=1}^{N} Q_i = C_T V.$$

Therefore
$$Q_T = \left\{\sum_{i=1}^{N} C_i\right\} V = C_T V,$$

and the total equivalent capacitance C_T of the parallel combination is

$$C_T = \sum_{i=1}^{N} C_i = C_1 + C_2 + \ldots + C_N.$$

(2) Now place the N condensers in series. The common element is the magnitude of the charge. We assume that the condensers

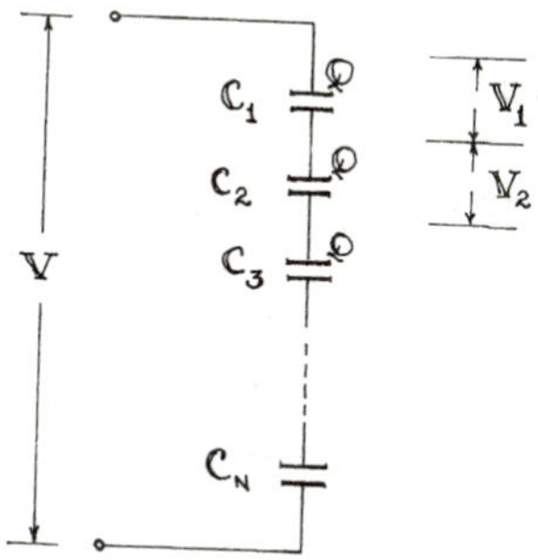

SERIES COMBINATION

are initially uncharged and then connected in series. After connecting, the system is charged. The total charge in a connecting branch between C_k and C_{k+1} is therefore *zero*. After a potential difference is imposed across the system, the *net charge* in an isolated branch remains zero; thus $Q_k = Q_{k+1}$, etc. This is a statement of the conservation of charge. Thus

$$Q_1 = Q_2 = Q_j = Q_N$$

and $$Q = C_1V_1 = C_2V_2 = \ldots = C_iV_i.$$

The total potential drop across the system is V_T and is the sum of the individual drops V_i,

$$V_T = \sum_{i=1}^{N} V_i = \sum_{i=1}^{N} \frac{Q}{C_i} = \left\{\sum_{i=1}^{N} \frac{1}{C_i}\right\} Q.$$

Because $$V_T = Q/C_T,$$

$$\frac{1}{C_T} = \sum_{i=1}^{N} \frac{1}{C_i}.$$

d. CAPACITIES FOR SPECIFIC CONFIGURATIONS.

(1) *The Parallel Plate Condenser.* Consider the field between the plates to be constant, i.e., neglect edge effects. According to the figure,

$$\mathscr{E} = \mathscr{E}_0 \mathbf{k}$$

$$V_z = \mathscr{E}_0 z = V_0 z/d.$$

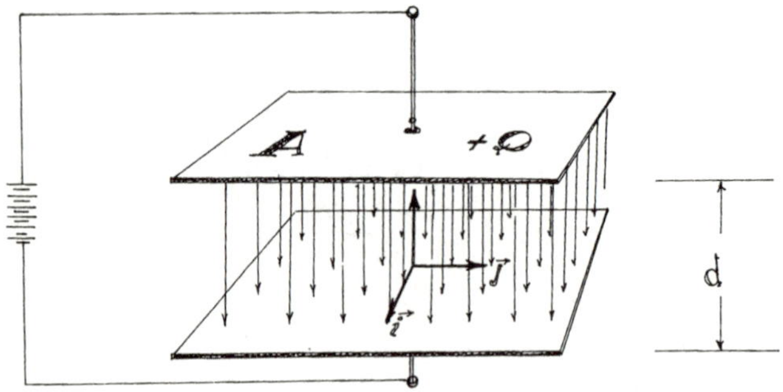

The potential difference between the plates, noting the plate separation as d, is

$$V_0 = -\int_0^d \mathscr{E} \cdot d\mathbf{r} = \int_0^d \mathscr{E}_0 \, dz = \mathscr{E}_0 d,$$

$$\mathscr{E}_0 = V_0/d.$$

The surface charge density is

$$\sigma = |\mathbf{D}| = \epsilon_0 |\mathscr{E}| = \epsilon_0 V_0/d.$$

The total charge per plate is

$$Q = \sigma A = \left(\frac{\epsilon_0 A}{d}\right) V_0 = CV.$$

Thus the capacity is

$$C = \epsilon_0 A/d,$$

where A is the area of either of the identical plates.

(2) *The Capacitance of The Isolated Sphere.* We calculate the capacitance of a sphere of radius R, with respect to ground at infinity. If the sphere carries a charge Q, the field at any distance r_p is

$$\mathscr{E}_p = \frac{Q}{4\pi\epsilon_0} \frac{\mathbf{r}_p}{r_p^3};$$

$$V_p = \frac{Q}{4\pi\epsilon_0} \frac{1}{r_p}.$$

At $r_p = R$

$$V_R = \frac{Q}{4\pi\epsilon_0 R},$$

or

$$C = \frac{Q}{V_R} = 4\pi\epsilon_0 R.$$

(3) *The Concentric Spherical Condenser.* For an inner sphere

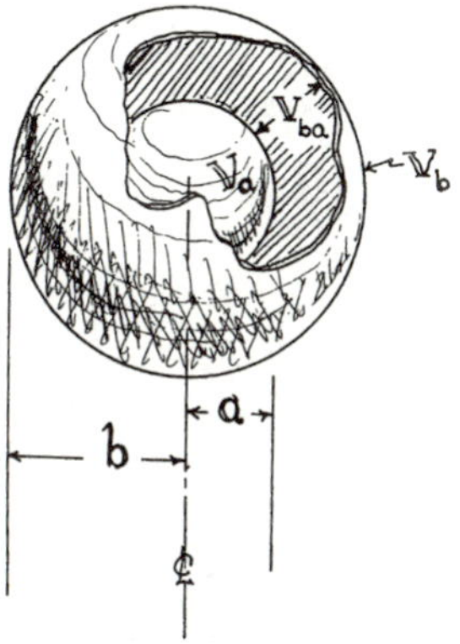

(radius a) carrying a charge of +Q and an outer sphere (of radius b) carrying a charge of −Q, we compute the *potential difference between* the spheres:

$$V_{ba} = -\int_b^a \mathscr{E} \cdot dr = -\int_b^a \frac{Q}{4\pi\epsilon_0}\frac{dr}{r^2} = \frac{Q}{4\pi\epsilon_0}\left\{\frac{1}{a} - \frac{1}{b}\right\}.$$

Then
$$C = \frac{Q}{V_{ba}} = \frac{4\pi\epsilon_0 ab}{(b - a)}.$$

(4) *The Capacity per Unit Length of a Concentric Cylindrical Condenser.* From Gauss's law the field between the cylinders is

$$\mathscr{E} = \frac{\lambda}{2\pi\epsilon_0}\frac{(x_p\mathbf{i} + y_p\mathbf{j})}{(x_p^2 + y_p^2)}$$

where
$$b^2 > a_p^2 = a_p^2 + y_p^2 > a^2.$$

If the cylinders carry $\pm\lambda$ charge per unit length respectively,

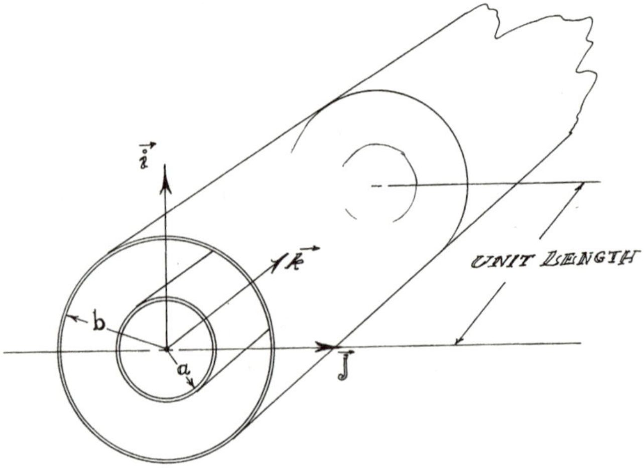

the potential difference between the cylinders is

$$V_{ab} = -\int_a^b \mathscr{E}_a \, da = \frac{\lambda}{2\pi\epsilon_0}\log\frac{b}{a};$$

then
$$C = \frac{\lambda}{V_{ab}} = \frac{2\pi\epsilon_0}{\log(b/a)},$$

which is the capacity per unit length.

d. MANY-BODY PROBLEMS.

(1) *The Parallel Plate Condenser with Unequal Charges on the Plates.* Bring two conducting plates, one carrying $+Q_a$ and the other carrying $-Q_b$, into a parallel configuration having a separation d. Assume $|Q_a| > |Q_b|$.

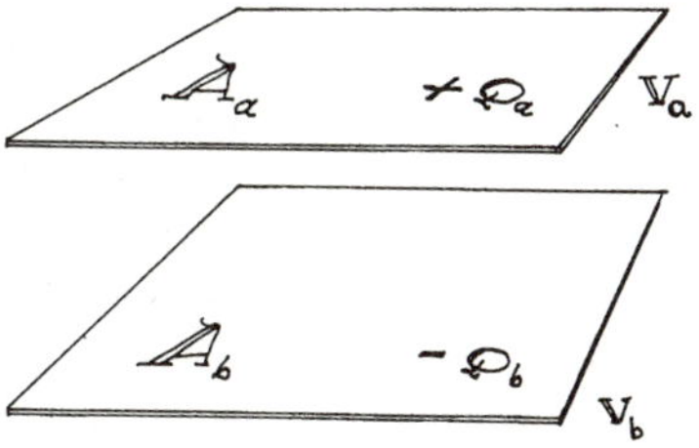

Since the lines of A_a (measured by Q_a) cannot all terminate on $A_b(|Q_b| < |Q_a|)$ *some of the lines must terminate at infinity* (a third body).

To achieve equilibrium, *the outside of the condenser must have a single potential with respect to infinity*; otherwise lines from one point outside would terminate at another outside point. Thus charge will have to move to the internal gap of the condenser. *The equilibrium arrangement is shown in the figure.* Since there is a capacitance

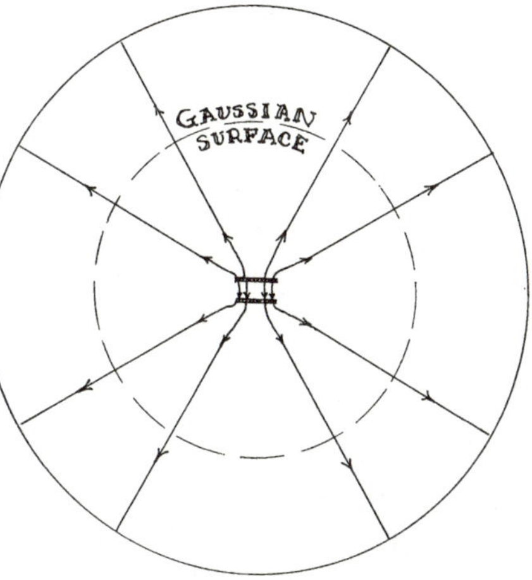

between the plates coupled with capacitance with respect to ∞, the definition of a *single* capacitance has no meaning. A Gaussian surface will measure $Q_a - Q_b$ lines or the total charge enclosed.

(2) *The Problem of Charging Two Condensers in Parallel; Isolating from the Battery; Disconnecting; and Then Connecting in Series.* The immediate reaction to such a problem is that the final voltage will be the sum of the voltages on the isolated condensers. Care must be taken in this problem, because whenever the inner branch in the series connection carries a nonzero charge, the system acts as a many-body system. In some cases the direct addition of the voltages may give a close approximation to the solution, but in general the application of the two-body concepts to a many-body problem is not correct.

(3) *The Concentric Spherical Condenser with the Inner Sphere Grounded.* With the inner sphere grounded, place a charge $+Q$ on the outer sphere. The outer sphere now has a potential V with respect to ground. However, the potential V must be consistent with both the zero potential at infinity and the zero potential of the inner sphere. Thus the outer sphere has lines going to infinity and lines terminating on the inner sphere.

Move a test charge from infinity to the surface of the outer sphere. Assume that the charge Q splits into mQ and $(1 - m)Q$,

$$V_{\infty \to a} = 0 = \frac{Q}{4\pi\epsilon_0}\left\{\frac{m}{b} - (1 - m)\left(\frac{b - a}{ab}\right)\right\}$$

where $$m = \frac{b - a}{b}$$

and $$(1 - m) = a/b.$$

Thus $$V_{b \to a} = \frac{-Q}{4\pi\epsilon_0}\left\{\frac{b - a}{b^2}\right\}.$$

The apparent capacity would be

$$\frac{4\pi\epsilon_0 b^2}{(b - a)}.$$

4. THE CONDENSER AND THE DIELECTRIC

a. FUNDAMENTAL EQUATIONS. When a dielectric is used to fill or partially fill the region intervening between the two conductors forming a condenser, the equations representing the capacitance, charge, energy stored, and force between the conductors are altered.

The relations which we shall use extensively are the following:

$$\mathbf{D} = \sigma_{\text{free}}\,\mathbf{n}$$

at the surface of a charged conductor;

$$\mathbf{D} = \epsilon_0\mathscr{E} + \mathbf{p}_v = \epsilon_0(1 + \chi)\mathscr{E};$$

$$V_b - V_a = -\int_a^b \mathscr{E}\cdot d\mathbf{r};$$

$$U = \tfrac{1}{2}\iiint_\tau \mathbf{D}\cdot\mathscr{E}\,d\tau$$

$$= \tfrac{1}{2}\iiint_\tau \epsilon_0\mathscr{E}^2\,d\tau + \tfrac{1}{2}\iiint_\tau (\mathbf{p}_v\cdot\mathscr{E})\,d\tau;$$

The pressure at a point on the surface of a conductor is,

$$\mathbf{P} = \tfrac{1}{2}\mathscr{E}\cdot\mathbf{D}\mathbf{n}.$$

b. PARALLEL PLATE CONDENSER WITH DIELECTRIC IN THE GAP. If the plate area is A and the total free charge on a plate is Q_f while the potential difference is V,

$$C = \frac{Q_f}{V} = \frac{\sigma_f A}{V}$$

where σ_f is the uniform surface density of charge. Assume that the dielectric filling the gap has a constant ϵ, and is also isotropic. At one plate

$$\sigma_f = |\mathbf{D}| = \epsilon_0\epsilon\,|\mathscr{E}|.$$

In a parallel plate condenser (neglecting edge effects) the electric field is constant. If the plate separation is d,

$$V = -\int_d^0 \mathscr{E}_0\,dx_3 = \mathscr{E}_0 d.$$

This integral was taken to make V positive. Then

$$\sigma_f = \epsilon_0\epsilon V/d,$$

and

$$C = \frac{\sigma_f A}{V} = \frac{\epsilon_0\epsilon A}{d} = \epsilon C_0.$$

Here C_0 is the capacitance of this condenser with a vacuum gap. We see that the dielectric increases the capacity of the condenser.

The energy stored, U, is increased by the dielectric for a fixed V (U_0 is the energy associated with a vacuum gap):

$$U = \tfrac{1}{2}CV^2 = \tfrac{1}{2}\epsilon C_0 V^2 = \epsilon U_0.$$

The additional energy supplied in charging the condenser with the dielectric is the energy required to polarize the dielectric,

$$\Delta U = U - U_0 = \chi U_0 = (\chi/\epsilon)U.$$

$$U = \tfrac{1}{2}\iiint_{\text{Gap}} \epsilon_0 \mathscr{E}^2 \, d\tau + \tfrac{1}{2}\iiint_{\text{Gap}} (\mathbf{p}_V \cdot \mathscr{E}) \, d\tau;$$

$$U = \tfrac{1}{2}\epsilon_0 \frac{V^2}{d^2} Ad + \tfrac{1}{2}\epsilon_0 \chi \frac{V^2}{d^2} Ad = U_0 + \chi U_0.$$

Thus the expressions for the volume integral show that the term ΔU is indeed the energy of polarization.

The force between the condenser plates is

$$\mathbf{F} = \iint_{\text{Plate Area}} \mathbf{P} \, dS = \tfrac{1}{2}(\mathbf{D} \cdot \mathscr{E})A\mathbf{n} = \tfrac{1}{2}\epsilon_0 \epsilon \frac{V^2}{d^2} A\mathbf{n},$$

where **n** is the surface normal so directed as to indicate that the force between the plates is attractive. Thus in magnitude,

$$|\mathbf{F}| = \tfrac{1}{2}\epsilon_0 \epsilon A \frac{V^2}{d^2}.$$

c. THE PARTIALLY FILLED GAP. If the dielectric segments are cut and arranged in a symmetric manner in the gap, one can usually represent the problem as a set of condensers in series and parallel.

The equivalent problems are indicated in the diagrams for three typical cases.

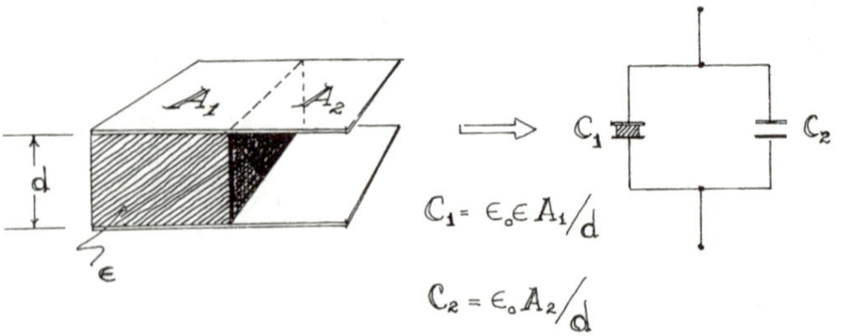

$$C_1 = \epsilon_0 A / d_1$$

$$C_2 = \epsilon \epsilon_0 A / d_2$$

$$C_1 = \epsilon_0 (A - a) / d$$
$$C_2 = \epsilon_0 a / (d - \delta)$$
$$C_3 = \epsilon_0 \epsilon \, a / \delta$$

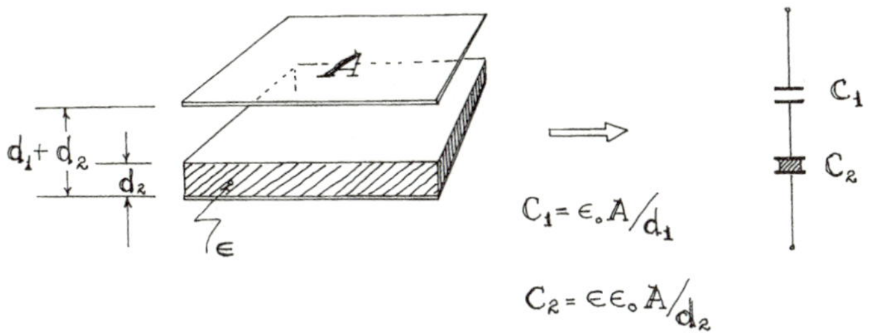

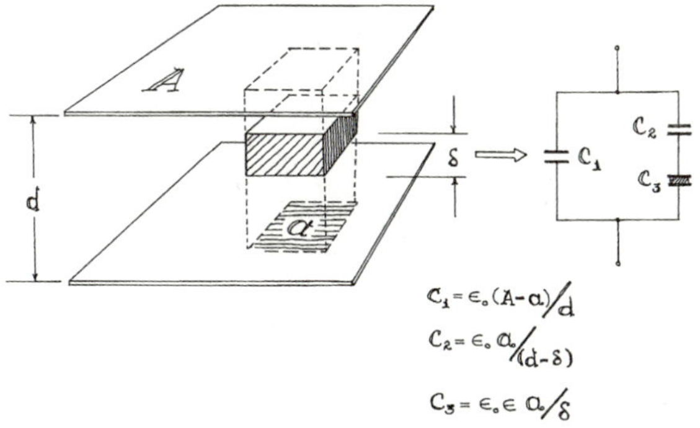

d. ENERGETICS OF PLACING A DIELECTRIC IN THE GAP OF A PARALLEL PLATE CONDENSER. For simplicity in the discussion to follow, a dielectric slab will be used which can *completely fill* the gap of a previously charged parallel plate condenser. There are *two* major types of problems:

(1) Inserting the dielectric at constant voltage, and
(2) Inserting the dielectric while keeping a constant charge on the condenser.

In either case the dielectric will be attracted into the gap. If we examine the field and polarization when the dielectric is just at the edge of the gap, it can be seen that the field is distorted at the edge, and the dipoles induced in the dielectric will be attracted in the direction of the field concentration.

Assume that the dielectric is restrained by an external force $\mathbf{F}_{ext}$ as it slides into the gap. Otherwise the dielectric, though starting from zero velocity and zero kinetic energy, would be pulled into the

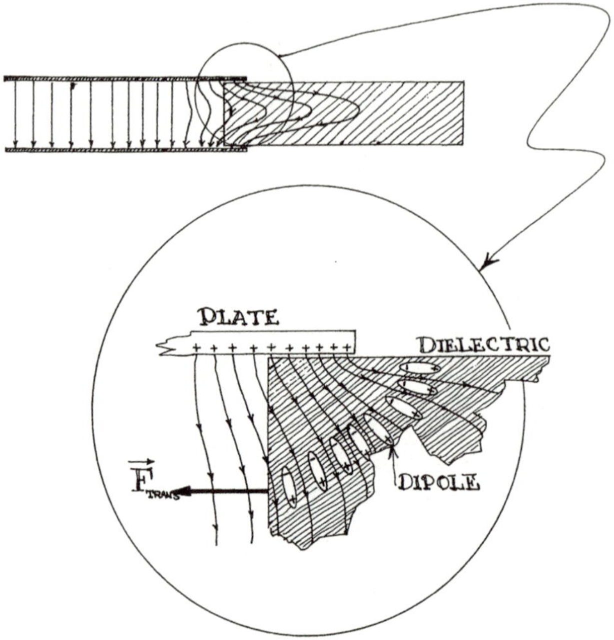

gap. At the center position it would have a zero translational force on it ($F_{trans} = 0$) or zero potential energy but it would have kinetic energy. The result would be that the dielectric would oscillate in and out of the gap.

Thus, in a problem involving a stationary dielectric, during insertion the dielectric must be restrained by an external force as it moves into the gap, the end result being that the condenser does work on an external system.

(1) *Constant Voltage* (*Varying Condenser Charge*). The condenser face area is A, the plate separation d. The following describe the condenser system before the dielectric is in the gap:

$V_i = V$ — P = pressure on the plates

$C_i = \epsilon_0 A/d$ — F = the force between the plates

$Q_i = C_i V$ — u = the energy density

$\sigma_i = Q_i/A = \epsilon_0 V/d$ — U = the stored energy

$U_i = \frac{1}{2} C_i V^2 = \epsilon_0 A V^2/2d$ — σ = the area charge density

$F_i = P_i A = u_i A = U_i/d = \epsilon_0 A V^2/2d^2$.

After the dielectric is in the gap:

$$
\begin{aligned}
V_f &= V \\
C_f &= \epsilon(\epsilon_0 A/d) = \epsilon C_i \\
Q_f &= C_f V_f = \epsilon Q_i \\
&= Q_i + \text{polarization charge } (\chi Q_i) \\
U_f &= \tfrac{1}{2} C_f V = \epsilon U_i \\
F_f = P_f A &= U_f/d = \epsilon F_i
\end{aligned}
$$

Thus as a result of inserting the dielectric at constant voltage, C, Q, U, and F increase by a factor equal to ϵ. Now let us examine carefully the energy stored in the condenser and the energy supplied by the battery:

$$Q_f - Q_i = \chi Q_i$$

$$U_f - U_i = \chi U_i = U_p, \quad \text{the energy of polarization.}$$

The energy supplied by the battery as the dielectric is moved into the gap is

$$(Q_f - Q_i)V = \chi Q_i V = W_{batt}.$$

However,

$$U_i = \tfrac{1}{2} Q_i V;$$

thus $W_{batt} = 2\chi U_i$ of which ONLY χU_i goes to polarize the dielectric.

Therefore the additional work done by the battery (other than U_p) goes into mechanical work against the external restraining force necessary to place a stationary dielectric in the gap:

$$W_{mech} = W_{batt} - U_p = \chi U_i.$$

The external mechanical work and internal energy of polarization are equal in the case of the constant voltage problem,

$$W_{mech} = U_p.$$

(2) *Constant Charge* (*Varying Condenser Voltage*). The condenser system before the dielectric is in the gap is described by:

$$
\begin{aligned}
Q_i &= Q \\
C_i &= \epsilon_0 A/d \\
V_i &= Q/C_i = Qd/\epsilon_0 A \\
Q_i &= \tfrac{1}{2} C_i V_i^2 = \tfrac{1}{2} Q^2/C_i = (d/2\epsilon_0 A)Q^2 \\
F_i &= U_i/d = Q^2/\epsilon_0 A.
\end{aligned}
$$

After the dielectric is in the gap:

$$Q_f = Q$$

$$C_f = \epsilon(\epsilon_0 A/d) = \epsilon C_i$$

$$V_f = Q/C_f = V_i/\epsilon$$

$$U_f = \tfrac{1}{2}Q^2/C_f = U_i/\epsilon$$

$$F_f = U_f/d = F_i/\epsilon.$$

Thus when the dielectric is inserted while the condenser charge is held constant, C increases by a factor ϵ but V, U, and F decrease by a factor $1/\epsilon$. The energy of polarization and mechanical work no longer have a simple relationship. The decrease in stored energy equals the mechanical work,

$$U_i - U_f = W_{mech}$$

$$U_i - U_f = \left(1 - \frac{1}{\epsilon}\right)U_i = (\chi/\epsilon)U_i.$$

The energy of polarization $U_p = (\chi/\epsilon)U_{final}$,

$$U_{final} = \text{field energy} + \text{energy of polarization.}$$

Since $$U_p = \frac{\chi}{\epsilon}U_f = \frac{\chi}{\epsilon^2}U_i,$$

therefore $$W_{mech} = \epsilon U_p.$$

We can conclude therefore that when the condenser system is forced to provide the mechanical work from its initial stored energy, a great part of the energy is lost to an external system.

V. Currents and Circuits

A. CURRENTS

The microscopic charge carriers were discussed in Section II-A when we defined charge. In metals, currents are constituted of moving electrons. The motion is not unobstructed as it would be for electrons in a vacuum; rather, the motion consists of a series of collisions between the electrons and the heavier atoms of the metal. Under the influence of an electric field the motion is not random but

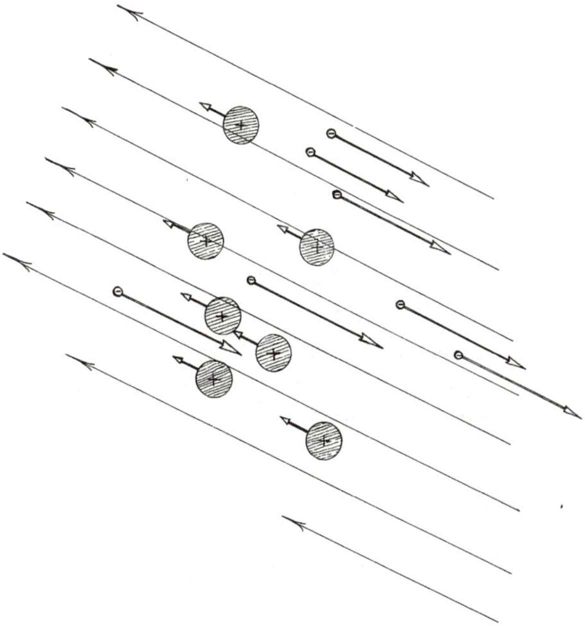

has a net vector displacement in a direction opposite to the field direction (because the electron carries a negative charge). In an electrolyte the currents are made up of positive and negative ions while in a gaseous discharge the conduction is produced by electrons plus the less mobile positive and negative ions.

Although the charge carriers in a metal are negative, it has been a convention to take the direction of positive charge flow as the direction of the positive current. Positive charge flow is a necessary fiction resulting from early concepts concerning the charge carriers.

In order to build up a picture of the current from its microscopic constituents, consider a volume density n_v of positive charge carriers

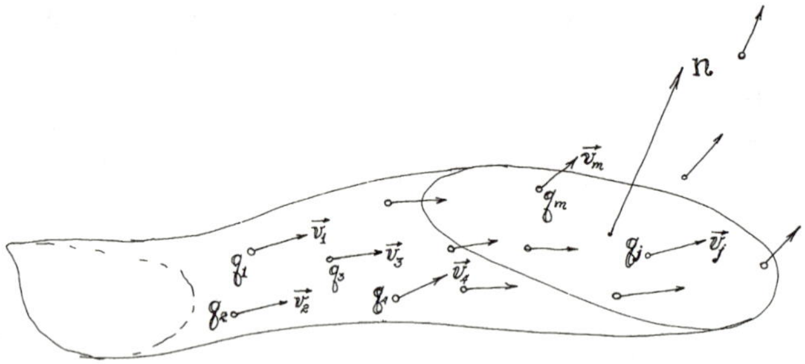

each of charge $+q$. At every point the velocity of the charge is specified by a velocity distribution $\mathbf{v}$.

In this case the current density $\mathbf{J}$ is

$$\mathbf{J}(x_1, x_2, x_3) = qn_v(x_1, x_2, x_3)\,\mathbf{v}(x_1, x_2, x_3)$$
$$= \rho\mathbf{v},$$

where ρ is the volume density of charge specified at every point.

If the moving charge distribution in a volume τ consists of N charges q_j moving with a velocity $\mathbf{v}_j$, then

$$\mathbf{J} = \frac{\sum_{n=1}^{N} q_n \mathbf{v}_n}{\tau}.$$

The element of current ΔI crossing a surface element $\Delta S\mathbf{n}$ is

$$\Delta I = \mathbf{J} \cdot \mathbf{n}\,\Delta S.$$

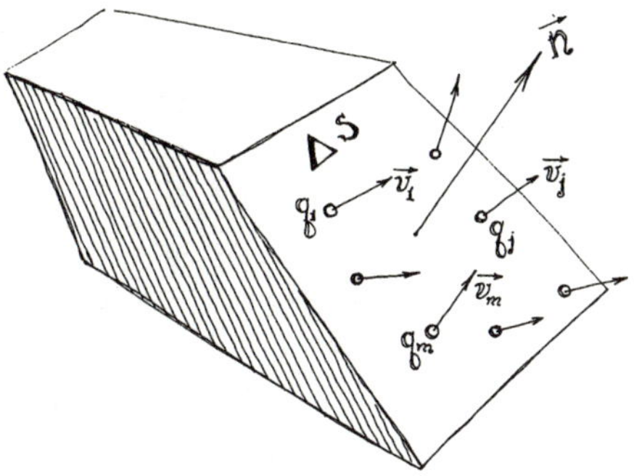

When circuits are involved it is usually sufficient to use the total current in a conductor. In such a conductor (a wire for instance) the cross-sectional area A is defined, and the total current flowing is

$$I = \iint_A \mathbf{J} \cdot \mathbf{n} \, dS,$$

while the total charge flowing across the area A in a time T is

$$Q_T = \int_0^T \frac{dQ}{dt} dt = \int_0^T I \, dt,$$

and

$$I = \frac{dQ}{dt}.$$

The current is the time rate of flow of charge through the cross-sectional area of a conductor.

B. CHARGE CONSERVATION AND THE CONTINUITY EQUATION

On the basis of the assumption that the net charge of the universe is constant, the build-up of charge in a volume τ enclosed by a surface S must be accounted for completely by the currents flowing across the surface.

The time rate of change of charge inside of S is equal to the net current in minus the net current out,

$$\frac{dQ_t}{dt} = I_{in} - I_{out} = -\iint_S \mathbf{J} \cdot \mathbf{n} \, dS.$$

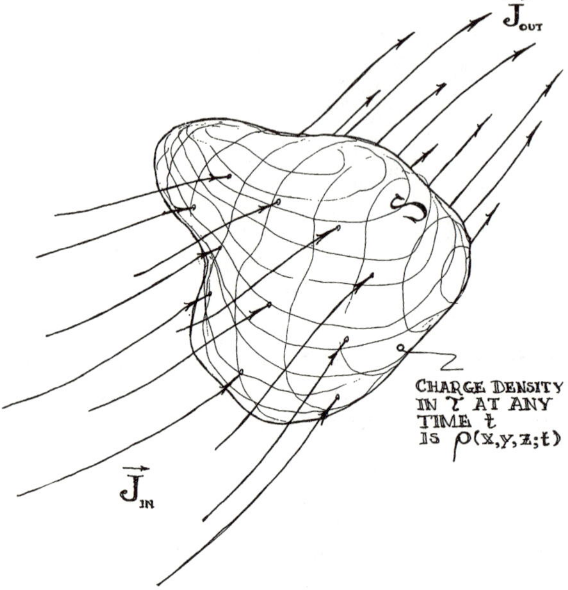

If the volume charge density at any time t is $\rho(x_1,x_2,x_3;\, t)$ then*

$$\frac{dQ}{dt} = \iiint_\tau \frac{\partial \rho}{\partial t}\, d\tau = -\iint_S \mathbf{J} \cdot \mathbf{n}\, dS.$$

Converting the surface integral to a volume integral via the divergence theorem,

$$\iiint_\tau \frac{\partial \rho}{\partial t}\, d\tau = -\iiint_\tau \operatorname{div} \mathbf{J}\, d\tau.$$

This result is independent of the volume τ, in other words it holds for $\tau \to 0$, therefore this equality is satisfied by the equation

$$\operatorname{div} \mathbf{J} + \frac{\partial \rho}{\partial t} = 0,$$

which is the *equation of continuity.*

* Because $dQ/dt = \partial Q/\partial t$, the partial derivative is appropriate when used with the volume integral.

C. ELECTROMOTIVE FORCE

Consider a region belonging to a current-carrying element. Such a region basically consists of charges moving under the influence of imposed fields. The work done by the field or against the field in moving a charge from a point a to another point b is

$$W_{ab} = \int_a^b \mathbf{F} \cdot d\mathbf{r} = q \int_a^b \mathscr{E} \cdot d\mathbf{r},$$

where $d\mathbf{r}$ represents the displacement of q at each point along the path of q. The integral is *positive* if the work is done by the electrical system and represents a conversion of electrical energy into kinetic energy. On the other hand, the work is *negative* if energy is supplied by an external system such as a chemical process or a mechanical operation. In this latter case the charge q moves against the electric field, and as a result the energy supplied by the external system appears as additional potential energy in the electrical system.

If the charge q is moving with a velocity $\mathbf{v}$ the instantaneous power developed is

$$P_q = \frac{dW}{dt} = q\mathscr{E} \cdot \mathbf{v}.$$

Again the positive sign represents a power loss and the negative sign a power gain. In a volume element $\Delta\tau$ the total power is

$$\Delta P = \mathscr{E} \cdot \mathbf{J}\, d\tau.$$

Here we have summed over all point charges times the associated velocities $q_i\mathbf{v}_i$ giving $\mathbf{J}\, d\tau$. The power associated with a volume τ is then

$$P = \iiint_\tau \mathscr{E} \cdot \mathbf{J}\, d\tau.$$

Steady state electrical currents are maintained in general by some nonelectrical agent such as a chemical reaction (battery), a mechanical device (d.c. generator), a thermal device (thermocouple), etc. Because nonelectrical energy is converted into electrical energy by these devices, they are considered to be nonconservative elements in the electrical network.

These devices will set up nonconservative fields $\mathscr{E}'$. A given circuit volume will then contain fields $\mathscr{E}$, set up by the charge carriers, and source (or battery) fields $\mathscr{E}'$. In a closed wire loop $\mathscr{L}$ the total

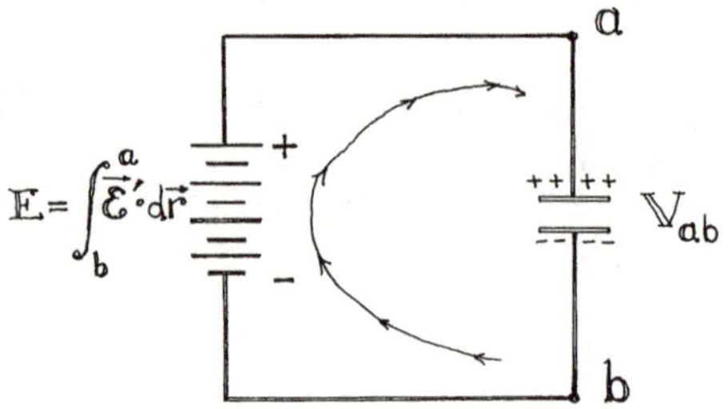

power associated with the loop is

$$P = \iiint \{\mathscr{E} + \mathscr{E}'\} \cdot \mathbf{J}\, d\tau$$

In elementary circuits we regard the current as being uniformly distributed across any cross section. Our volume integral then reduces to a loop integral over $\mathscr{L}$ by letting $\mathbf{J}\, d\tau$ go to I **dr**;

$$P = \oint_{\mathscr{L}} I\{\mathscr{E} + \mathscr{E}'\} \cdot d\mathbf{r}.$$

Because the field $\mathscr{E}$ is conservative the closed line integral about $\mathscr{L}$ vanishes, and the power developed in the loop is completely accounted for by the field $\mathscr{E}'$.

$$P = I \oint_{\mathscr{L}} \mathscr{E} \cdot d\mathbf{r} + I \oint_{\mathscr{L}} \mathscr{E}' \cdot d\mathbf{r} = I \oint \mathscr{E}' \cdot d\mathbf{r}.$$

We define the ELECTROMOTIVE FORCE, emf, of the loop as

$$E = \oint \{\mathscr{E} + \mathscr{E}'\} \cdot d\mathbf{r} = \oint \mathscr{E}' \cdot d\mathbf{r}.$$

As a result the total power delivered to the loop is the emf, E, times the current I;

$$P = EI.$$

The line integral between points a and b of a loop can be resolved into two components, a potential difference and an emf.

$$\oint_a^b \{\mathscr{E} + \mathscr{E}'\} \cdot d\mathbf{r} = \int_a^b \mathscr{E} \cdot d\mathbf{r} + \int_a^b \mathscr{E}' \cdot d\mathbf{r}$$
$$= -\{V_b - V_a\} + E_{ab}.$$

As we have defined the emf between points a and b, it is related only to the nonconservative forces $\mathscr{E}'$;

$$E_{ab} = \int_a^b \mathscr{E}' \cdot d\mathbf{r}.$$

The circuit branch a $\to$ b is a passive branch if E_{ab} is zero. On the other hand, if E_{ab} does not vanish in the element a $\to$ b the branch is said to contain a seat of emf and is therefore an active branch.

D. OHM'S LAW

In any conducting media such as a metal it is an *experimental fact* that the current density is proportional to the total field intensity, $\{\mathscr{E} + \mathscr{E}'\}$;

$$\mathbf{J} = \beta \mathscr{E}_{\text{total}} = \beta\{\mathscr{E} + \mathscr{E}'\}.$$

Here β is a scalar constant known as the conductivity of the medium. The units of β are mhos per meter, while the inverse β^{-1} is known as the resistivity of the medium in units of ohm-meters. The properties of conducting media are not always so simple. Many materials are found which have nonlinear conduction properties ($\beta = \beta(\mathbf{J})$), and in some cases are anisotropic. Anisotropic conductors can be represented by a tensor conductivity. In these materials

$$\mathbf{J} = \boldsymbol{\beta} \cdot \mathscr{E}_{\text{total}},$$

where $\boldsymbol{\beta}$ is a 3×3 symmetric matrix having elements β_{jk}.

In the discussion to follow we shall be primarily interested in linear isotropic conductors. Consider a circuit containing only linear isotropic conductors in series with various sources of emf. If we integrate Ohm's law between any two points a and b in the circuit, we find that

$$\int_a^b \mathscr{E} \cdot d\mathbf{r} + \int_a^b \mathscr{E}' \cdot d\mathbf{r} = \int_a^b \frac{\mathbf{J}}{\beta} \cdot d\mathbf{r},$$

or

$$-\{V_b - V_a\} + E_{ab} = \int_a^b \frac{\mathbf{J}}{\beta} \cdot d\mathbf{r}.$$

For simplicity assume that the resistive elements are cylindrical of length l and cross-sectional area A. Further assume that $\mathbf{J}$ is

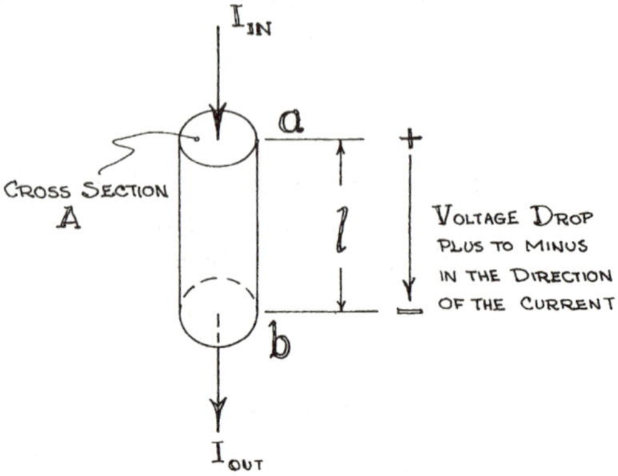

uniformly distributed across A and parallel to the length l (**J** is then perpendicular to A). Then

$$|\mathbf{J}| = \frac{I}{A},$$

where I is the total current flowing in the segment a → b. The line integral can be reduced to

$$-\{V_b - V_a\} + E_{ab} = \frac{Il}{\beta A}.$$

The quantity $\left\{\frac{l}{\beta A}\right\}$ is called the total resistance of the cylinder and is denoted by the symbol R;

$$R = \frac{l}{\beta A}.$$

If there are no seats of emf in the cylinder, $E_{ab} = 0$; and we obtain the relation between the voltage drop across the resistor and the current passing through the resistor

$$V_a - V_b = RI,$$

where $(V_a - V_b)$ is a positive quantity indicating that the entrance point a is at a higher potential than the exit point b.

The rules for series and parallel combinations of resistances can be established from the definition in terms of the conductivity.

For the *parallel* combination, the voltage drop which we shall

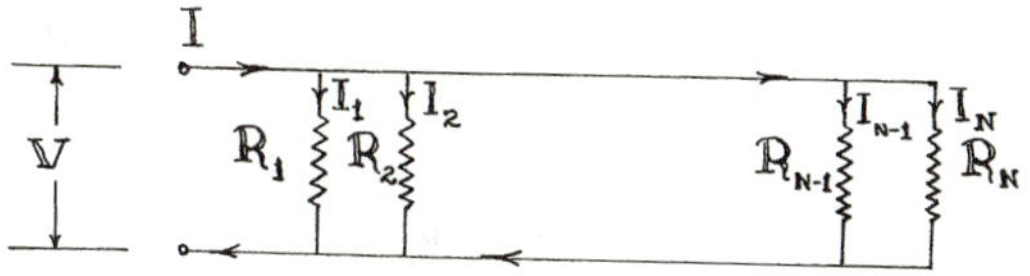

PARALLEL COMBINATION

label V is the common element. Consider N resistors R_j in parallel,

$$V = I_1R_1 = I_2R_2 = I_jR_j = \ldots = I_NR_N.$$

The total current is I_T,

$$I_T = \sum_{j=1}^{N} I_j = \sum_{j=1}^{N} V/R_j = \left\{\sum_{j=1}^{N} \frac{1}{R_j}\right\} V.$$

The equivalent total resistance of the parallel combination is R_T, where

$$I_T = V/R_T.$$

Therefore $$\frac{1}{R_T} = \sum_{j=1}^{N} \frac{1}{R_j}\,.$$

When the combination consists of a *series* connection of N

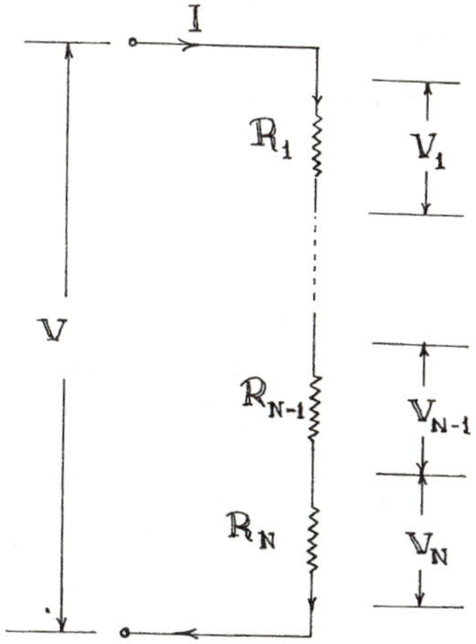

SERIES COMBINATION

resistors, the current I is common to all. Let V_j be the potential drop across the jth resistor R_j; then

$$I = \frac{V_1}{R_1} = \frac{V_2}{R_2} = \ldots = \frac{V_j}{R_j} = \ldots = \frac{V_N}{R_N}.$$

The total potential drop across the series combination is V_T,

$$V_T = \sum_{j=1}^{N} V_j = \sum_{j=1}^{N} R_j I = \left\{\sum_{j=1}^{N} R_j\right\} I.$$

Since the total resistance is given by $V = R_T I$,

$$R_T = \sum_{j=1}^{N} R_j = R_1 + R_2 + \ldots + R_N.$$

Series-parallel combinations can be compounded from these rules. Any odd-shaped resistor can be simulated by a series-parallel configuration. This is the basis of the statement that resistors regardless of their shape have an equivalent, calculable R.

The power balance can now be obtained from the integral

$$P = \iiint_{\substack{\text{Volume } \tau \\ \text{of the circuit}}} \{\mathscr{E} + \mathscr{E}'\} \cdot \mathbf{J}\, d\tau = \iiint \frac{J^2}{\beta}\, d\tau.$$

If we consider the resistance elements as cylinders of homogeneous isotropic material, $|\mathbf{J}| = I/A$, and

$$d\tau = A\,|d\mathbf{r}|,$$

where **J** and **dr** are parallel.

Then
$$P = I \oint \mathscr{E} \cdot d\mathbf{r} + I \oint \mathscr{E}' \cdot d\mathbf{r}$$
$$= I \oint \mathscr{E}' \cdot d\mathbf{r} = IE = \left\{\frac{l}{\beta A}\right\} I^2$$
$$= I^2 R;$$

or
$$P = EI = I^2 R.$$

This last equation is a statement that the power dissipated by the resistor is equal to the power delivered by the emf E.

E. STEADY STATE CIRCUIT THEORY

1. KIRCHHOFF'S LAWS

Combining the definition of the electromotive force with Ohm's law (for a linear circuit),

$$\beta(\mathscr{E} + \mathscr{E}') = \mathbf{J},$$

and $$E = \oint (\mathscr{E} + \mathscr{E}') \cdot dr = \oint \frac{J}{\beta} \cdot dr = \frac{I}{A}\frac{1}{\beta} = IR,$$

where R is the equivalent resistance of the circuit. This includes all resistances, even the internal battery resistance.

This equation plus the continuity equation make up Kirchhoff's laws.

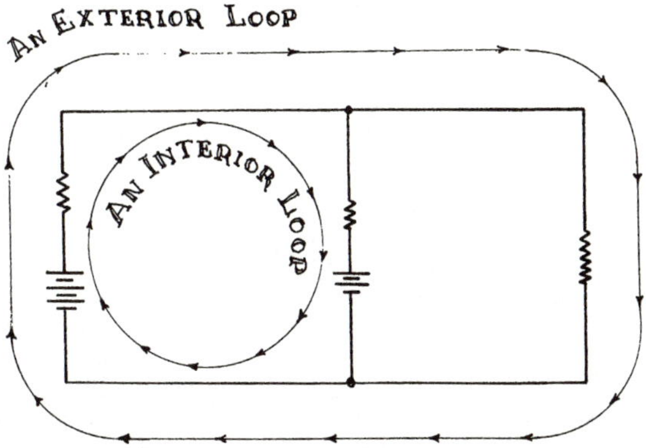

Kirchhoff's laws state that

(1) In a given loop the sum of the emfs (taken with appropriate signs) equals the sum of the IR drops,

$$\sum_{\text{all } j} E_j = \sum_{\text{all } n} I_n R_n$$

Here the maximum number of elements in each sum represents the number of adjoining loops, where the exterior region is considered as a loop. We also assume each segment reduced to an equivalent resistance.

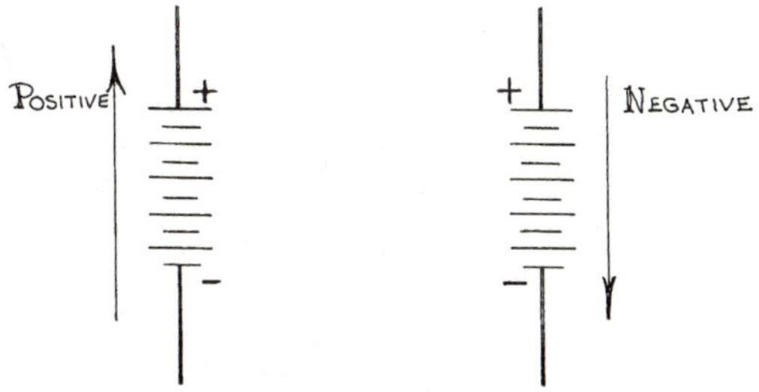

The sign of the emfs E_j is *positive* if the loop passes through the seat of emf from − to +. The sign of E_j is negative if the traversal is from + to −.

IR drops are positive when the path progresses from plus to minus in the resistor (i.e., in the direction of positive charge flow).

(2) The sum of the currents entering a branch point is ZERO. A branch point is a point at which more than two currents meet. A current leaving takes a negative sign. Let the current in the kth branch be i_k; at a branch

$$\sum_{\text{all } k} i_k = 0.$$

Since Kirchhoff's laws are well covered in the elementary texts and because they are extremely tedious to apply, we shall content ourselves with this brief statement.

2. THE MAXWELL LOOP METHOD

Of much more practical interest is the Maxwell loop method. This is a condensation of Kirchhoff's laws and lends itself readily to a matrix representation. A third method known as the method of linear superposition will be shown to be equivalent to the loop method in the matrix formalism.

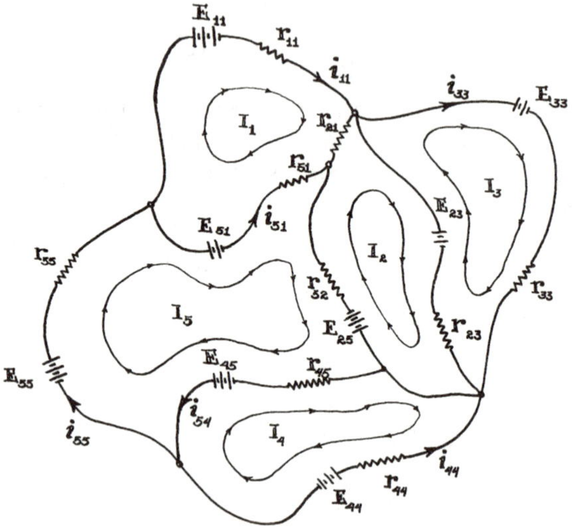

The loop method assigns a current to every loop. The current in any segment is then the vector sum of the adjacent or common loop currents.

Consider the diagram opposite; the segment (or branch) current labeled i_{12} is a superposition of loop currents I_1 and I_2 (with appropriate signs) because it is common to both loops. Taking I_1 and I_2 clockwise,

$$i_{12} = I_1 - I_2 = -i_{21}.$$

A reasonable convention is to take all loop currents in a clockwise direction. If the actual currents flow in the opposite direction, the appropriate direction will appear in the signs after solution.

We proceed in the same manner as was done with Kirchhoff's first law. However, the loop currents are now used instead of branch currents. This approach automatically includes the second law when we use the rule for computing the branch currents.

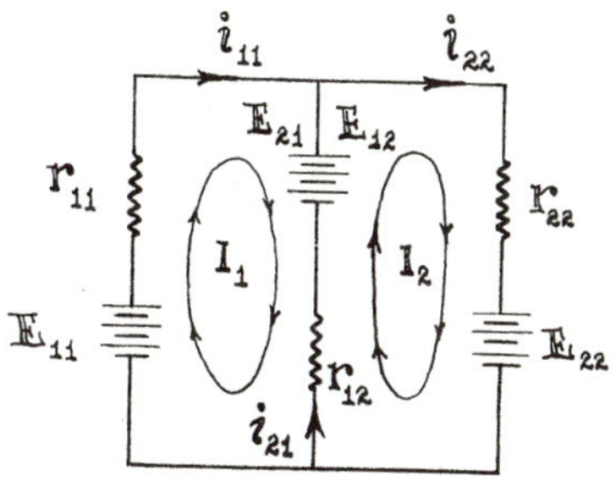

Consider the diagram above. Notice that all resistances r_{jk} bear a double subscript denoting the two loops common to a given branch. We use small r's in order that the matrix elements can be defined in terms of double subscripted capital R's. The r_{mn}'s are symmetric, $r_{mn} = r_{nm}$.

Using the loop currents, the sum of the emfs with appropriate signs equals the sum of the IR drops. The total current in the mn^{th} segment is the loop current I_m minus the loop current I_n. The minus sign appears consistently only if a clockwise loop convention is chosen; otherwise the signs must be entered in an appropriate manner.

Writing down the loop equations in our example,

$$r_{11}I_1 + r_{21}I_1 - r_{12}I_2 = E_{11} - E_{21}$$

$$-r_{21}I_1 + r_{12}I_2 + r_{22}I_2 = +E_{12} - E_{22}.$$

Remembering that

$$r_{12} = r_{21}$$

and

$$E_{12} = +E_{21},$$

we can write

$$R_{11}I_1 + R_{12}I_2 = F_1$$

$$R_{21}I_1 + R_{22}I_2 = F_2.$$

In this notation

$$R_{11} = r_{11} + r_{21}$$

$$R_{12} = -r_{12}$$

$$R_{21} = -r_{12} = -r_{21}$$

$$R_{22} = r_{12} + r_{22}$$

while

$$F_1 = E_{11} - E_{21}$$

$$F_2 = E_{12} - E_{22}.$$

In matrix notation,

$$\mathbf{R} \cdot \mathbf{I} = \mathbf{E},$$

or

$$\begin{pmatrix} R_{11} & R_{12} \\ R_{21} & R_{22} \end{pmatrix} \begin{bmatrix} I_1 \\ I_2 \end{bmatrix} = \begin{bmatrix} F_1 \\ F_2 \end{bmatrix}.$$

Usually the r_{jk}'s and E_{nm}'s are given, and we wish to find the I_j's. This is done by computing $\mathbf{R}^{-1}$. Multiplying the above equation by $\mathbf{R}^{-1}$,

$$\mathbf{I} = \mathbf{R}^{-1} \cdot \mathbf{E}.$$

It is clear then that once $\mathbf{R}^{-1}$ has been computed, ALL of the I_j solutions are available:

$$I_j = \sum_{k=1}^{2} R'_{jk}F_k$$

where R'_{jk} is an element of the inverse matrix $\mathbf{R}^{-1}$.

Although this method has not always been advocated there are many advantages associated with it. When the number of loops is greater than two the probability of error in a straightforward algebraic uncoupling of the algebraic equations is quite high.

In a standard algebraic uncoupling all of the elements of $\mathbf{R}^{-1}$ are essentially computed. However, they are never systematized. One great advantage of the matrix method is that once $\mathbb{R}^{-1}$ has been computed it can be immediately checked from the relation $\mathbb{R}^{-1} \cdot \mathbb{R} = \mathbb{I}$. When the number of loops is large this check can be of great value.

Going back to our 2×2 example, the elements of $\mathbb{R}^{-1}$ are

$$R'_{jk} = \frac{(-1)^{j+k} \text{ minor } R_{kj}}{|\mathbb{R}|}.$$

Thus
$$R'_{11} = \frac{+R_{22}}{(R_{11}R_{22} - R_{12}^2)},$$

and
$$R'_{12} = \frac{-R_{12}}{|\mathbb{R}|}, \quad \text{etc.}$$

To extend the formalism, consider the example of three loops as

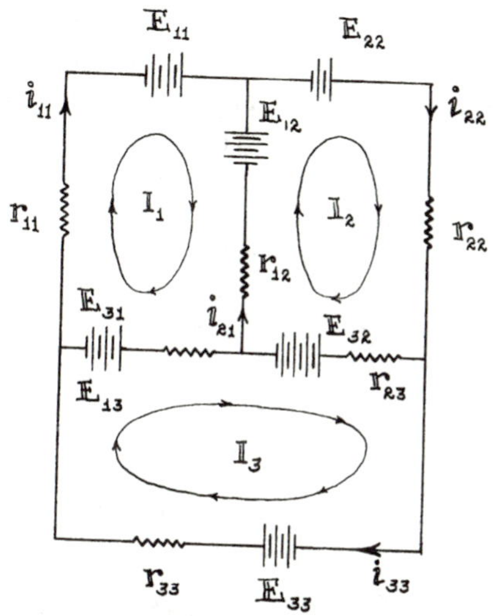

shown in the accompanying diagram. Then

$$r_{11}I_1 + r_{21}I_1 + r_{31}I_1 - r_{12}I_2 - r_{13}I_3 = E_{11} - E_{21} - E_{31};$$
$$-r_{21}I_1 + r_{12}I_2 + r_{12}I_2 + r_{32}I_2 - r_{23}I_3 = E_{12} + E_{22} - E_{32};$$
$$-r_{31}I_1 - r_{32}I_2 + r_{13}I_3 + r_{23}I_3 + r_{33}I_3 = E_{13} + E_{23} - E_{33}.$$

Defining $$R_{11} = r_{11} + r_{21} + r_{31} = \sum_{j=1}^{3} r_{j1}$$

$$R_{12} = -r_{12} = R_{21}$$

$$R_{13} = -r_{13} = R_{31}$$

$$R_{22} = r_{12} + r_{22} + r_{32} = \sum_{j=1}^{3} r_{j2}, \quad \text{etc.,}$$

and $$F_1 = E_{11} + E_{21} - E_{31} = \sum_{j=1}^{3} E_{j1}$$

(with appropriate signs on the E_{j1}), then

$$\begin{pmatrix} R_{11} & R_{12} & R_{13} \\ R_{21} & R_{22} & R_{23} \\ R_{31} & R_{32} & R_{33} \end{pmatrix} \begin{bmatrix} I_1 \\ I_2 \\ I_3 \end{bmatrix} = \begin{bmatrix} F_1 \\ F_2 \\ F_3 \end{bmatrix},$$

and $$\mathbf{I} = \mathbf{R}^{-1} \cdot \mathbf{E}.$$

3. THE METHOD OF SUPERPOSITION

In the method of linear superposition all batteries are shorted out but one. The currents are then computed in each branch; call them $i_j^{(mn)}$ if the mn^{th} battery is in the circuit. This is carried out for each battery. In the end the total current is the vector sum of all individual battery currents $i_j^{(mn)}$.

This can also be performed with individual loop currents for each battery. However, it can be demonstrated that the end result in matrix form is nothing more than the previous loop method.

Consider the original two-loop example. Now short out E_{12} and E_{22} and label the loop solutions $I_j^{(11)}$. Then

$$\begin{pmatrix} R_{11} & R_{12} \\ R_{21} & R_{22} \end{pmatrix} \begin{bmatrix} I_1^{(11)} \\ I_2^{(11)} \end{bmatrix} = \begin{bmatrix} E_{11} \\ 0 \end{bmatrix}$$

or $$\mathbf{I}^{(11)} = \mathbf{R}^{-1} \cdot \mathbf{E}^{(11)}.$$

In like manner $$\mathbf{I}^{(12)} = \mathbf{R}^{-1} \cdot \mathbf{E}^{(12)}$$

where $$\mathbf{E}^{(12)} = \begin{bmatrix} -E_{12} \\ +E_{12} \end{bmatrix}.$$

Finally $$\mathbf{I}^{(22)} = \mathbf{R}^{-1} \cdot \mathbf{E}^{(22)}$$

where $$\mathbf{E}^{(22)} = \begin{bmatrix} 0 \\ -E_{22} \end{bmatrix}.$$

The total current vector **I** for the system is

$$\mathbf{I} = \mathbf{I}^{(11)} + \mathbf{I}^{(12)} + \mathbf{I}^{(22)}$$
$$= \mathbb{R}^{-1} \cdot \{\mathbf{E}^{(11)} + \mathbf{E}^{(12)} + \mathbf{E}^{(22)}\}.$$

This however is merely our original solution by the Maxwell loop method,

$$\mathbf{I} = \mathbb{R}^{-1} \cdot \mathbf{E}.$$

Because the major effort in these problems consists in computing $\mathbb{R}^{-1}$, it would seem that the superposition method has little additional to offer in the matrix formalism.

Application of these techniques to special problems will be performed in the exercises.

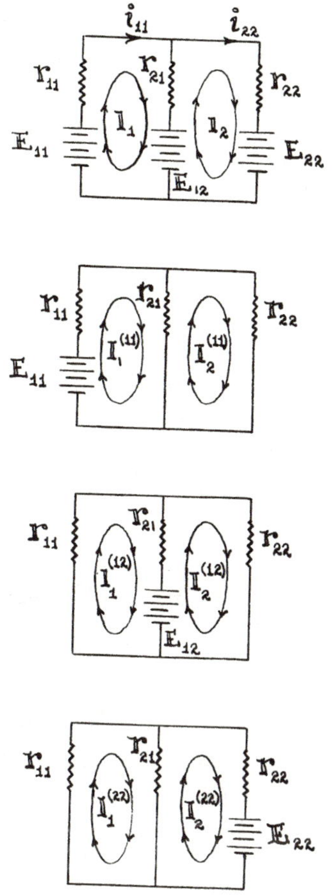

ELEMENTS OF THE METHOD OF SUPERPOSITION

4. THÉVENIN'S THEOREM

A flat mesh of (N – 1) loops which terminate in a load resistance R_L via a two-terminal connection can be replaced by an equivalent resistance R_e and an equivalent source of emf, E_e. When this equivalent series combination R_e and E_e is connected through the terminals to any load resistance R_L, the combination behaves as the original mesh of (N – 1) loops. This theorem is known as Thévenin's theorem.

From the statement of the theorem we see that the open circuit voltage E_{oc} ($R_L \to \infty$) is equal to the equivalent emf, while the short circuit current I_{sc} is equal to the open circuit voltage divided by the equivalent resistance R_e:

$$E_{oc} = E_e$$

$$R_e = E_{oc}/I_{sc}.$$

Experimentally then, given a two-terminal output from an unknown mesh we are able to replace the unknown mesh in terms of simple series equivalents by measuring the open circuit voltage and the short circuit current across the two terminals.

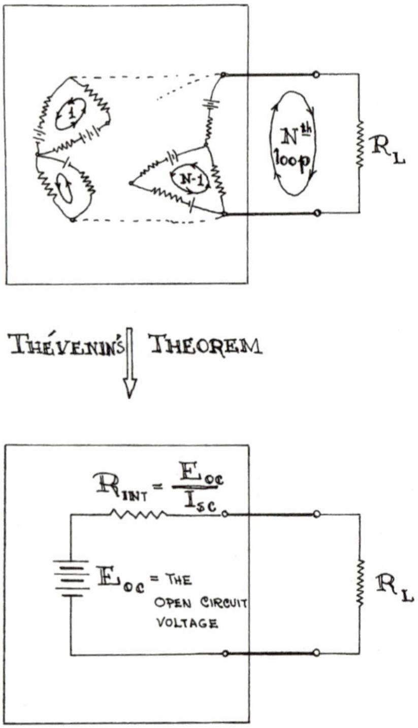

When a load resistor R_L is placed across the terminals, a current I_L flows through the load and this current is given by the relation

$$(R_e + R_L)I_L = E_{oc} = \left\{\frac{E_{oc}}{I_{sc}} + R_L\right\}I_L.$$

The proof of this theorem for an internal mesh of $(N - 1)$ loops can be performed readily using the matrix method.

Because the theorem also applies equally well to a.c. networks, the general proof is provided at the end of Section XII-F-2.

The reader is in a position to carry out this proof by referring to the appropriate section in Chapter XII. One merely substitutes the resistance matrix $\mathbb{R}$ for $\mathbb{Z}$ and $\mathbb{R}^{-1}$ for $\mathbb{Y}$ in that derivation.

VI. *The Lorentz Force and the Magnetic Field*

A. DEFINITION OF THE MAGNETIC INDUCTION FIELD

In defining the electrostatic field, the field quantities were specified in terms of forces upon charged bodies. By an analogous procedure the magnetic field is defined in terms of forces depending upon the charge on a body plus the velocity of that body.

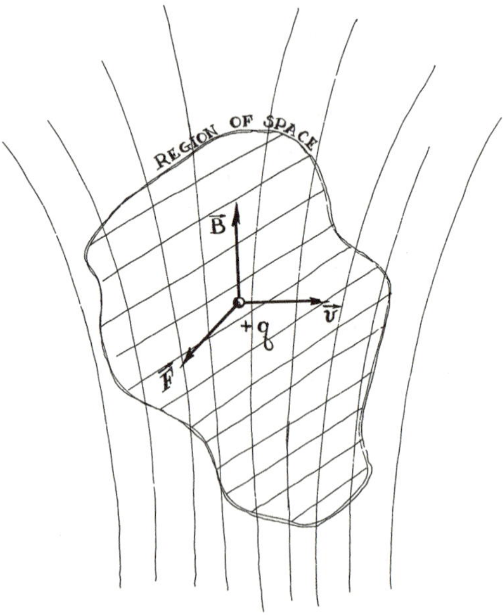

A magnetic field **B** *is said to be present in a specified region when a moving charge (i.e., current) experiences an electromagnetic force upon it which is observable only by virtue of the charge and the existence of the velocity of this charge.*

To describe the characteristics of the magnetic induction field **B**, a series of thought experiments can be performed which define **B** operationally. One can employ either a point charge moving with a velocity **v** or a current element I d$\boldsymbol{l}$ where the direction of d$\boldsymbol{l}$ is in the sense of the positive charge flow.

In order to simplify our development we shall employ a point charge q moving at any given instant with a velocity **v**. The experimental region contains no electric fields.

Assuming that **v** can be measured, the following experimental results are obtained.

(1) In the region in question, for some orientations of **v** a force $\mathbf{F}_B$ is observed directed perpendicular to **v**.

(2) If the magnitude of the charge is varied, $\mathbf{F}_B$ varies proportionally,

$$|\mathbf{F}_B| \propto q.$$

(3) For one particular orientation of **v** the force $\mathbf{F}_B$ is ZERO. This orientation of the velocity vector for zero $\mathbf{F}_B$ constitutes a line in space. The magnetic field is defined to be parallel or antiparallel to this line at the point of observation.

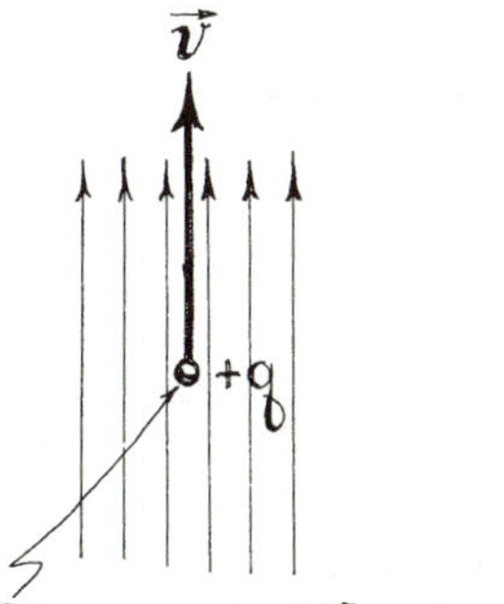

DIRECTION OF FIELD DEFINED BY $\vec{v}$ WHEN THE LORENTZ FORCE IS ZERO

(4) Furthermore (keeping q constant), $|\mathbf{F}_B|$ varies as the sine of the angle between **v** and the line defining the field direction,

$$|\mathbf{F}_B| \propto \sin \measuredangle^{\mathbf{B}}_{\mathbf{v}}.$$

(5) The magnitude of the magnetic field is determined by observing the maximum value of $\mathbf{F}_B$ for a fixed magnitude $|\mathbf{v}|$. This occurs when **v** is perpendicular to **B**:

$$|\mathbf{B}| = \frac{|\mathbf{F}_B(\max)|}{q|\mathbf{v}|}.$$

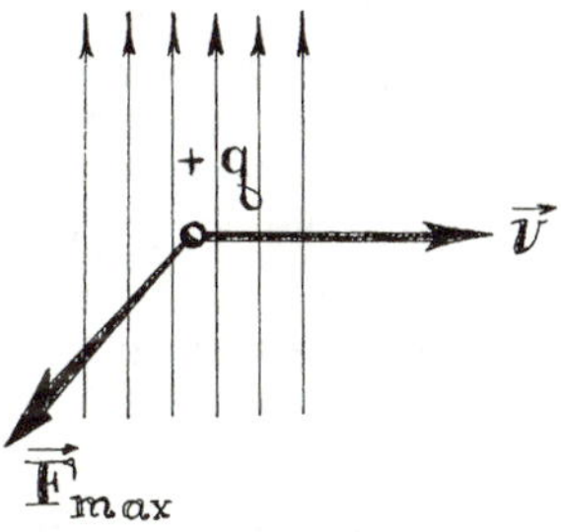

MAGNITUDE OF FIELD DETERMINED BY THE MAGNITUDE OF THE MAXIMUM VALUE OF THE LORENTZ FORCE

(6) The maximum force together with the velocity form a plane perpendicular to the zero line defining plus or minus the direction of **B**. The direction of **B** is taken in the direction of $\mathbf{F}_B(\max) \times \mathbf{v}$.

(7) Finally, by varying the magnitude of $\mathbf{v}$ we find $|\mathbf{F}_B| \propto |\mathbf{v}|$. Taking all observations and specifications into account,

$$\mathbf{F}_B = q\mathbf{v} \times \mathbf{B}.$$

This equation represents the magnetic component of the Lorentz force. The term $q\mathbf{v}$ is equivalent to a current element $I\,d\boldsymbol{l}$. Therefore the magnetic force exerted upon a current element $I\,d\boldsymbol{l}$ is

$$d\mathbf{F}_B = I\,d\boldsymbol{l} \times \mathbf{B}.$$

The units of $\mathbf{B}$ are newton-seconds per coulomb-meter and these define the units of the magnetic induction vector as weber/m^2,

$$1\,\frac{\text{newton-second}}{\text{coulomb-meter}} = 1\,\frac{\text{weber}}{(\text{meter})^2}.$$

B. THE LORENTZ FORCE

In general, a region of space in which a point charge q moves can contain both an external electric field $\mathscr{E}$ and an external magnetic field $\mathbf{B}$. An external field is assumed to be unaffected by the presence of the test charges. When both $\mathbf{E}$ and $\mathbf{B}$ are present at the position of the moving charge q, there is a vector force exerted upon q called the Lorentz force $\mathbf{F}$, where

$$\mathbf{F} = q(\mathscr{E} + \mathbf{v} \times \mathbf{B}).$$

There is no similar equation for a current element in a conducting medium because the net charge of the element is ordinarily zero (each volume element always carries equal numbers of positive and negative charges) and the interaction with the electric field can be neglected.

The total energy, kinetic plus potential, of a charge q moving in a magnetic field $\mathbf{B}$ is conserved. We employ conservative electric fields $\mathscr{E}$ such that $\mathscr{E} = -\text{grad } V$. Starting with Newton's second law, we multiply the Lorentz force from the left by $\mathbf{v} = d\mathbf{r}/dt$,

$$\frac{d}{dt}(m\mathbf{v}) = \mathbf{F} = q(\mathscr{E} + \mathbf{v} \times \mathbf{B})$$

where m is the mass of the particle. Then

$$\mathbf{v} \cdot \frac{d}{dt}(\mathbf{p}) = \frac{d}{dt}(T) = q\mathbf{v} \cdot \{-\text{grad } V + \mathbf{v} \times \mathbf{B}\}.$$

Now $$T = \tfrac{1}{2}mv^2 = p^2/2m$$

and $$(\text{grad } V) \cdot \mathbf{v} = \frac{dV}{dt}; \quad (V \neq V(t)).$$

Finally $$\mathbf{v} \cdot (\mathbf{v} \times \mathbf{B}) = 0;$$

thus $$\frac{d}{dt}\{T + V\} = 0,$$

and $$T + V = E = \text{a constant}.$$

This result could have been predicted from our initial definition of **B**. Since the force is perpendicular to the velocity, the power $\mathbf{v} \cdot \mathbf{F}_B$ is zero. Therefore the magnetic component of the force will not change the total energy of the particle.

We should indicate that the relativistic energy is also conserved. The momentum is $\mathbf{p} = m_0\mathbf{v}/\sqrt{1 - \beta^2}$ where $\beta = v/c$, and

$$\mathbf{v} \cdot \frac{d}{dt}\left\{\frac{m_0\mathbf{v}}{\sqrt{1 - \beta^2}}\right\} = q\mathbf{v} \cdot (\mathbf{v} \times \mathbf{B}) = 0;$$

with the result that

$$\mathbf{v} \cdot \frac{d}{dt}\left\{\frac{m_0\mathbf{v}}{\sqrt{1 - \beta^2}}\right\} = \frac{d}{dt}\left\{\frac{m_0c^2}{\sqrt{1 - \beta^2}}\right\} = 0,$$

thus $$E = \frac{m_0c^2}{\sqrt{1 - \beta^2}} = \text{a constant}.$$

C. THE MOTION OF CHARGED PARTICLES

1. MOTION IN A CONSTANT MAGNETIC FIELD

Consider a relativistic point charge q of mass m_0 moving in a constant magnetic field **B**. Take the direction of **B** along the x_3 (z) axis,

$$\mathbf{B} = B_0\mathbf{k}.$$

At any instant the position of q is given by **r** and the velocity is $\mathbf{v} = d\mathbf{r}/dt$. Designate the momentum as

$$\mathbf{p} = m_0\mathbf{v}/\sqrt{1 - \beta^2} \quad \text{where} \quad \beta^2 = v^2/c^2;$$

then $$\frac{d}{dt}\{\mathbf{p}\} = q\mathbf{v} \times \mathbf{B}.$$

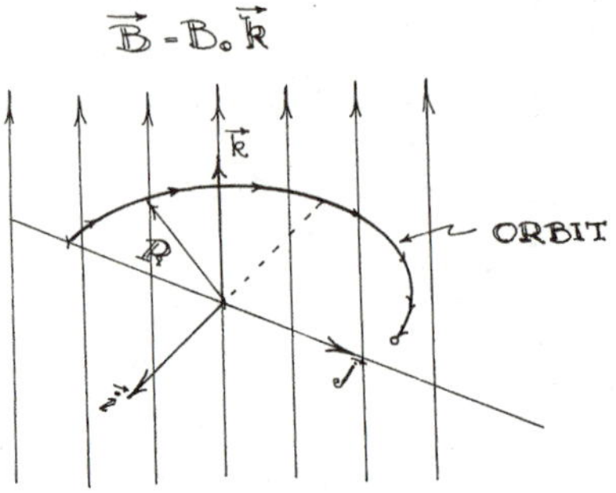

We have taken **B** as a constant; therefore

$$\mathbf{v} \times \mathbf{B} = \frac{d}{dt}(\mathbf{r} \times \mathbf{B}).$$

Transposing all terms in the Second Law to the left,

$$\frac{d}{dt}\{p - q\mathbf{r} \times \mathbf{B}\} = 0,$$

and

$$\mathbf{p} - q\mathbf{r} \times \mathbf{B} = \mathbf{P}.$$

The vector **P** is a constant of the motion and is the conjugate momentum appearing in the Hamiltonian of classical mechanics.

We have already shown that $|\mathbf{v}|^2$ is a constant. Therefore in both relativistic and nonrelativistic problems

$$|\mathbf{p}|^2 = \text{const.}$$

Since $\mathbf{r} \times \mathbf{B}$ is in the x_1, x_2 plane, $p_3 = P_3 = \text{const.}$ and can be extracted from the vector equation. Then

$$p_1^2 + p_2^2 = \text{const.} = [(q\mathbf{r} \times \mathbf{B}) + P_1\mathbf{i} + P_2\mathbf{j}]^2.$$

The left side of this equation is a statement that the magnitude of the projection of **p** on the x_1, x_2 plane is a constant. The right side shows that the orbit is a circle of radius R. Expanding the cross product and squaring, we find

$$\left(x_1 - \frac{P_2}{qB_0}\right)^2 + \left(x_2 + \frac{P_1}{qB_0}\right)^2 = R^2 = \left\{\frac{p_1^2 + p_2^2}{q^2B_0^2}\right\}.$$

This result is usually given in an abbreviated form. In many instances only the x_1, x_2 motion is considered and the result is written

$$p_\perp = \sqrt{p_1^2 + p_2^2} = qBR.$$

The angular velocity divided by 2π of the particle about the center of the circle is called the cyclotron frequency $\omega_c/2\pi$.

$$R\omega_c = v_{tan} = \frac{p_\perp}{m_0}\sqrt{1-\beta^2} = \frac{qBR}{m_0}\sqrt{1-\beta^2}.$$

When $\beta \ll 1$ (nonrelativistic), ω_c is essentially independent of β. As $\beta \to 1$, ω_c is a rapidly varying function of β;

$$\omega_c = \frac{qB_0}{m_0}\sqrt{1-\beta^2}.$$

The modifications of cyclotrons for proton energies above about 20 Mev are concerned with the problem of increasing $|\mathbf{B}|$ as the energy

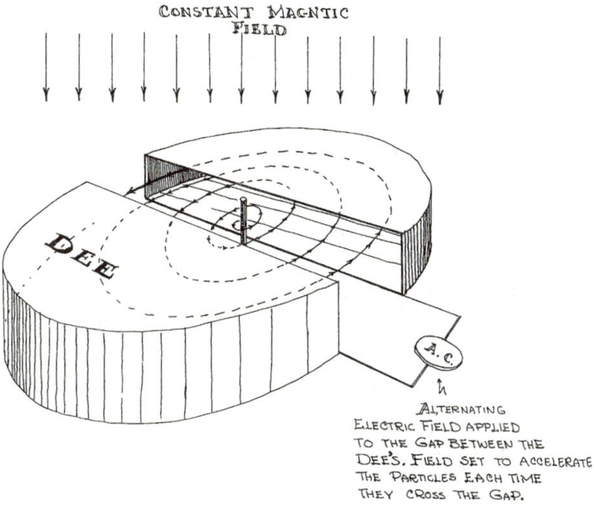
CONSTANT MAGNTIC FIELD
DEE
A.C.
ALTERNATING ELECTRIC FIELD APPLIED TO THE GAP BETWEEN THE DEE'S. FIELD SET TO ACCELERATE THE PARTICLES EACH TIME THEY CROSS THE GAP.

increases or with the possibility of decreasing the working frequency as the particle moves away from the center to correspond to the ω_c of the orbit.

2. MAGNETIC SPECTROMETERS

Many practical applications of the motion of charged particles in a magnetic field are available in the literature. The cyclotron has been mentioned briefly. The analysis of the momentum distributions of charged particles in a magnetic spectrometer serves as a convenient example of the application of the development presented in the preceding section.

There exists a large variety of magnetic spectrometers. Of these the 180-degree, constant field spectrometer is perhaps the simplest.

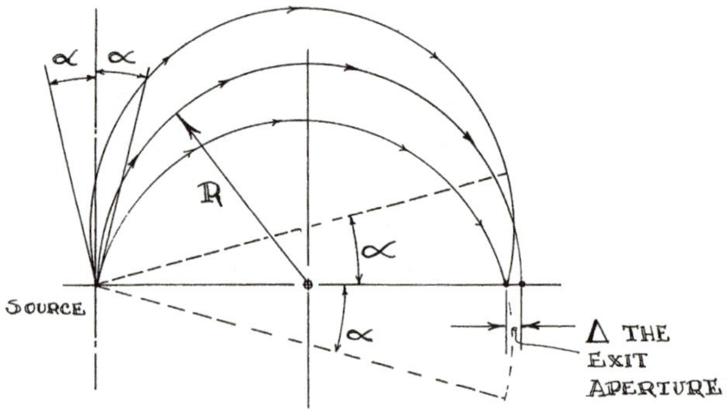

180° FOCUSING

A diagram of the spectrometer orbit is shown. This type of instrument selects particles initially moving in the orbital plane. In other words the **B** component of velocity of the entering particles must be small compared to the magnitude of **v**. If **B** is taken in the x_3 direction, the orbit is then approximately in the x_1,x_2 plane. The condition on v_3 is then

$$v_3 \ll |\mathbf{v}|.$$

The type of instrument shown is a constant radius spectrometer where the radius is R. The time which the particle spends in the 180° orbit after injection is approximately π/ω_c where ω_c is the 2π times the cyclotron frequency. In this time the particle moves a distance $(\pi/\omega_c)v_3$ along the x_3 axis. The maximum value of v_3 allowed is then determined by the height (in the **B** direction) of the *exit* aperture. Ordinarily the entering particles have an angular spread in the orbital plane of $\pm\alpha$ about the normal to the entrance window. In addition, the width of the window, δ, adds a spread to the exit beam.

The angular variation $\pm\alpha$ of the entering velocities causes a spread Δ in the orbits at the exit window

$$\Delta = 2R - 2R\cos\alpha = 2R(1 - \cos\alpha).$$

The total geometric spread at the exit window (in the x_1,x_2 plane) is $\Delta + \delta$. If the magnetic field is set to focus a central ray of momentum p where $p = qRB_0$, the exit window accepts a distribution of momenta Δp

$$\Delta p = qB_0\,\Delta R = qB_0\,\frac{(\Delta + \delta)}{2},$$

and $$\Delta p = qB_0\left\{R(1 - \cos\alpha) + \frac{\delta}{2}\right\};$$

whence $$\frac{\Delta p}{p} = \left\{(1 - \cos\alpha) + \frac{\delta}{2R}\right\}.$$

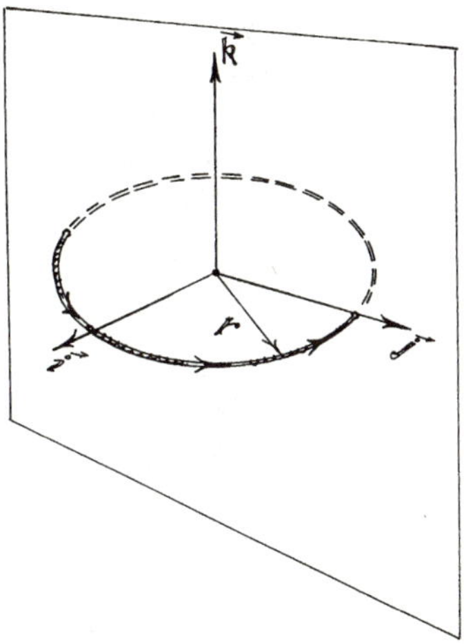

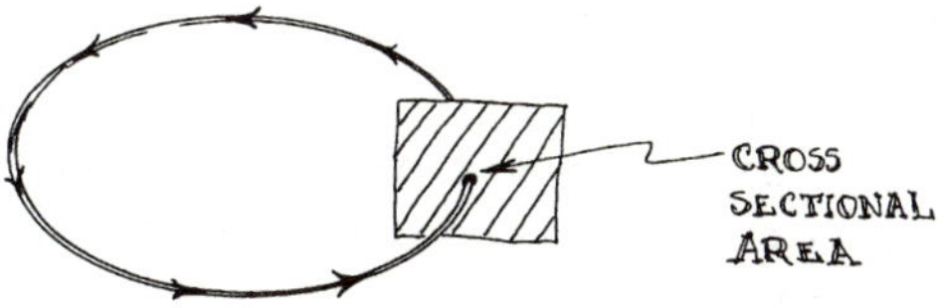

D. FORCES AND TORQUES ON CURRENT LOOPS

1. POTENTIAL ENERGY AND TRANSLATIONAL FORCE

Conceptually the current loop is a highly idealized system composed of a single filament of current. This filament is a closed curved line of zero cross-sectional area, with no breaks or sources of current included. In other words it is assumed to be a closed line of steady state current I. The current density is analogous to the Dirac delta function $\delta(\mathbf{r} - \mathbf{r}')$ associated with the volume density of charge for a point charge (see the figure at the bottom of page 185).

The current density $\mathcal{J}_f$ in a filament has an undefined magnitude. However, when one integrates across a small area element pierced by a filament, the total current is I.

$$\iint_{\text{cross section}} \mathcal{J}_f \cdot \mathbf{n}\, dA = I.$$

If the total current carried by a filament is an infinitesimal, then the current carried in a macroscopic wire is composed of a sum of the filaments. As in the case of the point charge problem the self-energy of an infinitesimal filament is defined as zero.

At present we can regard the filament merely as a very thin loop

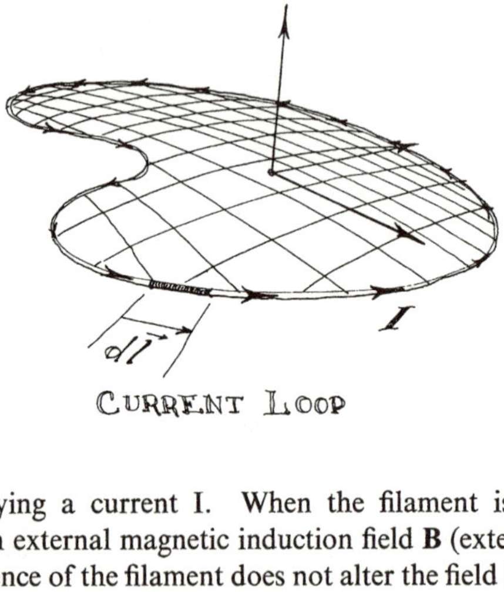

CURRENT LOOP

of wire carrying a current I. When the filament is in a region containing an external magnetic induction field **B** (external implying that the presence of the filament does not alter the field **B**), then every element of length d of the filament experiences a Lorentz force **dF** upon it,

$$\mathbf{dF} = \mathrm{I}\, \mathrm{d}\boldsymbol{l} \times \mathbf{B}$$

where **B** is specified at the position of d$\boldsymbol{l}$.

The total force exerted upon the loop is

$$\mathbf{F} = \oint_{\text{Loop}} \mathbf{dF} = \mathrm{I} \oint_{\text{Loop}} \mathrm{d}\boldsymbol{l} \times \mathbf{B}$$

A convenient method for finding **F** is to utilize the potential energy of the loop in the field.

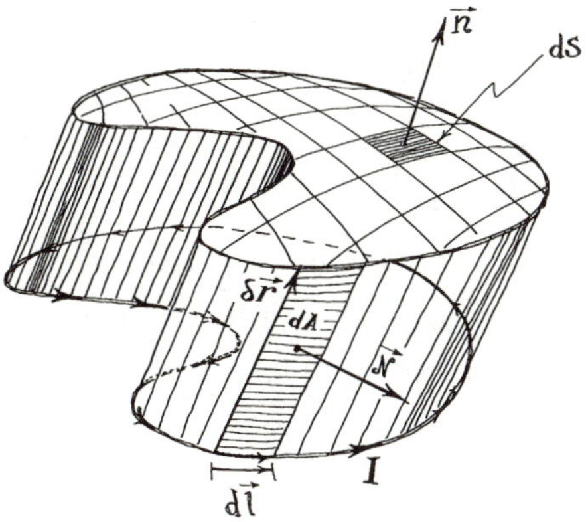

Vary the position of the loop by $\delta\mathbf{r}$; i.e., shift every d$\boldsymbol{l}$ upwards by $\delta\mathbf{r}$. The corresponding variation of energy δU is then

$$\begin{aligned}\delta \mathrm{U} &= -\oint_{\text{Loop}} \mathbf{dF} \cdot \delta\mathbf{r} \\ &= -\oint_{\text{Loop}} \mathrm{I}(\mathbf{dl} \times \mathbf{B}) \cdot \delta\mathbf{r} = +\mathrm{I} \oint_{\text{Loop}} (\mathbf{dl} \times \mathbf{r}) \cdot \mathbf{B}.\end{aligned}$$

When the loop is displaced by $\delta\mathbf{r}$, a volume is swept out. This volume has a ribbon-like side area made up of elements $(\mathbf{dl} \times \delta\mathbf{r})$;

$$(\mathbf{dl} \times \delta\mathbf{r}) = \mathcal{N}\, \mathrm{dA},$$

where $\mathcal{N}$ is a unit normal for the element of area on the side. Thus

$$\delta \mathrm{U} = \mathrm{I} \iint_{\text{side area}} \mathbf{B} \cdot \mathcal{N}\, \mathrm{dA}.$$

The integral of **B** over a surface area ΔA is a measure of the net magnetic flux linking that area,

$$\Phi_{\Delta A} = \iint_{\Delta A} \mathbf{B} \cdot \mathcal{N} \, dA.$$

The flux linking the cap surface S (i.e., the face area as opposed to the side area) of the filament loop is

$$\Phi_S = \iint_S \mathbf{B} \cdot \mathbf{n} \, dS.$$

As the position of the loop is varied, the flux Φ_S linking the cap surface is varied. The net flux passing through the side area represents the change $\delta\Phi_S$ in the flux linking S(cap). This assumes that there are no sources or sinks of **B** in the volume swept out, that is, that div **B** = 0 in this region. Later in the consideration of the sources of **B** it will be shown that in general **B** is a solenoidal field, and

$$\text{div}\,\mathbf{B} = 0, \quad \text{everywhere.}$$

Using the solenoidal property of **B** one can demonstrate analytically that the net flux penetrating the side area is the variation in Φ_S.

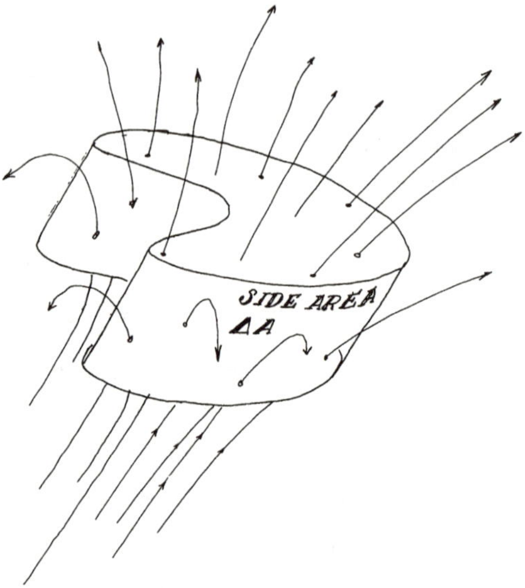

If div $\mathbf{B} = 0$, then

$$\iiint\limits_{\substack{\text{Volume}\\\text{swept out}}} \operatorname{div} \mathbf{B}\, d\tau = \iint\limits_{S(\delta r)} \mathbf{B}\cdot\mathbf{n}\, dS + \iint\limits_{\Delta A} \mathbf{B}\cdot\mathcal{N}\, dA - \iint\limits_{S(0)} \mathbf{B}\cdot\mathbf{n}\, dS$$

$$= 0 = \Phi_S(\delta r) - \Phi_S(0) + \Phi_{\Delta A}.$$

Now the variation in flux linking the loop is

$$\delta\Phi_S = \Phi_S(\delta r) - \Phi_S(0) = -\Phi_{\Delta A}.$$

Because the variation of the energy is proportional to the variation of flux linkage,

$$\delta U = I \iint\limits_{\Delta A} \mathbf{B}\cdot\mathcal{N}\, dA = -I\,\delta\Phi,$$

it follows that

$$U = -I\Phi_S = -I \iint\limits_{S(\text{cap})} \mathbf{B}\cdot\mathbf{n}\, dS.$$

If **B** is constant over S(cap) it can be removed from the integral, and

$$U = -\left\{ I \iint\limits_{S} \mathbf{n}\, dS \right\}\cdot\mathbf{B}.$$

From this result we define the *magnetic moment* **M** of a current loop as the current times the projection of the cap surface area. Here the projection of the cap surface means the surface integral of **n**. We can then define the magnetic moment as

$$\mathbf{M} = I \iint\limits_{S} \mathbf{n}\, dS.$$

If the loop lies in a plane and the area of the loop is S, then

$$\mathbf{M} = IS\mathbf{n}.$$

Having defined **M** we find that

$$U = -\mathbf{M}\cdot\mathbf{B},$$

again with the condition that **B** must be essentially constant over **S**. *Otherwise the more general expression must be used.*

The translational force on the loop, or on the magnetic moment, is then

$$\mathbf{F} = -\text{grad}\, U = \text{grad}\,(\mathbf{M}\cdot\mathbf{B}).$$

2. TORQUE

The force exerted upon a current element of the loop under consideration is

$$d\mathbf{F} = I\, d\boldsymbol{l} \times \mathbf{B}.$$

Let $\mathbf{r}$ be the position vector of $d\boldsymbol{l}$ with respect to an origin O; then the torque about O is

$$d\mathscr{T} = \mathbf{r} \times d\mathbf{F} = \mathbf{r} \times I(d\boldsymbol{l} \times \mathbf{B})$$
$$= I\{d\boldsymbol{l}(\mathbf{r} \cdot \mathbf{B}) - \mathbf{B} \cdot (\mathbf{r} \cdot d\boldsymbol{l})\}.$$

The total torque on the loop is

$$\mathscr{T} = I \oint_{\text{Loop}} d\boldsymbol{l}(\mathbf{r} \cdot \mathbf{B}) - I \oint_{\text{Loop}} \mathbf{B} \cdot (\mathbf{r} \cdot d\boldsymbol{l}).$$

The second term represents the torque relative to the origin. This torque component results from the translational force. If $\mathbf{B}$ is constant over the cap surface this term can be neglected.

The first integral transforms to a surface integral (see Appendix A):

$$I \oint d\boldsymbol{l}(\mathbf{r} \cdot \mathbf{B}) = -I \iint_{S(\text{cap})} \operatorname{grad}(\mathbf{r} \cdot \mathbf{B}) \times \mathbf{n}\, dS.$$

Now

$$\operatorname{grad}(\mathbf{r} \cdot \mathbf{B}) = (\mathbf{r} \cdot \nabla)\mathbf{B} + (\mathbf{B} \cdot \nabla)\mathbf{r} + \mathbf{r} \times (\nabla \times \mathbf{B}) + \mathbf{B} \times (\nabla \times \mathbf{r}) = \mathbf{B}.$$

Finally, considering only the torque about the centroid of the loop,

$$\mathscr{T} = -I \iint_{S(\text{cap})} \mathbf{B} \times \mathbf{n}\, dS \rightarrow \left\{I \iint_{S} \mathbf{n}\, dS\right\} \times \mathbf{B} = \mathbf{M} \times \mathbf{B}.$$

The last expression on the right can be employed when $\mathbf{B}$ is constant over S.

3. EXAMPLES

a. RECTANGULAR LOOP IN A CONSTANT FIELD $\mathbf{B}$. In the simplified geometry shown, it is apparent from the definition of the Lorentz force that

$$|\mathbf{F}_1| = I l B,$$

and

$$|\mathbf{F}_2| = -I l B.$$

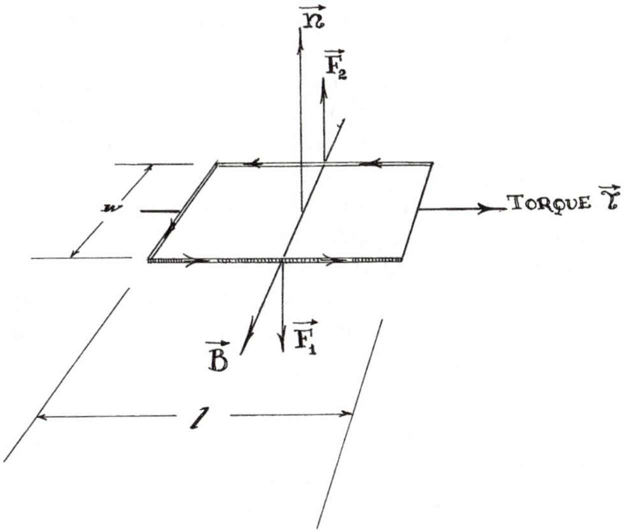

The magnitude of the torque on this system is

$$|\mathscr{T}| = \frac{w}{2}F_1 - \frac{w}{2}F_2 = I(lw)B = MB.$$

With the geometry shown, **n** and **B** are mutually perpendicular; thus our result is equivalent to $\mathscr{T} = \mathbf{M} \times \mathbf{B}$.

b. TORQUE ON A CIRCULAR CURRENT LOOP

$$\mathbf{R} = R(\cos\phi\mathbf{i} + \sin\phi\mathbf{j})$$

$$d\mathbf{R} = R(-\sin\phi\mathbf{i} + \cos\phi\mathbf{j})\,d\phi$$

$$\mathbf{B} = \sum_{j=1}^{3} B_j\boldsymbol{\epsilon}_j = \text{const.}$$

About an origin at the center,

$$\mathscr{T} = \oint_{\text{Loop}} \mathbf{R} \times d\mathbf{F} = \oint \mathbf{R} \times \{I\,d\mathbf{R} \times \mathbf{B}\}$$

$$= I(\pi R^2)\{-B_2\mathbf{i} + B_1\mathbf{j}\} = \mathbf{M} \times \mathbf{B}.$$

c. MOTION OF A MAGNETIC MOMENT IN AN INHOMOGENEOUS FIELD. Atomic systems may have an over-all magnetic moment. One of the original experiments which indicated the validity of the quantum theory consisted of a measurement of the deflection of an atomic magnetic moment in an inhomogeneous magnetic field.

A narrow beam of atoms is formed moving at a constant velocity $\mathbf{v} = v_0\mathbf{j}$ along the y axis. (This is an idealization because in an actual experiment the magnitude of the velocity of the beam ordinarily has a distribution about a mean value.) The beam is passed along the y axis in the symmetry plane of the inhomogeneous field. Let the field possess a z dependence only in this plane. Thus in the symmetry plane containing the beam axis, $\mathbf{B} = B_3(z)\mathbf{k}$.

If the magnetic moments $\mathbf{M}$ have a classical distribution of orientations relative to the z axis, in the beam the number of moments with orientations between θ and $\theta + d\theta$ is

$$dn(\theta) = N_0 e^{-MB\cos\theta/kT}\, 2\pi \sin\theta\, d\theta.$$

The result of such a distribution would give a continuous distribution of deflections in the z direction in the outgoing beam.

The force on a given magnetic moment is

$$\mathbf{F} = \text{grad}\,(\mathbf{M}\cdot\mathbf{B}) = M\cos\theta \left[\frac{\partial B_3(z)}{\partial z}\right]_{z=0} \mathbf{k}.$$

There are no changes in the orientations of the moments; therefore we assume that the quantity $M\cos\theta$ is constant. The inhomogeneity $B_3(z)$ can be constructed to be linear in z. This particular z dependence provides a constant force on the atom. If the mass of the particle is m,

$$m\frac{d^2z}{dt^2} = M\cos\theta\left[\frac{\partial B_3}{\partial z}\right].$$

Let $B_3 = B_0 z/L$; solving the differential equation above,

$$z(t) = \frac{1}{2}\left\{\frac{M\cos\theta}{m}\right\}\left[\frac{B_0}{L}\right]t^2,$$

where L is a constant having the units of a length. L describes the inhomogeneity of the field to first order.

After moving a distance Y in the inhomogeneous field, the deflection from the y axis is

$$z(Y) = \frac{1}{2}\left\{\frac{M\cos\theta}{m}\right\}\left[\frac{B_0}{L}\right]\frac{Y^2}{v_0^2}.$$

The probability of $M\cos\theta$ (or the relative number of atoms with an orientation between θ and $\theta + d\theta$) is $M\cos\theta\, dn(\theta)$ and the distribution in z would be obtained by integrating θ from 0 to π. We notice that variations in $\mathbf{v}_0$ also will give rise to further variations in the deflections z.

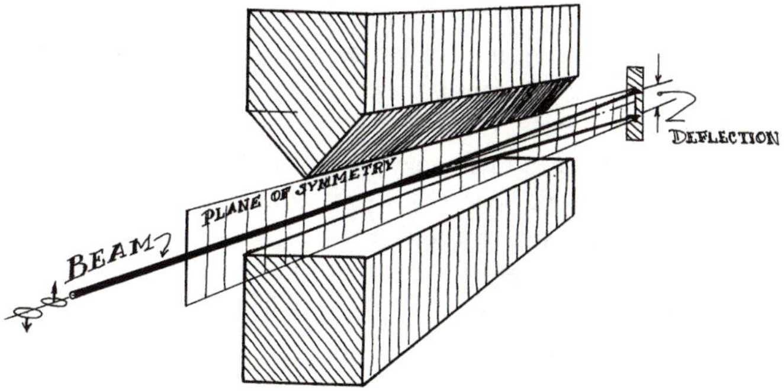

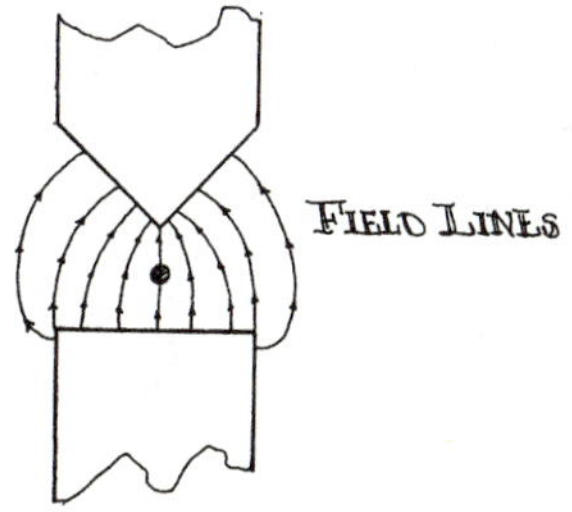

It is evident that a statistical distribution in orientations will produce a continuous distribution in the deflection z.

According to quantum theory, however, the magnetic moments are restricted to preferred orientations with respect to the z axis. The number of states of orientation is (2j + 1) where j is an integer or half integer. The angles of orientation are given by

$$\cos^{-1}\{m_j/\sqrt{j(j+1)}\}$$

where m_j can have any value from +j to −j in integral steps,

$$m_j = +j, j-1, j-2, \ldots, -j.$$

Since the orientations of the magnetic moments are quantized in this manner, the final deflected beam will have not a continuous distribution in the positive and negative z direction but rather (2j + 1) separate deflections given by

$$z(m_j) = \frac{1}{2}\left\{\frac{MB_0Y^2}{MLv_0^2}\right\}\frac{m_j}{\sqrt{j(j+1)}}.$$

The actual observation by Davisson and Germer of these discrete deflections in the case of silver atoms for which $j = \frac{1}{2}$ (giving two states) was one of the important verifications of the quantum description of the atom.

In actual practice such an experiment could be employed to measure the magnetic moment of an atom. An oven at absolute temperature T° is used to provide the atomic beam. The emitted beam is not monoenergetic but has a thermal distribution of velocities about a mean value $\overline{v_y^2}$, the most probable value of the square of the y component of velocity, and

$$\tfrac{1}{2}m\overline{v_y^2} = kT,$$

where k is the Boltzmann constant and T is the absolute temperature. This result together with the expression for the deflection in the case of $j = \frac{1}{2}$ gives the total separation between the maxima of the two lines as

$$\Delta = 2z(m_j = \tfrac{1}{2}) = \frac{M}{2}\left[\frac{\partial B}{\partial z}\right]\frac{Y^2}{kT}\frac{\frac{1}{2}}{\sqrt{\frac{3}{4}}},$$

or

$$M = \frac{2\sqrt{3}\,kT\,\Delta}{\left[\frac{\partial B}{\partial z}\right]Y^2}.$$

VII. The Magnetic Field of Steady Currents

A. AMPERE'S LAW

Early in the nineteenth century Oersted discovered that magnetic fields could be produced by electrical currents. Later the experimental work of Ampere supplied the specific relations between the magnetic fields and the currents.

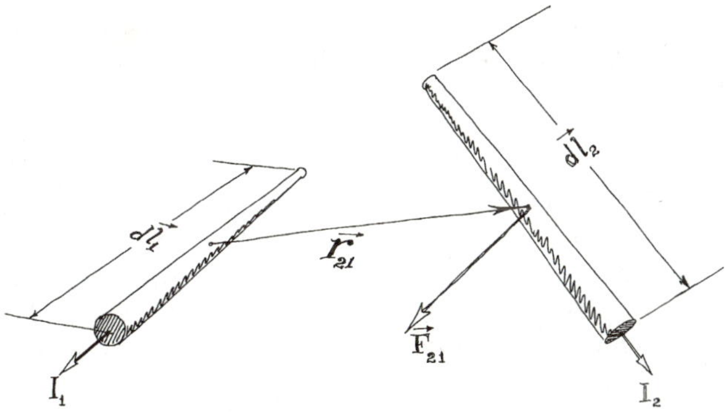

When two current-bearing elements $I_1\, dl_1$ and $I_2\, dl_2$ interact, the force exerted by element 1 upon element 2 is

$$\mathbf{dF}_{21} = \frac{\mu_0 I_2}{4\pi}\, dl_2 \times \left\{\frac{I_1\, dl_1 \times \mathbf{r}_{21}}{r_{21}^{\;3}}\right\}.$$

where $\mathbf{r}_{21} = \mathbf{r}_2 - \mathbf{r}_1$, $\mathbf{r}_j$ being the position vector of the jth element and μ_0 a constant of proportionality. From Newton's third law the force exerted by element 2 upon 1 is equal in magnitude to but oppositely directed from $d\mathbf{F}_{21}$,

$$d\mathbf{F}_{12} = -d\mathbf{F}_{21} = \frac{\mu_0}{4\pi} I_1\, dl_1 \times \left\{\frac{I_2\, dl_2 \times \mathbf{r}_{12}}{r_{12}^3}\right\}.$$

Because of the form of the Lorentz force, these equations define the magnetic fields set up by the two current elements. The magnetic field set up by $I_1\, dl_1$ at the position of element 2 is

$$d\mathbf{B}_{21} = \frac{\mu_0}{4\pi} I_1 \left\{\frac{dl_1 \times \mathbf{r}_{21}}{\mathbf{r}_{21}^3}\right\},$$

and $d\mathbf{B}_{12}$ is obtained by interchanging the indices.

The force law giving the interaction of the two current elements is a generalization of Ampere's experiment and is known as Ampere's law.

The constant μ_0 is the *magnetic permeability* of free space, established by definition as

$$\mu_0 = 4\pi(10)^{-7} \text{ henry/meter.}$$

Once defined, this constant serves as a determining factor in the definition of the absolute ampere. In practice the absolute ampere can be determined from a current balance in which the forces of attraction or repulsion of two circular coils are measured mechanically.

For convenience, we shall idealize this experiment and define the ampere in terms of the mutual force set up between two straight parallel wires one meter in length, separated a distance of one meter, and carrying currents of equal magnitude. If the current carried is one ampere, the magnitude of the force (attractive or repulsive) is $\mu_0/4\pi$ newtons. The unit of henries per meter is then newtons per (ampere)^2.

In Chapter II it was indicated that μ_0 is arbitrarily defined as $4\pi(10)^{-7}$ mks units. Then ϵ_0 is determined from the relation

$$\epsilon_0 = \frac{1}{\mu_0 c^2}$$

where c = the velocity of light in a vacuum in meters per second.

Having determined that the magnetic induction field $d\mathbf{B}_p$ set up at any point P by a current element I d$\boldsymbol{l}$ is

$$d\mathbf{B}_p = \frac{\mu_0 I}{4\pi}\left\{\frac{d\boldsymbol{l} \times (\mathbf{r}_p - \mathbf{r})}{|\mathbf{r}_p - \mathbf{r}|^3}\right\},$$

where **r** is the position vector of d$\boldsymbol{l}$, we can compute the total magnetic field at P by integrating over the closed current loop,

$$\mathbf{B}_p = \frac{\mu_0 I}{4\pi}\oint_{\text{Loop}} \frac{d\boldsymbol{l} \times (\mathbf{r}_p - \mathbf{r})}{|\mathbf{r}_p - \mathbf{r}|^3}.$$

B. THE SECOND MAXWELL EQUATION

The magnetic induction field can be *postulated* as a solenoidal field; however, Ampere's law serves actually to demonstrate that the fields set up by currents are solenoidal. If one examines the field of a long straight current-carrying wire, it is apparent that the lines of **B** are circles lying in a plane perpendicular to the wire and having their centers at the position of the wire. The lines do not terminate upon sources or sinks, but rather they form closed loops.

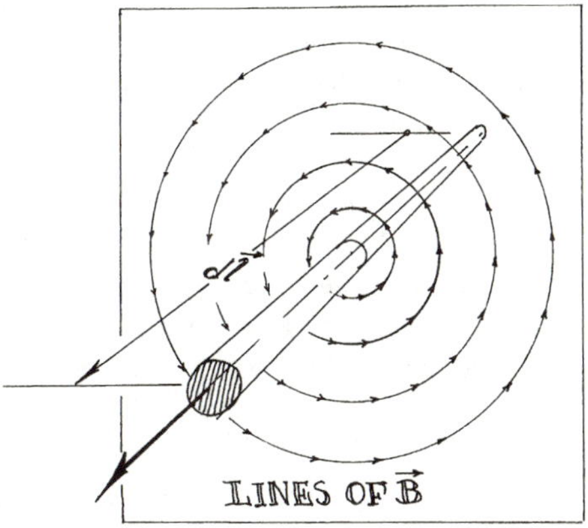

If a vector point function **B** can be generated by taking the curl of another vector point function **A**, then the generated function is solenoidal since the divergence of the curl of a vector is always zero. In other words, if

$$\mathbf{B} = \text{curl}\,\mathbf{A},$$

then

$$\text{div}\,\mathbf{B} = \text{div curl}\,\mathbf{A} = 0.$$

As a result, if we can find a vector point function $\mathbf{A}_p$ which satisfies the relation $B_p = \text{curl}_p\,\mathbf{A}_p$, then the $\mathbf{B}_p$ vector function is solenoidal. The reader will notice that the subscript p is emphasized because the curl is composed of variations of the coordinates of the point P at which the field is being examined, and the curl does not operate upon the coordinates of the current elements in the loop $\mathscr{L}$.

$$\text{curl}_p\left\{\frac{d\boldsymbol{l}}{|\mathbf{r}_p - \mathbf{r}|}\right\} = \text{grad}_p\left\{\frac{1}{|\mathbf{r}_p - \mathbf{r}|}\right\} \times d\boldsymbol{l} = \frac{d\boldsymbol{l} \times (\mathbf{r}_p - \mathbf{r})}{|\mathbf{r}_p - \mathbf{r}|^3}.$$

Therefore

$$\text{curl}_p \oint_{\mathscr{L}} \frac{\mu_0 I}{4\pi} \frac{d\boldsymbol{l}}{|\mathbf{r}_p - \mathbf{r}|} = \frac{\mu_0 I}{4\pi} \oint_{\mathscr{L}} \frac{d\boldsymbol{l} \times (\mathbf{r}_p - \mathbf{r})}{|\mathbf{r}_p - \mathbf{r}|^3}.$$

This demonstrates that the magnetic field of a current I is solenoidal.

We define the VECTOR MAGNETIC POTENTIAL $\mathbf{A}_p$ by the integral

$$\mathbf{A}_p = \frac{\mu_0 I}{4\pi} \oint_{\mathscr{L}} \frac{d\boldsymbol{l}}{|\mathbf{r}_p - \mathbf{r}|}.$$

The magnetic induction vector $\mathbf{B}_p$ then is obtained from

$$\mathbf{B}_p = \text{curl}_p\,\mathbf{A}_p.$$

This vector potential has the same usefulness in magnetic calculations as the scalar potential in electrostatics. In many problems it is much more convenient to compute $\mathbf{A}_p$ for a specific geometry $\mathscr{L}$ and then to generate $\mathbf{B}_p$ by application of the curl operator.

To review this section, we find that because $\mathbf{B}_p = \text{curl}\,\mathbf{A}_p$,

$$\text{div}_p\,\mathbf{B}_p = 0.$$

This expression is the second Maxwell equation, and it is a mathematical statement that $\mathbf{B}_p$ is solenoidal. The first Maxwell equation is the differential form of Gauss's law (div $\mathbf{D} = \rho$), representing a statement concerning the irrotational form of $\mathbf{D}$ in the presence of free or mobile charges.

The integral form of the second equation is a surface integral.

Using the divergence theorem,

$$\iint_{S(\text{closed})} \mathbf{B} \cdot \mathbf{n}\, dS = 0.$$

Because the flux is defined as $\Phi = \iint_S \mathbf{B} \cdot \mathbf{n}\, dS$, the integral form of the second Maxwell equation demonstrates that the net magnetic flux crossing the boundary of any closed surface is zero.

C. EXTENDED CURRENT DISTRIBUTIONS

1. THE MAGNETIC FIELD IN TERMS OF THE CURRENT DENSITY

Construct a current-carrying conductor of nonzero cross section.

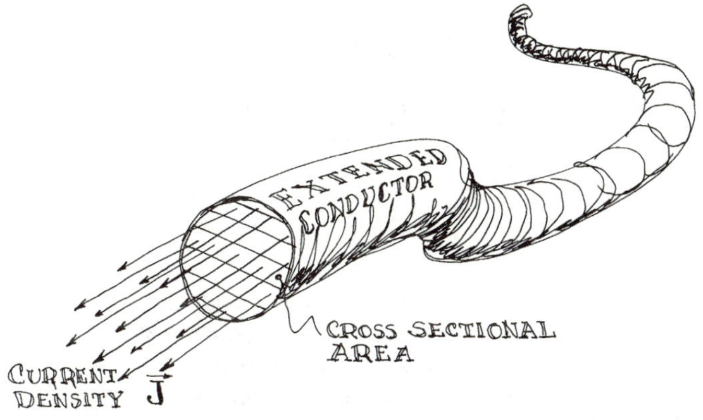

If the loop is closed, the total current I can be thought of as being made up of a sum of infinitesimal filaments of currents dI:

$$I_{\text{total}} = \iint_{\substack{\text{cross}\\ \text{section}}} \mathbf{J} \cdot \mathbf{n}\, dS.$$

The current density may now have a nonuniform distribution within the conductor. Under such conditions the loop integral

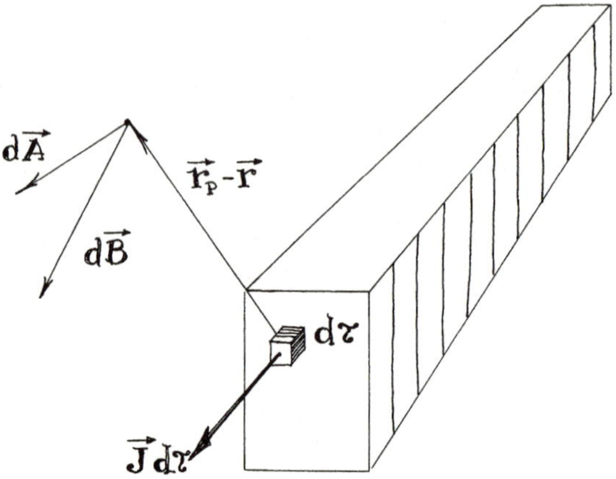

becomes a volume integral over the conducting volume τ_c containing the current density **J**. Let

$$I\,dl \rightarrow \mathbf{J}\,d\tau;$$

thus

$$\mathbf{A}_p = \frac{\mu_0}{4\pi}\iiint_{\tau_c} \frac{\mathbf{J}\,d\tau}{|\mathbf{r}_p - \mathbf{r}|},$$

and

$$\mathbf{B}_p = \frac{\mu_0}{4\pi}\iiint_{\tau_c} \frac{\mathbf{J}\times(\mathbf{r}_p - \mathbf{r})}{|\mathbf{r}_p - \mathbf{r}|^3}\,d\tau.$$

2. EXAMPLE CALCULATIONS

a. THE MAGNETIC INDUCTION FIELD OF A LONG STRAIGHT CURRENT FILAMENT. Consider the straight-line current filament I extending along the z axis from $-L/2$ to $+L/2$ and carrying current in the positive z direction. A current element I dz**k** at $\mathbf{r} = z\mathbf{k}$ contributes an infinitesimal field component

$$d\mathbf{B}_p = \frac{\mu_0}{4\pi}\frac{I(dz\mathbf{k})\times(\mathbf{r}_p - z\mathbf{k})}{|\mathbf{r}_p - z\mathbf{k}|^3}.$$

Let $L \rightarrow \infty$; then the magnetic induction field at P is

$$\mathbf{B}_p = \int_{-\infty}^{\infty} d\mathbf{B}_p = \frac{\mu_0 I}{2\pi}\frac{[-y_p\mathbf{i} + x_p\mathbf{j}]}{(x_p^2 + y_p^2)},$$

or

$$\mathbf{B}_p = \frac{\mu_0 I}{2\pi}\frac{(\mathbf{k}\times\mathbf{a}_p)}{a_p^2}.$$

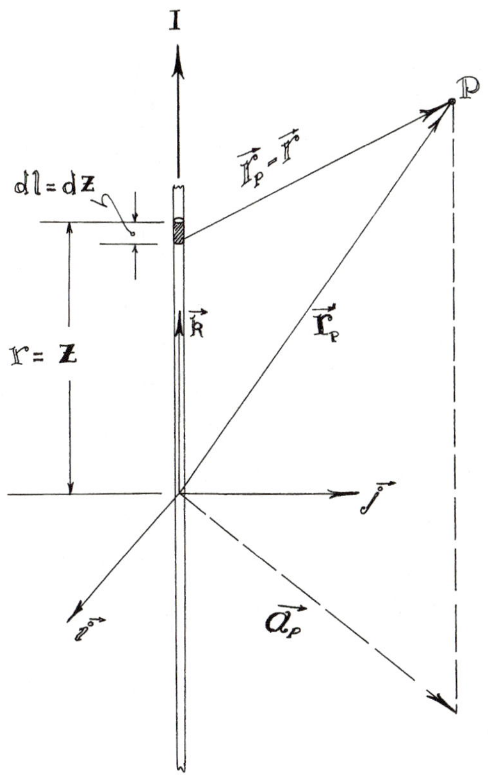

The corresponding vector magnetic potential is

$$\mathbf{A}_p = -\frac{\mu_0 I}{2\pi} \log\left(\frac{a_p}{C}\right)\mathbf{k}$$

where C is an arbitrary constant having the units of a length.

b. THE FIELD OF A CIRCULAR CURRENT LOOP. With the geometry shown in the diagram above,

$$\mathbf{R} = R(\cos\phi\mathbf{i} + \sin\theta\mathbf{j}),$$

and $$d\mathbf{R} = R(-\sin\theta\mathbf{i} + \cos\theta\mathbf{j})\, d\theta.$$

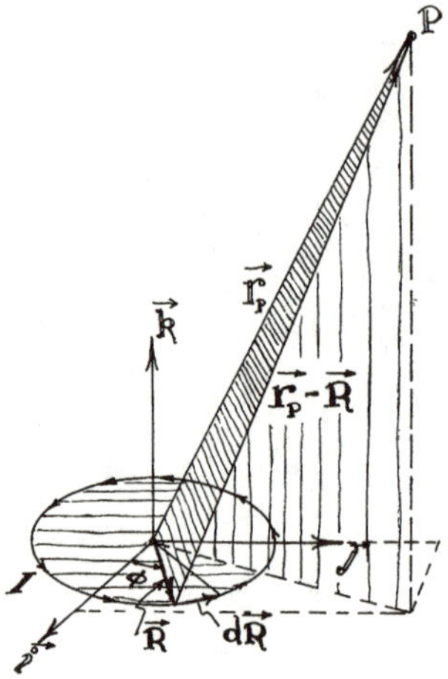

The vector magnetic potential is then

$$\mathbf{A}_p = \frac{\mu_0 I}{4\pi}\int_0^{2\pi} \frac{R(-\sin\theta\mathbf{i} + \cos\theta\mathbf{j})\,d\phi}{\{R^2 + r_p{}^2 - 2a_pR\cos(\phi - \zeta_p)\}^{1/2}},$$

where

$$a_p{}^2 = (x_p{}^2 + y_p{}^2)$$

and

$$\zeta_p = \tan^{-1}(y_p/x_p).$$

From an examination of the symmetry it is apparent that an evaluation of $\mathbf{A}_p$ at $\zeta_p = 0$ (or $y_p = 0$) will provide a result which

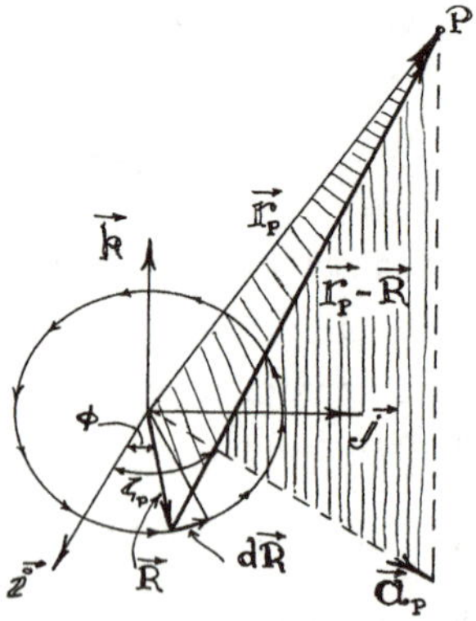

can be generalized for any ζ_p. Designate $\mathbf{A}_p(y_p = 0)$ as $\mathbf{A}_0$; then if*

$$k^2 = \frac{4Ra_p}{[(a_p + R)^2 + z_p{}^2]}$$

and $\quad |\mathbf{r}_p - \mathbf{R}| = \{(R + a_p)^2 + z_p{}^2\}^{1/2}(1 - k^2 \sin^2 \alpha)^{1/2}$,

where $\quad \sin^2 \alpha = \frac{1}{2}(1 + \cos \phi)$,

we can write the integral as

$$\mathbf{A}_0 = \frac{\mu_0 Ik}{4\pi}\left(\frac{R}{a_p}\right)^{1/2}\left\{\mathbf{i}\int_{+\pi/2}^{-\pi/2} \frac{2 \sin \alpha \cos \alpha \, d\alpha}{\sqrt{1 - k^2 \sin^2 \alpha}} - \mathbf{j}\int_{+\pi/2}^{-\pi/2} \frac{(2 \sin^2 \alpha - 1)}{\sqrt{1 - k^2 \sin^2 \alpha}}\, d\alpha\right\}.$$

The first integral (the coefficient of $\mathbf{i}$) is zero; therefore

$$\mathbf{A}_0 = \frac{\mu_0 Ik(R)^{1/2}}{2\pi(a_p)^{1/2}}\left\{\left(\frac{2}{k^2} - 1\right)K(k) - \frac{2}{k^2}E(k)\right\}\mathbf{j}.$$

Because $\mathbf{A}_0$ is evaluated at $y_p = 0$ we see that in general $\mathbf{A}_p$ will have a direction mutually perpendicular to $\mathbf{a}_p$ and $\boldsymbol{\epsilon}_3$, that is, the direction $\boldsymbol{\epsilon}_\phi = (\boldsymbol{\epsilon}_3 \times \mathbf{a}_p)/a_p$:

$$\mathbf{A}_p = \frac{\mu_0 Ik}{2\pi}\left(\frac{R}{a_p}\right)^{1/2}\left\{\left(\frac{2}{k^2} - 1\right)K(k) - \frac{2}{k^2}E(k)\right\}\frac{\boldsymbol{\epsilon}_3 \times \mathbf{a}_p}{a_p}.$$

K(k) and E(k) are the complete elliptic functions of the first and second kinds (see Appendix C).

Once the general form of $\mathbf{A}_p$ is at our disposal, the magnetic induction vector $\mathbf{B}_p$ can be obtained by taking the curl of $\mathbf{A}_p$. Remembering the techniques for differentiating K(k) and E(k) indicated in Section III-A-2-b, the reader should demonstrate as an exercise that in cylindrical coordinates

$$\mathbf{B}_p = \frac{\mu_0 Ik}{4\pi R^{1/2} a_p^{3/2}}\left\{-z_p\left[K(k) - \frac{1 - \dfrac{k^2}{2}}{1 - k^2}E(k)\right]\boldsymbol{\epsilon}_a + a_p\left[K(k) + \left(\frac{(R + a_p)k^2}{2a_p} - 1\right)\frac{E(k)}{(1 - k^2)}\right]\boldsymbol{\epsilon}_3\right\}.$$

At $z_p = 0$ or in the plane of the circle,

$$\mathbf{B}(z_p = 0) = \frac{\mu_0 I}{2\pi(R + a_p)}\left\{K(k_0) + \frac{(R + a_p)}{(R - a_p)}E(k_0)\right\}\boldsymbol{\epsilon}_3,$$

* Care must be exercised to differentiate between the function $k(a_p, z_p)$ and the base vector $\mathbf{k}$. $\boldsymbol{\epsilon}_3$ will be used whenever possible.

where $$k_0 = k(z_p = 0) = \sqrt{\frac{4Ra_p}{(R + a_p)}}\,.$$

The axial field ($a_p = 0$), which is the usual integral developed, is obtained directly remembering that

$$K(0) = \pi/2 \quad \text{and} \quad E(0) = \pi/2.$$

Therefore $$\mathbf{B}_{\text{axis}} = \frac{\mu_0 I}{2} \frac{R^2 \boldsymbol{\epsilon}_3}{(R^2 + z_p^2)^{3/2}}\,.$$

This last form can be obtained directly from the integral for **B** by allowing a_p to be zero.

C. THE FAR FIELD OF A CIRCULAR CURRENT LOOP. In the far field ($R \ll r_p$) the forms of $\mathbf{A}_p$ and $\mathbf{B}_p$ are obtained corresponding to a point magnetic moment $M = I(\pi R^2)\boldsymbol{\epsilon}_3$. The far field approximations can be developed from either the final expression for $\mathbf{A}_p$ in terms of K(k) and E(k), or as here, from the initial integral: Expanding

$$\mathbf{A}_p \xrightarrow[R \ll r_p]{} \frac{\mu_0 IR}{4\pi r_p} \int_0^{2\pi} (-\sin\phi\mathbf{i}\cos\phi\mathbf{j})\left\{1 + \left[\frac{Ra_p}{r_p^2}\right] \cos(\phi - \zeta_p)\right\} d\phi;$$

and $$\mathbf{A}_p \xrightarrow[R \ll r_p]{} \frac{\mu_0}{4\pi} \frac{\mathbf{M} \times \mathbf{r}_p}{r_p^3}\,.$$

Remembering that $\mathbf{M} = M\boldsymbol{\epsilon}_3$,

$$\mathbf{B}_p = \text{curl}\,\mathbf{A}_p \xrightarrow[R \ll r_p]{} \frac{\mu_0}{4\pi} \left\{\frac{3(\mathbf{M} \cdot \mathbf{r}_p)\mathbf{r}_p - r_p^2 \mathbf{M}}{r_p^5}\right\}.$$

This particular field form is familiar to us from the development of the field of a point electric dipole.

Some care must be exercised with this field. It can be generated by taking the curl of $\mathbf{A}_p$,

$$\mathbf{B}_p = \text{curl}_p \frac{\mu_0}{4\pi} \left\{\frac{\mathbf{M} \times \mathbf{r}_p}{r_p^3}\right\}.$$

Also, there exists a scalar potential function from which this $\mathbf{B}_p$ can be obtained:

$$\mathbf{B}_p = -\text{grad}_p \left\{\frac{\mu_0}{4\pi} \frac{\mathbf{M} \cdot \mathbf{r}_p}{r_p^3}\right\}.$$

This seems to indicate that the field $\mathbf{B}_p$ is both irrotational and solenoidal. This lack of uniqueness and pathological behavior is only apparent; one must keep in mind that both of these developments are for a far field region. This means that the near field behavior for nonzero configurations is specifically *not* given by this expression. In the case of point electric dipoles or magnetic moments the fields must be specified for *excluded sources and excluded sinks.* Unless the region about the point dipole is specifically excluded, the sources and sinks of the field lie at one point, and thus it is not clear whether the field is continuous or discontinuous at that point. Because of this lack of definition at the origin, the far field expression can be derived as a solenoidal or as an irrotational vector point function.

The scalar magnetic potential can be used if proper account is taken of the discontinuity at the source. Because $\mathbf{B}_p$ is solenoidal we shall prefer the vector point function.

Two magnetic moments 1 and 2 interact according to the relation already derived,

$$U_{21} = -\mathbf{M}_2 \cdot \mathbf{B}_{21} \quad \text{or} \quad U_{12} = -\mathbf{M}_1 \cdot \mathbf{B}_{12}.$$

Here $\mathbf{B}_{mn}$ is the magnetic field at the mth loop as established by the nth loop. Thus

$$U_{21} = U_{12} = -\frac{\mu_0}{4\pi}\left\{\frac{3(\mathbf{M}_2 \cdot \mathbf{r}_{21})(\mathbf{M}_1 \cdot \mathbf{r}_{21}) - r_{21}{}^2(\mathbf{M}_1 \cdot \mathbf{M}_2)}{r_{21}{}^5}\right\}.$$

If we assign one half the stored energy of a pair to each member of the pair, then for N magnetic moments

$$U = -\sum_{j=1}^{N} \tfrac{1}{2}\mathbf{M}_j \cdot \mathbf{B}_j$$

where $\mathbf{B}_j$ is the field at the jth magnetic moment set up by all of the other $(N - 1)$ moments.

Because permanent bar magnets behave like magnetic moments in the far field, the interaction of two bar magnets is described by this relation.

The potential energy is a minimum when $\mathbf{M}_1$ and $\mathbf{M}_2$ are antiparallel and when $\mathbf{r}$ lies in a plane perpendicular to the moments:

$$U = -\frac{\mu_0}{4\pi}\frac{M_1M_2}{r^3},$$

giving an attractive force.

The maximum potential energy occurs when the moments are parallel,

$$U = + \frac{\mu_0}{4\pi} \frac{M_1 M_2}{r^3},$$

corresponding to a repulsive force. Again the separation vector has been taken in a plane perpendicular to the direction of **M**.

The electrons in the 2p shell of hydrogen have an orbital angular momentum of $\hbar$* and therefore an orbital magnetic moment directed opposite to the angular momentum (because of negative charge of

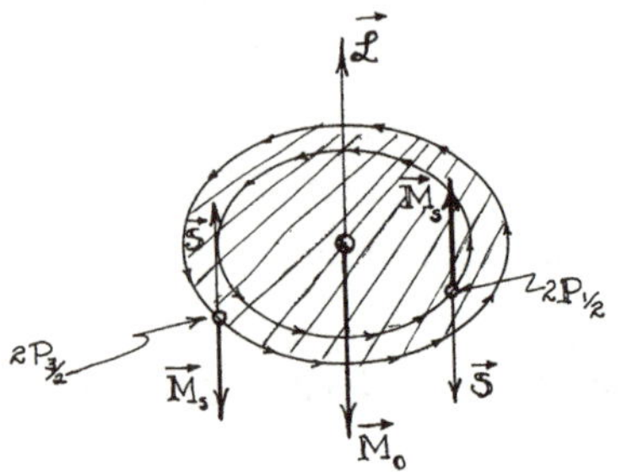

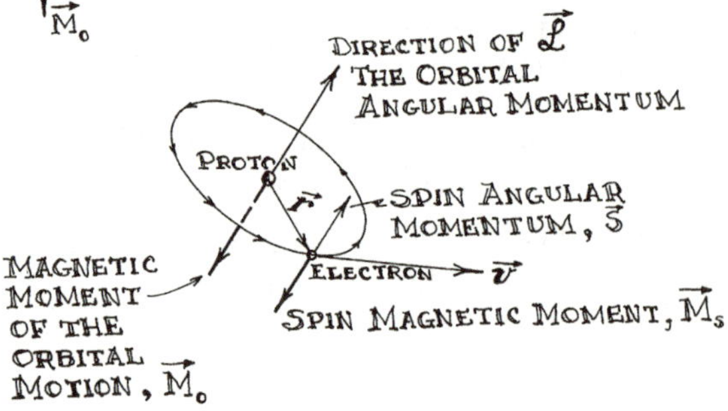

the electron). Besides the orbital magnetic moment, the electron has an intrinsic magnetic moment associated with its spin $\frac{1}{2}\hbar$. The intrinsic magnetic moment is directed opposite to the spin direction in a similar manner.

The 2p electrons have two possible total angular momenta: $\frac{3}{2}\hbar$ when the two magnetic moments associated with orbital and spin angular momenta are parallel, and $\frac{1}{2}\hbar$ when the orbital and spin magnetic moments are antiparallel.

* $\hbar$ is Planck's constant divided by 2π.

The most tightly bound 2p system is then the $2p_{1/2}$, while the energy of the $2p_{3/2}$ is the higher state.

d. THE FIELD OF A SOLENOID. A solenoid consists of an axially symmetric array of circular current loops. Suppose there are N loops (or turns) per unit length; this cylindrical current configuration can be simulated by a cylindrical current sheet of length $2l$, placed with its axis along the z axis and positioned symmetrically about the xy plane.

This problem is solved by superposing the magnetic fields of the

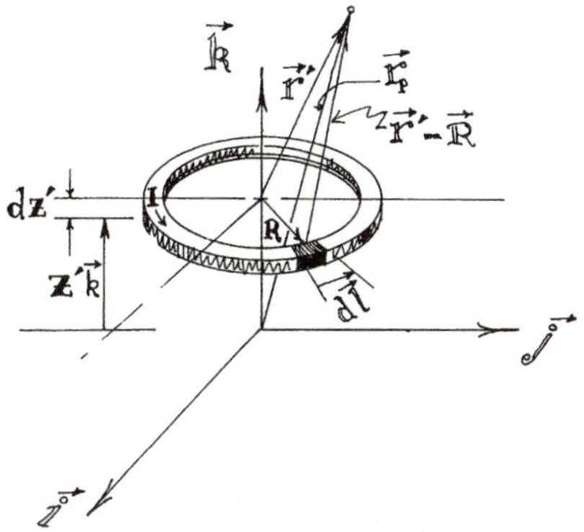

vertical cascade of circular loops. The field set up by the circular element of the current sheet between z and z + dz is

$$d\mathbf{B}_p = \frac{\mu_0}{4\pi}(\text{NI dz})\int_0^{2\pi}\frac{d\mathbf{R}\times(\mathbf{r}_p-\mathbf{R})}{|\mathbf{r}_p-\mathbf{R}|^3}.$$

At a point designated by $\mathbf{r}_p$ the vector potential of the circular loop dz is

$$d\mathbf{A}_p = \frac{\mu_0}{4\pi}(\text{NI dz})\int_0^{2\pi}\frac{d\mathbf{R}}{|\mathbf{r}_p-\mathbf{R}|}.$$

The integral for $\mathbf{B}_p$ can be approximated readily for a finite solenoid and can be computed exactly in the case of the infinite solenoid. After performing the integration over z from $-l$ to $+l$,

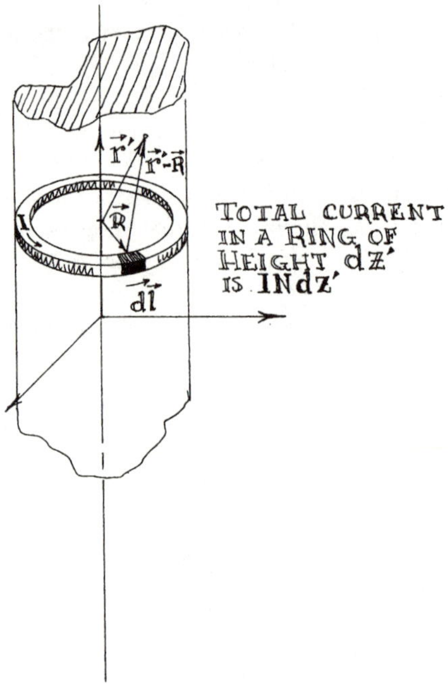

the individual components of **B** in cartesian bases are

$$B_1 = \frac{\mu_0 NIR}{4\pi}\int_0^{2\pi} \cos\phi \left\{\frac{1}{[b^2+(z_p-l)^2]^{1/2}} - \frac{1}{[b^2+(z_p+l)^2]^{1/2}}\right\} d\phi$$

$$B_2 = \frac{\mu_0 NIR}{4\pi}\int_0^{2\pi} \sin\phi \left\{\frac{1}{[b^2+(z_p-l)^2]^{1/2}} - \frac{1}{[b^2+(z_p+l)^2]^{1/2}}\right\} d\phi$$

$$B_3 = \frac{\mu_0 NIR}{4\pi}\int_0^{2\pi} (R - a_p\cos\xi) \left\{\frac{(z_p-l)}{[b^2+(z_p-l)^2]^{1/2}} - \frac{(z_p+l)}{[b^2+(z_p+l)^2]^{1/2}}\right\} d\phi$$

where $\qquad b^2 = R^2 + a_p{}^2 - 2Ra_p\cos\xi$

and $\qquad \xi = \phi - \tan^{-1}(y_p/x_p).$

If the length of the solenoid is infinite ($l \to \infty$), then the x and y components are zero and

$$B_3 = \mu_0 NI \quad \text{if} \quad a_p < R;$$
$$B_3 = 0 \quad \text{if} \quad R < a_p.$$

This is the well-known formula for the internal field of an infinite solenoid.

The field in the central region ($r_p \ll l$) of a finite solenoid can be computed by expanding the integrand of the components B_j in a Taylor's series:

$$B_1 \simeq \frac{3\mu_0 NIR^2 z_p}{2l^4} x_p ,$$

$$B_2 \simeq \frac{3\mu_0 NIR^2 z_p}{2l^4} y_p ,$$

and
$$B_3 \simeq \begin{cases} \dfrac{\mu_0 NIR}{2}\left\{\dfrac{2}{R} - \dfrac{R}{l^2}\right\} & \text{if} \quad a_p < R; \\ \dfrac{-\mu_0 NIR^2}{2l^2} & \text{if} \quad R < a_p. \end{cases}$$

By allowing the length $2l$ to be finite we find that a weak negative field begins to build up outside of the solenoid ($R < a_p$).

The vector potential of the solenoid is of some interest. Later when problems involving induced emfs are discussed we shall find it convenient to employ the vector potential.

Setting up the integral for the vector potential of the solenoid, we find that

$$\mathbf{A}_p = \mu_0 \frac{NIR}{4\pi} \int_0^{2\pi} \int_{-l}^{l} \frac{\{-\sin\phi\,\mathbf{i} + \cos\phi\,\mathbf{j}\}\,d\phi\,dz}{[b^2 + (z_p - z)^2]^{1/2}} .$$

This problem has cylindrical symmetry; thus to simplify the integration we set $y_p = 0$ and evaluate $\mathbf{A}_p$ at a point in the x,z plane. At $y_p = 0$ the component of the base vector $\mathbf{i}$ becomes zero. Then integrating over z

$$\mathbf{A}_p(y_p = 0) = \mu_0 \frac{NIR}{4\pi}\,\mathbf{j} \int_0^{2\pi} \log\left\{\frac{(z_p + l) + \sqrt{b^2 + (z_p + l)^2}}{(z_p - l) + \sqrt{b^2 + (z_p - l)^2}}\right\} \cos\phi\,d\phi .$$

If we evaluate this integral for $z_p \ll l$ (or let l become very large) we can set $z_p = 0$ for a fair approximation of $\mathbf{A}_p$. Then

$$\mathbf{A}_p(y_p = 0) = \mu_0 \frac{NIR}{4\pi}\,\mathbf{j} \int_0^{2\pi} \log\left\{\frac{\sqrt{1 + \delta^2} + 1}{\sqrt{1 + \delta^2} - 1}\right\} \cos\phi\,d\phi,$$

where
$$\delta = b^2/l^2,$$

and
$$b^2 = R^2 + x_p^2 - 2Rx_p \cos\phi.$$

The integral can be obtained in closed form for the infinite solenoid. Some care must be exercised to take the limit as $l \to \infty$

before integrating over ϕ. Taking the limit as $l \to \infty$, the vector potential for the infinite solenoid is then

$$\mathbf{A}_p(y_p = 0) = \mu_0 \frac{NIR}{4\pi} \mathbf{j} \int_0^{2\pi} \cos\phi \log\{R^2 + x_p^2 - 2Rx_p \cos\phi\}\, d\phi.$$

Integrating by parts we obtain

$$\mathbf{A}_p(y_p = 0) = \frac{\mu_0 NIx_p}{2}\mathbf{j}, \quad x_p \leqslant R;$$
$$= \frac{\mu_0 NIR^2}{2x_p}\mathbf{j}, \quad R \leqslant x_p.$$

This result can now be generalized to any point $\mathbf{a}_p$, giving

$$\mathbf{A}_p = \frac{\mu_0 NI}{2}(\mathbf{k} \times \mathbf{a}_p), \qquad a_p \leqslant R;$$
$$= \frac{\mu_0 NIR^2}{2}\frac{(\mathbf{k} \times \mathbf{a}_p)}{a_p^2}, \quad R \leqslant a_p.$$

These two forms are quite unique and useful. The interior solution is $\frac{1}{2}(\mathbf{B} \times \mathbf{a}_p)$ where $\mathbf{B}$ is the constant interior field of the solenoid. The reåder should check this by demonstrating that the vector potential of a constant field can always be represented in this manner.

The curl of the exterior solution is zero, which is the required value for the external field of an infinite solenoid. If one integrates the exterior solution about a closed path (say a circle for convenience) the path integral is not necessarily zero. Care must be exercised when Stokes' theorem is introduced. The function $(\mathbf{k} \times \mathbf{a}_p)/a_p^2$ has a zero curl at every point *except* the point at $a_p = 0$. The function is undefined at $a_p = 0$; however, the surface integral of the normal component of the curl (on the $\mathbf{k}$ axis) is nonvanishing and equal to 2π:

$$\oint_{\text{circle}} \frac{(\mathbf{k} \times \mathbf{a}_p)}{a_p^2} \cdot d\mathbf{r} = \iint_{\text{cap}} \text{curl}\left\{\frac{(\mathbf{k} \times \mathbf{a}_p)}{a_p^2}\right\} \cdot \mathbf{n}\, dS = 2\pi.$$

This integral plays a prominent role in the study of complex variables. The above consideration is academic in the case of the solenoid, because the singular point at $a_p = 0$ is excluded by the physical constraints of the problem. However this integral can be employed later to explain the induced emf in a loop external to the solenoid.

D. THE MAGNETIC QUADRUPOLE

Two antiparallel magnetic moments $\mathbf{M}_1$ and $\mathbf{M}_2$ separated by a vector $\boldsymbol{l}$ can produce a characteristic magnetic quadrupole field. If

$$\mathbf{M}_1 = \mathbf{M} = -\mathbf{M}_2,$$

then the vector magnetic potential in the far field goes as

$$\mathbf{A}_p \xrightarrow[r_p \text{ large}]{} \frac{\mu_0}{4\pi}\left\{\frac{\mathbf{M}\times[\mathbf{r}_p - (\boldsymbol{l}/2)]}{|\mathbf{r}_p - (\boldsymbol{l}/2)|^3} - \frac{\mathbf{M}\times[\mathbf{r}_p + (\boldsymbol{l}/2)]}{|\mathbf{r}_p + (\boldsymbol{l}/2)|^3}\right\}.$$

In the far field, $l \ll r_p$, by expanding the denominator we obtain

$$\mathbf{A}_p \xrightarrow[l \ll r_p]{} \frac{\mu_0}{4\pi}\left\{\frac{\mathbf{M}}{r_p^5}\right\} \times \{3(\mathbf{r}_p \cdot \boldsymbol{l})\mathbf{r}_p - r_p^2 \boldsymbol{l}\},$$

an expression characteristic of the vector potential of a magnetic quadrupole.

E. THE MULTIPOLE EXPANSION

The vector magnetic potential can be expanded in a series of multipoles just as its electrostatic counterpart was:

$$\mathbf{A}_p = \frac{\mu_0}{4\pi}\iiint_\tau \frac{\mathbf{J}\,d\tau}{|\mathbf{r}_p - \mathbf{r}|},$$

where the volume τ is sufficient to account for all the currents in the system.

In the same manner as employed in Chapters II and III, we expand $|\mathbf{r}_p - \mathbf{r}|^{-1}$ about the point $\mathbf{r}_p$. Then

$$\mathbf{A}_p = \frac{\mu_0}{4\pi}\frac{1}{r_p}\iiint_\tau \mathbf{J}\,d\tau - \frac{\mu_0}{4\pi}\iiint_\tau \mathbf{J}\left[\mathbf{r}\cdot\nabla_p\left(\frac{1}{r_p}\right)\right]d\tau$$
$$-\frac{\mu_0}{4\pi}\frac{1}{2!}\iiint_\tau \mathbf{J}\left\{\mathbf{r}_p \cdot \mathrm{T} \cdot \nabla_p\left(\frac{1}{3r_p^3}\right)\right\}d\tau + \text{h.t.}$$

where $$\mathrm{T}_{jk} = 3x_j x_k - \delta_{jk} r^2.$$

The first term in this expansion is zero if the current filaments ΔI_m all form closed loops. To show this, we write

$$\iiint_\tau \mathbf{J}\,d\tau = \sum_{\text{all m}} \Delta I_m \oint_{\mathscr{L}_m} d\mathbf{r}_m = 0$$

where ΔI_m is the current in the mth filament and $\mathscr{L}_m$ is the closed curve representing the filament.

The closed path integral is zero; therefore our first integral is zero (if the filaments form closed loops, i.e., if div $\mathbf{J} = 0$).

When the form of the second integral in the expansion is studied we find that the second term, as one would expect, is the potential contribution from the net magnetic moment of the current filaments ΔI_m.

$$-\iiint_\tau \mathbf{J}\left[\mathbf{r}\cdot\nabla_p\left(\frac{1}{r_p}\right)\right]d\tau = \sum_{\text{all m}} \Delta I_m \oint_{\mathscr{L}_m} d\mathbf{r}_m\left[\mathbf{r}_m\cdot\nabla_p\left(\frac{1}{r_p}\right)\right].$$

Using the expansion of $\nabla_p\left(\frac{1}{r_p}\right)\times(\mathbf{r}_m\times d\mathbf{r}_m)$ and of

$$d\left\{\mathbf{r}_m\left[\mathbf{r}_m\cdot\nabla_p\left(\frac{1}{r_p}\right)\right]\right\},$$

we find that

$$-\Delta I_m \oint_{\mathscr{L}_m} d\mathbf{r}_m\left[\mathbf{r}_m\cdot\nabla_p\left(\frac{1}{r_p}\right)\right] = -\left\{\frac{\Delta I_m}{2}\oint_{\mathscr{L}_m}\mathbf{r}_m\times d\mathbf{r}_m\right\}\times\nabla_p\left(\frac{1}{r_p}\right) + 0.$$

The zero comes from another closed integral of an exact differential.

The magnetic moment of the mth loop, however, is

$$\Delta\mathbf{M}_m = \frac{\Delta I_m}{2}\oint_{\mathscr{L}_m}\mathbf{r}_m\times d\mathbf{r}_m = \Delta I_m \iint_{S_m}\mathbf{n}_m\, dS_m.$$

The second term in the expansion is then

$$\frac{-\mu_0}{4\pi}\sum_{\text{all m}}\Delta\mathbf{M}_m\times\nabla_p\left[\frac{1}{r_p}\right] = \frac{\mu_0}{4\pi}\frac{\mathbf{M}\times\mathbf{r}_p}{r_p^3}$$

where $\mathbf{M}$ is the total magnetic moment of the current distribution. The breakdown into filaments is not necessary. If the volume integral is retained we find that the total magnetic moment is

$$\mathbf{M} = \tfrac{1}{2}\iiint_\tau \mathbf{r}\times\mathbf{J}\,d\tau$$

with the magnetic moment per unit volume $\mathbf{M}_v = \frac{1}{2}\mathbf{r}\times\mathbf{J}$.

The third term in the expansion describes the magnetic quadrupole contribution to $\mathbf{A}_p$. Without further simplification we can leave this term as

$$\sum_{j=1}^{3}\sum_{k=1}^{3}\frac{\mu_0}{4\pi}\frac{x_j^{(p)}x_k^{(p)}}{r_p^5}\iiint_\tau \mathbf{J}\{3x_jx_k - \delta_{jk}r^2\}\,d\tau.$$

One of the more important characteristics of this expansion is the vanishing of the first term. When contrasted with the expansion of the electrostatic potential we see that there is no term in the vector magnetic potential corresponding to the singlet charge term. Experimentally this is related to the fact that an isolated magnetic pole has never been observed. The reader should consider the magnetic field of a magnetic moment in the case in which the field is generated from a scalar potential. Under such a development the magnetic moment would be written as $\mathbf{M} = \mathrm{m}\boldsymbol{l}$ where m is an isolated magnetic pole. The magnetic moment can then be thought of as $+\mathrm{m}$ and $-\mathrm{m}$ separated by a distance l. Historically the concept of a pole strength m crept into usage because of the utility of the permanent bar magnet in physical experiments. The field of the bar magnet, however, is accounted for quite well by the vector potential, and the pole strength is an unnecessary concept.

F. AMPERE'S CIRCUITAL LAW

1. DERIVATION

The differential form of the circuital law is

$$\mathrm{curl}\,\mathbf{B} = \mu_0\mathbf{J}.$$

This relation can be proven by taking the curl of **B** as defined by the integral form of Ampere's law. Coupling this operation with the fact that div $\mathbf{J} = 0$ if this equation is to be satisfied, the form shown above is obtained.

Rather than approach the circuital law in this way, some physical insight is gained by examining the closed path integral of $\mathbf{B}_p$ when $\mathbf{B}_p$ is set up by a closed current filament.

At any point P on the path shown in the diagram on the next page,

$$\mathbf{B}_p = \frac{\mu_0 I}{4\pi}\oint_{\mathscr{L}} \frac{\mathbf{dl} \times \mathbf{r}}{r^3}.$$

If $\mathbf{B}_p$ is integrated about any closed path C,

$$\oint_C \mathbf{B}_p \cdot \mathbf{dr} = \oint_C \left\{\frac{\mu_0 I}{4\pi}\oint_{\mathscr{L}} \frac{(\mathbf{dl} \times \mathbf{r})}{r^3}\right\} \cdot \mathbf{dr}.$$

This particular integral represents a product of the component of $\mathbf{B}_p$ tangent to a point P lying on the path C and a variation **dr** along

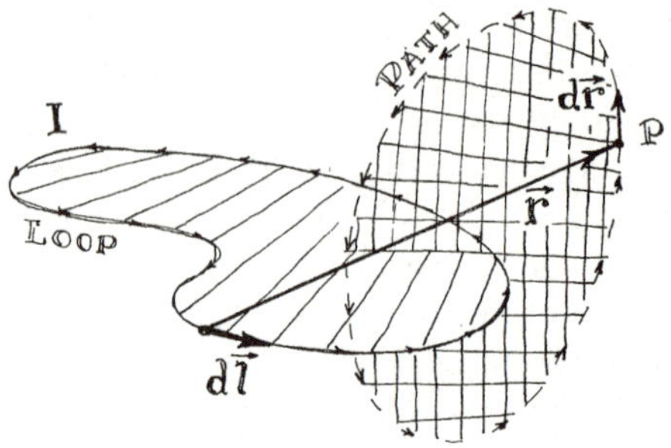

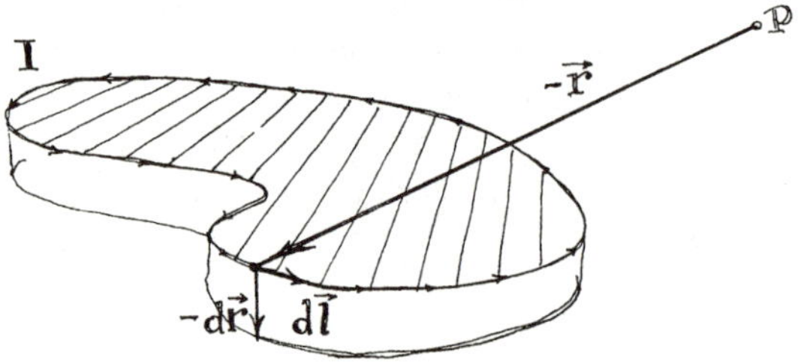

the path, accompanied by a sum over all $\Delta\mathbf{r}$ elements in the closed path C.

The same result can be obtained by holding the point fixed in space and varying the position of the loop by $-\mathbf{dr}$. As the loop is moved the spatial orientation of the loop must be maintained. In other words

$$\mathbf{n}' = \frac{1}{S}\iint_{S(\mathrm{cap})} \mathbf{n}\, dS$$

must be held fixed as the position of the loop is varied by $-\mathbf{dr}$.

When the loop position is varied, each element $\mathbf{dr}$ sweeps out an element of surface

$$\mathscr{N}\, dA' = -\mathbf{dr} \times \mathbf{dl}.$$

The radius vector locating the element of surface area relative to the fixed point P is $\mathbf{r}' = -\mathbf{r}$. Thus

$$\oint_C \mathbf{B}_p \cdot \mathbf{dr} = \oint_{C'} \frac{\mu_0 I}{4\pi} \oint_{\mathscr{L}} (-\mathbf{dr} \times \mathbf{dl}) \cdot \left(\frac{-\mathbf{r}}{r^3}\right)$$

$$= \frac{\mu_0 I}{4\pi} \iint_{A'} \left(\frac{\mathbf{r}'}{r'^3}\right) \cdot \mathscr{N}\, dA'$$

$$= \frac{\mu_0 I}{4\pi} \int_{\Omega_{A'}} d\Omega$$

After displacement about the closed mirror path C′ the loop sweeps out a set of closed surfaces. Because the orientation of the loop is maintained relative to the laboratory coordinates, the surfaces will be necked down at several points. The important aspect of this, however, is that all of these connected surfaces are closed.

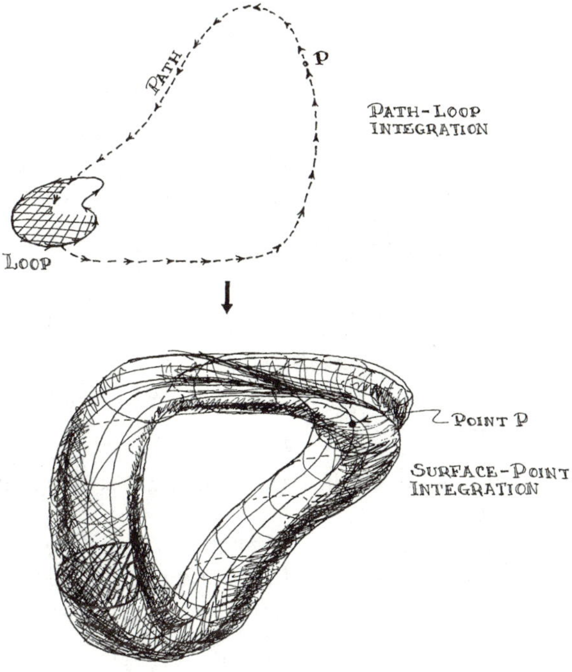

If the original path C passes through the cap surface of $\mathscr{L}$, the fixed point of the mirror operation lies INSIDE one of the closed surfaces, and

$$\int_{\Omega_{A'}} d\Omega = 4\pi; \quad \text{Path links cap surface of } \mathscr{L} \text{ or P lies inside surface A}'.$$

Under the condition that C does not penetrate the cap surface of $\mathscr{L}$, the fixed point of the mirror operation lies outside the surface

A′ swept out,

and $\int_{\Omega_{A'}} d\Omega = 0;$ Path does not link S(cap) or P lies outside A′.

Then $\oint_C \mathbf{B}_p \cdot d\mathbf{r} = \mu_0 I$ if C links $\mathscr{L}$;

$= 0$ if C does not link $\mathscr{L}$.

In general for any closed path C,

$$\oint \mathbf{B}_p \cdot d\mathbf{r} = \mu_0 I_{\text{linked}} = \mu_0 \text{ times the total current flowing through the cap surface of C.}$$

Now

$$I_{\text{linked}} = \iint_{S_C} \mathbf{J} \cdot \mathbf{n}\, dS$$

where S_C is the cap surface of the closed contour C. Using Stokes' theorem,

$$\oint_C \mathbf{B} \cdot d\mathbf{r} = \iint_{S_C} (\text{curl}\, \mathbf{B}) \cdot \mathbf{n}\, dS = \mu_0 \iint_{S_C} \mathbf{J} \cdot \mathbf{n}\, dS.$$

Since the choice of path C is arbitrary, it is necessary and sufficient that

$$\text{curl}\, \mathbf{B} = \mu_0 \mathbf{J}.$$

This result is conditioned by the fact that we utilized closed filaments, and thus

$$\text{div}\, \mathbf{J} = 0.$$

Later we shall find cases in which div **J** is not zero, and this law must be modified.

A direct development of the differential form of the circuital law can be performed in a short but elegant manner: by definition

$$\mathbf{B}_p = \text{curl}_p \frac{\mu_0}{4\pi} \iiint_\tau \frac{\mathbf{J}}{|\mathbf{r}_p - \mathbf{r}|}\, d\tau,$$

where the surface of the volume τ containing the currents **J** is extended such that $\mathbf{J} \to 0$ on the surface. Now remembering that

$$\text{curl curl} = \text{grad div} - \nabla^2,$$

$$\text{curl}_p\, \mathbf{B}_p = \text{curl}_p\, \text{curl}_p \left\{\frac{\mu_0}{4\pi} \iiint_\tau \frac{\mathbf{J}}{|\mathbf{r}_p - \mathbf{r}|}\, d\tau\right\}$$

$$= \frac{\mu_0}{4\pi} \text{grad}_p \iiint_\tau \vec{\mathbf{J}} \cdot \text{grad}_r \left\{\frac{1}{|\vec{\mathbf{r}}_p - \vec{\mathbf{r}}|}\right\} d\tau$$

$$- \frac{\mu_0}{4\pi} \iiint \mathbf{J}\, \nabla_r^2 \left\{\frac{1}{|\mathbf{r}_p - \mathbf{r}|}\right\} d\tau.$$

In both integrals we have let $\nabla_p \to -\nabla_r$. The first integral expands into two terms,

$$\text{grad}_p\left(\frac{-\mu_0}{4\pi}\iiint_\tau \text{div}_r\left\{\frac{\mathbf{J}}{|\mathbf{r}_p - \mathbf{r}|}\right\} d\tau + \frac{\mu_0}{4\pi}\iiint_\tau \frac{1}{|\mathbf{r}_p - \mathbf{r}|}\,\text{div}_r\,\mathbf{J}\,d\tau\right).$$

The second term in this expansion is zero because div $\mathbf{J} = 0$ for this problem. Utilizing the divergence theorem, the first integral becomes

$$\iint_S \frac{\mathbf{J}\cdot\mathbf{n}}{|\mathbf{r}_p - \mathbf{r}|}\,dS.$$

This term is zero in view of the preliminary boundary condition that $\mathbf{J} \to 0$ at the surface S of τ.

The remaining term has the form

$$\iiint_\tau \mathbf{J}\,\nabla_r^2\left\{\frac{1}{|\mathbf{r}_p - \mathbf{r}|}\right\} d\tau.$$

In Chapter II we defined the Dirac delta function $\delta(\mathbf{r}_p - \mathbf{r})$ by

$$\nabla_r^2\left\{\frac{1}{|\mathbf{r}_p - \mathbf{r}|}\right\} = -4\pi\delta(\mathbf{r} - \mathbf{r}_p).$$

Thus
$$\iiint_\tau \mathbf{J}(\mathbf{r})[-4\pi\delta(\mathbf{r} - \mathbf{r}_p)]\,d\tau = -4\pi\mathbf{J}(\mathbf{r}_p),$$

giving
$$\text{curl}_p\,\mathbf{B}_p = \mu_0\mathbf{J}(\mathbf{r}_p)$$

which is the differential form of Ampere's law.

The reader should keep in mind that the subscript p implies that these particular derivatives and functions involve the coordinates of the point P designated by $\mathbf{r}_p$.

All of these derivations have implicitly used the condition that div $\mathbf{A} = 0$. This can be seen by again examining the expansion of the first integral in $\text{curl}_p\,\mathbf{B}_p$. The condition div $\mathbf{A} = 0$ is known as an establishment of the GAUGE of the system. Later in a more general formulation we will see that the scalar and magnetic potentials are not unique, in that another potential function can be incorporated with them. This extra function is defined when the GAUGE is established, and the condition div $\mathbf{A} = 0$ does just this.

Using div $\mathbf{A} = 0$ we can reverse the derivation (dropping the subscript p):

$$\text{curl}\,\mathbf{B} = \text{curl curl}\,\mathbf{A} = \text{grad (div}\,\mathbf{A}) - \nabla^2\mathbf{A} = \mu_0\mathbf{J}.$$

Then $$\nabla^2 A_j = -\mu_0 J_j$$

with solutions (using the Green's function $G(\mathbf{r},\mathbf{r}') = (4\pi\,|r_p - r|)^{-1}$)

$$\mathbf{A} = \frac{\mu_0}{4\pi} \iiint_\tau \frac{\mathbf{J}}{|\mathbf{r}_p - \mathbf{r}|}\, d\tau,$$

demonstrating the consistency of our procedure.

2. APPLICATIONS OF THE CIRCUITAL LAW

The integral form of the circuital law proves to be quite useful in problems which have symmetry. The differential form has a greater utility because the solutions involve general expansions in complete sets of functions. These allow approximate general solutions in series.

a. THE MAGNETIC FIELD INSIDE AND OUTSIDE OF A LONG STRAIGHT CYLINDRICAL CURRENT-BEARING CONDUCTOR. Assume that the current density $\mathbf{J} = J\mathbf{k}$ across the face (or cross section) of the conductor is homogeneous. In reality this condition is slightly altered by the Lorentz attraction between straight current filaments. However, in practice the homogeneous assumption is quite good.

The external field has been obtained previously. In this application the solution is obtained from the cylindrical symmetry. This symmetry indicates that the external $\mathbf{B}$ vectors are tangent to circles whose centers lie on the axis of the conductor. The planes of these circles are perpendicular to this axis. The integral then becomes a product of scalars. Take a circle of radius $a_p > R$ (the radius of the conductor); then

$$\oint_{\text{Circle}} \mathbf{B} \cdot d\mathbf{r} = 2\pi a_p\,|\mathbf{B}_p| = \mu_0 \iint_{\text{Circle}} \mathbf{J} \cdot \mathbf{n}\, dS = \mu_0\,|\mathbf{J}|\,\pi R^2.$$

Then $$|\mathbf{B}_p| = \frac{\mu_0\,|\mathbf{J}|\,R^2}{2a_p},$$

and $$\mathbf{B}_p = \frac{\mu_0 R^2 \mathbf{J} \times \mathbf{a}_p}{2{a_p}^2},$$

where $${a_p}^2 = ({x_p}^2 + {y_p}^2).$$

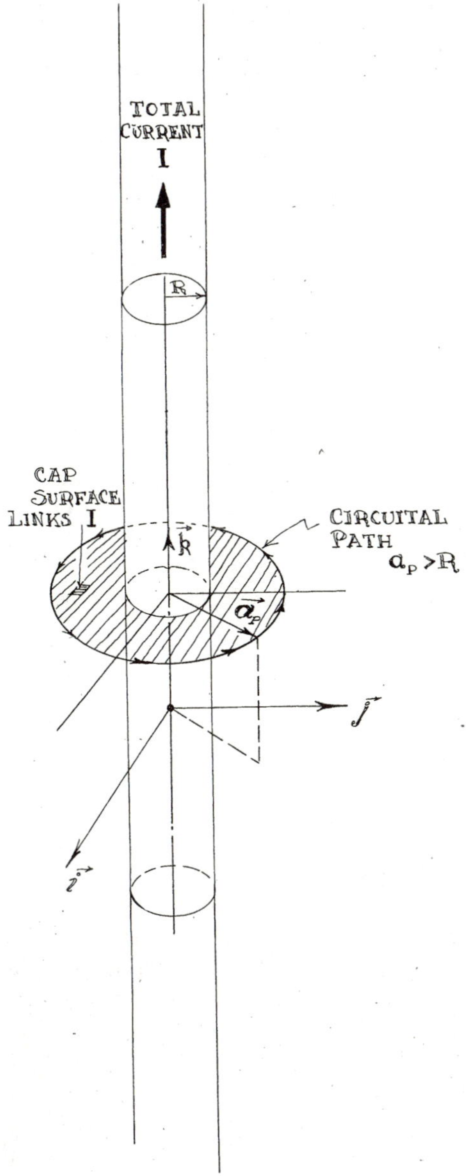
TOTAL
CURRENT
I
R
CAP
SURFACE
LINKS I
CIRCUITAL
PATH
ap >R

The internal field ($a_p' < R$) is obtained by the same method from

$$\oint \mathbf{B}_p \cdot d\mathbf{r} = 2\pi a_p' \, |\mathbf{B}_p| = \mu_0 \pi a_p{}^2 \, |\mathbf{J}|.$$

Then
$$\mathbf{B}_p = \frac{\mu_0 \mathbf{J} \times \mathbf{a}_p'}{2}.$$

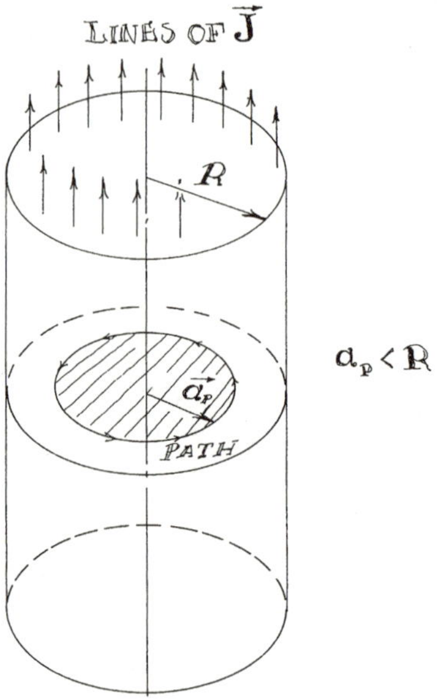

b. AXIALLY PARALLEL CYLINDRICAL HOLES. If a cylindrical cavity having an axis parallel to the conductor axis is cut in the conducting cylinder such that the entire cavity is enclosed within the surface of the conductor, assuming that the current density is constant $\mathbf{J} = J\mathbf{k}$, we find that the field in the cavity is homogeneous.

Let the radius of the cavity be b, and let the axis of the cavity be located by a vector $\mathbf{c}$ perpendicular to the axis of the conductor ($b + |\mathbf{c}| \leqslant R$). The cavity is simulated by the superposition of a negative current density $-\mathbf{J}$ in the cavity region.

Consider a point inside the cavity located by $\mathbf{a}_p'$ relative to the conductor axis and by $\mathbf{a}_p''$ relative to the cavity axis,

$$\mathbf{a}_p' = \mathbf{c} + \mathbf{a}_p''.$$

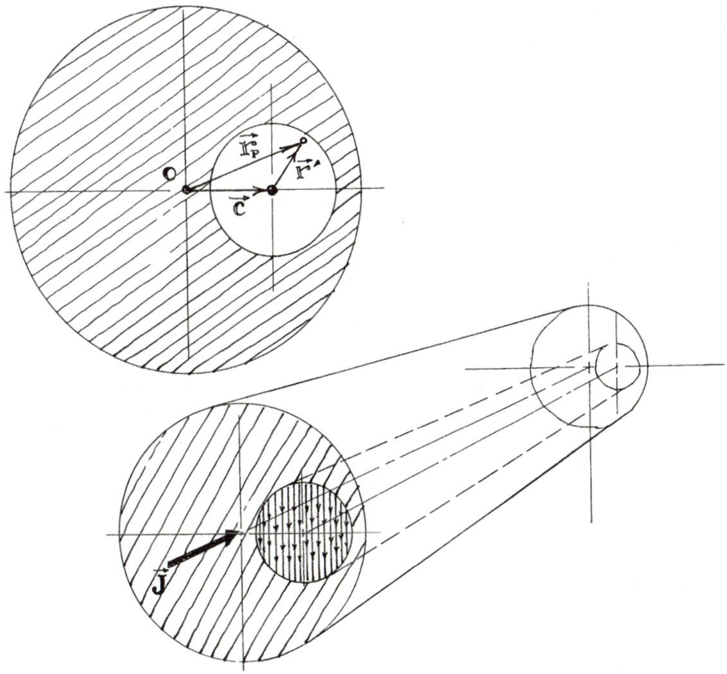

Inside the cavity

$$\mathbf{B}_p = \mu_0 \frac{\mathbf{J} \times \mathbf{a}'_p}{2} - \mu_0 \frac{\mathbf{J} \times \mathbf{a}''_p}{2}$$

$$= \frac{\mu_0 \mathbf{J}}{2} (\mathbf{a}'_p - \mathbf{a}''_p),$$

giving $\mathbf{B}_p = \mu_0 \dfrac{\mathbf{J} \times \mathbf{c}}{2}$ = a constant vector field.

G. THE ENERGY STORED IN A STEADY STATE CURRENT DISTRIBUTION

1. DERIVATIONS

During the preceding discussion we have been concerned with distributions of currents $\mathbf{J}$ for which div $\mathbf{J} = 0$, in other words current distributions composed of an ensemble of closed current loops. These loops $\mathscr{L}_m$ have thus far been considered as filaments of current I_m.

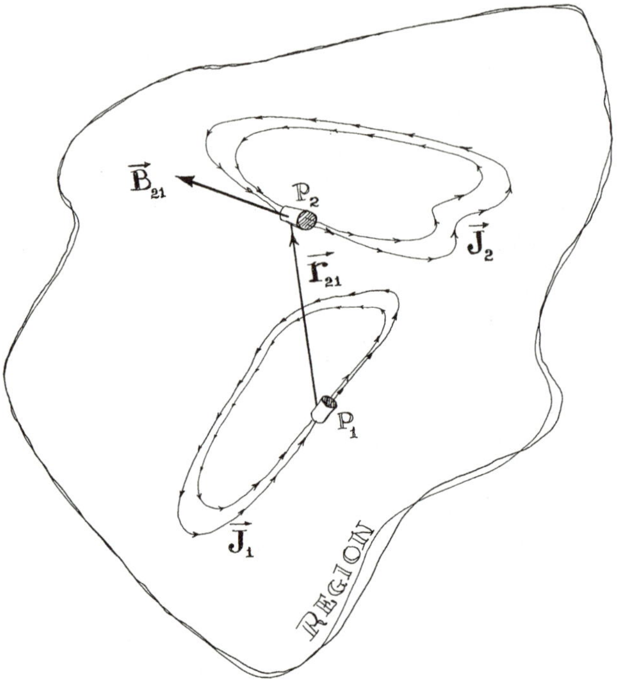

Although the filament picture is extremely useful, there exist inherent difficulties associated with it when energy storage is a problem. If the total current in a filament is kept constant while the cross-sectional area is allowed to become arbitrarily small, the current density in the filament becomes arbitrarily large, and as a result the "self-energy" of the line filament (like that of a point charge) is an undefined quantity. In the case of the point charge, the self-energy terms were the diagonal terms in the potential matrix $\mathbb{P}$ (i.e., the P_{jj}), and these self-energy coefficients were taken as zero. The self-energy coefficients of filaments must also be omitted if this concept is utilized. The final results for the stored energy, after appropriate sums, are the same as if closed tubes of current (finite **J**) of very small (not zero) cross section were employed.

The filament picture has the advantage that it provides some physical picture of the self-energy and self-inductance of a macroscopic current loop. In such a cross section the self-inductance can be viewed as stemming from the mutual interaction of all of the filaments which make up the single macroscopic loop.

Let us play it safe and content ourselves with closed current-carrying loops $\mathscr{L}_m$ having cross-sectional area $\Delta\mathscr{A}_m$ which is very small compared to the loop dimension. The **J** are then finite and the current of the mth loop is

$$I_m = |\mathbf{J}|\,\Delta\mathscr{A}_m$$

where **J** is evaluated at the coordinates of the mth loop.

Now consider two loops $\mathscr{L}_j$ and $\mathscr{L}_k$ and their interaction. U_{jk}, the energy stored is

$$U_{jk} = I_j\Phi_{jk} = I_j\frac{\mu_0}{4\pi}I_k\oint_{S_j}\mathbf{B}_{jk}\cdot\mathbf{n}_j\,dS_j$$

where $\mathbf{B}_{jk}$ is the field of the kth loop specified at every point on the cap surface of the jth loop;

$$\mathbf{B}_{jk} = \frac{\mu_0 I_k}{4\pi}\oint_{\mathscr{L}_k}\frac{d\boldsymbol{l}_k\times\mathbf{r}_{jk}}{r_{jk}^3},$$

with $$\mathbf{r}_{jk} = \mathbf{r}_j - \mathbf{r}_k.$$

If $j \neq k$ (i.e., the jth and k^{th} loops are not one and the same), the surface integral over S_j can be converted to a line integral on $\mathscr{L}_j$ by Stokes' theorem. Since

$$\mathbf{B}_{jk} = \text{curl}_j\,\mathbf{A}_{jk} = \text{curl}_j\left\{\frac{\mu_0 I_k}{4\pi}\oint_{\mathscr{L}_k}\frac{d\boldsymbol{l}_k}{r_{jk}}\right\},$$

$$\Phi_{jk} = \iint_{S_j}\mathbf{B}_{jk}\cdot\mathbf{n}_j\,dS_j = \oint_{\mathscr{L}_j}\mathbf{A}_{jk}\cdot d\boldsymbol{l}_j = \frac{\mu_0 I_k}{4\pi}\oint_{\mathscr{L}_j}\oint_{\mathscr{L}_k}\frac{d\boldsymbol{l}_j\cdot d\boldsymbol{l}_k}{r_{jk}}.$$

Then $$\Phi_{jk} = L_{jk}I_k,$$

with $$L_{jk} = L_{kj}.$$

The coefficients L_{jk} are known as the coefficients of MUTUAL INDUCTANCE, $j \neq k$, and they depend only upon the geometry of the loops.

When $j = k$ we retain the integration over the cap surface S_j.

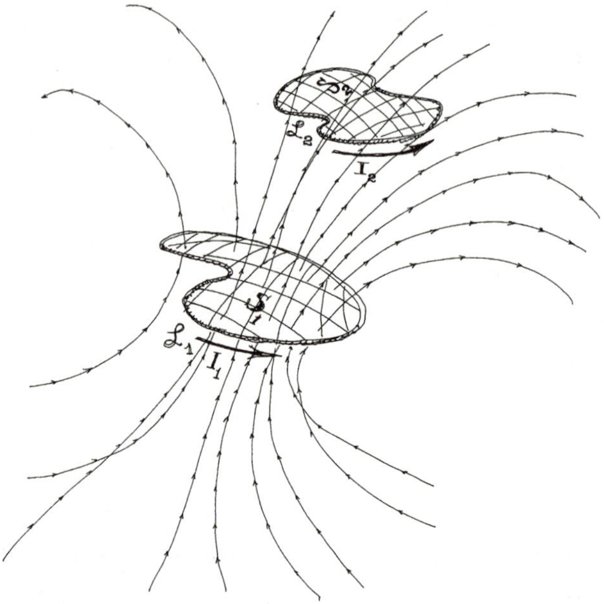

This precaution actually lets us keep this integral in a form most readily calculated.

$$\Phi_{jj} = \iint_{S_{j'}} \frac{\mu_0 I_j}{4\pi} \left\{ \oint_{\mathscr{L}_j} \frac{(d\boldsymbol{l}_j \times \mathbf{r}_{jj'})}{|\mathbf{r}_{jj'}|^3} \right\} \cdot \mathbf{n}'_j \, dS'_j = L_{jj} I_j.$$

$\mathbf{r}_j$ terminates in the cap surface while $\mathbf{r}_{j'}$ terminates on the loop contour.

The diagonal coefficients L_{jj} are called the coefficients of SELF-INDUCTANCE.

We have actually obtained forms useful for practical computations. The coefficients as defined by their integrals have the following form used in the case of microscopic loops:

$$L_{jk} = \frac{\mu_0}{4\pi} \oint_{\mathscr{L}_j} \oint_{\mathscr{L}_k} \frac{d\boldsymbol{l}_j \cdot d\boldsymbol{l}_k}{r_{jk}},$$

and
$$L_{jj} = \frac{\mu_0}{4\pi} \iint_{S_j} \left\{ \oint_{\mathscr{L}_{j'}} \frac{d\boldsymbol{l}_{j'} \times \mathbf{r}_{jj'}}{|\mathbf{r}_{jj'}|^3} \right\} \cdot \mathbf{n}_j \, dS_j.$$

Returning to our energy problem, when the jth and k^{th} loops are brought together in space, forces are exerted and energy is stored in the two-loop system. The energy is assumed to be shared equally; thus

$$\tfrac{1}{2}U_{jk} + \tfrac{1}{2}U_{kj} = \tfrac{1}{2}I_j\Phi_{jk} + \tfrac{1}{2}I_k\Phi_{kj}.$$

In terms of the inductances

$$\tfrac{1}{2}U_{jk} + \tfrac{1}{2}U_{kj} = \tfrac{1}{2}I_jL_{jk}I_k + \tfrac{1}{2}I_kL_{kj}I_j$$

where

$$L_{jk} = L_{kj}.$$

For N loops the total energy stored is

$$U = \sum_{m=1}^{N} \sum_{n=1}^{N} \tfrac{1}{2}\{I_mL_{mn}I_n\} = \sum_{m=1}^{N} \tfrac{1}{2}I_m\Phi_m$$

with

$$\Phi_m = \sum_{n=1}^{N} L_{mn}I_n.$$

Again (as in the case of the problem of the charged conducting bodies) we have obtained the energy of a discrete set of objects as a bilinear or quadratic form. The elements L_{jk} form a square $N \times N$ matrix $\mathbf{L}$ while the components I_m become the coefficients of a column vector $\mathbf{I}$ or a row vector $\tilde{\mathbf{I}}$.

$$\mathbf{L} = \begin{pmatrix} L_{11} & L_{12} & \dots & L_{1N} \\ L_{21} & & & \cdot \\ \cdot & & & \cdot \\ \cdot & & & \cdot \\ \cdot & & & \cdot \\ L_{N1} & & \dots & L_{NN} \end{pmatrix},$$

and

$$\mathbf{I} = \begin{bmatrix} I_1 \\ I_2 \\ \cdot \\ \cdot \\ \cdot \\ I_N \end{bmatrix}; \quad \text{or} \quad \tilde{\mathbf{I}} = [I_1, I_2, \dots, I_N].$$

Using this representation,

$$U = \tfrac{1}{2}\tilde{\mathbf{I}} \cdot \mathbf{L} \cdot \mathbf{I}.$$

The flux linking the nth loop now becomes the mth component Φ_m of a vector $\mathbf{\Phi}$

$$\Phi_m = \sum_{n=1}^{N} L_{mn} I_n,$$

or

$$\mathbf{\Phi} = \mathbb{L} \cdot \mathbf{I}.$$

In this form the matrix $\mathbb{L}$ is a geometric operator which transforms the vector $\mathbf{I}$ into the vector $\mathbf{\Phi}$.

If the I_m are added in groups to form various macroscopic loops, the Φ_m corresponding to each I_m are the same (the cross section of the wire is still very small compared with the dimensions of the loop). For M macroscopic loops, $\mathbb{L}$ contracts to an M $\times$ M matrix while $\mathbf{\Phi}$ and $\mathbf{I}$ contract to M-dimensional vectors.

In the same manner as the analogous problem in electrostatics, the eigenvalues and eigenvectors of $\mathbb{L}$ provide sets of currents which will produce fluxes at a given loop proportional to the current of that particular loop.

Of purely academic interest is the inversion of the flux equation. Because $|\mathbb{L}| \neq 0$ we can write

$$\mathbf{I} = \mathbb{L}^{-1} \cdot \mathbf{\Phi};$$

giving

$$U = \tfrac{1}{2}\tilde{\mathbf{\Phi}} \cdot \mathbb{L}^{-1} \cdot \mathbf{\Phi},$$

or

$$U = \tfrac{1}{2}\sum_{j=1}^{N}\sum_{k=1}^{N} \Phi_j L'_{jk} \Phi_k.$$

where L'_{jk} is an element of $\mathbb{L}^{-1}$.

This equation represents the expression for energy storage in discrete loops. If the current is continuously distributed in a volume τ with div $\mathbf{J} = 0$, we can separate out loops of current

$$I_m = \mathbf{J}_m \cdot \mathbf{n}_m \, \Delta\mathscr{A}_m,$$

where $\Delta\mathscr{A}_m$ is the cross sectional area of the loop.

Going back to our original expression,

$$U = \tfrac{1}{2} \sum_{\text{all } m} I_m \Phi_m$$

where

$$\Phi_m = \sum_{\text{all } k} L_{mk} I_k = \sum_{\text{all } k} \iint_{S_m} \mathbf{B}_{mk} \cdot \mathbf{n}_m \, dS_m$$

$$= \sum_{\text{all } k} \oint_{\mathscr{L}_m} \mathbf{A}_{mk} \cdot d\boldsymbol{l}_m;$$

$$\Phi_m = \oint_{\mathscr{L}_m} \mathbf{A}_m \cdot d\boldsymbol{l}_m.$$

Because $I_m = \mathbf{J}_m \cdot \mathbf{n}_m \, \Delta\mathscr{A}_m$, and because

$$\mathbf{J}_m/J_m = \boldsymbol{l}_m/l_m,$$

$$U = \tfrac{1}{2}\sum_{\text{all m}} \oint \mathbf{J}_m \cdot \mathbf{A}_m \, dl_m \, \Delta\mathscr{A}_m.$$

Taking the sum over m to an integral $dl_m \, d\mathscr{A}_m = d\tau$,

$$U = \tfrac{1}{2}\iiint_\tau \mathbf{J} \cdot \mathbf{A} \, d\tau.$$

τ is the volume containing **J**. In all cases **J** vanishes on the closed surface at infinity; otherwise we would have to deal with systems having an undefined total current and total energy.

In terms of the fields which can be developed in the absence of magnetic media,

$$\mu_0 \mathbf{J} = \text{curl}\, \mathbf{B};$$

then

$$\mathbf{A} \cdot \mathbf{J} = \frac{\mathbf{A}}{\mu_0} \cdot \text{curl}\, \mathbf{B} = \mu_0^{-1}\{\text{div}\,(\mathbf{B} \cdot \mathbf{A}) + \mathbf{B} \cdot (\text{curl}\, \mathbf{A})\}.$$

Inserting this expansion into our volume integral and extending the surface of τ to infinity (where the fields vanish),

$$U = \tfrac{1}{2} \iiint_{\text{All Space}} \mathbf{A} \cdot \mathbf{J} \, d\tau = \frac{\mu_0^{-1}}{2} \iiint_{\text{All Space}} \text{div}\,(\mathbf{B} \times \mathbf{A}) \, d\tau + \frac{1}{2\mu_0} \iiint_{\text{All Space}} \mathbf{B} \cdot \mathbf{B} \, d\tau.$$

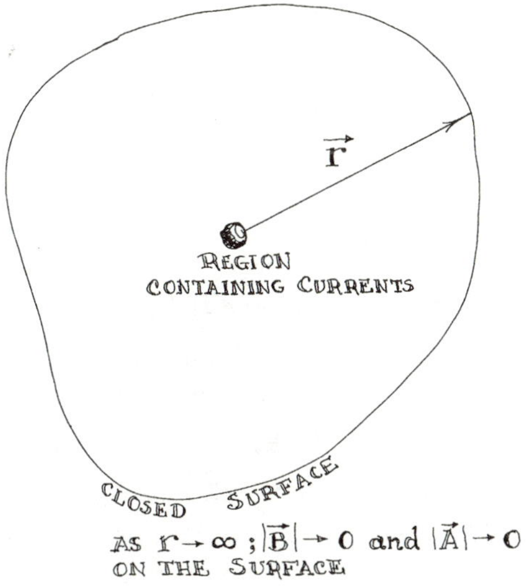

The first integral converts to a surface integral by Stokes' theorem and vanishes because **B** and **A** are zero on the surface at infinity. Finally, in the absence of magnetic media

$$U = \frac{1}{2\mu_0} \iiint_{\text{All Space}} B^2 \, d\tau.$$

The quantity $B^2/2\mu_0$ is an energy density analogous to energy density in electrostatics. Viewing the energy storage in this manner we obtain a representation in which every volume element in space has a portion of the energy storage associated with it. This viewpoint will become an aid in understanding the negative sign in Faraday's law, to be taken up in the next chapter.

2. EXAMPLE CALCULATIONS

a. THE MUTUAL INDUCTANCE OF TWO COAXIAL CIRCULAR LOOPS. Let the radius of one loop be R_1 and that of the other R_2. The

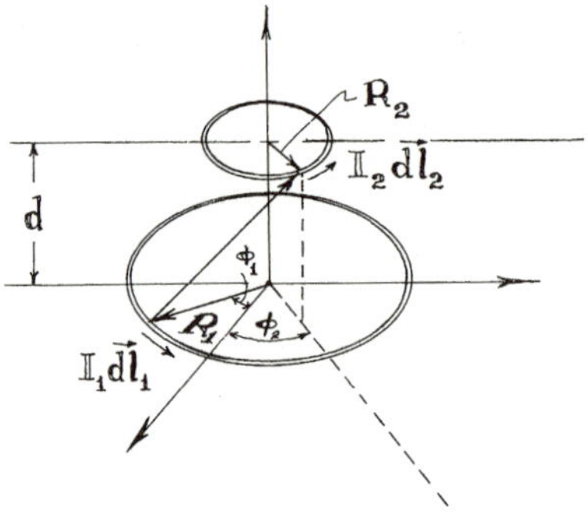

separation of the planes of the loops is d.

$$L_{12} = \frac{\mu_0}{4\pi} \oint_{\mathscr{L}_1} \oint_{\mathscr{L}_2} \frac{d\boldsymbol{l}_1 \cdot d\boldsymbol{l}_2}{|\mathbf{r}_1 - \mathbf{r}_2|}$$

where $$d\boldsymbol{l}_1 \cdot d\boldsymbol{l}_2 = |d\boldsymbol{l}_1| \, R_2 \, d\phi_2 \cos(\phi_1 - \phi_2)$$

and $$r_{12} = \{R_1{}^2 + R_2{}^2 + d^2 - 2R_1R_2 \cos(\phi_1 - \phi_2)\}^{1/2}.$$

Because of the symmetry, all positions of $d\boldsymbol{l}_1$ will receive the same contribution from the integration over loop 2. Thus the final result is $2\pi R_1$ times the integration over ϕ_2 with ϕ_1 set equal to zero. Write ϕ_2 as ϕ; then

$$L_{12} = \frac{\mu_0 R_1}{2}\int_0^{2\pi} \frac{R_2 \cos\phi\, d\phi}{\{d^2 + R_1^{\,2} + R_2^{\,2} - 2R_1R_2\cos\phi\}^{1/2}};$$

giving

$$L_{12} = \mu_0(R_1R_2)\left\{\left(\frac{2}{k_0} - k_0\right)K(k_0) - \frac{2}{k_0}E(k_0)\right\}.$$

$K(k_0)$ and $E(k_0)$ are complete elliptic functions of the first and second kind, while

$$k_0^{\,2} = \frac{4R_1R_2}{(R_1 + R_2)^2 + d^2}.$$

Special approximations of this expression can be quite useful. When $d \gg R_1$ and $d \gg R_2$,

$$L_{12} \simeq \frac{\mu_0\pi}{16}(R_1R_2)^{1/2}k_0^{\,3}.$$

This result can also be obtained using the axial field of either loop. If $d \ll R_1$ or R_2, then (taking $R_1 \sim R_2$)

$$L_{12} \simeq \mu_0 R_1\left\{\log\left[\frac{8R_1}{\{(R_1 - R_2)^2 + d^2\}^{1/2}}\right] - 2\right\}.$$

This last result gives the approximate self-inductance of a circular turn of wire where the radius of the wire cross section is much less than R:

$$L_{11} \simeq \mu_0 R\left\{\log\left(\frac{8R}{d}\right) - 2\right\},$$

where d is approximately the radius of the wire.

To see this, take two loops carrying currents I_1 and I_2 respectively.

$$U = \tfrac{1}{2}L_{11}I_1^{\,2} + \tfrac{2}{2}L_{12}I_1I_2 + \tfrac{1}{2}L_{22}I_2^{\,2}.$$

Let $L_{11} = L_{22}$ and $I = I_1 + I_2$; then

$$U = \tfrac{1}{2}\{L_{11}I_1^{\,2} + 2L_{12}I_1I_2 + L_{11}I_2^{\,2}\}.$$

In this form L_{11} is the self-inductance and L_{12} is close to becoming a self-inductance.

Thus with

$$L_{12} = L_{11},$$

$$U = \tfrac{1}{2}L_{11}(I_1 + I_2)^2,$$

showing that L_{12} is a good approximation for L_{11}.

Some criticism might be leveled at this approximation. The result is certainly valid if the wire radius d is not too small. We can show that, contrary to the seeming divergence exhibited, as $\delta \to 0$ in a real problem the energy goes to zero. δ is the distance from the center of the wire. Unless $|\mathbf{J}|$ is allowed to be infinite as $\delta \to 0$ the integrand goes as $\frac{1}{2}J\pi\delta \times |\mathbf{A}(R - \delta)|2\pi R$. The term $\delta\,|\mathbf{A}|$ goes to zero in the limit as $\delta \to 0$. Thus there are no spurious divergences to invalidate the argument.

b. THE ENERGY STORED PER UNIT LENGTH IN A LONG SOLENOID. In the central region of a long solenoid where **B** is approximately constant (in the z direction),

$$\mathbf{B} \simeq \mu_0 N I \mathbf{k}.$$

N is the number of turns per unit length and I the current per turn. Then

$$\mathbf{A} = -\tfrac{1}{2}\{\mathbf{a}_p \times \mathbf{B}\},$$

where $\mathbf{a}_p = x_p\mathbf{i} + y_p\mathbf{j}$.

At the position of the current, taking the radius of the solenoid as R, **A** is tangent to the circular turns. The energy per unit length, $U_{u.l.}$ then is

$$\begin{aligned} U_{u.l.} &= \tfrac{1}{2}NI\,|\mathbf{A}(R)|\,2\pi R \\ &= \pi RNI\{\tfrac{1}{2}\mu_0 NIR\} \\ &= \tfrac{1}{2}\mu_0(\pi R^2)N^2I^2. \end{aligned}$$

The self-inductance per unit length is then

$$L_{u.l.} = \mu_0(\pi R^2)N^2.$$

If the total length is l, the approximate total self-inductance (neglecting end effects) is

$$L \simeq \mu_0(\pi R^2)N^2 l.$$

VIII. Induced Electric Fields

A. FARADAY'S LAW

From experience it is found that a time variation of a magnetic field **B** produces an electric field $\mathscr{E}$. The electric field induced is not irrotational but rather a rotational field.

Because this law is developed from experiment, we can postulate it in one of several forms. Customarily Faraday's law is quoted in terms of the emf E developed in a conducting loop when the magnetic flux Φ linking this loop is changed:

$$\mathrm{E} = -\frac{d\Phi}{dt}.$$

Attempts have been made to develop this equation from the energy storage equation. However, although one can demonstrate that conservation of energy is consistent with Faraday's law, conservation considerations alone do not uniquely determine this law. Also, the presence of magnetic media can introduce a nonlinear dependence upon the currents.

As stated above, Faraday's law accounts for stationary current loops with varying fields and for moving conductors in magnetic fields.

The emf E is related to the electric field $\mathscr{E}$ set up by the time variation of **B** by

$$\mathrm{E} = \oint_{\mathscr{L}} \mathscr{E} \cdot dl,$$

and

$$\mathrm{E} = \oint_{\mathscr{L}} \mathscr{E} \cdot dl = \iint_{S(\text{cap})} (\text{curl}\, \mathscr{E}) \cdot \mathbf{n}\, dS.$$

From Faraday's law,

$$\mathrm{E} = -\frac{d\Phi}{dt} = -\frac{d}{dt}\left\{ \iint_{S(\text{cap})} \mathbf{B} \cdot \mathbf{n}\, dS \right\}.$$

Thus
$$\iint_S (\text{curl}\, \mathscr{E}) \cdot \mathbf{n}\, dS = -\frac{d}{dt} \iint_S \mathbf{B} \cdot \mathbf{n}\, dS.$$

The total time derivative operates both upon **B** and upon the geometry of the loop. As a result, in those cases in which the loop moves, the limits of integration are explicitly time dependent.

This problem can be handled by using coordinates which are stationary with respect to the loop. The induced emf must be expressed relative to the loop; therefore this transformation to the loop coordinates is quite reasonable.

Coordinates stationary relative to the loop are analogous to the body coordinates of a rigid body in classical mechanics.

We consider the induced emf E' in the rest frame of the loop where an element of length on the loop is $d\mathbf{l}'$. The problem of transformation is applied now to the total time derivative. The general motion of a rigid loop can be quite complicated, involving translations plus rotations relative to the laboratory coordinates. Rotations specifically would require transformation of all time derivatives, as discussed at the end of Appendix A. Simple translations of the loop will serve to illustrate the approach. A nonrelativistic or Galilean transformation of the total time derivative to a frame of reference moving with a velocity **v** relative to the laboratory coordinates is

$$\frac{d}{dt} = \mathbf{v} \cdot \nabla + \frac{\partial}{\partial t}.$$

In the rest frame of the loop the induced emf is

$$E' = \oint_{\mathscr{L}'} \mathscr{E}' \cdot d\mathbf{l}' = -\frac{d}{dt} \iint_{S'} \mathbf{B} \cdot \mathbf{n}'\, dS' = -\iint_{S'} \left\{ (\mathbf{v} \cdot \nabla)\mathbf{B} \cdot \mathbf{n}' + \frac{\partial \mathbf{B}}{\partial t} \cdot \mathbf{n}' \right\} dS',$$

where the primes indicate that the coordinate variables are those of the rest frame of the loop. Using the vector relation

$$\text{curl}\, (\mathbf{v} \times \mathbf{B}) = \mathbf{v} \cdot \text{div}\, \mathbf{B} - \mathbf{B} \cdot \text{div}\, \mathbf{v} + (\mathbf{B} \cdot \nabla)\mathbf{v} - (\mathbf{v} \cdot \nabla)\mathbf{B},$$

and because $\text{div}\, \mathbf{B} = 0$ and **v** is not an explicit function of **r** we find that

$$(\mathbf{v} \cdot \nabla)\mathbf{B} = -\text{curl}\, (\mathbf{v} \times \mathbf{B}).$$

Substituting and employing Stokes' theorem for the first term in the surface integral,

$$\oint_{\mathscr{L}'} \{\mathscr{E}' - \mathbf{v} \times \mathbf{B}\} \cdot d\mathbf{l} = -\iint_{S'} \left\{\frac{\partial \mathbf{B}}{\partial t}\right\} \cdot \mathbf{n}'\, dS'.$$

We write this as

$$\oint_{\mathscr{L}'} \mathscr{E} \cdot d\boldsymbol{l}' = -\iint_{S'} \left\{\frac{\partial \mathbf{B}}{\partial t}\right\} \cdot \mathbf{n}'\, dS',$$

which is the form the equation must take for a stationary loop. Then

$$\mathscr{E}' = \mathscr{E} + \mathbf{v} \times \mathbf{B}.$$

The last equation provides the Galilean transformation of the electric field $\mathscr{E}$ when the loop moves relative to the laboratory coordinates.

Our general relation is then

$$\oint_{\mathscr{L}} \mathscr{E} \cdot d\boldsymbol{l} = -\iint_{S} \left\{\frac{\partial \mathbf{B}}{\partial t}\right\} \cdot \mathbf{n}\, dS$$

where $$\mathscr{E} = \mathscr{E}' - \mathbf{v} \times \mathbf{B}.$$

Both sides of the equation can be expressed as surface integrals,

$$\oint_{\mathscr{L}} \mathscr{E} \cdot d\boldsymbol{l} = \iint_{S} \{\text{curl}\, \mathscr{E}\} \cdot \mathbf{n}\, dS = -\iint_{S} \left\{\frac{\partial \mathbf{B}}{\partial t}\right\} \cdot \mathbf{n}\, dS.$$

Because the geometry of the loop is arbitrary,

$$\text{curl}\, \mathscr{E} = -\frac{\partial \mathbf{B}}{\partial t}.$$

This equation is the fourth Maxwell equation.

To return to the transformation of $\mathscr{E}$, it can now be demonstrated that the Galilean transformation of (d/dt) served to incorporate the Lorentz force into our definition of the induced electric field.

If $\mathbf{B}$ is time independent, then $\partial \mathbf{B}/\partial t$ is zero, and the field $\mathscr{E}$ is zero. When the loop is moving, however, the field $\mathscr{E}'$ does not vanish. Under these conditions

$$\frac{\partial \mathbf{B}}{\partial t} = 0,$$

$$\mathscr{E} = 0,$$

and $$\mathscr{E}' = \mathbf{v} \times \mathbf{B}.$$

Multiplying by the charge q present on each charge carrier, we find that

$$q\mathscr{E}' = q\mathbf{v} \times \mathbf{B} = \mathbf{F}.$$

As one would expect, the Lorentz force associated with the motion of a conductor in an external magnetic field can be interpreted as an electric field $\mathscr{E}'$ acting upon the charges q in the conductor.

In closing, the reader should remember that the preceding development is nonrelativistic. This implies that the magnitude of the velocity of the loop is much less than the speed of light. Actually the transformation of $\mathscr{E}$ as we have developed it holds only up to terms of the order v/c.

The fourth Maxwell equation implies a relation between the electric field $\mathscr{E}$ and the time variations of the vector potential function **A**.

Since
$$\text{curl}\,\mathscr{E} = -\frac{\partial \mathbf{B}}{\partial t}$$

and
$$\mathbf{B} = \text{curl}\,\mathbf{A},$$

we can write
$$\text{curl}\left\{\mathscr{E} + \frac{\partial \mathbf{A}}{\partial t}\right\} = 0.$$

A solution for this differential equation is

$$\mathscr{E} + \frac{\partial \mathbf{A}}{\partial t} = -\text{grad V}$$

where V(x,y,z) is a scalar point function. Then

$$\mathscr{E} = -\text{grad V} - \frac{\partial \mathbf{A}}{\partial t}.$$

This last equation demonstrates that $\mathscr{E}$ can be generated from a scalar and a vector potential function.

B. SYSTEMS OF STATIONARY LOOPS

1. INDIVIDUAL CURRENT LOOPS

Faraday's law transforms into Lentz's law when we express Φ_j, the flux linking the jth loop, in terms of the currents of the N loops making up a closed system.

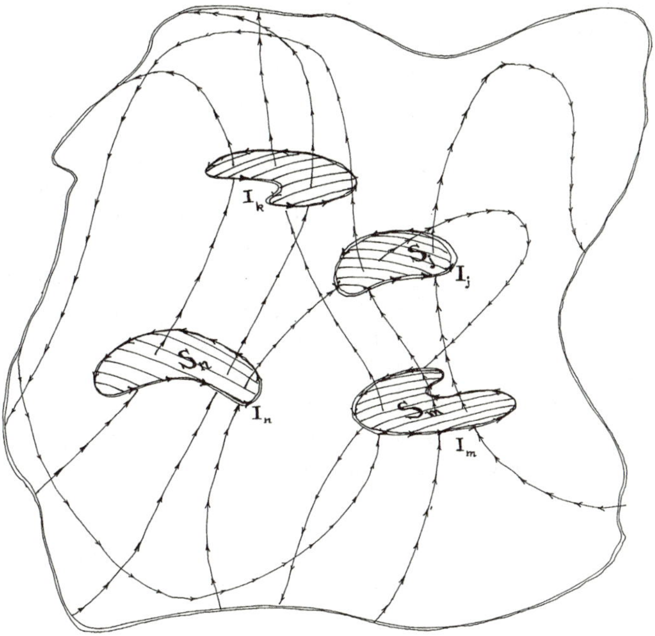

Previously we developed the forms

$$\Phi_j = \sum_{k=1}^{N} L_{jk} I_k.$$

In the case of stationary loops, $dL_{jk}/dt = 0$, and

$$\frac{d\Phi_j}{dt} = \sum_{k=1}^{N} L_{jk}\left(\frac{dI_k}{dt}\right)$$

with $$E_j = -\sum_{k=1}^{N} L_{jk}\left(\frac{dI_k}{dt}\right) \quad \text{(Lentz's law).}$$

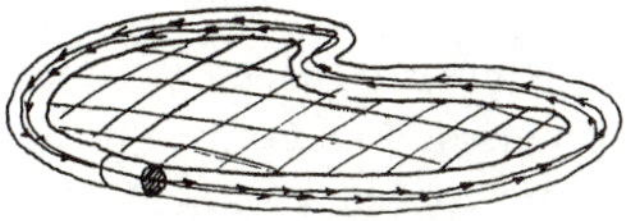

MUTUAL INTERACTION OF FILAMENTS IN THE SAME CONDUCTOR PROVIDES THE SELF INDUCTANCE

When the system consists of a single loop $\mathscr{L}_1$ the only inductance in the problem is the self-inductance L_{11}, and

$$E_1 = -L_{11}\frac{dI_1}{dt} \quad \text{(for an isolated circuit).}$$

It has been the practice to withhold the definition of the L_{mn}'s until Faraday's law has been postulated. This can be unfortunate because the inductance coefficients are geometrical terms which in turn are a direct consequence of Ampere's law. Since we have realized the form of the L_{mn}'s in the discussion of Ampere's law, the fact that these coefficients must have a consistent definition in the transition between Ampere's law and Faraday's law appears quite natural.

In Chapter IX the effects of magnetic media will be introduced. These considerations will indicate that the flux Φ_j in the presence of magnetic media need not be a linear function of the currents. Under such circumstances the emf E_j developed in the jth circuit can be related to the time variation of currents in all the loops with the restriction that the inductance matrix now can be a function of the currents:

$$\Phi_j = \sum_{k=1}^{N} \Phi_{jk}$$

where, as before, Φ_{jk} is the mutual flux linking the jth and k^{th} loops; then

$$E_j = -\frac{d\Phi_j}{dt} = -\sum_{k=1}^{N}\frac{d\Phi_{jk}}{dI_k}\left(\frac{dI_k}{dt}\right).$$

A more general definition of the elements L_{mn} of $\mathbb{L}$ is therefore

$$L_{mn} = \frac{d\Phi_{mn}}{dI_n}.$$

From this definition we again obtain

$$E_j = -\sum_{k=1}^{N} L_{jk}\left(\frac{dI_k}{dt}\right).$$

We can perceive the manner in which the energy relations enter into this problem. Assume that an emf $E_m^{(b)}(t)$ is *imposed* upon the circuit (say by a battery and a switch, or some other external source of emf). The power supplied at any instant by the mth battery, $I_m E_m^{(b)}$, plus the power supplied by the induced emf, $-I_m\dot{\Phi}_m$,* must

* $\dot{\Phi}_m$ is shorthand for $d\Phi_m/dt$.

be equal to the instantaneous power dissipated in the resistance R_m of the mth circuit,

$$I_m E_m^{(b)} - I_m \dot{\Phi}_m = I_m{}^2 R_m.$$

Rearranging, and expanding $\dot{\Phi}_m$,

$$I_m E_m^{(b)} = I_m R_m I_m + I_m \sum_{k=1}^{N} L_{mk}\left(\frac{dI_k}{dt}\right).$$

This relation contains the common factor I_m. Extracting this factor we obtain the usual relation that the sum of the emfs is equal to the voltage drop across the resistance of the circuit. Again the induced emf $-\dot{\Phi}_m$ is transferred to the right-hand side, giving

$$E_m^{(b)} = R_m I_m + \sum_{k=1}^{N} L_{mk} \dot{I}_k.$$

The total energy of N loops is given by a power balance equation

$$\sum_{j=1}^{N} I_j E_j^{(b)} - \sum_{j=1}^{N} \sum_{k=1}^{N} I_j L_{jk} \dot{I}_k = \sum_{j=1}^{N} I_j R_j I_j.$$

To obtain some insight into the negative sign associated with the induced emf let us consider the relatively simple two-loop configuration shown in the diagram on the following page.

Assume that loop 1 is initially connected to a battery and that a steady state current $I_1(0)$ flows in loop 1. Loop 2 is a closed loop. When I_1 is constant, I_2 is zero. Using a shorting switch we remove the battery from the loop of circuit 1.

The initial state of the system has a steady state magnetic field set up by $I_1(0)$. This field exists throughout all space and energy is stored in space because of this field. A volume element $\Delta\tau$ located a distance $\mathbf{r}$ from the first loop has an energy associated with it

$$\Delta U = \frac{B^2(r)}{2\mu_0} \Delta\tau.$$

When the current source in loop 1 is removed by the short, the energy of the field of the loop must be dissipated either by radiation or by absorption in the loop resistances. Actually both of these energy sinks are available; however, the dissipation of the energy by radiation can be shown to be small.

Viewing the energy as stored in terms of the fields we are left with an energy transport problem in which, as the field changes, the energy initially stored in the magnetic field is absorbed by the conducting loops and dissipated by the resistances.

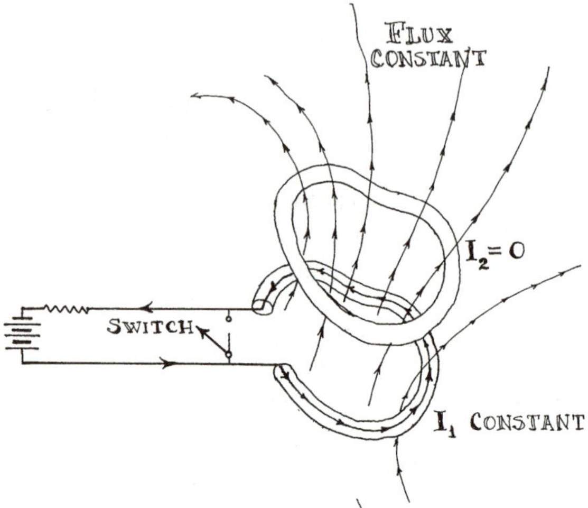

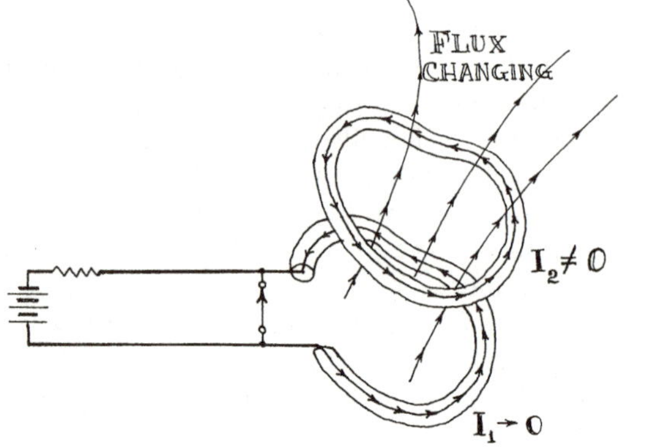

If the transport velocity of the field were infinite, the absorption would consist of an instantaneous impulse. However, the transport velocity is finite and thus the field **B** cannot collapse instantaneously.

Faraday's law accounts for this quite conveniently. As the **B** field begins to collapse, the induced emfs $-\dot{\Phi}_j$ set up currents *which tend to oppose the change*. These induced currents die out exponentially as the field energy is dissipated in I^2R losses.

In the two-loop problem under consideration, as the field starts to decrease the induced emf sets up a current in such a direction as to oppose the decrease. Thus the induced currents $I_1(t)$ and $I_2(t)$ produce field components in the direction of the initial field. This accounts for the minus sign. The emf is directed to oppose $\dot{I}_1$ (or $\dot{\Phi}$).

When the field **B** is increased, emfs are developed which create currents opposing the increase.

The behavior of inductances in circuits will be covered in more detail in Chapter XII.

2. EXTENDED CONDUCTORS: SKIN EFFECT

In a conducting medium, if **J** varies in time the associated magnetic field varies also. As a result an induced electric field $\mathscr{E}$ is set up. This induced electric field acts upon the charge carriers in such a manner as to alter the current density.

In a nonmagnetic conductor

$$\text{curl}\,\mathbf{B} = \mu_0\mathbf{J}.$$

If the magnetic properties of the conductor are isotropic and homogeneous, curl **B** is still proportional to **J**; however, the curl is scaled by a factor μ,

$$\text{curl}\,\mathbf{B} = \mu_0\mu\mathbf{J}.$$

The factor μ is called the relative magnetic permeability of the medium. Comparing the two equations it is apparent that the magnetic permeability of a nonmagnetic conductor is 1.

If the conductor is stationary at every point,

$$\text{curl}\,\mathscr{E} = -\frac{\partial\mathbf{B}}{\partial t}.$$

Assuming that Ohm's law holds in the conductor,

$$\mathbf{J} = \beta\mathscr{E}.$$

Taking the curl of Faraday's law we substitute Ampere's law and Ohm's law,

$$\text{curl curl}\,\mathscr{E} = \text{curl curl}\left[\frac{\mathbf{J}}{\beta}\right]$$

$$= -\frac{\partial}{\partial t}(\text{curl}\,\mathbf{B}) = -\frac{\partial}{\partial t}(\mu\mu_0\mathbf{J}).$$

Expanding curl curl and assuming that **J** is solenoidal (div **J** = 0),

$$-\nabla^2\left[\frac{\mathbf{J}}{\beta}\right] = -\mu\mu_0\frac{\partial\mathbf{J}}{\partial t},$$

or

$$\nabla^2\mathbf{J} = \beta\mu\mu_0\frac{\partial\mathbf{J}}{\partial t}.$$

This is the equation governing eddy currents in conductors, and it can be solved by the method of separation of variables.

An example can be given briefly in the form of an idealized problem. Assume that a current varying as $e^{i\omega t}$ is present in a

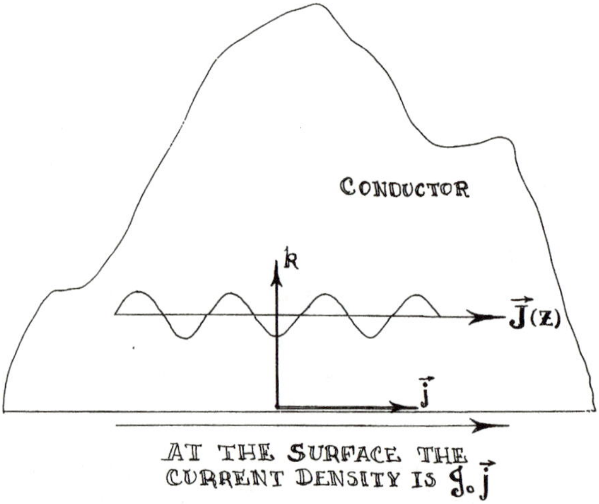

conductor β which fills the region above z = 0. In addition we will assume that, near the origin, **J** is unidirectional along the y axis. The region z < 0 is a vacuum and **J**(z < 0) is zero.

To maintain finite **J** we assume that as $z \to +\infty$, $|\mathbf{J}| \to 0$.

The symmetry in this problem is rectangular, so we take

$$\mathbf{J} = \mathscr{J}_0 Z(z)e^{i\omega t}\mathbf{j}.$$

Then

$$\frac{d^2Z}{dz^2} = i\omega\beta\mu\mu_0 Z(z) = +\alpha^2 Z(z),$$

giving

$$\alpha = (1 + i)\left\{\frac{\beta\omega\mu\mu_0}{2}\right\}^{1/2}$$

and

$$Z = Ae^{\alpha z} + B^{-\alpha z}.$$

If $|\mathbf{J}|$ is to be finite as $z \to \infty$, the coefficient A must be zero. The constant $\mathscr{J}_0$ is the magnitude of the current density at the surface, i.e., at $z = 0$. The complete expression for $\mathbf{J}$ is then

$$\mathbf{J} = \mathscr{J}_0 e^{-\alpha z} e^{i\{\omega t - \alpha z\}} \mathbf{j}$$

where

$$\alpha = \sqrt{\beta\omega\mu_0\mu/2}.$$

We find that $\mathbf{J}$ decreases rapidly as z increases. The depth at which the current is down by a factor e^{-1} is $z' = 1/\alpha$. As the frequency $\omega/2\pi$ increases, the penetration depth z' decreases as $1/\omega^{1/2}$; thus at high frequencies the current flows mainly on the surface, increasing the effective resistance of the medium.

The phase of the current wave lags as z increases. At the depth z' the phase lag is 1 radian.

C. MOTIONAL EMFS

The Lorentz force exerted upon the mobile charge in a conductor which moves in a magnetic field is in effect an induced emf.

1. SLIDING BAR PROBLEMS

Consider a long cylindrical conductor of length l moving relative to a magnetic field $\mathbf{B}$. The electromotive force developed from a to b

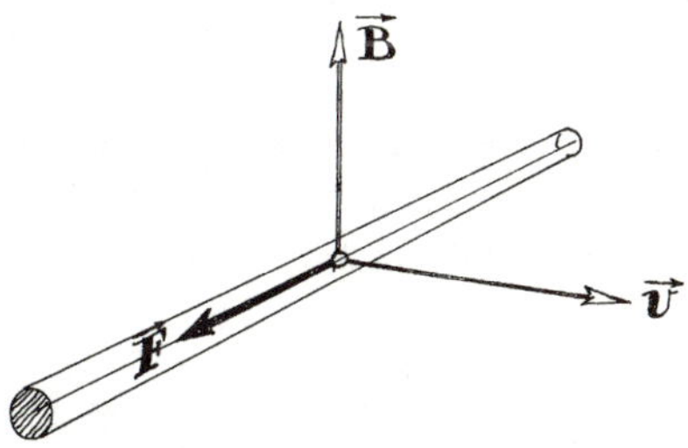

across the length of the conductor is

$$E_{ab} = \int_a^b \frac{\mathbf{F}}{q} \cdot d\mathbf{r} = \int_a^b (\mathbf{v} \times \mathbf{B}) \cdot d\mathbf{r}.$$

If **B** is constant (remembering that **v** is independent of **r**),

$$E_{ab} = \boldsymbol{l} \cdot (\mathbf{v} \times \mathbf{B}) = -\mathbf{B} \cdot (\mathbf{v} \times \boldsymbol{l}) = -\mathbf{B} \cdot \mathbf{n} \frac{dS}{dt}$$

$$= -d\Phi/dt = -(\text{rate at which flux is swept out}).$$

As an elementary example $\boldsymbol{l}$, **v**, and **B** can be taken mutually perpendicular and then $E_{ab} = Blv$.

Because **F** is perpendicular to **B** and **v**, there is zero power associated with **F**. However, once a current flows because of the induced emf, an electromagnetic reaction sets in. In other words the induced emf sets up a current I which then interacts with **B** providing a reaction force **f**.

To illustrate this reaction force, allow the conducting cylinder to slide along two rails which are perpendicular to $\boldsymbol{l}$ and parallel to **v**,

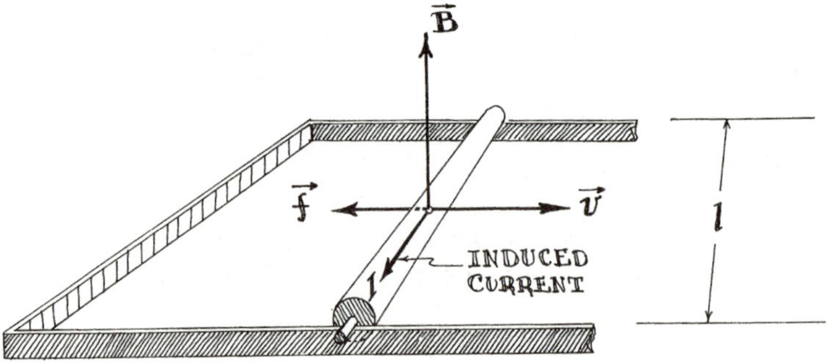

for simplicity. At one end the rails are connected to close the circuit. If the total resistance of the circuit at any time t is R(t), neglecting inductance, then the current at t is

$$I(t) = \frac{E_{ab}}{R(t)}.$$

I(t) interacts with **B** to give a force of reaction

$$\mathbf{f} = I(t)\, \boldsymbol{l} \times \mathbf{B},$$

and
$$f = \frac{[\boldsymbol{l} \cdot (\mathbf{v} \times \mathbf{B})](\boldsymbol{l} \times \mathbf{B})}{R(t)}.$$

The mechanical power required to move the conductor along the rail is

$$P_f = \mathbf{f} \cdot \mathbf{v} = \frac{[\boldsymbol{l} \cdot (\mathbf{v} \times \mathbf{B})][\mathbf{v} \cdot (\boldsymbol{l} \times \mathbf{B})]}{R(t)}$$

or $$P_f = -\frac{[\boldsymbol{l} \cdot (\mathbf{v} \times \mathbf{B})]^2}{R(t)} = -\frac{E_{ab}{}^2}{R(t)}.$$

2. THE ROTATING COIL

Consider a rectangular coil of N turns having a length $\boldsymbol{l}$ parallel to the axis of rotation ($\boldsymbol{\omega}$) and a dimension d perpendicular to that axis. The rotation axis $\boldsymbol{\omega}$ bisects the dimension d.

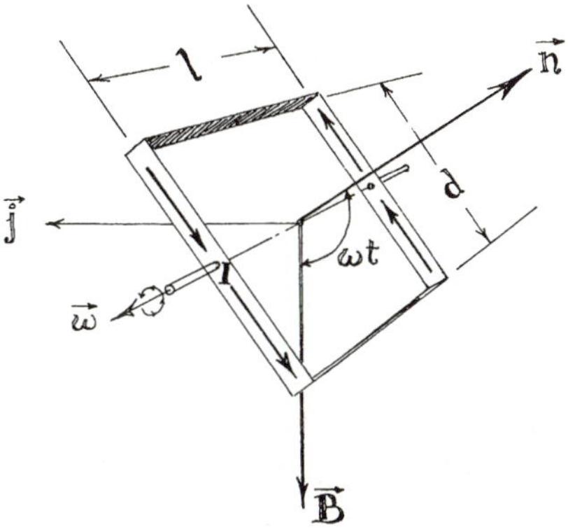

If the coil rotates at a constant angular velocity $\boldsymbol{\omega}$ in a constant magnetic field $\mathbf{B}$ which is perpendicular to $\boldsymbol{\omega}$, the emf set up in a single turn is

$$E = 2[\boldsymbol{l} \cdot (\mathbf{v} \times \mathbf{B})]$$

where $\boldsymbol{l}$ is in the direction of the current flow. $\mathbf{v}$ is parallel or anti-parallel to the unit vector $\mathbf{n}$ normal to the plane cap surface of the rectangle. The velocity of the arms is

$$\mathbf{v} = \pm\boldsymbol{\omega} \times \mathbf{d}/2 = \pm\frac{\omega d}{2}\,\mathbf{n}.$$

If $\boldsymbol{\omega}$ is constant,

$$\mathbf{n} = -\sin\omega t\,\mathbf{j} + \cos\omega t\,\mathbf{k}$$

for the geometry shown. Thus for one turn

$$E = 2B\frac{l\omega d}{2}\sin\omega t = B(ld)\omega\sin\omega t.$$

The cap surface area of the loop is ld, and the maximum flux linkage is

$$\Phi_{max} = B(ld).$$

The total emf developed for N turns is

$$E_N = N\,\Phi_{max}\sin\omega t.$$

For this simplified generator, the general characteristic is that the total emf developed depends upon the speed of rotation and the magnitude of the impressed field.

3. A POINT CHARGE q OUTSIDE AN INFINITE SOLENOID

Consider for the moment an infinite solenoid of radius R and N turns per unit length, carrying a current of I amperes per turn, and oriented to have the symmetry axis along the z axis.

Assume that a point charge q is located at a point designated by $\mathbf{a}_p$ outside of the solenoid. The current is now varied in time, $I = I(t)$.

If a conducting loop of wire linked the solenoid, the presence of a changing magnetic field (the interior field of the solenoid) across the cap surface of the loop would imply that an electric field should be induced in the loop. The infinite solenoid is an interesting case since the external magnetic field is *always* zero. This seeming paradox is quickly resolved when it is remembered that the infinite solenoid has a nonvanishing VECTOR POTENTIAL in the exterior region:

$$\mathbf{A}_p(\mathbf{a}_p;t) = \frac{\mu_0 NIR^2}{2}\frac{(\mathbf{k}\times\mathbf{a}_p)}{a_p^2}, \quad R \leqslant a_p.$$

Because a time-varying vector potential function gives rise to an electric field $\mathscr{E}$, the force on a charge q outside of an infinite solenoid having a time-varying current is

$$\mathbf{F}_q = q\mathscr{E}_p = -q\frac{\partial\mathbf{A}_p}{\partial t} = -\left\{\frac{q\mu_0 NR^2}{2}\frac{(\mathbf{k}\times\mathbf{a}_p)}{a_p^2}\right\}\frac{dI}{dt}.$$

The exterior vector potential function could have been deduced from Faraday's law by examining the induced emf in a circular loop of radius a_p. From the symmetry

$$E = \oint_{\text{Circle}} \mathscr{E} \cdot d\mathbf{l} = |\mathscr{E}|\, 2\pi a_p = -\frac{d\Phi}{dt} = -\frac{d}{dt}\{\mu_0 N I \pi R^2\}.$$

The magnitude of the exterior electric field is then

$$|\mathscr{E}| = -\left\{\frac{\mu_0 N R^2}{2a_p}\right\}\frac{dI}{dt} = -\left|\frac{\partial \mathbf{A}}{\partial t}\right|.$$

The force on a charge q outside the infinite solenoid would then be perpendicular to the position vector $\mathbf{a}_p$. The resulting orbit is a spiral.

IX. Magnetic Media

~~~~~~

A. THE MAGNETIC FIELD OF A CONTINUOUS DISTRIBUTION OF MOMENTS

If a region τ contains a volume distribution of point magnetic moments $\mathbf{m_v}$ (analogous to the volume distribution of electric dipoles), then the vector magnetic potential at any point P is

$$\mathbf{A_p} = \frac{\mu_0}{4\pi} \int\!\!\!\int\!\!\!\int_\tau \mathbf{m_v} \times \frac{(\mathbf{r_p} - \mathbf{r})}{|\mathbf{r_p} - \mathbf{r}|^3} \, d\tau,$$

or

$$\mathbf{A_p} = \frac{\mu_0}{4\pi} \int\!\!\!\int\!\!\!\int_\tau \mathbf{m_v} \times \mathrm{grad}_r \left\{ \frac{1}{|\mathbf{r_p} - \mathbf{r}|} \right\} \, d\tau.$$

The integrand can be expanded into two terms:

$$\mathbf{m_v} \times \mathrm{grad}_r \left\{ \frac{1}{|\mathbf{r_p} - \mathbf{r}|} \right\} = -\mathrm{curl}_r \left\{ \frac{\mathbf{m_v}}{|\mathbf{r_p} - \mathbf{r}|} \right\} + \frac{\mathrm{curl}\ \mathbf{m_v}}{|\mathbf{r_p} - \mathbf{r}|}.$$

The first term in the expansion can be converted from a volume integral to a closed surface integral,

$$\int\!\!\!\int\!\!\!\int_\tau \mathrm{curl}_r \left\{ \frac{\mathbf{m_v}}{|\mathbf{r_p} - \mathbf{r}|} \right\} \, d\tau = \int\!\!\!\int_S \frac{(\mathbf{m_v} \times \mathbf{n})}{|\mathbf{r_p} - \mathbf{r}|} \, dS.$$

Then

$$\mathbf{A_p} = \frac{\mu_0}{4\pi} \int\!\!\!\int\!\!\!\int_\tau \left\{ \frac{\mathrm{curl}_r\ \mathbf{m_v}}{|\mathbf{r_p} - \mathbf{r}|} \right\} \, d\tau + \frac{\mu_0}{4\pi} \int\!\!\!\int_S \frac{(\mathbf{m_v} \times \mathbf{n})}{|\mathbf{r_p} - \mathbf{r}|} \, dS.$$

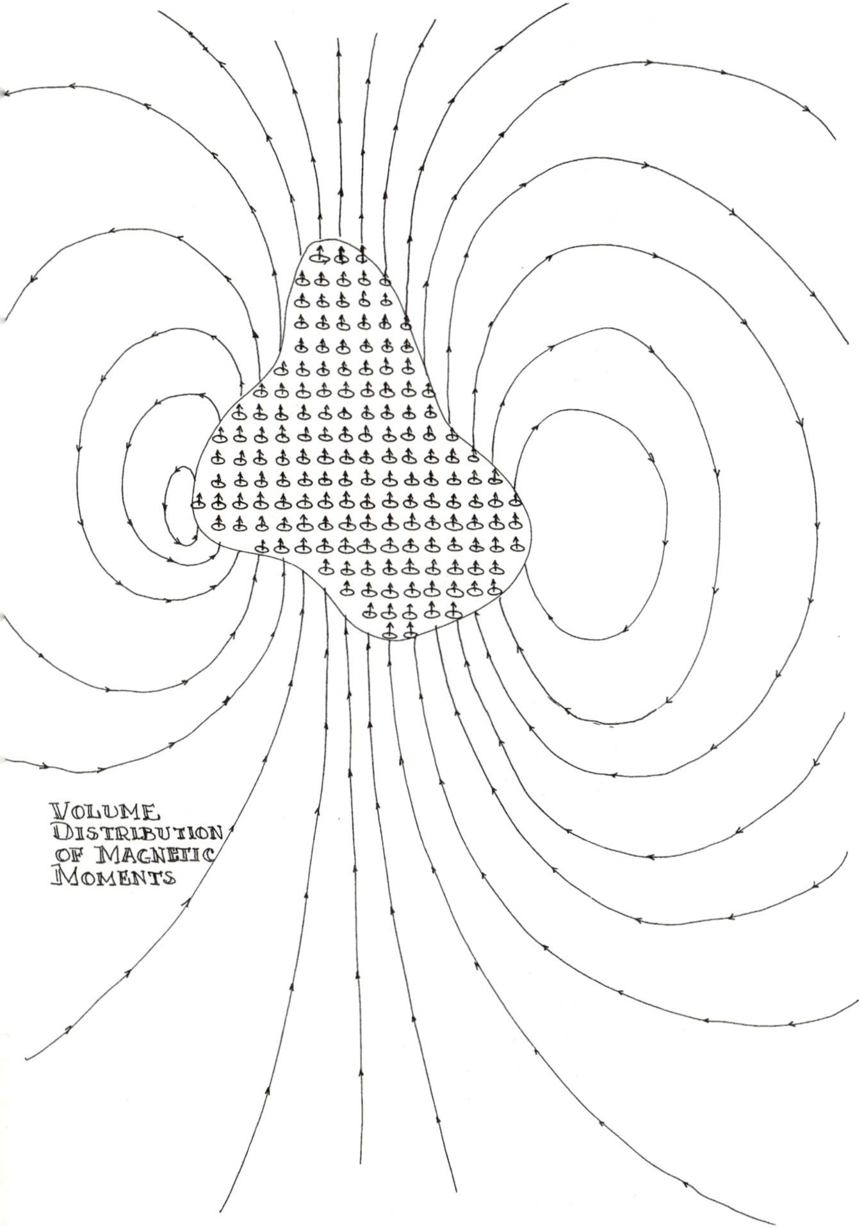
VOLUME
DISTRIBUTION
OF MAGNETIC
MOMENTS

This particular form corresponds to the definition of $\mathbf{A}_p$ in terms of AMPERIAN CURRENTS:

A Volume Distribution $\mathbf{J}_A = \text{curl } \mathbf{m}_V$,

A Surface Current $\mathbf{I}_A = \mathbf{m}_V \times \mathbf{n}$.

The surface current $\mathbf{I}_A$ is the current per unit length ($\mathbf{I}_A$ is perpendicular to the length) measured on the surface in a shell of thickness δ

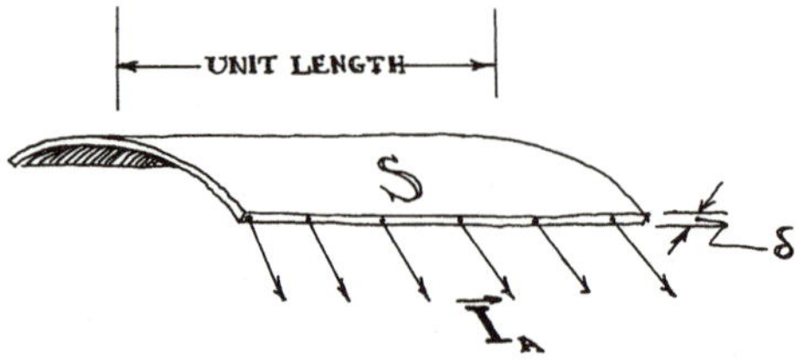

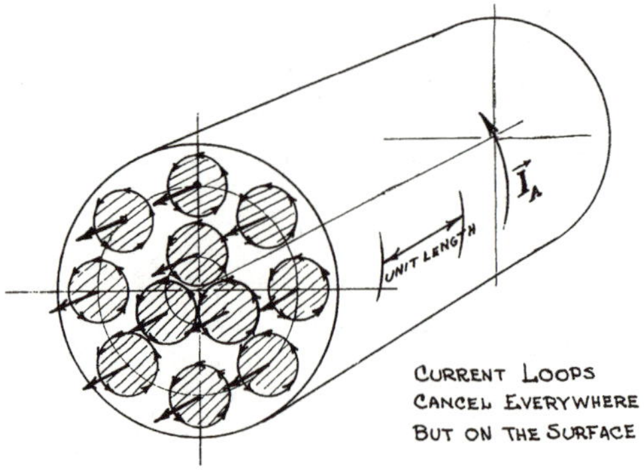

where δ is arbitrarily small. In the surface shell the equivalent current density is $\mathbf{J'}_A = \mathbf{I}_A/\delta$.

The diagrams above and opposite illustrate (a) the relation between the curl of $\mathbf{m}_V$ and $\mathbf{J}_A$ and (b) a homogeneous distribution of constant $\mathbf{m}_V$ resulting in a surface current on the cylindrical surface.

To proceed with the method of Amperian currents, the magnetic induction field at a point P is

$$\mathbf{B}_p = \text{curl}_p \left\{ \frac{\mu_0}{4\pi} \iiint_\tau \frac{\mathbf{J}_A}{|\mathbf{r}_p - \mathbf{r}|} d\tau + \frac{\mu_0}{4\pi} \iint_S \frac{\mathbf{I}_A}{|\mathbf{r}_p - \mathbf{r}|} dS \right\}.$$

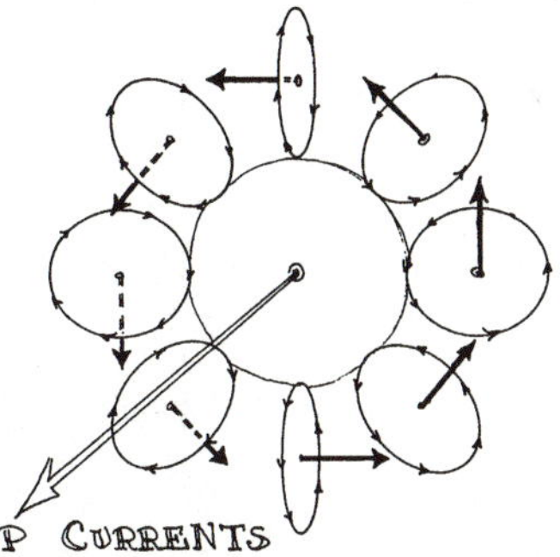
A DISTRIBUTION
$\vec{m}_v$ WITH A NON-ZERO
CURL
LOOP CURRENTS
SUPERPOSE AT CENTER
TO PRODUCE AN INDUCED
CURRENT DENSITY $\vec{J}_A$

Using relations of the type

$$\text{curl}_p \left\{\frac{\mathbf{J}_A(\mathbf{r})}{|\mathbf{r}_p - \mathbf{r}|}\right\} = \frac{1}{|\mathbf{r}_p - \mathbf{r}|} \text{curl}_p\, \mathbf{J}_A(\mathbf{r}) + \text{grad}_p \left\{\frac{1}{|\mathbf{r}_p - \mathbf{r}|^3}\right\} \times \mathbf{J}_A$$

$$= \nabla_p \left\{\frac{1}{|\mathbf{r}_p - \mathbf{r}|}\right\} \times \mathbf{J}_A = \mathbf{J}_A \times \frac{(\mathbf{r}_p - \mathbf{r})}{|\mathbf{r}_p - \mathbf{r}|^3},$$

it is possible to incorporate $\{\mathbf{I}_A/\delta\}\, dS$ into $\mathbf{J}_A\, d\tau$ to simplify the form, and then

$$\mathbf{B}_p = \frac{\mu_0}{4\pi} \iiint_\tau \left\{\mathbf{J}_A \times \frac{(\mathbf{r}_p - \mathbf{r})}{|\mathbf{r}_p - \mathbf{r}|^3}\right\} d\tau.$$

Using the method employed previously to derive the differential form of Ampere's law, the magnetic field of a volume distribution of magnetic moments is given in terms of the Amperian current density, $\mathbf{J}_A$, as

$$\text{curl}\, \mathbf{B}_m = \mu_0 \mathbf{J}_A = \mu_0\, \text{curl}\, \mathbf{m}_v$$

(for magnetic media in the absence of external free currents). We designate the magnetic induction field of the magnetic moments as $\mathbf{B}_m$. Then

$$\text{curl}\, \{\mathbf{B}_m - \mu_0 \mathbf{m}_v\} = 0,$$

and $$\mathbf{B}_m - \mu_0 \mathbf{m}_v = -\text{grad}\, W,$$

where W is a scalar potential.

In other words, given the fact that curl $\{\mathbf{B}_m - \mu_0\mathbf{m}_v\}$ is zero, the result shown above is the most general solution. Curl grad W is zero if W is defined at every point of interest.

The only remaining problem then is to discover the form of the scalar function W when this particular representation of $\mathbf{B}_p$ is to be employed. To accomplish this we take the curl of $\mathbf{A}_p$,

$$\mathbf{B}_p = \text{curl}\,\mathbf{A}_p = \text{curl}_p \frac{\mu_0}{4\pi} \iiint_\tau \mathbf{m}_v \times \frac{(\mathbf{r}_p - \mathbf{r})}{|\mathbf{r}_p - \mathbf{r}|^3}\, d\tau.$$

The integrand can be expanded using the following relation (see Appendix A):

$$\text{curl}_p\left\{\mathbf{m}_v \times \frac{\mathbf{r}'}{r'^3}\right\} = \mathbf{m}_v \text{div}_p\left(\frac{\mathbf{r}'}{r'^3}\right) - \frac{\mathbf{r}'}{r'^3}\text{div}_p\,\mathbf{m}_v - (\mathbf{m}_v\cdot\nabla_p)\left(\frac{\mathbf{r}'}{r'^3}\right) + \left(\frac{\mathbf{r}'\cdot\nabla_p}{r'^3}\right)\mathbf{m}_v$$

where $$\mathbf{r}' = \mathbf{r}_p - \mathbf{r},$$

and ∇_p operating upon $m_v(\mathbf{r})$ is zero. Thus

$$\mathbf{B}_p = \frac{\mu_0}{4\pi}\iiint_\tau \text{curl}_p\left\{\mathbf{m}_v \times \frac{\mathbf{r}'}{r'^3}\right\} d\tau$$

$$= \frac{\mu_0}{4\pi}\iiint_\tau \mathbf{m}_v\,\text{div}_p\left(\frac{\mathbf{r}'}{r'^3}\right) d\tau - \frac{\mu_0}{4\pi}\iiint_\tau \{\mathbf{m}_v\cdot\nabla_p\}\left(\frac{\mathbf{r}'}{r'^3}\right) d\tau.$$

Recall that $\text{div}_p\,(\mathbf{r}'/r'^3)$ is zero everywhere except at $\mathbf{r}' = 0$ where the operation ordinarily might be undefined. In Chapter II it was demonstrated that $\text{div}_p\,\{(\mathbf{r}_p - \mathbf{r})/|\mathbf{r}_p - \mathbf{r}|^3\}$ appearing under an integral sign behaves as the Dirac delta function $+4\pi\delta(\mathbf{r} - \mathbf{r}_p)$. This was demonstrated in several ways, one of which was a conversion to a surface integral with the singularity excluded by a small sphere. As the radius of the sphere went to zero the surface integral surrounding the singular point took on the value 4π, indicating that the volume integral of $\text{div}\,(\mathbf{r}'/r'^3)$ was 4π when the singular point $\mathbf{r}' = 0$ lay within the volume. Thus

$$\mathbf{B}_p = \frac{\mu_0}{4\pi}\iiint_\tau \mathbf{m}_v(\mathbf{r})4\pi\delta(\mathbf{r} - \mathbf{r}_p)\, d\tau - \frac{\mu_0}{4\pi}\iiint_\tau \{\mathbf{m}_v\cdot\nabla_p\}\left(\frac{\mathbf{r}'}{r'^3}\right) d\tau$$

$$= \mu_0\mathbf{m}_v(\mathbf{r}_p) - \frac{\mu_0}{4\pi}\iiint \{\mathbf{m}_v\cdot\nabla_p\}\left(\frac{\mathbf{r}'}{r'^3}\right) d\tau.$$

The last integral in this expression can be converted to the gradient of a scalar by the expansion (see Appendix A)

$$\mathrm{grad_p}\left\{\mathbf{m}_V\cdot\frac{\mathbf{r}'}{r'^3}\right\}=\{\mathbf{m}_V\cdot\nabla_p\}\left(\frac{\mathbf{r}'}{r'^3}\right)+\left\{\frac{\mathbf{r}'}{r'^3}\cdot\nabla_p\right\}\mathbf{m}_V$$
$$+\,\mathbf{m}_V\,\mathrm{curl_p}\left(\frac{\mathbf{r}'}{r'^3}\right)+\left(\frac{\mathbf{r}'}{r'^3}\right)\times\mathrm{curl_p}\,\mathbf{m}_V.$$

Inside the integral $\mathbf{m}_V$ is not a function of $\mathbf{r}_p$; therefore we can write

$$-\iiint_\tau\{\mathbf{m}_V\cdot\nabla_p\}\left(\frac{\mathbf{r}'}{r'^3}\right)d\tau=\iiint_\tau\mathbf{m}_V\times\mathrm{curl_p}\left(\frac{\mathbf{r}'}{r'^3}\right)d\tau$$
$$-\iiint_\tau\mathrm{grad_p}\left\{\mathbf{m}_V\cdot\frac{\mathbf{r}'}{r'^3}\right\}d\tau.$$

Once again we have a term which involves the differentiation of the singular function $(\mathbf{r}_p-\mathbf{r})/|\mathbf{r}_p-\mathbf{r}|^3$. The curl of this term can be taken as zero everywhere *except at the point* $\mathbf{r}'=0$ or $\mathbf{r}_p=\mathbf{r}$, because

$$\mathrm{curl_p}\,\frac{\mathbf{r}'}{r'^3}=-\mathrm{curl_p}\,\mathrm{grad_p}\,\frac{1}{r'}\,.$$

The integral can be converted to $\mathrm{curl_r}$ and

$$\iiint_\tau\mathrm{curl_p}\left(\frac{\mathbf{r}'}{r'^3}\right)d\tau=-\iiint_\tau\mathrm{curl_r}\left(\frac{\mathbf{r}'}{r'^3}\right)d\tau=\iint_S\mathbf{n}\times\frac{\mathbf{r}'}{r'^3}\,dS.$$

To evaluate this integral, surround the singular point by a small spherical surface S_R connected to the external surface S by a tube of negligible area. Define the volume included between S and S_R

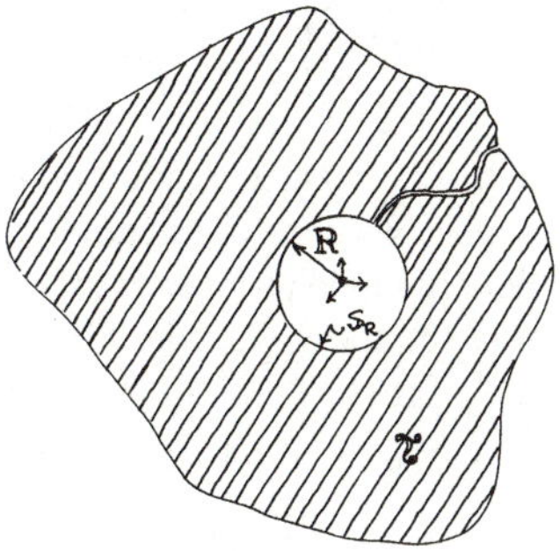

(excluding $r = r_p$) as τ'. In the limit as the radius R of the sphere goes to zero, τ' approaches τ.

In the region external to r_p, $\text{curl}_r\,(\mathbf{r}'/r'^3) = 0$. Therefore

$$\iiint_{\tau'} \text{curl}_r\left(\frac{\mathbf{r}'}{r'^3}\right) d\tau = 0 = \iint_{S_\tau} \mathbf{n} \times \frac{\mathbf{r}'}{r'^3}\, dS + \iint_{S_R} \mathbf{n}_R \times \frac{\mathbf{r}'}{r'^3}\, dS_R.$$

$$\lim_{R\to 0} \iint \mathbf{n}_R \times \frac{\mathbf{r}'}{r'^3}\, dS_R = \lim_{R\to 0} \left\{\frac{-\mathbf{R}}{R} \times \frac{\mathbf{R}}{R^3} 4\pi R^2\right\} = 0 = \lim_{R\to 0} - \iint_S \mathbf{n} \times \frac{\mathbf{r}'}{r'^3}\, dS,$$

and therefore $$\iiint_\tau \text{curl}_p\left(\frac{\mathbf{r}'}{r'^3}\right) d\tau = 0.$$

It may appear that a great amount of effort has been expended to demonstrate that the integral of the undefined quantity $\text{curl}_p\,(\mathbf{r}'/r'^3)$ is indeed zero. The purpose of this demonstration is to impress upon the reader the great care which must be exercised when using the differential operators at undefined points, particularly when integrals of the type shown are used. Although the integral of this curl term vanishes, this is certainly not, though often so stated, a consequence of the operation of the curl on the gradient operator, since these operations cannot be clearly defined at singular points.

Substituting our expansion into the integral of $(\mathbf{m}_v \cdot \nabla_p)(\mathbf{r}'/r'^3)$, we obtain

$$\mathbf{B}_p = \mu_0 \mathbf{m}_v(\mathbf{r}_p) - \text{grad}_p \left\{\frac{\mu_0}{4\pi} \iiint_\tau \mathbf{m}_v \cdot \frac{\mathbf{r}'}{r'^3}\, d\tau\right\},$$

and the scalar potential function W is defined by this last integral:

$$\mathbf{B}(\mathbf{r}_p) = \mu_0 \mathbf{m}_v(\mathbf{r}_p) - \text{grad}_p\, W(\mathbf{r}_p);$$

giving

$$W(\mathbf{r}_p) = \frac{\mu_0}{4\pi} \iiint_\tau \frac{\mathbf{m}_v \cdot \mathbf{r}'}{r'^3}\, d\tau.$$

For a volume distribution of magnetic moments the distribution $\mathbf{m}_v$ may be irrotational. Because the magnetic induction field $\mathbf{B}_m$ is solenoidal a vector field is added to $\mathbf{m}_v$ to satisfy the equality. Outside the volume containing the moments, the magnetic induction field is governed completely by $\text{grad}_p\, W$.

The second Maxwell equation is

$$\text{div}_p\, \mathbf{B}_p = 0;$$

therefore $\mu_0 \operatorname{div} \mathbf{m}_v - \nabla^2 W = 0$ inside τ

$$-\nabla^2 W = 0 \quad \text{outside } \tau.$$

Suppose that we are given an example of a spherical volume of radius R containing a constant dipole moment per unit volume $\mathbf{m}_v = m_v\mathbf{k}$ throughout. The field $\mathbf{B}_p$ at any point P can be computed using the Amperian currents or by utilizing the scalar potential W_p.

The vector magnetic potential at any point P in terms of the Amperian currents is

$$\mathbf{A}_p = \frac{\mu_0}{4\pi}\int_0^{\pi}\int_0^{2\pi} \frac{m_v R^2 \sin^2\theta\,[\sin\phi\,\mathbf{i} - \cos\phi\,\mathbf{j}]\,d\theta\,d\phi}{\{R^2 + r_p^2 - 2Rr_p(\cos\theta_p\cos\theta + \sin\theta_p\sin\theta\cos\phi)\}^{1/2}}.$$

This is a rather formidable integral. To illustrate the usefulness of the scalar function W_p, we solve the problem from the form

$$\mathbf{B} = \mu_0\mathbf{m}_v - \operatorname{grad} W,$$

remembering the two conditions on **B** used to obtain this particular equation,

$$\operatorname{div}\mathbf{B} = 0$$

and

$$\operatorname{curl}\{\mathbf{B} - \mu_0\mathbf{m}_v\} = 0.$$

The first condition (from the divergence theorem) implies that the *normal component*, B_r, is conserved across the boundary of the sphere. Using spherical coordinates,

$$\boldsymbol{\epsilon}_r \cdot \mathbf{B}_{in}(r_p = R) = \boldsymbol{\epsilon}_r \cdot \mathbf{B}_{out}(R).$$

The second condition states that the tangential component (coefficient of $\boldsymbol{\epsilon}_\theta$ in this example) of $\{\mathbf{B} - \mu_0\mathbf{m}_v\}$ is conserved at $r = R$:

$$\boldsymbol{\epsilon}_\theta \cdot \{\mathbf{B}_{in} - \mu_0\mathbf{m}_v\} = \boldsymbol{\epsilon}_\theta \cdot \{\mathbf{B}_{out} - 0\}.$$

The fact that W(r) is a solution of Laplace's equation is demonstrated by the first condition:

$$\operatorname{div}\mathbf{B} = \operatorname{div}\{\mu_0\mathbf{m}_v - \operatorname{grad} W\} = 0.$$

Thus $\nabla^2 W = 0 \quad (r < R \text{ and } r > R),$

for since $\mathbf{m}_v = m_v\mathbf{k} = m_v(\cos\theta\,\boldsymbol{\epsilon}_r - \sin\theta\,\boldsymbol{\epsilon}_\theta)$, $\operatorname{div}\mathbf{m}_v = 0$ inside the sphere.

The boundary conditions on the r and θ components enable us to handle div $\mathbf{m}_v$ at the discontinuity.

$$\mathbf{B}_{in} = \mu_0\mathbf{m}_v - \operatorname{grad} W_{in}.$$

$$\mathbf{B}_{out} = -\operatorname{grad} W_{out}.$$

The interior potential W_{in} must be well behaved at $r = 0$; therefore, as in previous potential problems using spherical coordinates,

$$W_{in} = \sum_{n=0}^{\infty} a_n r^n P_n(\cos\theta)$$

W_{out}, the exterior solution, must be well behaved for large r; thus

$$W_{out} = \sum_{n=0}^{\infty} b_n r^{-(n+1)} P_n(\cos\theta) + a_0 + a_1 r P_1(\cos\theta).$$

When r is large, the system of magnetic dipoles near the origin produces the standard dipole field accompanied by a dipole scalar potential:

$$W_{out} \xrightarrow[\text{r large}]{} \frac{\mu_0}{4\pi}\left\{\frac{4\pi}{3} R^3 m_v\right\} \frac{\cos\theta}{r^2}.$$

Equating the exterior expansion to this form at large r and equating like coefficients of $\cos\theta$, we find that the only nonvanishing b_n is b_1, and

$$b_1 = \frac{\mu_0}{3} m_v R^3,$$

with
$$W_{out} = \frac{\mu_0}{3} m_v \frac{R^3}{r^2} P_1(\cos\theta).$$

$$\mathbf{B}_{out} = -\text{grad } W_{out} = -\sum_{k=1}^{3} \frac{\boldsymbol{\epsilon}_k}{h_k} \frac{\partial W_{out}}{\partial \xi_k}$$

$$= \frac{\mu_0}{3} m_v \frac{R^3}{r^3} \{2\cos\theta\, \boldsymbol{\epsilon}_r + \sin\theta\, \boldsymbol{\epsilon}_\theta\}.$$

In the equations above ξ_k is a generalized coordinate.

The interior solution is of the form (taking $\mathbf{m}_v$ and grad W_{in})

$$\mathbf{B}_{in} = \mu_0 m_v(\cos\theta\, \boldsymbol{\epsilon}_r - \sin\theta\, \boldsymbol{\epsilon}_\theta)$$
$$- \boldsymbol{\epsilon}_r \sum_{n=1}^{\infty} a_n n r^{n-1} P_n - \frac{\boldsymbol{\epsilon}_\theta}{r} \sum_{n=0}^{\infty} a_n r^n \left(\frac{dP_n}{d\theta}\right).$$

At $r = R$ the equality of the normal components of $\mathbf{B}$ is

$$\mu_0 m_v \cos\theta - \{a_1 \cos\theta + a_2 2RP_2 + \ldots\} = \frac{\mu_0}{3} 2\cos\theta,$$

giving
$$a_1 = \frac{\mu_0}{3} m_v,$$

with
$$a_2 = a_3 = \ldots = 0.$$

The relation for the conservation of the tangentiaı component of $(\mathbf{B} - \mu_0\mathbf{m}_V)$ at $r = R$ is

$$-\frac{1}{R}\left\{a_1 R \frac{d(\cos\theta)}{d\theta} + a_2 R^2 \frac{dP_2}{d\theta} + \ldots\right\} = \frac{\mu_0}{3} m_V \sin\theta.$$

This relation merely shows that our result is consistent, namely that

$$a_1 = \frac{\mu_0}{3} m_V, \qquad \text{and} \qquad a_2 = a_3 = \ldots = 0.$$

Finally,

$$\mathbf{B}_{in} = \mu_0\mathbf{m}_V - \frac{\mu_0}{3} m_V\{\cos\theta\, \boldsymbol{\epsilon}_r - \sin\theta\, \boldsymbol{\epsilon}_\theta\},$$

or

$$\mathbf{B}_{in} = \tfrac{2}{3}\mu_0\mathbf{m}_V.$$

The quantity $(\mathbf{B} - \mu_0\mathbf{m}_V)$ proves exceedingly useful for computational purposes and is called the MAGNETIC INTENSITY **H** times μ_0:

$$\mu_0\mathbf{H} = \mathbf{B} - \mu_0\mathbf{m}_V.$$

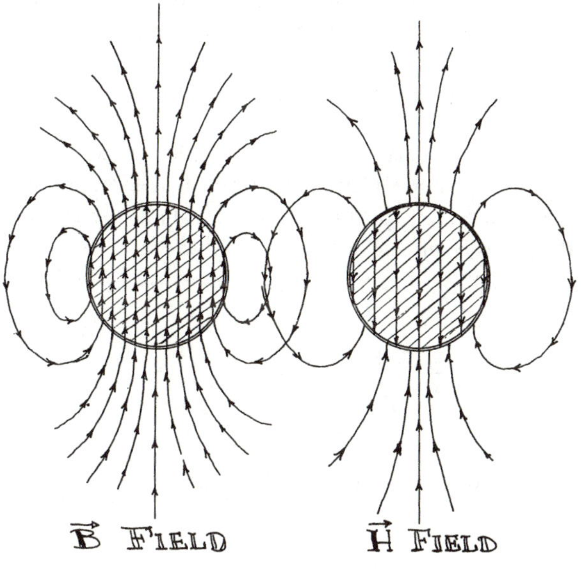

The implications of this will become apparent in the next section when free currents (as opposed to the dipole currents or Amperian currents) are superposed on the magnetic medium or volume distribution of moments.

Before terminating this discussion we should remark that fictitious magnetic pole densities can be associated with the scalar potential W(r) in the same manner as the induced volume and surface charges were developed from the scalar potential of a distribution of electric dipoles. This interpretation will be employed later in a discussion of ferromagnetism.

B. EXPANSION OF B IN TERMS OF MICROSCOPIC AND MACROSCOPIC CURRENTS

By definition microscopic currents are to be considered as those currents associated with the Amperian currents of the microscopic moments of order one, and also those currents associated with the equivalent currents set up by magnetic quadrupoles and higher moments. In the same manner as the analogous problem in electrostatics, all of the currents associated with moments higher than that of the dipole can be shown to produce an equivalent volume distribution of $\mathbf{m}'_v$.

From now on we shall include in $\mathbf{m}_v$ the first-order moments plus all the equivalent moments produced by the higher multipoles. With this qualification the expression becomes completely general.

In the presence of true or conventional currents $\mathbf{J}$ plus a magnetic moment distribution $\mathbf{m}_v$, producing Amperian currents $\mathbf{J}_A$, the vector potential at any point P is

$$\mathbf{A}_p = \frac{\mu_0}{4\pi} \iiint_\tau \frac{\mathbf{J}}{|\mathbf{r}_p - \mathbf{r}|} d\tau + \frac{\mu_0}{4\pi} \iiint_\tau \frac{\mathbf{J}_A}{|\mathbf{r}_p - \mathbf{r}|} d\tau.$$

Here

$$\mathbf{J}_A = \text{curl } \mathbf{m}_v.$$

For computational convenience we can utilize our expansion in terms of $\mathbf{m}_v$ and the scalar potential,

$$\mathbf{B}_p = \text{curl}_p \mathbf{A}_p,$$

and in one form

$$\mathbf{B}_p = \frac{\mu_0}{4\pi} \iiint_\tau \frac{\mathbf{J} \times (\mathbf{r}_p - \mathbf{r})}{|\mathbf{r}_p - \mathbf{r}|^3} d\tau + \mu_0 \mathbf{m}_v - \text{grad } W_p(\mathbf{r}_p).$$

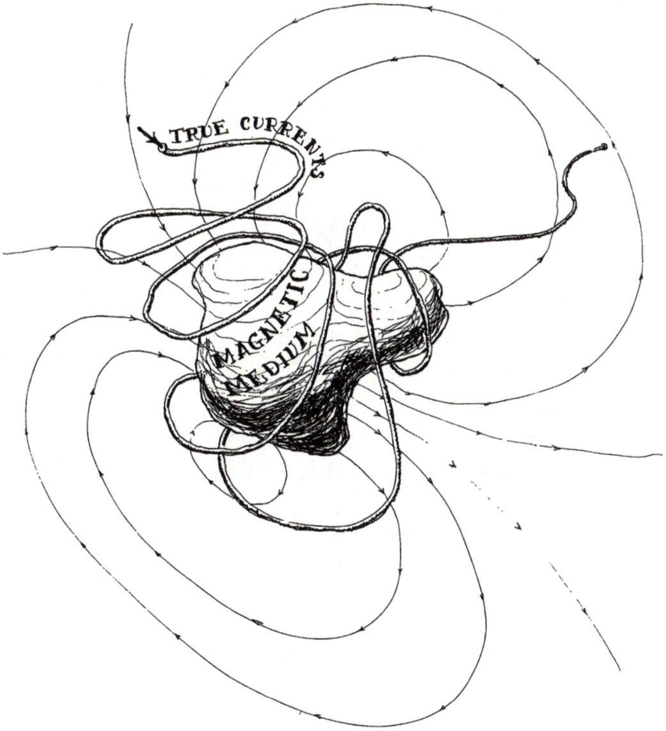

Then taking the curl of both sides,

$$\text{curl}\ \mathbf{B}_p = \mathbf{J} + \mu_0\ \text{curl}\ \mathbf{m}_v$$

or

$$\text{curl}_p\ \{\mathbf{B}_p - \mu_0 \mathbf{m}_v\} = \mathbf{J}.$$

We should notice the consistency that occurs in describing $\mathbf{B}_p$ as

$$\mathbf{B}_p = \frac{\mu_0}{4\pi} \iiint \frac{(\mathbf{J} + \mathbf{J}_A) \times (\mathbf{r}_p - \mathbf{r})}{|\mathbf{r}_p - \mathbf{r}|^3}\, d\tau,$$

which then gives

$$\text{curl}_p\ \mathbf{B}_p = \mathbf{J} + \mu_0\ \text{curl}\ \mathbf{m}_v.$$

The general solution for this last equation includes the term $-\nabla_p W(\mathbf{r}_p)$ in the definition of **B**. The volume τ must include all regions containing **J** and the magnetic media.

We have noticed previously that the quantity $(\mathbf{B}_p - \mu_0 \mathbf{m}_v)$ is important in that the tangential component of this term is conserved across a boundary *when* **J** *is zero.*

The *magnetic intensity* **H** has been defined as

$$\mu_0 \mathbf{H} = \mathbf{B} - \mu_0 \mathbf{m}_V;$$

then

$$\text{curl}\, \mathbf{H} = \mathbf{J},$$

and the curl of **H** is determined completely by the true currents **J**.

Returning to the expansion of **B** in terms of $\mathbf{m}_V$,

$$\mathbf{B}_P = \mu_0(\mathbf{H}_P + \mathbf{m}_V(\mathbf{r}_P)),$$

where

$$\mathbf{H}_P = \frac{1}{4\pi}\iiint \frac{\mathbf{J} \times (\mathbf{r}_P - \mathbf{r})}{|\mathbf{r}_P - \mathbf{r}|^3}\, d\tau - \frac{1}{\mu_0}\,\text{grad}_P\, W(\mathbf{r}_P).$$

We find in this result that **H** is not completely determined by **J**. The presence of the scalar potential produces a nonzero **H** in permanent magnets (an example is the homogeneous volume distribution of $\mathbf{m}_V$ discussed here in Section A).

In conclusion we can summarize our fundamental equations:

$$\text{div}\, \mathbf{B} = 0,$$

$$\text{curl}\, \mathbf{H} = \mathbf{J},$$

$$\mathbf{B} = \mu_0(\mathbf{H} + \mathbf{m}_V),$$

and

$$\mathrm{H} = \frac{1}{4\pi}\iiint_\tau \frac{\mathbf{J} \times (\mathbf{r}_P - \mathbf{r})}{|\mathbf{r}_P - \mathbf{r}|^3}\, d\tau - \frac{1}{\mu_0}\,\text{grad}_P\, W(\mathbf{r}_P),$$

where

$$W_P = \frac{\mu_0}{4\pi}\iiint_\tau \frac{\mathbf{m}_V \cdot (\mathbf{r}_P - \mathbf{r})}{|\mathbf{r}_P - \mathbf{r}|^3}\, d\tau.$$

C. THE ENERGY STORED IN A SYSTEM CONTAINING CURRENTS AND MAGNETIC MOMENTS

Previously we derived the energy of a system of interacting currents as

$$U = \tfrac{1}{2}\iiint_{\text{All Space}} \mathbf{J} \cdot \mathbf{A}\, d\tau$$

where **J** is assumed to include surface currents.

Now consider a system of magnetic moments $\mathbf{m}_V$ (bound Amperian currents $\mathbf{J}_A$) and external true currents **J**. The stored energy is varied by δU if the true currents are varied by $\delta \mathbf{J}$. At the same time the total vector magnetic potential is varied by $\delta\mathbf{A}$; then

$$\delta U = \tfrac{1}{2}\iiint_{\text{All Space}} \{\delta\mathbf{J}\cdot\mathbf{A} + \mathbf{J}\cdot\delta\mathbf{A}\}\, d\tau.$$

We must now examine the exact nature of **J**. It is clear that the variation of **J** is external to the microscopic system. In other words the current density in this equation represents the true currents which can be varied arbitrarily. We recognize that an alteration in **J** gives rise to a variation in $\mathbf{m}_V$ and therefore $\mathbf{J}_A$. This variation is taken into account by $\delta\mathbf{A}$. Thus

$$U = \tfrac{1}{2}\iiint_{\text{All Space}} \mathbf{J}\cdot\mathbf{A}\, d\tau.$$

This is the "action at a distance" form of U. To obtain the expansion in terms of fields we use the expansion

$$\mathbf{A}\cdot\mathbf{J} = \mathbf{A}\cdot\text{curl}\,\mathbf{H} = \text{div}\,\{\mathbf{H}\times\mathbf{A}\} + \mathbf{H}\cdot\text{curl}\,\mathbf{A}.$$

Here we have used the relation curl $\mathbf{H} = \mathbf{J}$; we then substitute $\mathbf{B} = \text{curl}\,\mathbf{A}$, and

$$U = \tfrac{1}{2}\iiint_{\text{All Space}} \text{div}\,\{\mathbf{H}\times\mathbf{A}\}\, d\tau + \tfrac{1}{2}\iiint_{\text{All Space}} \mathbf{H}\cdot\mathbf{B}\, d\tau.$$

Because the currents are bounded in a finite volume, **H** and **A** vanish on the closed surface at infinity, S_∞. Therefore the volume integral of the divergence of $\{\mathbf{H}\times\mathbf{A}\}$ vanishes, and letting u be the energy density,

$$U = \tfrac{1}{2}\iiint_{\text{All Space}} \mathbf{H}\cdot\mathbf{B}\, d\tau = \tfrac{1}{2}\iiint_{\text{All Space}} u\, d\tau$$

This particular form is not unique. We see this if the variations δJ and δA are maintained in the expansion. The presence of the surface integral in the variational form allows the final form to have added to it any surface integral of the form $\iint \delta G\cdot\mathbf{n}\, dS = 0$.

By the relation $\mathbf{B} = \mu_0(\mathbf{H} + \mathbf{m}_V)$ we can separate U into the field interaction and the magnetic moment interaction with the fields:

$$U = \tfrac{1}{2}\iiint_{\text{All Space}} \mathbf{H}\cdot\mathbf{B}\, d\tau = \frac{\mu_0}{2}\iiint_{\text{All Space}} \{\mathbf{H}\cdot\mathbf{H} + \mathbf{H}\cdot\mathbf{m}_V\}\, d\tau.$$

D. THE MAGNETIC PERMEABILITY AND THE MAGNETIC SUSCEPTIBILITY

The magnetic fields **B** and **H** are related by the true currents and magnetic moments. As we have seen, the **B** field is directly related to the magnetic moments. In turn, the moment density $\mathbf{m}_V$ is a function of the external excitation **H** coupled with the interactions with all moments in the neighborhood.

In the same manner as the electric dipole density depends on the internal field $\mathscr{E}$, the magnetic moment density in many substances is a linear transformation of **H**, or a linear transformation of **B**. Relating $\mathbf{m}_V$ to **B** is somewhat more realistic; however, convention has traditionally related $\mathbf{m}_V$ to **H**.

Then
$$\mathbf{m}_V = \mathbb{X}_m \cdot \mathbf{H}$$
where $\mathbb{X}_m$ is a symmetric 3 × 3 matrix.* Then

$$\mathbf{B} = \mu_0 \mathbf{H} + \mu_0 \mathbb{X}_m \cdot \mathbf{H} = \mu_0(\mathbb{I} + \mathbb{X}_m) \cdot \mathbf{H}.$$

The relative magnetic permeability tensor $\boldsymbol{\mu}$ is defined as

$$\boldsymbol{\mu} = \mathbb{I} + \mathbb{X}_m,$$

and
$$\mathbf{B} = \mu_0 \boldsymbol{\mu} \cdot \mathbf{H}.$$

This development is quite similar to the analogous problem in electrostatics. At a boundary between any two anisotropic magnetic materials of permeabilities $\boldsymbol{\mu}^{(1)}$ and $\boldsymbol{\mu}^{(2)}$, the general relations between the fields in the two media can be expressed with the $\mathbb{N}$ and $\mathbb{T}$ matrices of Chapter V. If there are no true currents at the interface,

$$\text{curl}\, \mathbf{H} = 0,$$

and the tangential component of **H** (or $\mathbf{B} - \mu_0 \mathbf{m}_V$, see Section A of this chapter) is conserved. If the $\boldsymbol{\epsilon}_3$ cartesian base (i.e., **k**) is normal to the interface, the conservation of the tangential component of H is represented by

$$\mathbb{T} \cdot \mathbf{H}^{(1)} = \mathbb{T} \cdot \mathbf{H}^{(2)},$$

where
$$\mathbb{T} = \begin{pmatrix} 1 & 0 & 0 \\ 0 & 1 & 0 \\ 0 & 0 & 0 \end{pmatrix}$$

* The fact that $\boldsymbol{\mu}$ and $\mathbb{X}$ are symmetric is based upon the commutation of the inner product of **B** and **H**: $u = \frac{1}{2}(\mathbf{B} \cdot \mathbf{H}) = \frac{1}{2}(\mathbf{H} \cdot \mathbf{B})$.

B is solenoidal, and div **B** = 0 indicates the conservation of the normal component of **B** at the interface,

$$\mathrm{N}\cdot\mathbf{B}^{(1)} = \mathrm{N}\cdot\mathbf{B}^{(2)}.$$

$$\mathrm{N} = \begin{pmatrix} 0 & 0 & 0 \\ 0 & 0 & 0 \\ 0 & 0 & 1 \end{pmatrix}$$

Using $$\mathbf{B}^{(m)} = \mu_0 \boldsymbol{\mu}^{(m)} \cdot \mathbf{H}^{(m)}$$

we can solve for $\mathbf{B}^{(2)}$ in terms of $\mathbf{B}^{(1)}$ or $\mathbf{H}^{(2)}$ in terms of $\mathbf{H}^{(1)}$. Because **B** is the fundamental force vector we will solve for the B fields. Inverting the last equation,

$$\mathbf{H}^{(j)} = \frac{1}{\mu_0} \boldsymbol{\mu}^{(j)^{-1}} \cdot \mathbf{B}^{(j)}$$

where the elements μ'_{mn} of $\boldsymbol{\mu}^{-1}$ are

$$\mu'_{mn} = \frac{(-1)^{m+n} \text{minor } \mu_{nm}}{|\boldsymbol{\mu}|}.$$

Using the inverted form

$$\mathrm{T}\cdot\{\boldsymbol{\mu}^{(1)}\}^{-1}\cdot\mathbf{B}^{(1)} = \mathrm{T}\cdot\{\boldsymbol{\mu}^{(2)}\}^{-1}\cdot\mathbf{B}^{(2)}$$

with $$\mathrm{N}\cdot\mathbf{B}^{(1)} = \mathrm{N}\cdot\mathbf{B}^{(2)}$$

and adding,

$$\{\mathrm{N} + \mathrm{T}\cdot(\boldsymbol{\mu}^{(1)})^{-1}\}\cdot\mathbf{B}^{(1)} = \{\mathrm{N} + \mathrm{T}\cdot(\boldsymbol{\mu}^{(2)})^{-1}\}\cdot\mathbf{B}^{(2)},$$

we have $$\mathbf{B}^{(2)} = \{\mathrm{N} + \mathrm{T}\cdot(\boldsymbol{\mu}^{(2)})^{-1}\}^{-1}\cdot\{\mathrm{N} + \mathrm{T}\cdot(\boldsymbol{\mu}^{(1)})^{-1}\}\cdot\mathbf{B}^{(1)}.$$

If region 1 is a vacuum,

$$\mathbf{B}^{(2)} = \{\mathrm{N} + \mathrm{T}\cdot(\boldsymbol{\mu}^{(2)})^{-1}\}^{-1}\cdot\mathbf{B}^{(1)}.$$

The methods for computing the refraction at the surface were covered in detail in the discussion of dielectrics.

$$\mathrm{N} + \mathrm{T}\cdot\boldsymbol{\mu}^{-1} = \begin{pmatrix} 0 & 0 & 0 \\ 0 & 0 & 0 \\ 0 & 0 & 1 \end{pmatrix} + \begin{pmatrix} 1 & 0 & 0 \\ 0 & 1 & 0 \\ 0 & 0 & 0 \end{pmatrix} \begin{pmatrix} \mu'_{11} & \mu'_{12} & \mu'_{13} \\ \mu'_{21} & \mu'_{22} & \mu'_{23} \\ \mu'_{31} & \mu'_{32} & \mu'_{33} \end{pmatrix}.$$

If $$\mathbf{B}^{(2)} = \begin{bmatrix} \mathrm{B}_1^{(2)} \\ \mathrm{B}_2^{(2)} \\ \mathrm{B}_3^{(2)} \end{bmatrix},$$

then
$$\mathbf{B}^{(1)} = \begin{bmatrix} \sum_{j=1}^{3} \mu'_{1j} B_j^{(2)} \\ \sum_{j=1}^{3} \mu'_{2j} B_j^{(2)} \\ B_3^{(2)} \end{bmatrix}.$$

Because
$$\tan \alpha^{(j)} = \frac{\{(B_1^{(j)})^2 + (B_2^{(j)})^2\}^{1/2}}{B_3^{(j)}},$$

$$\frac{\tan \alpha^{(2)}}{\tan \alpha^{(1)}} = \frac{\{(B_1^{(2)})^2 + (B_2^{(2)})^2\}^{1/2}}{\{(\sum_j \mu'_{1j} B_j^{(2)})^2 + (\sum_j \mu'_{2j} B_j^{(2)})^2\}^{1/2}}.$$

In the case of an isotropic magnetic medium, $\mathbb{X}_m$ is a scalar χ_m times the unit matrix and $\boldsymbol{\mu}$ is a scalar μ times $\mathbb{I}$,

$$\boldsymbol{\mu} = (\mathbb{I} + \chi_m \mathbb{I}) = \mu \mathbb{I}.$$

Thus in this case
$$\mathbf{B} = \mu_0 \mu \mathbf{H}$$

where μ is a dimensionless scalar function.

In this special situation, $\boldsymbol{\mu}^{-1} = (1/\mu)\mathbb{I}$, and

$$\mathbf{B}^{(2)} = \left\{\mathrm{N} + \frac{1}{\mu^{(2)}} \mathrm{T}\right\}^{-1} \cdot \left\{\mathrm{N} + \frac{1}{\mu^{(1)}} \mathrm{T}\right\} \cdot \mathbf{B}^{(1)}.$$

and
$$\mathbf{B}^{(1)} = \begin{bmatrix} \frac{\mu^{(1)}}{\mu^{(2)}} B_1^{(2)} \\ \frac{\mu^{(1)}}{\mu^{(2)}} B_2^{(2)} \\ B_3^{(2)} \end{bmatrix}.$$

Finally, for the scalar susceptibility,

$$\frac{\tan \alpha^{(2)}}{\tan \alpha^{(1)}} = \frac{\mu^{(2)}}{\mu^{(1)}}.$$

Much more information can be obtained by similar calculations. For instance we could have solved for $\mathbf{H}^{(1)}$ in terms of $\mathbf{H}^{(2)}$ or we could have found the deviation of $\mathbf{m}_v$ from $\mathbf{B}$. These exercises will be left as problems.

The reader should be cautioned that we have discussed homogeneous and anisotropic media. The only inhomogeneities introduced were the boundaries which are given as surface discontinuities in $\mathbf{m}_v$. The problems of general inhomogeneous media are in most cases extremely difficult to handle, and such problems will not be treated here.

E. THE MAGNETIC PROPERTIES OF MATTER

1. THE TOROIDAL COIL OF RECTANGULAR CROSS SECTION

To introduce the effect of magnetic media upon the fields of a system of true currents I, we envisage a toroidal coil having rectangular cross section.

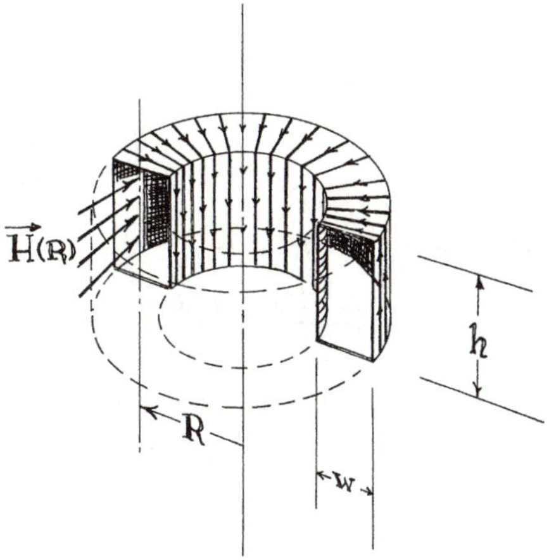

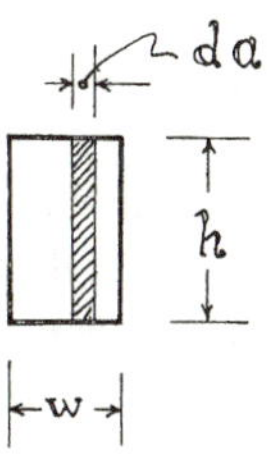

The dimensions of the rectangle are w and h; the inside radius of the toroid is R − (w/2) while the external radius is R + (w/2). Assume that the coil has a total of N turns and carries a current of I amperes per turn. The system has cylindrical symmetry about the z axis. According to the circuital law the field **H**, inside the toroid and on a circle of radius a relative to the symmetry axis, is tangent to that circle; and

$$\oint_{\text{circle a}} \mathbf{H}\cdot d\mathbf{r} = |\mathbf{H}|\, 2\pi a = I_{\text{linked}} = NI, \qquad \left(R - \frac{w}{2}\right) \leqslant a \leqslant \left(R + \frac{w}{2}\right).$$

This is a restatement of curl $\mathbf{H} = \mathbf{J}$ in the integral form. Then the interior field is

$$\mathbf{H} = \frac{NI}{2\pi}\frac{(\mathbf{k}\times\mathbf{a})}{a^2}$$

where

$$\mathbf{a} = x\mathbf{i} + y\mathbf{j}.$$

The total flux linking a single turn is

$$\Phi_{\text{per turn}} = \mu_0 \int_{R-(W/2)}^{R+(W/2)} \frac{NIh}{2\pi a}\, da = \frac{\mu_0 NIh}{2\pi}\log\left(\frac{2R + w}{2R - w}\right).$$

The total flux linked by N turns is

$$N\Phi_{\text{turn}} = \left\{\frac{\mu_0 N^2 h}{2\pi}\log\left(\frac{2R + w}{2R - w}\right)\right\} I.$$

Inside the braces is the geometrical coefficient which is the *self-inductance* L of the system,

$$L = \frac{\mu_0 N^2 h}{2\pi}\log\left(\frac{2R + w}{2R - w}\right).$$

As we shall discover in the discussion of time-varying currents, there are many experimental techniques to measure L. One simple technique is to wire several external secondary turns N_2 about the primary coil. Just a few secondary turns are used in order to reduce the back reaction on the primary coil of N_1 turns.

If the current I_1 in the primary coil is changed, an induced emf is set up in the secondary turns which is a measure of Φ_1, the initial flux in the primary:

$$E_2 = -N_2\dot{\Phi}_1 = -N_2 L_{21}\dot{I}_1.$$

In this case

$$L_{21} = \frac{N_2}{N_1} L_{11} \quad (N_2 \ll N_1).$$

To discover the gross properties of magnetic media, magnetic material is introduced into the core of the toroid (region of Φ) and completely fills the core.

If the relative magnetic permeability of the core is a scalar μ, the value of Φ and therefore L_{11} is altered in that

$$L_{11}(\mu) = \frac{\mu_0\mu N^2 h}{2\pi} \log\left(\frac{2R + w}{2R - w}\right) = \mu L_{11}(0)$$

where $L_{11}(0)$ is the self-inductance of the toroid with a vacuum core.

Experimentally it has been found that there are two types of material media. With magnetic media in the core, the inductance $L(\mu)$ can be smaller than $L(0)$ or greater than $L(0)$.

(1) Media for which $L(\mu) < L(0)$ have $\mu < 1$. Therefore the susceptibility χ_m is negative. Such media are called DIAMAGNETIC. Examples of diamagnetic substances are bismuth, copper, nitrogen, etc.

(2) When $L(\mu) > L(0)$, then $1 < \mu$, and χ_m is positive. This class of media is called PARAMAGNETIC. Also in this class of materials we incorporate a series of nonlinear anisotropic materials which are called *ferromagnetic.*

Actually all media are diamagnetic. The presence of paramagnetism usually produces a relatively large effect, and therefore the diamagnetism is neglected in these cases.

2. SEMICLASSICAL DESCRIPTION OF THE DIAMAGNETIC EFFECT

When an atomic system is placed in an external magnetic field, the magnetic moment associated with each electronic orbit will precess about the field direction with a classical frequency known as the Larmor precession frequency.

The precession of all orbits is in such a direction as to produce a net magnetic field opposite to the imposed field. This electromagnetic reaction is merely an expression of the Lorentz force on the moving electron in its orbit.

If the magnetic moment of an electron in the jth orbit is $\mathbf{M}_j$, then the torque $\mathscr{T}$ exerted upon $\mathbf{M}_j$ by an external magnetic induction field $\mathbf{B}$ is

$$\mathscr{T}_j = \mathbf{M}_j \times \mathbf{B}.$$

Classically the magnetic moment produced by a charge moving in a closed orbit is related to the angular momentum $\mathscr{L}_j$ in the following manner.

At any instant $\mathscr{L}_j = \mathbf{r}_j \times \mathbf{p}_j$, where $\mathbf{p}_j$ is the linear momentum of the electron,

$$\mathbf{p}_j = m\mathbf{v}_j = m(\boldsymbol{\omega}_j \times \mathbf{r}_j + \mathbf{v}_r).$$

Here $\mathbf{v}_r$ is the radial component of velocity and m is the reduced mass of the electron.

Then $$\mathscr{L}_j = m\mathbf{r}_j \times (\boldsymbol{\omega}_j \times \mathbf{r}_j) + 0.$$

$\boldsymbol{\omega}$ is the instantaneous angular velocity. Expanding the triple vector product,

$$\mathscr{L}_j = m(r_j^2)\boldsymbol{\omega}_j.$$

We assume that the forces producing the motion are central forces, giving a plane orbit. In such an orbit $\mathbf{v}$ and $\mathbf{r}$ lie in the plane;

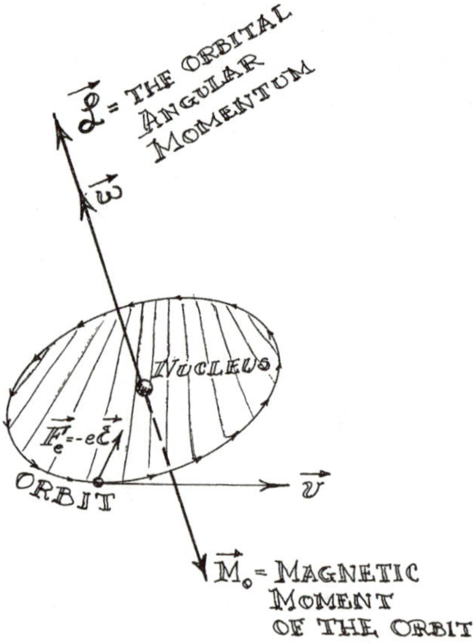

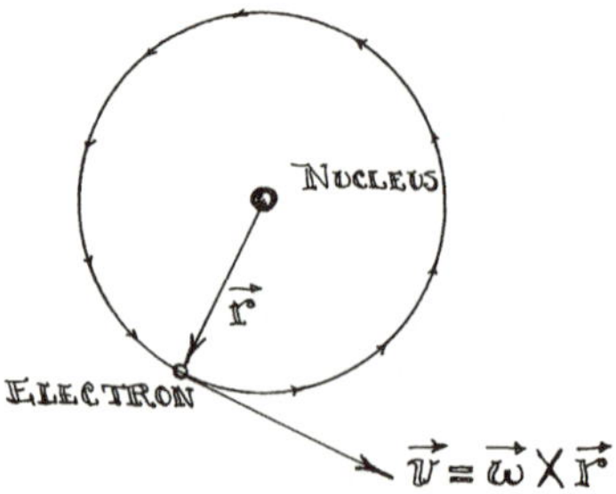

thus $\mathscr{L}_j$ is perpendicular to the plane, requiring that $\mathscr{L}_j$ have no radial components. To satisfy this condition $\boldsymbol{\omega}_j \cdot \mathbf{r}_j$ must be zero, indicating that $\boldsymbol{\omega}_j$ is perpendicular to the plane of the orbit and parallel to $\mathscr{L}_j$.

$\mathbf{r}_j$ varies in direction and magnitude; therefore we must take $\mathbf{r}_j$ and $\boldsymbol{\omega}_j$ as instantaneous values. Only $\mathscr{L}_j$ is conserved. When the orbits are circular, $\boldsymbol{\omega}_j$ can be a constant of the motion.

According to our definition of the magnetic moment,

$$\mathbf{M} = \mathrm{I}\mathbf{A}\mathbf{n} = \oint_{\text{Orbit}} \tfrac{1}{2}\mathrm{I}\mathbf{r} \times \mathrm{d}\mathbf{r};$$

then at any instant

$$\mathrm{d}\mathbf{M} = \tfrac{1}{2}\mathrm{I}\mathbf{r} \times \mathrm{d}\mathbf{r} = \tfrac{1}{2}\mathrm{dq}\,\mathbf{r} \times \mathbf{v}.$$

If dq is constant and equal to the charge on the electron $-e$, then $\mathrm{d}\mathbf{M} \to \mathrm{M}$,

$$\mathbf{M}_j = \frac{-e}{2m}(\mathbf{r}_j \times \mathbf{p}_j) = \frac{-e\mathscr{L}_j}{2m}.$$

As stated initially, the torque on the jth electron is $\mathbf{M}_j \times \mathbf{B}$, or

$$\mathscr{T}_j = \frac{-e\mathscr{L}_j}{2m} \times \mathbf{B} = \frac{+e}{2m}(\mathbf{B} \times \mathscr{L}_j).$$

Newton's second law then gives us the equations of motion for $\mathscr{L}_j$:

$$\frac{\mathrm{d}\mathscr{L}_j}{\mathrm{d}t} = \mathscr{T}_j,$$

and

$$\frac{\mathrm{d}\mathscr{L}_j}{\mathrm{d}t} = \frac{+e}{2m}(\mathbf{B} \times \mathscr{L}_j).$$

Transforming the equations of motion to a coordinate system which precesses relative to the laboratory coordinates with an angular velocity $\boldsymbol{\omega}_L$ (see Appendix A, Section 10),

$$\left[\frac{\mathrm{d}\mathscr{L}_j}{\mathrm{d}t}\right]_{\text{lab}} \to \left[\frac{\mathrm{d}\mathscr{L}'_j}{\mathrm{d}t}\right]_{\text{rot}} + \boldsymbol{\omega}_L \times \mathscr{L}'_j,$$

and

$$\frac{\mathrm{d}\mathscr{L}'_j}{\mathrm{d}t} + \boldsymbol{\omega}_L \times \mathscr{L}'_j = \frac{+e}{2m}(\mathbf{B} \times \mathscr{L}'_j).$$

This equation has a trivial solution

$$\frac{d\mathscr{L}_j'}{dt} = 0$$

or $$\mathscr{L}_j' = \text{const.}$$

when $$\boldsymbol{\omega}_L = \frac{+e}{2m}\mathbf{B}.$$

We can view the motion then as the original motion having a constant angular momentum $\mathscr{L}_j'$ in a frame of reference precessing

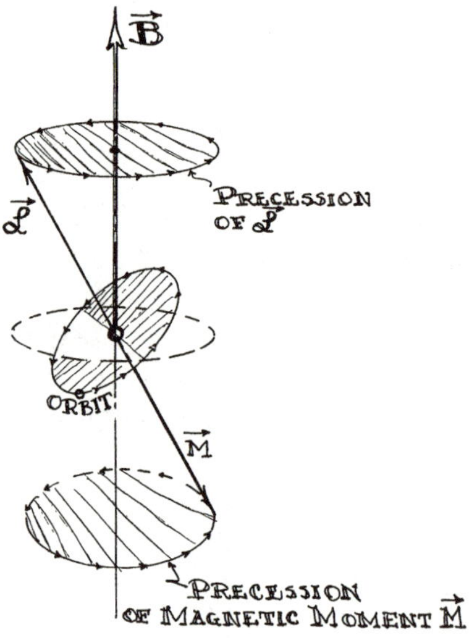

at the Larmor frequency or at an angular velocity $\boldsymbol{\omega}_L$ relative to the laboratory coordinates.

The net magnetic effect of a charged particle precessing at an average square radius $\overline{a_j^2}$ with a constant angular velocity $\boldsymbol{\omega}_L$ is to produce a magnetic moment via $\mathscr{L}_L = \overline{ma_j^2}\boldsymbol{\omega}_L$,

$$\mathbf{M}_L = \frac{-e\mathscr{L}_L}{2m} = \frac{-e\overline{a_j^2}}{2}\boldsymbol{\omega}_L.$$

If there are Z electronic orbits in the atomic system, the total diamagnetic moment per atom is

$$\mathbf{M}_{\text{dia}} = \sum_{j=1}^{Z} \left\{ \frac{-ea_j^2}{2} \right\} \omega_L = \frac{-e\omega_L}{2} \sum_{j=1}^{Z} a_j^2.$$

Let

$$ZR^2 = \sum_{j=1}^{Z} a_j^2$$

where R^2 is an effective atomic radius; thus

$$\mathbf{M}_{\text{dia}} = \frac{-ZeR^2}{2} \omega_L = \frac{-Ze^2R^2}{4m} \mathbf{B}.$$

The diamagnetic moment is NEGATIVE.

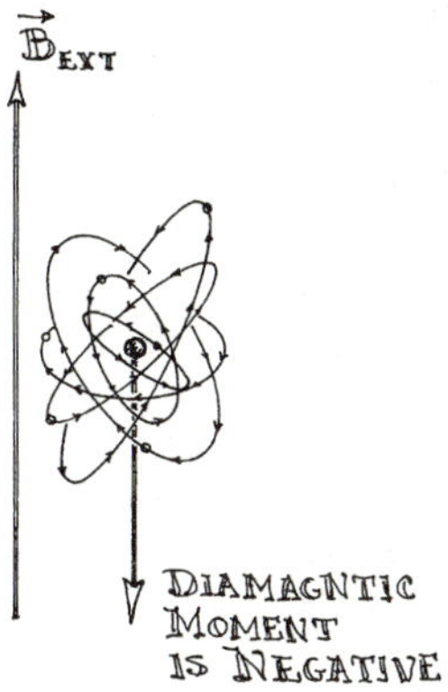

All microscopic systems exhibit a diamagnetic reaction in the presence of a magnetic field. If the net magnetic moment of an atom or molecule is zero, the observed magnetic effect is diamagnetic. When the net moment is not zero the paramagnetic orientation of this moment *in the direction of* **B** aids the field and is usually an effect larger than the diamagnetic reaction.

Assuming that a given atomic system could be truly isolated, the net effect of the total torque on the system would consist of a precession about **B** which in turn sets up microscopic fields in a direction opposite to **B**. Real media are made up of mutually interacting atomic systems. Therefore if the microscopic system possesses a net magnetic moment $\mathbf{M}_T$, the orientations of $\mathbf{M}_T$ relative to **B** will assume a statistical distribution with the greater number oriented in the direction of the magnetic field. The statistical tendency of the atomic moments to assume orientations parallel to **B** produces a net field parallel to and enhancing **B**. This constitutes the paramagnetic effect.

The magnetic moments of atomic systems are larger than the induced Larmor moments which produce the diamagnetic effect. Therefore *if an atomic system possesses a net nonzero angular momentum* $\mathscr{L}_{\mathrm{T}}$, that is, if

$$\sum_{j=1}^{Z} \mathscr{L}_j = \mathscr{L}_{\mathrm{T}}$$

and

$$\mathrm{M_T} \neq 0,$$

then the paramagnetic orientation effects predominate. Although the diamagnetic reaction is always present, the medium is characterized as paramagnetic.

Diamagnetic media consist of microscopic systems for which the total permanent magnetic moment is *zero*; summing over the Z electron orbits this requirement is

$$\mathbf{M}_{\mathrm{T}} = 0 = \sum_{j=1}^{Z} \mathbf{M}_j.$$

In such a case the magnetic behavior of the medium arises from the diamagnetic reaction.

Assume that we are dealing with diamagnetic media; then if n_v is the number of atoms or molecules per unit volume,

$$\mathbf{m}_v = n_v \mathbf{M}_{dia} = n_v \left\{ \frac{Ze^2R^2}{4m} \right\} \mathbf{B}_{eff}.$$

We notice that $\mathbf{B}_{eff}$ is the effective magnetic induction field at the position of the atom. If the atom in question is at the center of a spherical cavity of radius R, the effective field at the center is given by the **B** in the medium minus the effect of the cavity. A cavity in a polarized medium $\mathbf{m}_v$ is equivalent to the superposition of the **B** field of the medium plus the field inside a spherical volume distribution of magnetic moments $-\mathbf{m}_v$ (polarized in a direction antiparallel to $\mathbf{m}_v$). The calculation of the field of a magnetized sphere was performed in Section A of this chapter; thus

$$\begin{aligned}\mathbf{B}_{eff} &= \mathbf{B} - \tfrac{2}{3}\mu_0 \mathbf{m}_v.\\ &= \mu_0\{\mathbf{H} + \tfrac{1}{3}\mathbf{m}_v\};\end{aligned}$$

giving $$\mathbf{m}_V = -n_V \frac{Ze^2R^2}{4m} \mu_0\{\mathbf{H} + \tfrac{1}{3}\mathbf{m}_V\}.$$

Because $$\mathbf{m}_V = \chi_m \mathbf{H}$$

the diamagnetic susceptibility is

$$\chi_m = \frac{-3n_V Ze^2R^2\mu_0}{\{12m + n_V Ze^2R^2\mu_0\}}.$$

3. THE PARAMAGNETIC SUSCEPTIBILITY

As described in the previous section, the Lorentz torque which acts upon a microscopic magnetic moment in the presence of a magnetic force field **B** produces two effects. The first effect is the precession about the field, giving the diamagnetic moment of the system.

The second effect refers to the energy of orientation. As stated, an *isolated* magnetic moment would maintain its initial orientation relative to the **B** field; however, it would precess about **B**. If the medium is made up of many interacting systems (interactions such as collisions in a gas) the moments will tend to assume a direction corresponding to the minimum potential energy of orientation. The energy of a magnetic moment **M** in a field **B** is

$$U = -\mathbf{M} \cdot \mathbf{B}.$$

Thus the minimum energy occurs when **M** is *parallel* to **B**. The magnetic moment considered here is a permanent atomic or molecular moment and is not an induced moment.

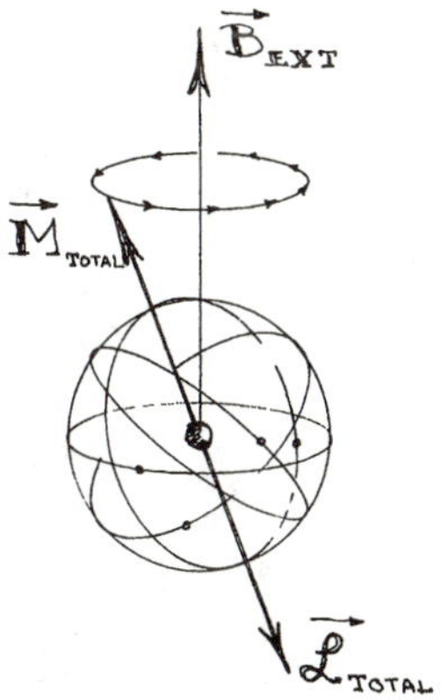

To illustrate a typical paramagnetic effect we can consider a volume of gas having n_v particles per unit volume and having a total moment $\mathbf{M}_T$ per particle.

The calculation of the number of particles dn for which the orientation of $\mathbf{M}_T$ lies between θ and $\theta + d\theta$ relative to $\mathbf{B}$ is dependent upon a Maxwell-Boltzmann distribution of energies. The reader is referred to Chapter IV, Section C-3-d where an equivalent calculation was performed for molecules possessing a permanent electric dipole moment.

In the same manner as before, the probability of a projection of $\mathbf{M}$, M cos θ in the direction of the field $\mathbf{B}$ is

$$dM = M \cos\theta \, dn(\theta),$$

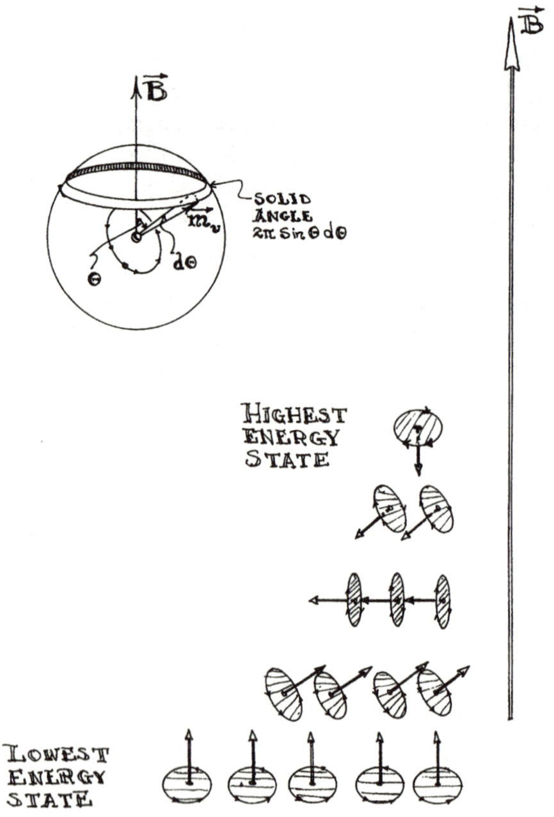

where again

$$dn = \left\{\frac{n_V(MB/kT)}{4\pi \sinh (MB/kT)}\right\} e^{(MB \cos \theta)/kT} 2\pi \sin \theta \, d\theta.$$

Integrating from O to π and multiplying by the number of particles per unit volume,

$$m_V = Mn_V\left\{\left(\coth \frac{MB}{kT}\right) - \frac{kT}{MB}\right\}.$$

When MB/kT is small we can expand the hyperbolic cotangent, and

$$\mathbf{m}_V \simeq \frac{M^2 n_V}{3kT} \mathbf{B}_{\text{eff}}.$$

Notice that the effective force field $\mathbf{B}_{\text{eff}} \simeq \mu_0(\mathbf{H} + \frac{1}{3}\mathbf{m}_V)$, and

$$\mathbf{m}_V \simeq \frac{\mu_0 M^2 n_V}{3kT\left\{1 - \dfrac{M^2 n_V \mu_0}{9kT}\right\}} \mathbf{H}.$$

The paramagnetic susceptibility of this simple system is

$$\chi_{\text{para}} \simeq \frac{3\mu_0 n_V M^2}{(9kT - \mu_0 n_V M^2)} = \frac{\mu_0 n_V M^2}{3kT\{1 - (\mu_0 n_V M^2/9kT)\}}.$$

The term in the denominator $(-M^2 n_V \mu_0/9kT)$ will be small in most cases and can be neglected. This result is known as the Curie-Langevin function, and describes the dependence of χ_{para} upon the absolute temperature T of the medium.

At small fields and fixed temperature such that $\xi \ll 1$, where $\xi = MB/kT$, the magnetic moment density is relatively linear. However, in the presence of very strong fields when we must use the complete Langevin formula, $\mathbf{m}_V$ is a nonlinear function of $\mathbf{H}$. A plot of m_V vs. ξ demonstrates this. In the general case paramagnetic materials are not only nonlinear in $\mathbf{H}$ but exhibit a saturation phenomenon. This is not surprising because the orientation effect itself must show saturation.

4. FERROMAGNETISM

From the relations governing the paramagnetic susceptibility, the conditions for ferromagnetic effects are in evidence. In a ferromagnetic material it is possible to align permanently the microscopic magnetic moments.

The unmagnetized ferromagnet is made up of randomly oriented domains. A domain is a volume element very much larger than the volume of a single magnetic moment. The molecular or atomic moments in a given domain are parallel, and the domain may contain many billions of moments.

In order for a permanent magnetic moment to exist, the field produced by neighboring moments must be greater than the minimum required field $\frac{1}{3}\mu_0 m_v$ at a given moment position. From our discussion of the paramagnetic effect in a semi-infinite medium, if the true currents are zero, the effective field at any position is

$$\mathbf{B}_{\text{eff}} = \gamma\mu_0 \mathbf{m}_v$$

where

$$\gamma \geqslant \tfrac{1}{3}.$$

The magnetic moment per unit volume from the Langevin formula is

is
$$m_v = Mn_v\left\{\left(\coth\frac{MB_{\text{eff}}}{kT}\right) - \frac{kT}{MB_{\text{eff}}}\right\}.$$

Let
$$\xi = \frac{MB_{\text{eff}}}{kT};$$

then
$$\xi = \frac{M}{kT}\{\gamma\mu_0 m_v\}.$$

From our approximation with $\xi \ll 1$,

$$m_v \simeq \frac{M^2 n_v}{3kT}\mu_0(\gamma m_v),$$

setting a lower limit on γ.

Therefore

$$\frac{\gamma M^2 n_v \mu_0}{3kT} \geqslant 1.$$

This implies that m_v is much less than $n_v M$.

Taking the equality it is found that γ must be of the order of 10^3 (here M is assumed to be the order of the Bohr magneton $e\hbar/2m$). The customary classical nearest neighbor calculation of $\gamma \simeq \frac{1}{3}$ obviously does not account for this large magnitude of γ. Heisenberg has demonstrated that quantum mechanical electron exchange forces are commensurate with this order of γ.

To provide for permanent magnetization under the Langevin formula two conditions must be satisfied by m_v:

$$m_v(T) = Mn_v\{(\coth \xi) - \xi^{-1}\},$$

and

$$m_v(T) = \frac{kT}{\gamma\mu_0 M}\,\xi(T).$$

If we solve for kT in terms of $\xi(T)$,

$$kT = \gamma\mu_0 n_v M^2 \frac{1}{\xi}\{(\cot \xi) - \xi^{-1}\}.$$

As $\xi(T)$ approaches zero, T goes to a constant known as the Curie temperature T_c. One can observe graphically that kT is a monotonically decreasing function of ξ, and that at $\xi = 0$ the field providing permanent magnetization is effectively zero (since T remains non-zero).

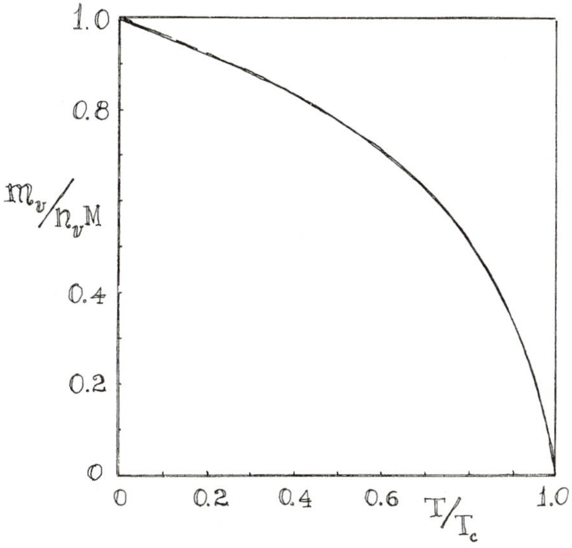

This value of T, labeled T_c, is found at the point $\xi = 0$ by expanding coth ξ in a power series, giving

$$\lim_{\xi\to 0} \frac{1}{\xi}\left\{(\coth \xi) - \frac{1}{\xi}\right\} = \frac{1}{3},$$

and $$kT_c = \tfrac{1}{3}\gamma\mu_0 n_v M^2.$$

A plot of m_v/n_vM vs. T/T_c is shown in the accompanying diagram.

At temperatures above the Curie temperature one expects the medium to cease behaving as a ferromagnetic medium and exhibit paramagnetic characteristics.

At a fixed temperature below the Curie point, ferromagnetic materials exhibit a nonlinear dependence upon $|\mathbf{H}|$, so that we must regard $\chi_m(H)$ and $\mu(H)$ as nonlinear operators.

The fact that the relative magnetic permeability μ is a nonlinear function of $|\mathbf{H}|$ should not be surprising in view of the fact that a paramagnetic medium in general can exhibit nonlinear behavior at high fields. The nonlinearity is quite different for ferromagnetic media and will exhibit a hysteresis effect. In other words, if a magnetic field is applied to the initially unmagnetized medium, after

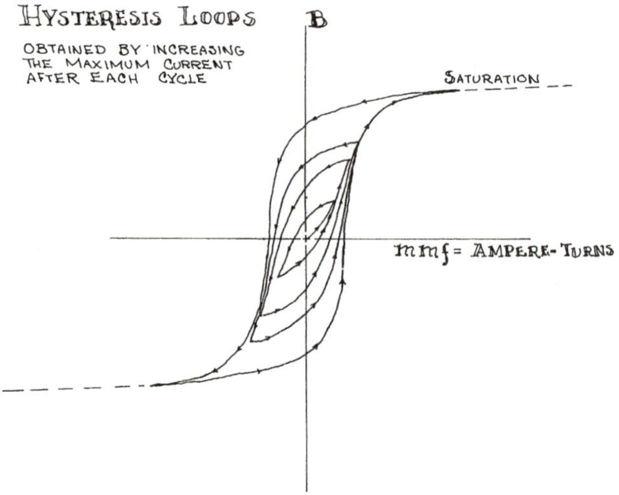

the exciting field is removed the medium will retain a permanent magnetization.

It will become obvious from the examples to follow that although we can think of the permeability as being a function of $|\mathbf{H}|$, this viewpoint can be misleading when applied to complicated magnetic circuits. We consider for simplicity the toroid discussed earlier, with a ferromagnetic core. The magnetic moment density is homogeneous.

A coil of N turns is wound about a section of the toroid, the section having a length l_N.

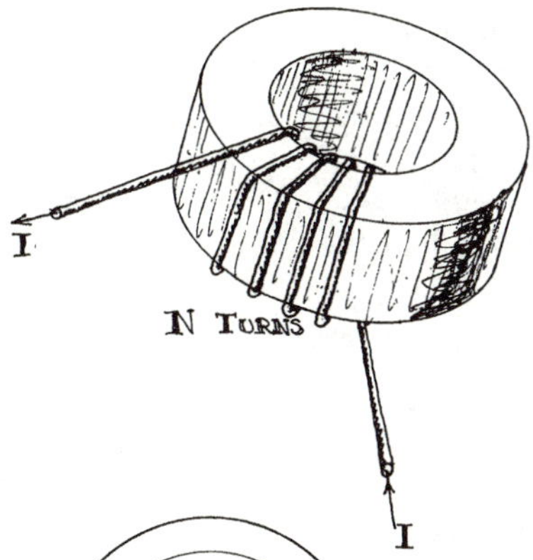

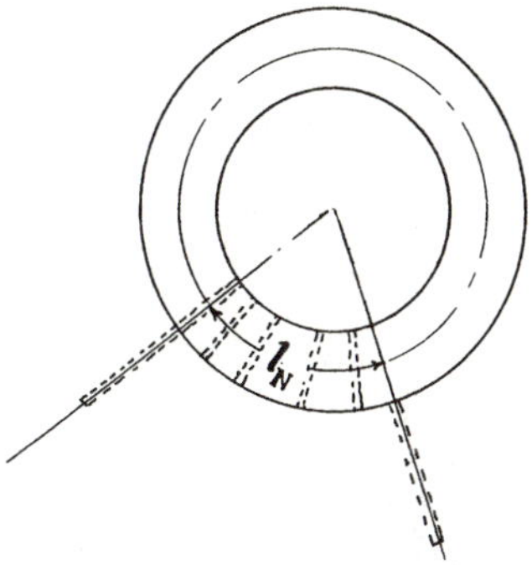

Because the ring is closed (i.e., no air gaps), **B**, inside the toroid, is everywhere tangent to circles centrally oriented with respect to the axis of symmetry of the toroid. IN THIS SPECIAL CASE both **B** and $\mathbf{m}_v$ are solenoidal, therefore grad W = 0. This particular geometry then is a very *particular* case in which

$$H = NI/l_N.$$

In other words, for this unique problem H is a measure of the true currents only.

It is worthwhile in a discussion of permanent magnetization to investigate further the content of the scalar potential function W, where

$$W_p = \frac{\mu_0}{4\pi} \iiint\limits_{\substack{\text{Volume}\\ \text{of magnetic}\\ \text{material}}} \frac{\mathbf{m}_V \cdot (\mathbf{r}_P - \mathbf{r})}{|\mathbf{r}_P - \mathbf{r}|^3}\, d\tau.$$

By the same methods as employed in the discussion of the potential of a volume distribution of electric dipoles, we can reduce this integral to a form which exhibits an induced surface and volume distribution (σ_M and ρ_M) of magnetic monopoles. As stated previously, this is strictly a mathematical convenience since there is no experimental evidence for the singlet magnetic pole. By expanding

$$+ \left\{ \mathrm{m}_V \cdot \nabla_r \left(\frac{1}{|\mathbf{r}_P - \mathbf{r}|} \right) \right\}$$

we get $$W = \frac{\mu_0}{4\pi} \iint\limits_S \frac{\sigma_M}{|\mathbf{r}_P - \mathbf{r}|}\, dS + \frac{\mu_0}{4\pi} \iiint\limits_\tau \frac{\rho_M}{|\mathbf{r}_P - \mathbf{r}|}\, d\tau$$

where $$\sigma_M = \mathbf{m}_V \cdot \mathbf{n}$$

and $$\rho_M = -\mathrm{div}\, \mathbf{m}_V.$$

Returning to the problem of the toroid; when the circuit is closed, $\mathbf{m}_V$ is solenoidal (i.e., div $\mathbf{m}_V = 0$), and $\mathbf{m}_V$ is everywhere perpendicular to the surface normals $\mathbf{n}$ (thus $\mathbf{m}_V \cdot \mathbf{n} = 0$).

This development of W demonstrates that H is given by NI/l_N in this particular problem.

Most magnetic circuits do not have such a convenient result. The reader should be cautioned since most $|\mathbf{B}|$, $|\mathbf{H}|$ curves are implied to be $|\mathbf{B}|$ vs. NI/l curves. Only in the cases of the type under discussion will this be true. This experimental arrangement proves to be quite convenient for our purposes since the plot of B vs. NI/l provides a proper measure of $\mu(H)$.

This plot can then be used for magnetic circuits involving gaps or other magnetic media, by utilizing a shift in the origin.

Before becoming concerned with permanent magnets we can develop the general approach to the theory of magnetic circuits. Consider a small gap cut in the toroid of the previous example. The faces of the gap are parallel to radial lines extending from the z axis.

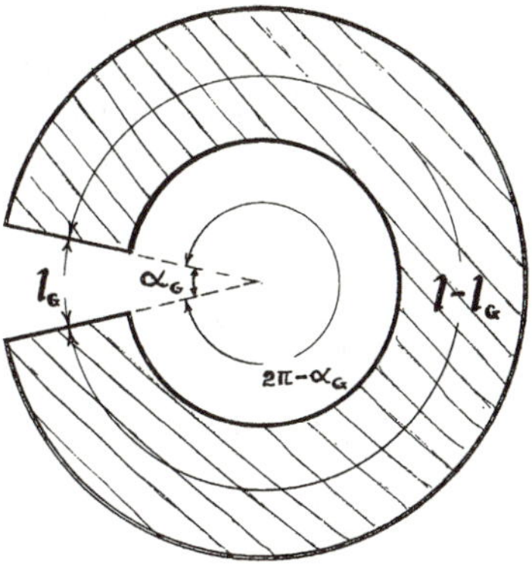

Introducing a gap or another magnetic medium results in an inhomogeneity in the permeability of the entire toroid. With the geometry shown, div $\mathbf{m}_v = 0$ in every region except at the faces of the gap. At these faces $\mathbf{m}_v \cdot \mathbf{n}$ is nonzero, thus $\mathbf{H}$ varies between the gap region and the region of magnetic material. In fact if the magnitude of $\mathbf{m}_v$ is sufficiently large as to dominate the term NI/l, then $\mathbf{H}$ will reverse direction in the medium as shown. This over-concern with H as a real physical quantity is misguided. If we consistently concern ourselves with $\mathbf{B}$ and $(\mathbf{B} - \mu_0\mathbf{m}_v)$ there will be less confusion in the discussion of magnetic media.

Let the gap width be designated as l_G. The circuital law allows us to compute $\mathbf{B}$ and $\mathbf{m}_v$ everywhere as a function of NI. Consider a circular path of radius r where $[R - (w/2)] \leqslant r \leqslant [R + (w/2)]$. According to the circuital law, taking advantage of the symmetry,

$$\oint_{\text{Circle r}} \mathbf{H} \cdot d\mathbf{r} = \oint_{\text{Circle}} \frac{(\mathbf{B} - \mu_0\mathbf{m}_v)}{\mu_0} \cdot d\mathbf{r} = NI$$

$$(2\pi r - l_G)(B_r^{(\mu)} - \mu_0 m_r) + l_G B_r^{(A)} = \mu_0 NI.$$

The subscript r above refers to the fact that the subscripted function has been evaluated at a point r meters from the axis of symmetry.

In constructing a hysteresis curve of $|\mathbf{B}|$ vs. μ_0NI, it is relatively simple to plot also $|\mathbf{m}_v(r)|$ as a function of μ_0NI. These graphs are in a sense much more realistic than straight graphs of B vs. H, particularly when the media involved are ferromagnetic.

If B is measured as a function of μ_0NI we can find m_V. Again because of the symmetry, the magnitude of the induction field in the medium, $B_r^{(\mu)}$, and the magnitude of the induction field in the air gap, $B_r^{(A)}$, are equal (neglecting fringing) for any specified r; thus

$$B_r - \left\{1 - \frac{l_G}{2\pi r}\right\}\mu_0 m_r = \frac{\mu_0 NI}{2\pi r}.$$

For a given cycle on the hysteresis curve, B_r and m_r are related by

$$m_r = \zeta(NI)B_r.$$

A trivial manipulation shows that $\zeta = \chi/\mu$. Then

$$B_r = \frac{\mu_0 NI}{2\pi r}\left\{(1-\zeta) + \frac{l_G}{2\pi r}\zeta\right\}^{-1}.$$

By permitting $l_G \to 0$ we see that the proper result is obtained:

$$B_r \xrightarrow[l_G \to 0]{} \frac{\mu\mu_0 NI}{2\pi r}.$$

When one considers magnetic circuits with many segments of different μ, the same approach can be utilized. Suppose we maintain the simple geometry of the toroid of rectangular cross section. This time we shall have four regions, three containing magnetic media μ_1, μ_2, μ_3, plus an air gap $\mu_A = 1$. The lengths in the four regions are

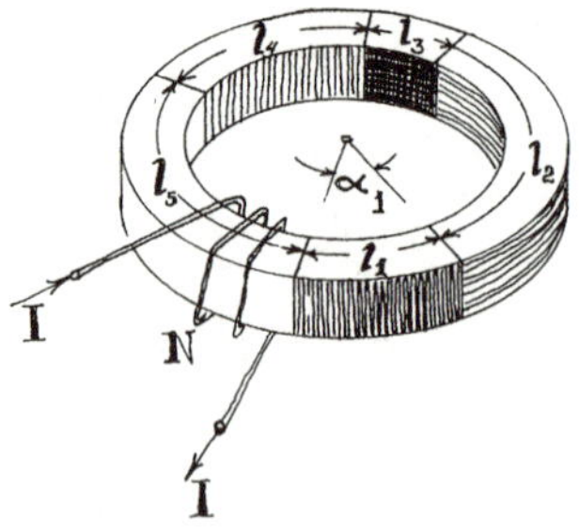

l_1, l_2, l_3, and l_G. If α_1, α_2, and α_3 are the angles subtended at the center by the three regions μ_j, then

$$l_j = \alpha_j r.$$

On a circular path of radius r

$$\sum_{j=1}^{3} (B_r - \mu_0 m_j) l_j + B_r l_G = \mu_0 NI.$$

If

$$\mu_0 m_j = \zeta_j(NI)B,$$

then

$$B_r\left\{\sum_{j=1}^{3} (1 - \zeta_j) l_j + l_G\right\} = \mu_0 NI.$$

If the air gap region is considered as region 4 with $\zeta_4 = 0$, substituting for l_j

$$\mathbf{B}_r\left\{\sum_{j=1}^{4} (1 - \zeta_j)\alpha_j\right\} r = \mu_0 NI = \mathbf{B}_r \sum_{j=1}^{4} \frac{1}{\mu_j}.$$

The invariant quantity in a magnetic circuit when the cross-sectional area varies from one medium to the next is the total flux Φ:

$$\Phi = \iint_{\substack{\text{Cross}\\ \text{section}}} \mathbf{B} \cdot \mathbf{n}\, dS.$$

In our simplified problem with all cross sections the same

$$\Phi = \int_{R-(w/2)}^{R+(w/2)} \frac{\mu_0 NIh\, dr}{\{\sum_j (1 - \zeta_j)\alpha_j\} r}$$

$$= \frac{\mu_0 NIh}{\{\sum_j (1 - \zeta_j)\alpha_j\}} \log\left\{\frac{2R + w}{2R - w}\right\}.$$

We define the MAGNETOMOTIVE FORCE of the problem as

$$\text{mmf} = \mu_0 NI;$$

the RELUCTANCE $\mathscr{R}$ of the magnetic circuit is defined by the relation

$$\mathscr{R} = \frac{\text{mmf}}{\Phi}.$$

In the problem above then

$$\mathscr{R} = \frac{\sum_{j=1}^{4} (1 - \zeta_j)\alpha_j}{\log\left(\dfrac{2R + w}{2R - w}\right)} = \frac{\sum_{j=1}^{4} \alpha_j/\mu_j}{\log\left(\dfrac{2R + w}{2R - w}\right)}.$$

Although this is not the standard result, it is instructive to see that the reluctance can be approximated by the standard formulas shown below only in very special cases.

Even in this problem we have neglected leakage flux from one gap face to various portions of the circuit other than the opposite gap face. Also, the reader should notice immediately that we have taken the interfaces between the media perpendicular to **B**. Thus in our problem there is no refraction of **B** at an interface. One should be aware, however, that an arbitrary geometry can present a formidable problem.

To investigate a common situation and by far a simpler problem, assume that B is relatively constant in the cross sections $\mathscr{A}_j$ of each of the four media. If this is true then

$$\oint \mathbf{H} \cdot d\mathbf{r} \simeq \left\{ \sum_{j=1}^{4} (1 - \zeta_j) \frac{l_j}{\mathscr{A}_j} \right\} \Phi = \mu_0 NI.$$

From our definition of the reluctance

$$\mathscr{R} = \sum_{j=1}^{4} \frac{l_j}{\mu_j \mathscr{A}_j}.$$

This is the most widely used formula and one sees that under these conditions the relation between mmf, Φ, and $\mathscr{R}$ is analogous to the relations between E, I, and $\mathscr{R}$ in a d.c. circuit.

As a matter of fact series and parallel magnetic circuits can be handled, in approximation, in the same manner as series and parallel d.c. circuits.

The last and in many cases the most interesting problem in magnetic circuitry evolves when permanent magnets are considered. These can be misleading when the magnetomotive force is zero. Much of the confusion, however, arises in efforts to utilize B vs. H curves when NI (i.e., the ampere-turns) is zero. An absence of true currents in no way implies that the **H** field is zero.

First regard the original toroid *without* an air gap. When NI → 0, $\mathbf{B} = \mu_0 \mathbf{m}_V$.

If, however, a small gap of separation l_G is cut in the toroid, the circuital law tells us that (again NI = 0)

$$(2\pi r - l_G)(B_r - \mu_0 m_V) + l B_r = 0.$$

Solving for $\mathbf{B}_r$,

$$\mathbf{B}_r = \left\{1 - \frac{l_G}{2\pi r}\right\} \mu_0 \mathbf{m}_r(r),$$

and

$$\mu_0 \mathbf{H}_r^{(\mu)} = \mathbf{B}_r - \mu_0' \mathbf{m}_V(r) = \left(\frac{-l_G}{2\pi r} \mu_0 \mathbf{m}_V \right)$$

inside the medium, and

$$\mu_0 \mathbf{H}_r^{(A)} = \mathbf{B}_r$$

in the air gap.

From this example it is seen that H_r reverses direction at the interface. This reversal is accounted for by the $\mathbf{m}_V \cdot \mathbf{n}$ term at the faces of the gap.

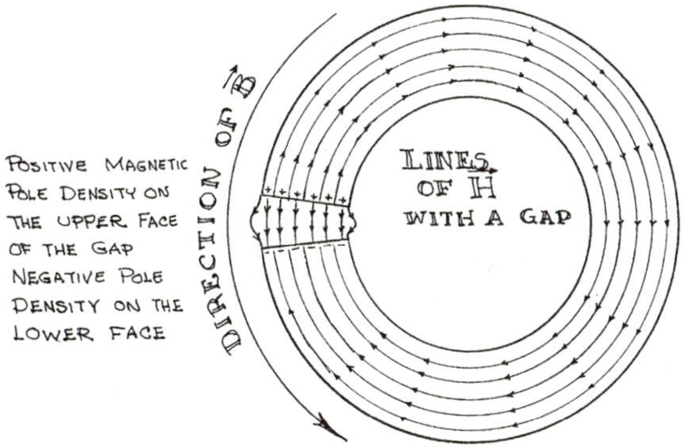

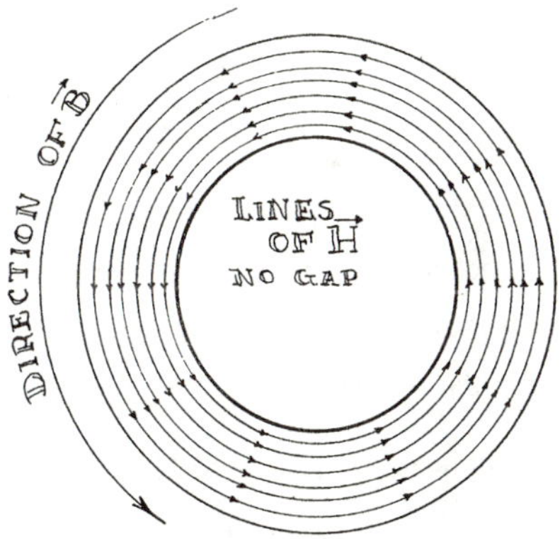

In a case in which **B** is relatively constant in the medium but varies from one medium to another because of changes of cross section, we can use the fact that Φ is conserved in the circuit. Then if the cross-sectional areas $\mathscr{A}_\mu$ vary,

$$(2\pi r - l_G)\left\{\frac{\Phi}{\mathscr{A}_\mu} - \mu_0 m_V\right\} + l_G \frac{\Phi}{\mathscr{A}_G} = 0,$$

and inside the magnet

$$\left\{1 - \frac{l_G}{l_T}\left(1 - \frac{\mathscr{A}_\mu}{\mathscr{A}_G}\right)\right\}B^{(\mu)} \simeq \left(1 - \frac{l_G}{l_T}\right)\mu_0 m_V$$

where

$$l_r = 2\pi r.$$

$$\mathscr{A}_M = \text{cross section in the magnet}$$

$$\mathscr{A}_G = \text{cross section in the air gap}$$

If one has only a B vs. $\mu_0 H (= \mu_0 NI)$ curve available, the operating

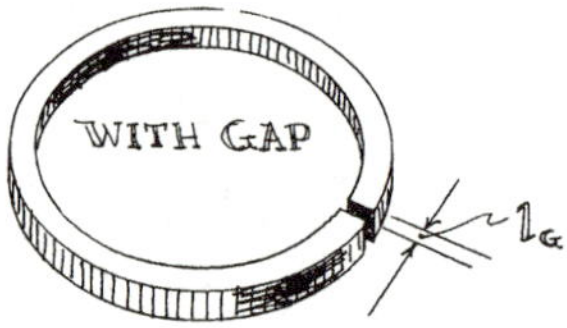

point of the permanent "C" magnet with a gap can be found from the relation

$$\mu_0 H' = (B - \mu_0 m_V) = \frac{l_G}{l_T}\left\{\frac{B}{\mu} - \frac{\mathscr{A}_\mu}{\mathscr{A}_G}B\right\}.$$

μ is usually large compared to l; therefore

$$B \simeq \left\{\frac{l_T \mathscr{A}_G}{l_G \mathscr{A}_\mu}\right\}\mu_0 H'.$$

This linear plot of B vs. H intersects the hysteresis curve at the operating point.

X. The Maxwell Equations

A. THE DISPLACEMENT CURRENT

The Maxwell equations have been developed in their complete form with the exception of the third law, which has been given up to now as a differential form of Ampere's law:

$$\text{curl}\,\{\mathbf{B} - \mu_0 \mathbf{m}_V\} = \mathbf{J};$$

or

$$\text{curl}\,\mathbf{H} = \mathbf{J}.$$

That this particular equation is not complete is apparent when we recall the fact that a vector, in this case **J**, which is derived from the curl of another vector is solenoidal. In many instances of time-varying currents, problems must be considered in which the circuit of **J** is broken, say by a condenser, and **J** in fact becomes an irrotational vector. In such cases the flow of a time-varying current density still provides an external field intensity **H**. Thus our relation curl **H** = **J** must be modified.

Before Maxwell, Faraday and Mossotti developed a way of overcoming this dilemma for dielectric media: It can be shown that the time variation of current is related to the time variation of the *induced* volume density of charge $\rho_1 = -\text{div}\,\mathbf{p}_V$. This particular problem has a reasonable physical interpretation. In this situation the induced charge is certainly *displaced* across a surface element in a particular time interval.

If we define a displacement current produced by the time variation of the induced dipole moment in order to continue the external lines of $\mathbf{J}_{\text{true}}$, a partial condition is that in the dielectric the effective current from $-\text{div}\,\mathbf{p}_V$ is

$$\text{div}\,\mathbf{J}_{\text{diel}} = -\frac{\partial \rho_1}{\partial t} = +\text{div}\,\frac{\partial \mathbf{p}_V}{\partial t}.$$

The electric field in the dielectric is related to dipole moment by

$$\mathbf{p}_v = \chi \epsilon_0 \mathscr{E}_{\text{diel}}.$$

Thus

$$\text{div } \mathbf{J}_d = +\text{div } \frac{\partial}{\partial t} \{\chi \epsilon_0 \mathscr{E}_d\}$$

or

$$\mathbf{J}_d = \frac{\partial}{\partial t} \{\chi \epsilon_0 \mathscr{E}_d\} = \frac{\partial}{\partial t} \left\{ \frac{\mathbf{D}_d}{1 + \chi^{-1}} \right\}.$$

When Ampere's law is modified to include this current of displaced charge produced by the polarization of the dielectric, we find

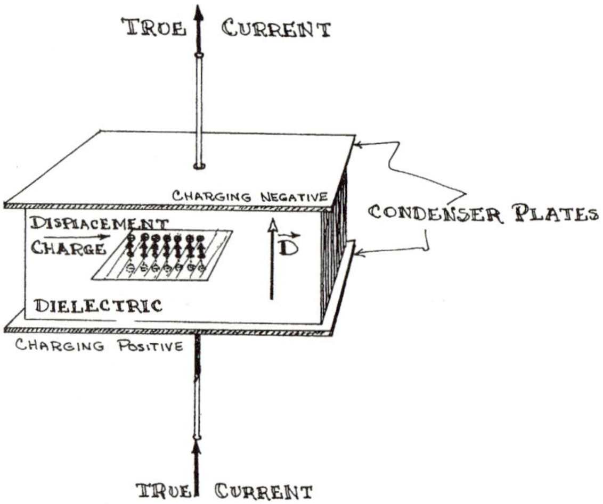

that curl **H** must in fact also be generated by a time-varying electric field in the dielectric.

Maxwell was aware that this derivation omitted a significant time-varying charge density. This was the time variation of the singlet or free charge density in the region in question. Because the mathematical formulation of electromagnetic theory was at that time somewhat more cumbersome, Maxwell invoked a vacuum which possessed an infinite subsceptibility. This approach led to a great deal of controversy concerning the initial proposal of the modification.

This sequence of events, however, illustrates vividly the reason for the name DISPLACEMENT CURRENT, since the name originally was applied to actual induced displacement currents in a dielectric.

To perform a complete derivation, consider a set of condenser plates, 1 and 2, with current density $\mathbf{J}_1$ flowing into plate 1 and $\mathbf{J}_2$ flowing away from plate 2. On plate 1 charge is building up. The region between plates is nonconducting; it can however contain a vacuum or a dielectric.

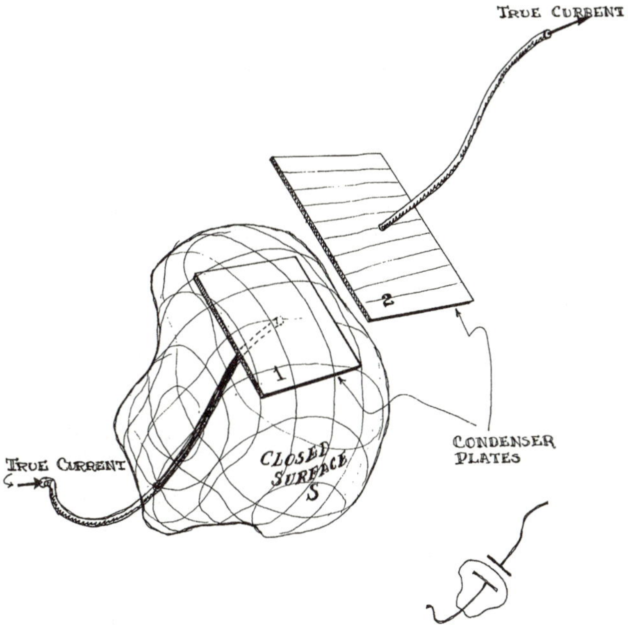

If plate 1 is surrounded by a closed surface S, the *law of conservation of charge* demands that

$$\text{div}\,\mathbf{J} + \frac{\partial \rho}{\partial t} = 0$$

at every point in the volume enclosed. The volume charge density ρ_f in this region enclosed by S must be related to the electric fields and dipolar moments by the first Maxwell equation

$$\text{div}\,\mathbf{D} = \rho_f.$$

Then
$$\iiint_\tau \operatorname{div} \mathbf{J}\, d\tau = -\iiint_\tau \operatorname{div} \frac{\partial \mathbf{D}}{\partial t}\, d\tau,$$

or
$$\iint_{S(\text{closed})} \left(\mathbf{J} + \frac{\partial \mathbf{D}}{\partial t}\right) \cdot \mathbf{n}\, dS = 0.$$

This is a statement that the quantity

$$\left\{\mathbf{J} + \frac{\partial \mathbf{D}}{\partial t}\right\}$$

is solenoidal. A further term consisting of the curl of an arbitrary vector can be added to this *effective* current. However, this term can be considered zero or contained in superposition with **J**. One can demonstrate that such a term is interpreted as the flow of free or mobile charge across the gap.

We now define a new current vector $\mathbf{J}_{\text{total}} = \mathbf{J}_{\text{true}} + \dfrac{\partial \mathbf{D}}{\partial t}$.

Ampere's law is then related to the total current flow in this region. We now have

$$\operatorname{curl} \mathbf{H} = \mathbf{J}_{\text{total}} = \mathbf{J} + \frac{\partial \mathbf{D}}{\partial t}$$

where **J** is considered as the true current and **D** is the electric displacement field. We see immediately that the Faraday-Mossotti hypothesis is contained as a part of this expression. Expanding **D**,

$$\operatorname{curl} \mathbf{H} = \mathbf{J} + \epsilon_0 \frac{\partial \mathscr{E}}{\partial t} + \frac{\partial \mathbf{p}_v}{\partial t}.$$

The third term on the right represents the original Faraday-Mossotti term. We find in addition that a time variation of inhomogeneous higher moments also contributes. These higher terms can however be incorporated in a modified $\mathbf{p}_v$ vector.

B. A SUMMARY OF THE ELECTROMAGNETIC EQUATIONS

We now are in a position to summarize the Maxwell equations plus the equations relating the fields.

(1) *The First Maxwell Equation*

$$\text{div } \mathbf{D} = \rho_{\text{free}}.$$

A statement of the inverse square law.

(2) *The Second Maxwell Equation*

$$\text{div } \mathbf{B} = 0.$$

This equation represents a statement that **B**, the basic magnetic force field vector is solenoidal with $\mathbf{B} = \text{curl } \mathbf{A}$.

(3) *The Third Maxwell Equation*

$$\text{curl } \mathbf{H} = \mathbf{J} + \frac{\partial \mathbf{D}}{\partial t}.$$

(4) *The Fourth Maxwell Equation*

$$\text{curl } \mathscr{E} = -\frac{\partial \mathbf{B}}{\partial t}.$$

An expression of Faraday's law.

(5) *Interrelations*

$$\mathbf{D} = \epsilon_0 \mathscr{E} + \mathbf{p}_v + \text{h.t.} = \epsilon_0 \boldsymbol{\epsilon} \cdot \mathscr{E};$$
$$\mathbf{B} = \mu_0(\mathbf{H} + \mathbf{m}_v) = \mu_0 \boldsymbol{\mu} \cdot \mathbf{H}.$$

In practice the higher terms in the expansion of **D** and **B** can be incorporated into equivalent dipole moments per unit volume.

(6) *The Lorentz Force*

$$\mathbf{F} = q(\mathscr{E} + \mathbf{v} \times \mathbf{B}).$$

This expression defines the electric field intensity $\mathscr{E}$ and the magnetic induction vector **B** in terms of the forces on a charged particle.

C. THE POTENTIALS AND THE GAUGE

By analogy with the methods used in electrostatics and for static magnetic fields, we can show that the $\mathscr{E}$ and **B** vectors can be generated from scalar and vector potentials.

Because B is solenoidal,

$$\mathbf{B} = \text{curl } \mathbf{A}$$

where $\quad$ **A** is the magnetic vector potential.

By definition

$$\mu_0 \mathbf{H} = \mathbf{B} - \mu_0 \mathbf{m}_V = \text{curl } \mathbf{A} - \mu_0 \mathbf{m}_V.$$

Faraday's law or the fourth Maxwell equation implies that the electric field can be associated with both a scalar and a vector potential. The relation

$$\mathscr{E} = -\text{grad V} - \frac{\partial \mathbf{A}}{\partial t}$$

satisfies the Maxwell equations.

Usually the potentials are specified in the absence of magnetic media or for regions in which

$$\mu_0 \text{H} = \boldsymbol{\mu}^{-1} \cdot \mathbf{B}.$$

Then

$$\mathbf{H} = \frac{1}{\mu_0} \boldsymbol{\mu}^{-1} \cdot (\text{curl } \mathbf{A}).$$

Further, if the electrical matter considered is isotropic such that

$$\boldsymbol{\mu}^{-1} = \frac{1}{\mu} \mathbf{I},$$

then

$$\mathbf{H} = \frac{1}{\mu_0 \mu} \text{curl } \mathbf{A}.$$

For *homogeneous isotropic media* this is the expression for **H**, and

$$\mathscr{E} = -\text{grad V} - \frac{\partial \mathbf{A}}{\partial t}.$$

At first thought one might conclude that this specification of the potentials is unique. This is not the case however. $\mathscr{E}$ and **H** alone do not uniquely determine V and **A**.

Notice that if we transform **A** and **V** to

$$\mathbf{A} = \mathbf{A}' - \text{grad } \psi$$

and

$$\text{V} = \text{V}' + \frac{\partial \psi}{\partial t},$$

where **A**′ is a vector and V′ is a scalar function,

then

$$\mathbf{H} = \frac{1}{\mu_0 \mu} \text{curl } \mathbf{A}'$$

and

$$\mathscr{E} = -\text{grad V}' - \frac{\partial \mathbf{A}'}{\partial t}.$$

Because **A** and V are not unique, any physical law involving **A** and V must be unaffected by a gauge transformation of the type shown. In other words, the final calculation of forces and energies must not be changed by the incorporation of a GAUGE POTENTIAL ψ.

In many cases, to insure consistency an equation of constraint which determines ψ is specified.

From the third Maxwell equation we can establish an interrelation between **A** and V. Let

$$\mathbf{H} = \frac{1}{\mu_0\mu}\,\text{curl}\,\mathbf{A},$$

and

$$\mathbf{D} = -\epsilon_0\epsilon\left\{\text{grad V} + \frac{\partial \mathbf{A}}{\partial t}\right\}.$$

Then

$$\text{curl}\,\mathbf{H} = \mathbf{J} + \frac{\partial \mathbf{D}}{\partial t}$$

becomes

$$\text{grad}\left\{\text{div}\,\mathbf{A} + \frac{\mu\epsilon}{c^2}\frac{\partial V}{\partial t}\right\} - \left\{\nabla^2 - \frac{\mu\epsilon}{c^2}\frac{\partial^2}{\partial t^2}\right\}\mathbf{A} = \mu_0\mu\mathbf{J}.$$

In this equation $c^2 = (\epsilon_0\mu_0)^{-1}$.

If $\mathbf{J} = \beta\mathscr{E} + \mathbf{J}_b$, where $\mathbf{J}_b$ represents the forced battery currents,

$$\text{grad}\left\{\text{div}\,\mathbf{A} + \frac{\mu\epsilon}{c^2}\frac{\partial V}{\partial t} + \beta\mu_0\mu V\right\} - \left\{\nabla^2 - \frac{\mu\epsilon}{c^2}\frac{\partial^2}{\partial t^2} - \beta\mu_0\mu\frac{\partial}{\partial t}\right\}\mathbf{A} = \mu_0\mu\mathbf{J}_b.$$

The *Lorentz condition* is specified by requiring the first term to be zero,

$$\text{div}\,\mathbf{A} + \frac{\mu\epsilon}{c^2}\frac{\partial V}{\partial t} + \beta\mu_0\mu V = 0.$$

Then

$$\left\{\nabla^2 - \frac{\mu\epsilon}{c^2}\frac{\partial^2}{\partial t^2} - \beta\mu_0\mu\frac{\partial}{\partial t}\right\}\mathbf{A} = -\mu_0\mu\mathbf{J}_b.$$

Using the first Maxwell equation we obtain the equation defining V,

$$\left\{\nabla^2 - \frac{\mu\epsilon}{c^2}\frac{\partial^2}{\partial t^2} - \beta\mu_0\mu\frac{\partial}{\partial t}\right\}V = -\rho/\epsilon_0\epsilon.$$

Thus V and **A** are defined by inhomogeneous differential equations in terms of the sources of the field $\mathbf{J}_b$ and ρ.

If we insert a gauge transformation into the Lorentz condition and further require that $\mathbf{A}'$ and $\mathbf{V}'$ satisfy the same conditions as $\mathbf{A}$ and $\mathbf{V}$, then ψ must be a solution of

$$\left\{\nabla^2 - \frac{\mu\epsilon}{c^2}\frac{\partial^2}{\partial t^2} - \beta\mu_0\mu\frac{\partial}{\partial t}\right\}\psi = 0.$$

In other words ψ is a solution of the homogeneous parts of the equations defining A and V. The existence of a gauge transformation is not surprising when we notice that ψ is a solution of the homogeneous equation. The gauge transformation implies that we are employing the condition that to a solution of the inhomogeneous equation, any number of additional solutions can be manufactured by adding solutions of the homogeneous equation.

D. THE ENERGY STORED IN THE ELECTROMAGNETIC SYSTEM

The energy stored in the electrostatic field and the magnetostatic field have been developed previously. During these developments it was shown that the energy could be represented in terms of quantities involving "action at a distance" or in terms of the fields themselves: the storage in terms of the fields is

$$U_{\text{elec}} = \tfrac{1}{2}\iiint \mathbf{D}\cdot\mathscr{E}\, d\tau,$$

and

$$U_{\text{mag}} = \tfrac{1}{2}\iiint \mathbf{B}\cdot\mathbf{H}\, d\tau.$$

These integrals extend over all space.

The Maxwell equations in their final form contain time-varying radiation fields as well as static fields. These equations can be manipulated into a form containing the previous relations associated with the energy stored plus additional terms which must be interpreted.

The third equation is multiplied by $\mathscr{E}$, giving

$$\mathscr{E}\cdot(\text{curl}\,\mathbf{H}) = \mathscr{E}\cdot\mathbf{J} + \mathscr{E}\cdot\frac{\partial\mathbf{D}}{\partial t},$$

while the fourth is multiplied by **H**,

$$\mathbf{H} \cdot (\text{curl}\, \mathscr{E}) = -\mathbf{H} \cdot \frac{\partial \mathbf{B}}{\partial t}.$$

Before proceeding further let us investigate the forms

$$\mathscr{E} \cdot \frac{\partial \mathbf{D}}{\partial t} \quad \text{and} \quad \mathrm{H} \cdot \frac{\partial \mathbf{B}}{\partial t}.$$

If the medium is linear and bilateral (bilateral implies that $\boldsymbol{\epsilon}$ is symmetric), then

$$\mathbf{D} = \epsilon_0 \boldsymbol{\epsilon} \cdot \mathscr{E}$$

or

$$\mathrm{D}_{\mathrm{j}} = \epsilon_0 \sum_{\mathrm{m}=1}^{3} \epsilon_{\mathrm{jm}} \mathscr{E}_{\mathrm{m}},$$

while

$$\mathbf{B} = \mu_0 \boldsymbol{\mu} \cdot \mathbf{H}$$

or

$$\mathbf{B}_{\mathrm{j}} = \mu_0 \sum_{\mathrm{m}=1}^{3} \mu_{\mathrm{jm}} \mathrm{H}_{\mathrm{m}}.$$

$\boldsymbol{\epsilon}$ and $\boldsymbol{\mu}$ are constant operators independent of $\mathscr{E}$ and **H**. Keep in mind that these relations hold only for a special class of media. There exist many examples of nonlinear media, for instance the ferromagnetic materials and even the paramagnetic media for large **B** fields. In addition there exist in nature materials which are not bilateral. The operators $\boldsymbol{\epsilon}$ and $\boldsymbol{\mu}$ are not symmetric in the case of nonbilateral media.

Because such a large number of practical problems involve linear bilateral media, and because an extension of our formalism to include examples not in this class would take us beyond the stated limits of this text, we shall confine our discussion of radiation problems to cases for which $\boldsymbol{\epsilon}$ and $\boldsymbol{\mu}$ are linear symmetric operators. Later we will restrict our discussion to scalar operators.

For this class of problems the permittivity and permeability are not functions of the coordinates or the time;

$$\frac{\partial \mathrm{D}_{\mathrm{m}}}{\partial t} = \epsilon_0 \sum_{\mathrm{n}=1}^{3} \epsilon_{\mathrm{mn}} \frac{\partial \epsilon_{\mathrm{n}}}{\partial t},$$

or

$$\frac{\partial \mathbf{D}}{\partial t} = \epsilon_0 \boldsymbol{\epsilon} \cdot \frac{\partial \mathscr{E}}{\partial t};$$

and

$$\frac{\partial \mathbf{B}}{\partial t} = \mu_0 \boldsymbol{\mu} \cdot \frac{\partial \mathbf{H}}{\partial t}.$$

The mechanical analogue to the problem of a nonlinear $\boldsymbol{\epsilon}$ or relative permeability $\boldsymbol{\mu}$ is the problem of the rotation of a nonrigid body, i.e., a mechanical system for which the inertia tensor is a function of the total angular velocity of the system. There are many mechanical examples of this, such as the rotation of a sponge rubber ball.

Returning to the formalism in electromagnetics, if $\boldsymbol{\epsilon}$ and $\boldsymbol{\mu}$ are symmetric operators and constant in time,

$$\mathscr{E} \cdot \frac{\partial \mathbf{D}}{\partial t} = \frac{1}{2}\frac{\partial}{\partial t}\{\mathscr{E} \cdot \mathbf{D}\}$$

and
$$\mathbf{H} \cdot \frac{\partial \mathbf{B}}{\partial t} = \frac{1}{2}\frac{\partial}{\partial t}\{\mathbf{H} \cdot \mathbf{B}\}.$$

We can now write

$$\mathscr{E} \cdot (\text{curl } \mathbf{H}) = \frac{1}{2}\frac{\partial}{\partial t}\{\mathscr{E} \cdot \mathbf{D}\} + \mathscr{E} \cdot \mathbf{J}$$

$$\mathbf{H} \cdot (\text{curl } \mathscr{E}) = \frac{1}{2}\frac{\partial}{\partial t}\{\mathbf{H} \cdot \mathbf{B}\}.$$

Subtracting the first equation from the second and using the expansion

$$\text{div}\,(\mathscr{E} \times \mathbf{H}) = \mathbf{H} \cdot (\text{curl } \mathscr{E}) - \mathscr{E} \cdot (\text{curl } \mathbf{H}),$$

we find that
$$\text{div } \mathbf{S} = -\frac{\partial u}{\partial t} - \mathscr{E} \cdot \mathbf{J},$$

where
$$\mathbf{S} = \mathscr{E} \times \mathbf{H} = \text{the Poynting vector}$$

and
$$u = \tfrac{1}{2}\{\mathscr{E} \cdot \mathbf{D} + \mathbf{H} \cdot \mathbf{B}\}.$$

As yet the quantities **S** and u are symbolic definitions. No attempt has been made to identify them with particular physical quantities.

We first notice that u is the sum of the electric and magnetic energy densities. The volume integral of u over a region τ bounded by a closed surface* S is a measure of the electromagnetic energy stored in the volume.

If the total electric field $\mathscr{E}_T$ of the system is made up of an external electric field $\mathscr{E}$ and an effective battery or generator field $\mathscr{E}_b$ (one must include $\mathscr{E}_b$ in order to set up currents **J** via some source), then the total current **J** is defined by the superposition of the field $\mathscr{E}$

* The scalar capital S represents a surface. The vector **S** is the Poynting vector.

and the battery field $\mathscr{E}_b$:

$$\mathbf{J} = \boldsymbol{\beta}(\mathscr{E} + \mathscr{E}_b).$$

Here β is the conductivity of the medium. The power dissipated per unit volume is given by

$$\tilde{\mathscr{E}}_b \cdot \mathbf{J} - \tilde{\mathbf{J}} \cdot \boldsymbol{\beta}^{-1} \cdot \mathbf{J} = -\tilde{\mathscr{E}} \cdot \mathbf{J}.$$

Ordinarily we will contend with isotropic conductors; then

$$\boldsymbol{\beta} = \beta \mathbf{I}$$

and

$$\mathscr{E}_b \cdot \mathbf{J} - \beta^{-1}\mathbf{J}^2 = -\mathscr{E} \cdot \mathbf{J}.$$

This equation then identifies the term $-\mathscr{E} \cdot \mathbf{J}$ as the energy expended by the battery *minus* the heat loss $\beta^{-1}\mathbf{J}^2$. In other words $\left\{-\iiint_\tau \mathscr{E} \cdot \mathbf{J}\, d\tau\right\}$ represents the rate at which *electromagnetic energy* is fed into a volume τ.

With this description we can attempt to understand our power balance equation above by integrating over the volume of interest τ surrounded by the closed surface S,

$$-\iiint_\tau \mathscr{E} \cdot \mathbf{J}\, d\tau = \iiint_\tau u\, d\tau + \iint_S \mathbf{S} \cdot \mathbf{n}\, dS.$$

For the last term the divergence theorem has been used to convert the volume integral to a closed surface integral. The power balance equation then can be interpreted in a straightforward manner, viz.:

$$\left\{\begin{matrix}\text{Admission of}\\ \text{electromagnetic}\\ \text{power into } \tau\end{matrix}\right\} = \left\{\begin{matrix}\text{Rate at which energy}\\ \text{is stored in } \tau\end{matrix}\right\} + \left\{\begin{matrix}\text{Rate of flow of energy}\\ \text{across the surround-}\\ \text{ing surface S.}\end{matrix}\right\}$$

The Poynting vector $\mathbf{S} = \mathscr{E} \times \mathbf{H}$ has the units rate of energy flow across a unit area. This identification of $\mathbf{S}$ as the surface density of power flow CANNOT be utilized without the divergence operator. The reason for this is obvious. Consider two independent fields, one $\mathscr{E}_S$ electrostatic and the other $\mathbf{H}_S$ magnetostatic. Although $\mathscr{E}_S \times \mathbf{H}_S$ can be nonzero, the physical problem does not entail the surface flow of electromagnetic energy at any point. Taking the divergence of this static problem we then find that $\operatorname{div}(\mathscr{E}_S \times \mathbf{H}_S) = 0$.

A better interpretation of $\mathbf{S} = \mathscr{E} \times \mathbf{H}$ is then obtained by regarding div $\mathbf{S}$ or the closed surface integral of $\mathbf{S}$. Div $\mathbf{S}$ is of course the rate of power radiated per unit volume.

E. THE FIELD EQUATIONS

1. GENERAL

The equations governing the radiation fields $\mathscr{E}$ and **H** are contained in the four Maxwell equations. The results desired first are separated differential equations specifying $\mathscr{E}$ and **H**. If the regions involved are anisotropic it is difficult to separate the equations formally. The field equations also are inhomogeneous in that the singlet charge densities and true currents which represent the sources of the fields are included.

Contrary to the usual practice, we will first develop the complexity of the equations in an inhomogeneous anisotropic region, both in terms of the susceptibilities and in terms of the volume densities of electric and magnetic dipole moments. When simplifications are made, the reader should have a greater perspective concerning the difficulty of solutions in media which are not isotropic and homogeneous.

From the first and second Maxwell equations,

$$\operatorname{div} \mathbf{D} + \epsilon_0 \operatorname{div} \mathscr{E} + \operatorname{div} \mathbf{p}_V = \rho_f,$$

and

$$\operatorname{div} \mathbf{B} = \mu_0 \operatorname{div} \mathbf{H} + \mu_0 \operatorname{div} \mathbf{m}_V = 0.$$

Then

$$\operatorname{div} \mathscr{E} = \frac{1}{\epsilon_0}\{\rho_f - \operatorname{div} \mathbf{p}_V\},$$

and

$$\operatorname{div} \mathbf{H} = -\operatorname{div} \mathbf{m}_V.$$

Here $\mathbf{p}_V$ includes real moments plus all induced moments from higher order terms such as $-\frac{1}{6} \operatorname{div} \widetilde{\mathbf{Q}}_V$. $\mathbf{m}_V$ also contains real plus induced moments.

The separation of field vectors is achieved by taking the curl of the third and fourth equations. Taking the curl of the third equation,

$$\operatorname{curl} \operatorname{curl} \mathbf{H} = \operatorname{curl} \mathbf{J} + \frac{\partial}{\partial t}\{\operatorname{curl} \mathbf{D}\}.$$

Now

$$\operatorname{curl} \frac{\mathbf{D}}{\epsilon_0} = \operatorname{curl} \mathscr{E} + \frac{1}{\epsilon_0} \operatorname{curl} \mathbf{p}_V;$$

using the fourth equation

$$\operatorname{curl} \boldsymbol{\epsilon} \cdot \mathscr{E} = -\frac{\partial \mathbf{B}}{\partial t} + \frac{1}{\epsilon_0} \operatorname{curl} \mathbf{p}_V$$

$$= -\mu_0 \frac{\partial}{\partial t}\{\boldsymbol{\mu} \cdot \mathbf{H}\} + \frac{1}{\epsilon_0} \operatorname{curl} \mathbf{p}_V.$$

Incorporating this expansion in the equation for {curl curl **H**},

$$\text{curl curl H} + \frac{1}{c^2}\frac{\partial^2}{\partial t^2}\{\boldsymbol{\mu}\cdot\mathbf{H}\} = \text{curl } \mathbf{J} + \text{curl}\frac{\partial}{\partial t}\mathbf{p}_v,$$

where $$\mu_0\epsilon_0 = 1/c^2.$$

These equations are coupled in the coefficients H_k and are inhomogeneous, the true current density plus the electric dipole moment representing the inhomogeneity. Actually the presence of the term $\mathbf{p}_v$ indicates that the separation of the equations for $\mathscr{E}$ and H may not be complete if the dielectric is anisotropic and or inhomogeneous. We are not able to use the fourth equation to convert this term to a form containing **H**. Unless $\mathbf{X}_e$ is independent of x,y,z and proportional to a unit matrix (i.e., a homogeneous isotropic dielectric), the susceptibility tensor cannot be extracted from the curl.

If $\mathbf{X}_e$ is homogeneous, the time derivatives of the left-hand side of the equation can be uncoupled by a space rotation $\mathbb{S}$. The problem can be set up in bases parallel to the principal axes of the magnetic susceptibility $\mathbf{X}_m$ (or $\boldsymbol{\mu}$); then the left side contains components

$$[\text{curl curl } \mathbf{H}]_k + \frac{1}{c^2}\mu_{kk}\frac{\partial^2 H_k}{\partial t^2}.$$

In practice this equation is still complicated. We expand the operator (curl curl) into (grad div) minus ∇^2. This brings in a quantity div **H**. Previously we equated div **H** to $-$div $\mathbf{m}_v$. Thus in addition to requiring $\mathbf{p}_v$ we now include $\mathbf{m}_v$. Carrying out this expansion of **B** into **H** and $\mathbf{m}_v$ throughout, we obtain

$$-\Box\mathbf{H} = \text{curl } \mathbf{J} + \left\{\text{grad div} - \frac{1}{c^2}\frac{\partial^2}{\partial t^2}\right\}\mathbf{m}_v + \text{curl}\frac{\partial}{\partial t}\mathbf{p}_v,$$

where the d'Alembertian operator

$$\Box = \nabla^2 - \frac{1}{c^2}\frac{\partial^2}{\partial t^2}.$$

2. HOMOGENEOUS ISOTROPIC MEDIA

In a homogeneous isotropic dielectric and a homogeneous isotropic magnetic medium, this equation simplifies quite readily.

Then

$$\text{div}\,\mathbf{B} = \mu_0\mu\,\text{div}\,\mathbf{H} = 0$$

and

$$\text{div}\,\mathbf{m}_\text{v} = 0.$$

The term in $\mathbf{p}_\text{v}$ becomes

$$\text{curl}\,\frac{\partial}{\partial t}\,\mathbf{p}_\text{v} = -\frac{\chi_\text{e}\mu}{c^2}\frac{\partial^2 \text{H}}{\partial t^2}\,;$$

also

$$-\frac{1}{c^2}\frac{\partial^2}{\partial t^2}\,\text{m}_\text{v} = -\frac{\chi_\text{m}}{c^2}\frac{\partial^2 \mathbf{H}}{\partial t^2}.$$

Thus in the case of a homogeneous isotropic medium

$$-\Box\mathbf{H} + \frac{(\chi_\text{e}\mu + \chi_\text{m})}{c^2}\frac{\partial^2 \mathbf{H}}{\partial t^2} = \text{curl}\,\mathbf{J}$$

or

$$\left\{-\nabla^2 + \frac{\epsilon\mu}{c^2}\frac{\partial^2}{\partial t^2}\right\}\mathbf{H} = \text{curl}\,\mathbf{J}.$$

If the true currents $\mathbf{J}$ are not sources of the field but are only induced by the $\mathscr{E}$ field, we can employ Ohm's law, assuming of course that our medium is linear and conducting:

$$\mathbf{J} = \beta\boldsymbol{\mathscr{E}}.$$

Then

$$\text{curl}\,\mathbf{J} = \beta\,\text{curl}\,\boldsymbol{\mathscr{E}} = -\beta\mu_0\mu\,\frac{\partial \mathbf{H}}{\partial t}.$$

Finally, when the medium is homogeneous, isotropic, and conducting (containing no source terms curl $\mathbf{J}_\text{b}$), we have

$$\left\{-\nabla^2 + \beta\mu_0\mu\,\frac{\partial}{\partial t} + \frac{\epsilon\mu}{c^2}\frac{\partial^2}{\partial t^2}\right\}\mathbf{H} = 0.$$

The equation which describes the electric field is developed in a similar manner by taking the curl of the fourth equation:

$$\text{curl curl}\,\boldsymbol{\mathscr{E}} = -\frac{\partial}{\partial t}\,\text{curl}\,\mathbf{B} = -\mu_0\frac{\partial}{\partial t}\,\text{curl}\,\mathbf{H} - \mu_0\frac{\partial}{\partial t}\,\text{curl}\,\mathbf{m}_\text{v}.$$

Substituting the third equation on the right side we obtain

$$\text{curl curl}\,\boldsymbol{\mathscr{E}} + \frac{1}{c^2}\frac{\partial^2}{\partial t^2}(\boldsymbol{\epsilon}\cdot\boldsymbol{\mathscr{E}}) = -\mu_0\frac{\partial \mathbf{J}}{\partial t} - \mu_0\,\text{curl}\,\frac{\partial}{\partial t}\,\mathbf{m}_\text{v}.$$

Finally, expanding $\boldsymbol{\epsilon}\cdot\boldsymbol{\mathscr{E}}$ and then div $\boldsymbol{\mathscr{E}}$,

$$-\Box\boldsymbol{\mathscr{E}} = -\mu_0\frac{\partial \mathbf{J}}{\partial t} - \frac{1}{\epsilon_0}\,\text{grad}\,\rho_\text{f} + \frac{1}{\epsilon_0}\left\{\text{grad div} - \frac{1}{c^2}\frac{\partial^2}{\partial t^2}\right\}\mathbf{p}_\text{v} - \mu_0\,\text{curl}\,\frac{\partial}{\partial t}\,\mathbf{m}_\text{v}.$$

Once again we take the special case in which the medium is homogeneous, isotropic, and conducting; then the terms involving $\mathbf{p}_v$, curl $\mathbf{m}_v$, and $\mathbf{J}$ reduce to terms in $\mathscr{E}$.

For the homogeneous and isotropic (h.i.) medium,

$$\left\{-\nabla^2 + \frac{\mu\epsilon}{c^2}\frac{\partial^2}{\partial t^2}\right\}\mathscr{E} = -\mu_0\mu\frac{\partial \mathbf{J}}{\partial t} - \frac{1}{\epsilon_0\epsilon}\operatorname{grad}\rho_f.$$

In an h.i. conducting medium for which grad $\rho = 0$ and $\mathbf{J} = \beta\mathscr{E}$,

$$\left\{-\nabla^2 + \frac{\mu\epsilon}{c^2}\frac{\partial^2}{\partial t^2} + \beta\mu_0\mu\frac{\partial}{\partial t}\right\}\mathscr{E} = 0.$$

Thus to summarize the preceding development, the Maxwell equations for homogeneous isotropic media are

$$\epsilon_0\epsilon \operatorname{div}\mathscr{E} = \rho_f$$

$$\operatorname{div}\mathbf{H} = 0$$

$$\operatorname{curl}\mathbf{H} = \mathbf{J} + \epsilon_0\epsilon\frac{\partial\mathscr{E}}{\partial t}$$

and

$$\operatorname{curl}\mathscr{E} = -\mu_0\mu\frac{\partial\mathbf{H}}{\partial t}.$$

Taking the curl of the last two equations and then employing the original equations again, we obtained

$$\left\{-\nabla^2 + \frac{1}{v^2}\frac{\partial^2}{\partial t^2}\right\}\mathbf{H} = \operatorname{curl}\mathbf{J}$$

and

$$\left\{-\nabla^2 + \frac{1}{v^2}\frac{\partial^2}{\partial t^2}\right\}\mathscr{E} = -\mu_0\mu\left\{\frac{\partial\mathbf{J}}{\partial t} - v^2\operatorname{grad}\rho_f\right\}.$$

Here

$$v^2 = c^2/\epsilon\mu = 1/\epsilon_0\epsilon\mu_0\mu.$$

Once again, in an isotropic conducting medium in which $\mathbf{J} = \beta\mathscr{E}$ and grad $\rho = 0$,

$$\left\{\nabla^2 - \beta\mu_0\mu\frac{\partial}{\partial t} - \frac{1}{v^2}\frac{\partial^2}{\partial t^2}\right\}\mathscr{E}\ (\text{or } \mathbf{H}) = 0.$$

One can view these equations in two ways (keeping in mind that not all solutions are acceptable since any solution must satisfy the Maxwell equations). First we can treat the transient solution which would be obtained with an input pulse. In other words such a solution will be damped out in a given time after the initial impulse. If a direct separation of variables is performed such a solution with a damping constant $2\epsilon_0\epsilon/\beta$ will be present.

When we confine our discussion to monochromatic plane waves in a steady state configuration, the time dependence of the wave T(t) in the conductor behaves as $e^{-i\omega t}$, where we consider the real parts of the amplitudes as the physically acceptable quantities. Setting

$$\mathscr{E}(\mathbf{r},t) = \mathbf{F}(\mathbf{r})e^{-i\omega t},$$

the equation in the conducting medium reduces to

$$\left\{\nabla^2 + i\omega\beta\mu_0\mu + \frac{\omega^2}{v^2}\right\}\mathbf{F}(\mathbf{r}) = 0.$$

We define the wave number squared γ^2 as a complex number,

$$\gamma^2 = i\omega\beta\mu_0\mu + \frac{\omega^2}{v^2}.$$

From this we see that plane waves in the conductor will be exponentially damped. Because γ is complex, a term behaving as $e^{-\alpha z}$ is present (z is the direction of propagation). The damping constant α is

$$\alpha = \left(\frac{\omega}{v}\right)\frac{\sin \frac{1}{2}\xi}{\sqrt{\cos \xi}}$$

where

$$\xi = \tan^{-1}\left\{\frac{\beta}{\omega\epsilon_0\epsilon}\right\}.$$

If β is zero, $\xi \to 0$, and if β is infinite $\xi \to \pi/2$. The infinite conductivity causes the wave to vanish in the medium. Thus if a plane wave is incident upon a conducting surface for which $\beta \to \infty$, the refracted wave is zero.

Returning to the form of α we see that when ω is very large compared to $\beta/\epsilon_0\epsilon$,

$$\alpha \xrightarrow[\omega \text{ large}]{} \left\{\frac{\beta}{v2\epsilon_0\epsilon}\right\},$$

and the higher the value of β the smaller the penetration of depth α^{-1}.

XI. Radiation

A. WAVES IN NONCONDUCTING MEDIA

1. THE FIELD EQUATIONS

In a homogeneous isotropic nonconducting source-free region, the equations governing $\mathscr{E}$ and H assume a particularly simple form:

$$\Box_v \mathscr{E} = 0,$$

and

$$\Box_v \mathbf{H} = 0,$$

where

$$\Box_v = \left\{\nabla^2 - \frac{1}{v^2}\frac{\partial^2}{\partial t^2}\right\}$$

and

$$v^{-2} = \epsilon_0 \epsilon \mu_0 \mu.$$

As has been already indicated these equations do not uniquely determine $\mathscr{E}$ and **H**, since the field quantities must also satisfy the Maxwell equations

$$\text{curl } \mathbf{H} = \epsilon_0 \epsilon \frac{\partial \mathscr{E}}{\partial t}$$

and

$$\text{curl } \mathscr{E} = -\mu_0 \mu \frac{\partial \mathbf{H}}{\partial t}.$$

We shall work with a symmetric set of equations. This is done by scaling the **H** field.

By a separation of variables we find that the four equations above are satisfied by a steady state time dependence $e^{-i\omega t}$. This provides monochromatic (single frequency) solutions. No generality is lost since any wave packet (or distorted wave) can be formed by a linear combination of the resulting complete set of monochromatic solutions. Writing $\mathscr{E}$ and **H** in a separated form,

$$\mathscr{E}(\mathbf{r},t) = \mathbf{F}(\mathbf{r})e^{-i\omega t},$$

and $$\mathbf{H}(\mathbf{r},t) = -i\left\{\frac{\epsilon_0\epsilon}{\mu_0\mu}\right\}^{1/2}\mathbf{M}(\mathbf{r})e^{-i\omega t}.$$

By substituting into the four equations specified previously, the following equations become the basis for the analysis to follow:

$$\nabla^2\mathbf{F}(\mathbf{r}) + k^2\mathbf{F}(\mathbf{r}) = 0$$
$$\nabla^2\mathbf{M}(\mathbf{r}) + k^2\mathbf{M}(\mathbf{r}) = 0,$$

also $$\text{curl}\,\mathbf{M}(\mathbf{r}) = k\mathbf{F}(\mathbf{r})$$
$$\text{curl}\,\mathbf{F}(\mathbf{r}) = k\mathbf{M}(\mathbf{r}),$$

where $$kv = \frac{k}{\sqrt{\epsilon_0\epsilon\mu_0\mu}} = \omega.$$

Thus by scaling **H** we obtain a symmetric set of relations. Furthermore, we need only two of these equations. If **F** is determined from $(\nabla^2 + k^2)\mathbf{F} = 0$ then **M** is specified uniquely by curl $\mathbf{F} = k\mathbf{M}$.

2. PLANE WAVES IN NONCONDUCTING MEDIA

We can initiate a solution for either **F** or **M**; we will begin the analysis with **F**(**r**). Plane wave solutions are obtained when the Laplacian in cartesian coordinates is employed. Writing x, y, and z as x_1, x_2, and x_3 we assume a solution for the components of the first equation, the mth component of **F** is

$$F_m(\mathbf{r}) = X_{m1}(x_1)X_{m2}(x_2)X_{m3}(x_3)$$
$$= \prod_{j=1}^{3} X_{mj}(x_j).$$

Substituting into the equation $(\nabla^2 + k^2)F_m = 0$ and dividing by F_m, we obtain

$$\sum_{n=1}^{3} \frac{1}{X_{mn}}\frac{d^2X_{mn}}{dx_n^2} = -k^2.$$

Separating off the first term in the sum,

$$\frac{1}{X_{m1}}\frac{d^2X_{m1}}{dx_1^2} = -\sum_{n=2}^{3}\frac{1}{X_{mn}}\frac{d^2X_{mn}}{dx_n^2} - k^2 = -k_1^2.$$

If these two differential equations are equal they must equal a constant $-k_1^2$, giving

$$X_{m1} = e^{+ik_1x_1}.$$

Separating off the second term and then the third with separation constants k_2 and k_3,

$$F_m(\mathbf{r}) = F_{om}e^{+ik_1x_1}e^{+ik_2x_2}e^{+ik_3x_3}$$

or

$$F_m(\mathbf{r}) = F_{om}e^{i\mathbf{k}\cdot\mathbf{r}},$$

where

$$\mathbf{k} = k_1\boldsymbol{\epsilon}_1 + k_2\boldsymbol{\epsilon}_2 + k_3\boldsymbol{\epsilon}_3 \quad \text{with} \quad k^2 = k_1^2 + k_2^2 + k_3^2.$$

k is a vector*; therefore it accounts for the possibility of either sign.

The *plane wave condition* requires that the components k_1, k_2, and k_3, the separation constants, *be the same* for all F_m.

Then

$$\mathbf{F}(\mathbf{r}) = \mathbf{F}_0e^{i(\mathbf{k}\cdot\mathbf{r})},$$

and

$$\mathscr{E}(\mathbf{r},t) = \mathbf{F}_0e^{i(\mathbf{k}\cdot\mathbf{r}-\omega t)}.$$

The quantity $\mathbf{F}_0$ is the amplitude vector of the electric plane wave. If a point of constant phase ϕ is followed, where the phase is $\{\mathbf{k}\cdot\mathbf{r} - kvt\}$, then

$$\frac{d\phi}{dt} = 0 = \frac{\mathbf{k}}{k}\cdot\mathbf{v} - v$$

or

$$\frac{\mathbf{k}}{k}\cdot\mathbf{v} = v,$$

and **k** is parallel to the velocity of the wave front.

In the absence of free or mobile charges ρ_f, the field is transverse. Under this condition

$$\operatorname{div}\mathbf{D} = \epsilon_0\epsilon\operatorname{div}\mathscr{E} = 0$$

and

$$\operatorname{div}\mathbf{F}(\mathbf{r}) = 0.$$

Taking the plane wave form of **F(r)**, with amplitude components F_{0m}

$$\mathbf{F} = \sum_{m=1}^{3} F_{0m}e^{i\mathbf{k}\cdot\mathbf{r}}\boldsymbol{\epsilon}_m$$

$$\operatorname{div}\mathbf{F}(\mathbf{r}) = \{ik_1F_{01} + ik_2F_{02} + ik_3F_{03}\}e^{i\mathbf{k}\cdot\mathbf{r}} = 0.$$

Thus

$$\mathbf{k}\cdot\mathbf{F} = 0.$$

Using the fact that **k** is in the direction of **v**, the relation above demonstrates that the electric vector $\mathscr{E}$ is perpendicular to **k** and

* The $\boldsymbol{\epsilon}_j$ representation of the unit axes is used here in order to lessen the confusion which can arise in distinguishing between **k** the wave vector and the unit base along the z axis. The reader must keep in mind that $\boldsymbol{\epsilon}_j$ the base vector is completely different from ϵ the dielectric constant.

therefore transverse. Further,

$$\text{curl } \mathbf{F}(\mathbf{r}) = k\mathbf{M} = i\mathbf{k} \times \mathbf{F}(\mathbf{r}),$$

giving

$$\mathbf{M}(\mathbf{r}) = i\frac{\mathbf{k}}{k} \times \mathbf{F}(\mathbf{r}).$$

This relation demonstrates that the **H** vector is also transverse. Finally

$$\mathbf{H}(\mathbf{r},t) = -i\left(\frac{\epsilon_0\epsilon}{\mu_0\mu}\right)^{1/2}\mathbf{M}(\mathbf{r})e^{-i\omega t} = \sqrt{\frac{\epsilon_0\epsilon}{\mu_0\mu}}\left(\frac{\mathbf{k}}{k} \times \mathbf{F}_0\right)e^{i(\mathbf{k}\cdot\mathbf{r}-\omega t)},$$

or

$$\mathbf{H}(\mathbf{r},t) = \frac{1}{\omega\mu_0\mu}(\mathbf{k} \times \mathbf{F}_0)e^{i(\mathbf{k}\cdot\mathbf{r}-\omega t)}.$$

3. REFLECTION AND REFRACTION

By establishing the appropriate boundary conditions for **F**(**r**) and **M**(**r**) we can develop the equations governing the reflection and refraction of a plane wave incident upon a plane boundary between two nonconducting media specified by the parameters $\epsilon^{(1)},\mu^{(1)}$, and $\epsilon^{(2)},\mu^{(2)}$.

Before listing the boundary conditions it is perhaps wise to note the various relations between field vectors which will be used:

$$\mathscr{E} = \mathbf{F}_0 e^{(i\mathbf{k}\cdot\mathbf{r}-\omega t)}$$

$$\mathbf{H} = -i\sqrt{\frac{\epsilon_0\epsilon}{\mu_0\mu}}\,\mathbf{M}(\mathbf{r})e^{-i\omega t}$$

$$\mathbf{B} = \mu_0\mu\mathbf{H} = -i\frac{k}{\omega}\mathbf{M}(\mathbf{r})e^{-i\omega t}$$

$$\mathbf{M}(\mathbf{r}) = i\left\{\frac{\mathbf{k}}{k} \times \mathbf{F}(\mathbf{r})\right\}$$

$$\omega = kv = \frac{k}{(\epsilon_0\epsilon\mu_0\mu)^{1/2}}$$

$$\omega = k_0 c = \frac{kc}{(\epsilon\mu)^{1/2}}$$

$$k = (\epsilon\mu)^{1/2}k_0.$$

The fields **F** and **M**, besides satisfying the wave equation, must satisfy the Maxwell equations. The reader will find that the use of the symmetric form $\mathbf{M}(r)$ introduces some additional constants into the equations which express the boundary conditions.

It is assumed that the wave vector $\mathbf{k}$ is incident from medium 1 upon a plane surface between media 1 and 2. The reflected wave vector (also in medium 1) is $\mathbf{k}''$, while the refracted wave vector (in medium 2) is $\mathbf{k}'$.

The first Maxwell equation relates the normal components of the incident, refracted, and reflected waves $\mathbf{F}$, $\mathbf{F}'$, and $\mathbf{F}''$. As in the electrostatic problem we orient the base vector $\boldsymbol{\epsilon}_3$ along the outward normal to the surface of 2. A pillbox of negligible thickness with its faces parallel to the surface is oriented about the point at which $\mathbf{k}$ intersects the surface. Because there is no mobile charge on the boundary,

$$\text{div } \mathbf{D} = 0$$

and

$$\iint_{\substack{\text{Pillbox}\\ \text{surface}}} \epsilon \mathbf{F} \cdot \mathbf{n}\, dS = 0.$$

Thus the normal component of $\mathbf{F}$ is conserved. This can be represented, using the matrix $\mathbb{N}$, as

$$\epsilon^{(1)}\mathbb{N} \cdot (\mathbf{F} + \mathbf{F}'') = \epsilon^{(2)}\mathbb{N} \cdot \mathbf{F}',$$

where

$$\mathbb{N} = \begin{pmatrix} 0 & 0 & 0 \\ 0 & 0 & 0 \\ 0 & 0 & 1 \end{pmatrix}.$$

The second Maxwell equation implies that the normal component of $k\mathbf{M}$ is conserved. It is assumed that the frequency of the wave is the same in both regions.

$$\text{div } \mathbf{B} = 0$$

gives

$$\iint_{\text{Pillbox}} k\mathbf{M} \cdot \mathbf{n}\, dS = 0,$$

or

$$k\mathbb{N} \cdot (\mathbf{M} + \mathbf{M}'') = k'\mathbb{N} \cdot \mathbf{M}'.$$

We have made use of the fact that the magnitude of $\mathbf{k}$ is the same as the magnitude of $\mathbf{k}''$, i.e., $|\mathbf{k}| = |\mathbf{k}''|$.

The third equation relates the tangential component of $\mathbf{M}$ to the normal component of $\mathbf{F}$. Because the conductivity is zero, the integral over a cap surface bounded by a rectangular path with sides

$\pm\boldsymbol{l}$ and sides $\pm(\delta_1 + \delta_2)\mathbf{n}$ is

$$\iint_{\text{Cap}} (\text{curl } \mathbf{H}) \cdot \mathbf{n} \, dS = \oint_{\text{Rectangle}} \mathbf{H} \cdot d\mathbf{r}$$

$$= \mathbf{H}_1 \cdot \boldsymbol{l}_1 - \mathbf{H}_2 \cdot \boldsymbol{l}_2 - \delta_1 \mathbf{n} \cdot \mathbf{H}_1 + \delta_1 \mathbf{n} \cdot \mathbf{H}_1 - \delta_2 \mathbf{n} \cdot \mathbf{H}_2 + \delta_2 \mathbf{n} \cdot \mathbf{H}_2$$

$$= \iint_{\text{Cap}} \frac{\partial \mathbf{D}}{\partial t} \cdot \mathbf{n} \, dS.$$

As $\delta_1 \to 0$ and $\delta_2 \to 0$ the cap surface $A \to 0$ with the net result that the integral over $\partial\mathbf{D}/\partial t$ goes to zero (assuming that $\mathbf{D}$ is continuous and bounded). As a result the tangential component of $\mathbf{H}$ is conserved,

$$\left\{\frac{\epsilon^{(1)}}{\mu^{(1)}}\right\}^{1/2} \mathbf{T} \cdot (\mathbf{M} + \mathbf{M}'') = \left\{\frac{\epsilon^{(2)}}{\mu^{(2)}}\right\}^{1/2} \mathbf{T} \cdot \mathbf{M}';$$

where

$$\mathbf{T} = \begin{pmatrix} 1 & 0 & 0 \\ 0 & 1 & 0 \\ 0 & 0 & 0 \end{pmatrix}.$$

Substituting $\mathbf{M} = i\left\{\frac{\mathbf{k}}{k} \times \mathbf{F}\right\}$ into the equation for $\mathbf{M}$ we get

$$\frac{1}{\mu^{(1)}} \mathbf{T} \cdot \{\mathbf{k} \times \mathbf{F} + \mathbf{k}'' \times \mathbf{F}''\} = \frac{1}{\mu^{(2)}} \mathbf{T} \cdot \{\mathbf{k}' \times \mathbf{F}'\}.$$

The conservation of the tangential component of $\mathscr{E}$ is obtained from the fourth Maxwell equation, taking the same rectangular cap surface perpendicular to the boundary and allowing the sides parallel to $\mathbf{n}$ (the boundary normal) to approach zero. Then

$$\iint_{\text{Cap}} (\text{curl } \mathscr{E}) \cdot \mathbf{n} \, dS \to \mathscr{E}_1 \cdot \mathbf{l} - \mathscr{E}_2 \cdot \mathbf{l},$$

and

$$\mathbf{T} \cdot (\mathbf{F} + \mathbf{F}'') = \mathbf{T} \cdot \mathbf{F}'.$$

The last two forms, showing the conservation of the tangential components of $\mathbf{F}$ and $\sqrt{\epsilon/\mu}\,\mathbf{M}$, are the most useful in the present application.

Finally, to achieve the matching equations at the boundary between the two media, the phases of all three waves must be the same at the boundary surface. For simplicity choose the origin of the coordinates in the plane of the boundary. Then let $\mathbf{r}_0$ be a vector locating any point in the boundary plane relative to the origin. With this set of coordinates $\mathbf{r}_0$ is perpendicular to $\mathbf{n}$, i.e.,

$$\mathbf{r}_0 \cdot \mathbf{n} = 0.$$

Matching the phases of the three waves at the boundary,

$$\mathbf{k} \cdot \mathbf{r}_0 = \mathbf{k}'' \cdot \mathbf{r}_0 = \mathbf{k}' \cdot \mathbf{r}_0.$$

Subtracting $\mathbf{k} \cdot \mathbf{r}_0$ we get

$$(\mathbf{k}'' - \mathbf{k}) \cdot \mathbf{r}_0 = 0,$$

$$(\mathbf{k}' - \mathbf{k}) \cdot \mathbf{r}_0 = 0,$$

and

$$(\mathbf{k}'' - \mathbf{k}') \cdot \mathbf{r}_0 = 0,$$

indicating that the wave vector differences or transfers are perpendicular to the boundary surface. Because these *wave vector transfers* are parallel (or antiparallel) to $\mathbf{n}$,

$$\mathbf{n} \times (\mathbf{k}'' - \mathbf{k}) = 0$$

$$\mathbf{n} \times (\mathbf{k}' - \mathbf{k}) = 0$$

$$\mathbf{n} \times (\mathbf{k}'' - \mathbf{k}') = 0.$$

(k′ − k) and (k″ − k) are colinear; therefore the three vectors $\mathbf{k}$, $\mathbf{k}''$, and $\mathbf{k}'$ must be coplanar.

These last relations give the law of reflection:

$$|\mathbf{k}| = |\mathbf{k}''|$$

and

$$\mathbf{n} \times \mathbf{k} = \mathbf{n} \times \mathbf{k}'',$$

giving

$$k \sin \theta = k \sin \theta'',$$

or

$$\theta = \theta''.$$

The angle of incidence $\{\sphericalangle^{\mathbf{n}}_{-\mathbf{k}}\}$ equals the angle of reflection $\{\sphericalangle^{\mathbf{n}}_{\mathbf{k}''}\}$.

The equation

$$\mathbf{n} \times \mathbf{k} = \mathbf{n} \times \mathbf{k}'$$

represents Snell's law

$$k \sin \theta = k' \sin \theta'.$$

To develop the amplitudes of the refracted and reflected wave, resolve the electric field amplitude into its components along two unit vectors perpendicular to the wave vector. First let the x_2 base vector, $\boldsymbol{\epsilon}_2$, lie in the plane of incidence and on the boundary. With this orientation (ϵ_3 along the surface normal $\mathbf{n}$) the base ϵ_1 is perpendicular to the plane of incidence.

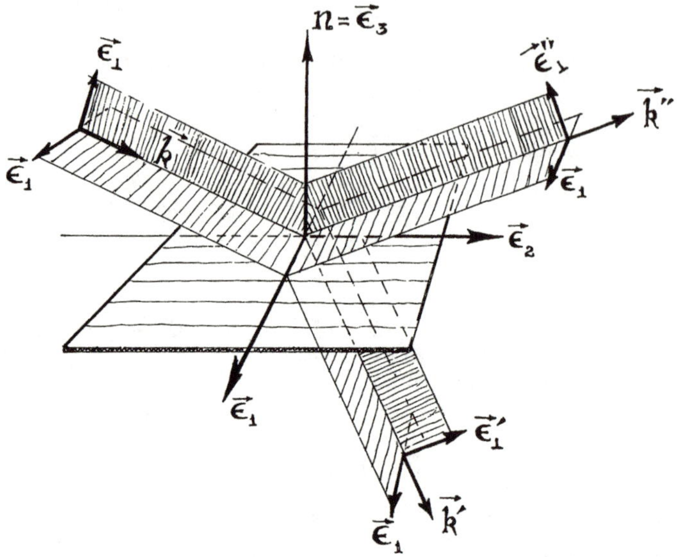

REFLECTION AND REFRACTION AT A PLANE BOUNDARY

ϵ_1 in this orientation is perpendicular to each wave vector and is a convenient choice for one of the bases. The second base is designated $\epsilon_\perp$ and is defined by

$$\epsilon_\perp = \epsilon_1 \times \frac{\mathbf{k}}{k}.$$

Because ϵ_1 and $\mathbf{k}$ are orthogonal, the system ϵ_1, $\mathbf{k}/k$, and $\epsilon_\perp$ form a convenient right-handed system.

The transverse vector amplitude $\mathbf{F}$ can be expanded in terms of ϵ_1 and $\epsilon_\perp$,

$$\mathbf{F} = F_1\epsilon_1 + F_\perp\epsilon_\perp.$$

Now
$$\mathbf{k} = k(\sin\theta\,\epsilon_2 - \cos\theta\,\epsilon_3),$$

$$\mathbf{k}'' = k(\sin\theta\,\epsilon_2 + \cos\theta\,\epsilon_3),$$

and
$$\mathbf{k}' = k'(\sin\theta'\,\epsilon_2 - \cos\theta'\,\epsilon_3).$$

We can then find the base $\boldsymbol{\epsilon}_\perp$ in terms of $\boldsymbol{\epsilon}_1$, $\boldsymbol{\epsilon}_2$, and $\boldsymbol{\epsilon}_3$:

$$\boldsymbol{\epsilon}_\perp = \boldsymbol{\epsilon}_1 \times \frac{\mathbf{k}}{k} = \{\cos\theta\,\boldsymbol{\epsilon}_2 + \sin\theta\,\boldsymbol{\epsilon}_3\},$$

$$\boldsymbol{\epsilon}''_\perp = \boldsymbol{\epsilon}''_1 \times \frac{\mathbf{k}''}{k''} = \{-\cos\theta\,\boldsymbol{\epsilon}_2 + \sin\theta\,\boldsymbol{\epsilon}_3\},$$

and

$$\boldsymbol{\epsilon}'_\perp = \boldsymbol{\epsilon}'_1 \times \frac{\mathbf{k}'}{k'} = \{\cos\theta'\,\boldsymbol{\epsilon}_2 + \sin\theta'\,\boldsymbol{\epsilon}_3\}.$$

Finally, the amplitudes of the electric vectors (in terms of $\boldsymbol{\epsilon}_1$, $\boldsymbol{\epsilon}_2$, and $\boldsymbol{\epsilon}_3$) are

$$\mathbf{F}_0 = \begin{bmatrix} F_1 \\ F_\perp \cos\theta \\ F_\perp \sin\theta \end{bmatrix}; \qquad \mathbf{F}''_0 = \begin{bmatrix} F''_1 \\ -F''_\perp \cos\theta \\ F''_\perp \sin\theta \end{bmatrix}; \qquad \mathbf{F}'_0 = \begin{bmatrix} F'_1 \\ F'_\perp \cos\theta \\ F'_\perp \sin\theta \end{bmatrix}.$$

The equation governing the tangential component of $\mathscr{E}$ gives

$$\mathbf{T}\cdot(\mathbf{F} + \mathbf{F}'') = \mathbf{T}\cdot\mathbf{F}',$$

with

$$F_1 + F''_1 = F'_1$$

and

$$(F_\perp - F''_\perp)\cos\theta = F'_\perp \cos\theta'.$$

The conservation of the tangential component of **H**,

$$\frac{1}{\mu^{(1)}}\mathbf{T}\cdot(\mathbf{k}\times\mathbf{F} + \mathbf{k}''\times\mathbf{F}'') = \frac{1}{\mu^{(2)}}\mathbf{T}\cdot(\mathbf{k}'\times\mathbf{F}'),$$

gives

$$(F_1 - F''_1) = \frac{\mu^{(1)}k'\cos\theta'}{\mu^{(2)}k\cos\theta}F'_1 = aF'_1$$

and

$$(F_\perp + F''_\perp) = \frac{\mu^{(1)}k'}{\mu^{(2)}k}F'.$$

These four equations allow solutions for the reflected and refracted amplitudes in terms of the incident amplitude.

The amplitude components of the *reflected wave* are

$$F''_1 = \frac{(1-a)}{(1+a)}F_1,$$

and

$$F''_\perp = \frac{(1-b)}{(1+b)}F_\perp;$$

where

$$a = \frac{\mu^{(1)}k'\cos\theta'}{\mu^{(2)}k\cos\theta}$$

and

$$b = \frac{\mu^{(2)}k\cos\theta'}{\mu^{(1)}k'\cos\theta}.$$

Thus the component F_1 parallel to the plane appears in its standard form which is one of the Fresnel equations

$$F_1'' = \left\{\frac{\cos\theta - (v_1/v_2)\cos\theta'}{\cos\theta + (v_1/v_2)\cos\theta'}\right\} F_1.$$

The amplitude components of the *refracted wave* are

$$F_1' = \frac{2F_1}{(1 + a)}$$

and

$$F_\perp' = \frac{2\left\{\dfrac{\mu^{(2)}k}{\mu^{(1)}k'}\right\}}{(1 + b)} F_\perp.$$

The cross product with **k** produces the magnetic components

$$M_\perp = -iF_1$$

and

$$M_1 = iF_\perp, \quad \text{etc.}$$

These results can be checked by demonstrating that the sum of the normal components of the reflected and refracted Poynting vectors equals the normal component of the incident Poynting vector. The real parts of the amplitudes must be used. Further, it is convenient to employ the time average of the Poynting vector. In Chapter XII we will show that the time average of the product of two complex vectors is one half the real part of the product of the complex conjugate of one, times the second. Therefore the time average of **S** is

$$\langle \mathbf{S} \rangle = \langle \mathscr{R}\mathscr{E} \times \mathscr{R}\mathbf{H} \rangle = \tfrac{1}{2}\mathscr{R}(\mathscr{E}^* \times \mathbf{H})$$

where $\mathscr{E}^*$ is the complex conjugate of $\mathscr{E}$, and $\mathscr{R}\mathscr{E}$ means the real part of the complex quantity $\mathscr{E}$. Then

$$\langle \mathbf{S} \rangle = \tfrac{1}{2}\mathscr{R}\left\{\mathbf{F}^*(\mathbf{r}) \times \left(-i\sqrt{\frac{\epsilon_0\epsilon}{\mu_0\mu}}\,\mathbf{M}(\mathbf{r})\right)\right\};$$

and substituting for **M**,

$$\langle \mathbf{S} \rangle = \frac{(\mathbf{F}_0^* \cdot \mathbf{F}_0)}{2\omega\mu_0\mu}\mathbf{k}$$

where $\mathbf{F}_0 = F_1\boldsymbol{\epsilon}_1 + F_\perp\boldsymbol{\epsilon}_\perp$.

Since we are free to choose $\mathbf{F}_0$, the amplitude of the plane wave, we can assume that $\mathbf{F}_0$ is real, and $\mathbf{F}_0^* \cdot \mathbf{F}_0 = \mathbf{F}_0{}^2$; then

$$\langle \mathbf{S} \rangle = \frac{(F_1{}^2 + F_\perp{}^2)}{2\omega\mu_0\mu}\mathbf{k}.$$

The vector **S** is an energy flow per unit area; therefore the energy crossing a surface **n** ΔS is given by **S** · **n** ΔS.

The *coefficient of reflection* R is

$$R = \frac{\langle S'' \rangle \cdot \mathbf{n}}{\langle -S \rangle \cdot \mathbf{n}} = \left\{ \frac{\left(\frac{1-a}{1+a}\right)^2 F_{\parallel}^2 + \left(\frac{1-b}{1+b}\right)^2 F_{\perp}^2}{F_{\parallel}^2 + F_{\perp}^2} \right\}.$$

The *coefficient of transmission* $\mathscr{T}$ is

$$\mathscr{T} = \frac{\langle S' \rangle \cdot \mathbf{n}}{\langle S \rangle \cdot \mathbf{n}} = 4a \left\{ \frac{\frac{F_{\parallel}^2}{(1+a)^2} + \frac{k^2 F_{\perp}^2}{k'^2(1+b)^2}}{(F_{\parallel}^2 + F_{\perp}^2)} \right\}.$$

With these expressions we find that

$$R + \mathscr{T} = 1.$$

B. REFLECTIONS FROM CONDUCTING SURFACES

To compute the equations governing the reflection of plane waves from a plane conducting surface of conductivity β, the boundary conditions must be changed to include the surface charge densities σ and currents **J**. We shall list the Maxwell equations and the appropriate constraints on the field components at the boundary. Again we use a pillbox with faces $\mathscr{A}$ parallel to the conductive surface.

(1) $$\operatorname{div} \mathbf{D} = \rho_f \rightarrow \iint_{\text{Pillbox}} \mathbf{D} \cdot \mathbf{n}\, dS = \sigma \mathscr{A};$$

giving $$\epsilon_0 \mathrm{N} \cdot (\epsilon^{(1)} \mathbf{F} + \epsilon^{(1)} \mathbf{F}'' - \epsilon^{(2)} \mathbf{F}') = \sigma.$$

(2) $$\operatorname{div} \mathbf{B} = 0,$$

or $$\mathrm{N} \cdot [k(\mathbf{M} + \mathbf{M}'') - k'\mathbf{M}'] = 0.$$

(3) $$\operatorname{curl} \mathbf{H} = \mathbf{J} + \frac{\partial \mathbf{D}}{\partial t},$$

giving $$\frac{k}{\mu^{(1)}} \mathrm{T} \cdot (\mathbf{M} + \mathbf{M}'') = \frac{k'}{\mu^{(2)}} \mathrm{T} \cdot \mathbf{M}' \cdot$$

We have assumed a finite conductivity such that the **J** contribution vanishes.

(4)
$$\text{curl}\,\mathscr{E} = -\frac{\partial \mathbf{B}}{\partial t}$$

with
$$\mathbf{T}\cdot(\mathbf{F}+\mathbf{F}'') = \mathbf{T}\cdot\mathbf{F}'.$$

(5) Conservation of charge:
$$\text{div}\,\mathbf{J} = -\frac{\partial \rho}{\partial t},$$

or
$$\mathbf{N}\cdot(\mathbf{J}_1-\mathbf{J}_2) = -\frac{\partial \sigma}{\partial t}\boldsymbol{\epsilon}_3.$$

$\mathbf{J}_m$ is the current density in region m.
(6) Ohm's law:
$$\mathbf{J} = \beta\mathscr{E};$$

current conservation at the surface gives
$$\mathbf{N}\cdot\{\beta_1(\mathscr{E}+\mathscr{E}'')-\beta_2\mathscr{E}'\} = -\frac{\partial \sigma}{\partial t} = i\omega\sigma,$$

where we have assumed that the charge density varies in time as $e^{-i\omega t}$
Combining (6) and (1) to eliminate the surface charge,
$$(i\omega\epsilon_0\epsilon^{(1)}-\beta_1)\mathbf{N}\cdot(\mathbf{F}+\mathbf{F}'') = (i\omega\epsilon_0\epsilon^{(2)}-\beta_2)\mathbf{N}\cdot\mathbf{F}'.$$

If both β_1 and β_2 are zero, the two media are dielectrics and $\sigma = 0$. When $\beta_1 = 0$ and β_2 is nonzero and finite we have
$$\text{curl}\,\mathbf{H}' = (\beta_2 - i\omega\epsilon_0\epsilon^{(2)})\mathscr{E}'.$$

Thus when β_2 becomes very large, $\mathscr{E}' \to 0$ and $\mathbf{H}' \to 0$.
To solve the problem of an incident plane wave on a conducting plane of conductivity β_2 we can write the defining equations as
$$\{\nabla^2 + k^2\}\mathbf{F}\ (\text{or } \mathbf{F}'') = 0, \quad \text{in region 1;}$$

k^2 is real and is equal to ω^2/c^2 if region 1 is a vacuum. The amplitude in region 2 is obtained from
$$\{\nabla^2 + i\omega\beta_2\mu_0\mu^{(2)} + \omega^2\epsilon_0\epsilon^{(2)}\mu_0\mu^{(2)}\}\mathbf{F}' = 0,$$

or
$$\{\nabla^2 + K^2\}\mathbf{F}' = 0.$$

K^2 is complex. Solving for K we find

$$K = \frac{\omega}{v_2} \sec^{1/2}\xi \, e^{i\xi/2},$$

where $$\xi = \tan^{-1}\left[\frac{\beta_2}{\omega\epsilon_0\epsilon^{(2)}}\right].$$

The real and imaginary parts of K can be designated as k_c and α where

$$K = k_c + i\alpha,$$

and $$k_c = \frac{\omega}{v_2}\left\{\frac{\cos(\xi/2)}{\sqrt{\cos\xi}}\right\},$$

with $$\alpha = \frac{\omega}{v_2}\left\{\frac{\sin(\xi/2)}{\sqrt{\cos\xi}}\right\}.$$

For a good conductor, $\beta^2/\omega\epsilon_0\epsilon^{(2)} \gg 1$; thus to a good approximation

$$k_c \simeq \frac{\omega}{v_2}\sqrt{\frac{\beta_2}{2\omega\epsilon_0\epsilon^{(2)}}} = k'\sqrt{\frac{\beta^2}{2\omega\epsilon_0\epsilon^{(2)}}},$$

$$\alpha \simeq k'\sqrt{\frac{\beta_2}{2\omega\epsilon_0\epsilon^{(2)}}} = mk',$$

and $$K \simeq (1 + i)mk';$$

where $$m = \sqrt{\frac{\beta_2}{2\omega\epsilon_0\epsilon^{(2)}}}.$$

We are now in a position to utilize the same analysis in terms of F_1 and $F_\perp$ as that which was employed for the nonconducting media of the preceding section. We define new parameters (the a and b functions were defined in the discussion nonconducting media):

$$a_c = (1 + i)ma,$$

and $$b_c = \frac{b}{(1 + i)m}.$$

The components of the amplitude of the electric vector in our present problem can be shown to be again

$$F_1'' = \left\{\frac{1 - a_c}{1 + a_c}\right\}F_1,$$

and $$F_\perp'' = \left\{\frac{1 - b_c}{1 + b_c}\right\}F_\perp.$$

Because the quantities a_c and b_c are complex, the products $\mathbf{F}_0^* \cdot \mathbf{F}_0$ must be written

$$\mathbf{F}_0^* \cdot \mathbf{F}_0 = F_1^* F_1 + F_\perp^* F_\perp = |F_1|^2 + |F_\perp|^2.$$

The reflection coefficient then becomes

$$R = \left\{ \frac{\left|\frac{1 - a_c}{1 + a_c}\right|^2 |F_1|^2 + \left|\frac{1 - b_c}{1 + b_c}\right|^2 |F_\perp|^2}{|F_1|^2 + |F_\perp|^2} \right\}.$$

Once again the choice of a real amplitude for **F** does not alter the results since we are only interested in relative phases.

When the wave is incident normal to the plane (i.e., $\theta = 0$),

$$R_0 \simeq 1 - \frac{1}{2}\sqrt{\frac{2\omega\epsilon_0\mu^{(2)}}{\beta_2}}.$$

Notice that $\epsilon^{(1)}$ and $\mu^{(1)}$ have been taken as 1.

The development above demonstrates that care must be exercised when the ($\theta = 0$) reflection coefficients are employed for randomly polarized waves incident at nonzero angles. The reflection coefficient in such a case is a function of *both* the polarization and the angle of incidence.

C. GUIDED WAVES

As an elementary approach to wave guides, consider a plane wave being reflected back and forth between two semi-infinite conducting planes parallel to the xz plane. Assume that the conductivity of the plates is infinite such that the $\mathscr{E}$ and **H** fields vanish inside the conducting planes. Further, place the planes at $\pm$a on the y axis as shown opposite.

Assume that the wave is propagating almost parallel to the y axis and that the wave vector makes an angle δ relative to the y axis.

The wave vector traveling to the right is

$$\mathbf{k}_1 = k(\cos\delta\,\mathbf{j} + \sin\delta\,\mathbf{k});$$

while the wave vector traveling to the left is

$$\mathbf{k}_2 = k(-\cos\delta\,\mathbf{j} + \sin\delta\,\mathbf{k}).$$

The two plane waves corresponding to $\mathbf{k}_1$ and $\mathbf{k}_2$ superpose and form a standing wave,

$$\mathscr{E} = F_1 e^{i(\mathbf{k}_1\cdot\mathbf{r} - \omega t)} + F_2 e^{i(\mathbf{k}_2\cdot\mathbf{r} - \omega t)}, \quad (-a < y < a).$$

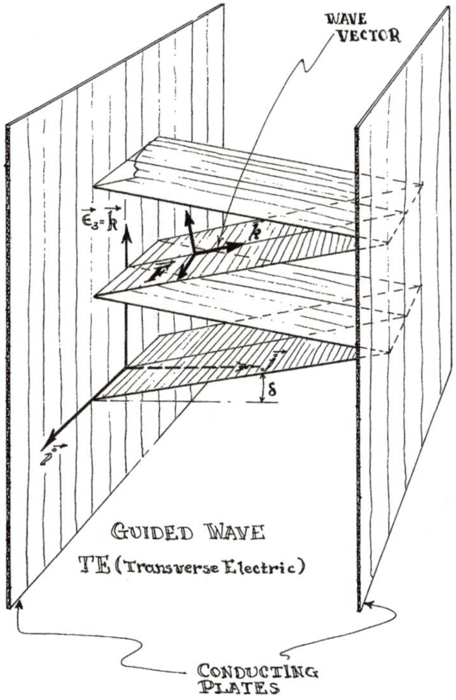

Conservation of the tangential component of $\mathscr{E}$ at either conducting surface is expressed by

$$\mathrm{T}\cdot(\mathscr{E}_1+\mathscr{E}_2)=0.$$

Taking the polarization along the x axis, (a TE, transverse electric, mode),

$$\mathbf{F}_n(\mathbf{r})=\boldsymbol{\epsilon}_1 F_n\cos\left\{\frac{(n+1)}{2a}\pi y-\frac{n\pi}{2}\right\}e^{ik_n z\sin\delta}$$

where $$\frac{(n+1)}{2a}\pi=k_n\cos\delta.$$

Notice that n is an integer ranging from 0 to arbitrarily large values. This form of $k\cos\delta$ satisfies the boundary conditions at $|y|=a$.

In the case of a periodic but nonsinusoidal wave, a Fourier

representation in the region $|y| < a$ is valid and

$$\mathbf{F}(\mathbf{r}) = \boldsymbol{\epsilon}_1 \sum_{n=1}^{\infty} F_n \cos\left\{\frac{(n+1)}{2a}\pi y - \frac{n\pi}{2}\right\} e^{ik_n z \sin\delta}.$$

We see that the components of the wave vector k_n for the nth harmonic have a unique relationship:

$$k_{ny} = k_n \cos\delta = \frac{(n+1)}{2a}\pi,$$

$$k_{nz} = k_n \sin\delta = \frac{(n+1)}{2a}\pi \tan\delta,$$

and

$$k_{ny}^2 + k_{nz}^2 = k_n^2 = \omega_n^2/c^2 = \left\{\frac{(n+1)}{2a}\pi\right\}^2 (1 + \tan^2\delta).$$

If $\omega_n^2/c^2 < \left\{\frac{(n+1)}{2a}\pi\right\}^2$, then $\tan^2\delta$ must be negative and $\sin\delta \rightarrow i \sin\delta'$. This causes the propagation in the z direction to be damped, and as a result the wave does not propagate.

The frequency corresponding to zero propagation is called the cut-off frequency, $\omega_c/2\pi$. From the equation above with $n = 1$

$$\frac{\omega_c}{c} = \frac{\pi}{2a}.$$

The lowest harmonic ($n = 0$) has been taken because the cut-off for $n = 0$ requires a cut-off for all higher harmonics. The reverse, however, is not the case since the nth harmonic can be cut off while all the lower harmonics propagate with different velocities. As a result of this situation a system operating in the region of frequencies near cut-off will produce large distortions in pulses which are traveling between the plates.

The z component of velocity for the nth harmonic is

$$v_{nz} = \frac{\omega_n}{k_{nz}} = \frac{C}{\sin\delta} = \text{the phase velocity } v_p.$$

The y component of velocity is c.

Although the phase velocity is arbitrarily large at cut-off, the special theory of relativity is not violated in view of the fact that the energy of the wave is transported with a group velocity v_g less than c.

The group velocity is the time average of the Poynting vector **S** integrated between $-a$ and a, divided by the time average of the energy density u in the same interval:

$$\langle u \rangle = \tfrac{1}{4}\mathscr{R}(\mathscr{E}^* \cdot \mathbf{D} + \mathbf{B}^* \cdot \mathbf{H})$$
$$= \tfrac{1}{4}\left\{\epsilon_0 + \frac{k_n^2}{\omega_n^2 \mu_0}\right\}\mathscr{R}(\mathbf{F}^* \cdot \mathbf{F})$$

while
$$|\langle \mathbf{S} \rangle| = \frac{1}{2\omega_n \mu_0}\,\mathscr{R}(\mathbf{F}^* \cdot \mathbf{F}) k_{nz}.$$

Integrating over y from $-a$ to a and taking the ratio of $\langle \mathbf{S} \rangle$ to $\langle u \rangle$ we obtain

$$v_g = \frac{\overline{|\langle \mathbf{S} \rangle|}}{\overline{\langle u \rangle}} = \frac{k_{nz}\omega}{k_n^2} = c \sin \delta.$$

Thus $\qquad v_g v_p = c^2 \qquad$ for bounded problems.

Notice that the bar over the time average indicates a space integration. The magnitude signs imply that we have taken the magnitude of the vector.

D. SPHERICAL WAVES

The wave equation in spherical coordinates is particularly useful. Because the solutions can be obtained in a form emphasizing the spherically diverging wave, this coordinate representation is well suited to problems in which the sources of radiation are at the origin.

The time-independent field equations have the same form as before. The Laplacian operator ∇^2, however, must now be considered in spherical variables:

$$(\nabla^2 + k^2)\mathbf{F}(\mathbf{r}) = 0$$
$$(\nabla^2 + k^2)\mathbf{M}(\mathbf{r}) = 0$$
$$\text{curl } \mathbf{F} = k\mathbf{M}$$
$$\text{curl } \mathbf{M} = k\mathbf{F},$$

and
$$\langle \mathbf{S} \rangle = \tfrac{1}{2}\mathscr{R}(\mathscr{E}^* \times \mathbf{H}) = -\frac{i}{2}\left(\frac{\epsilon_0 \epsilon}{\mu_0 \mu}\right)^{1/2} \mathscr{R}(\mathbf{F}^* \times \mathbf{M}).$$

As before, the angular brackets indicate a time average.

The medium is considered to be a homogeneous and isotropic; therefore

$$k^2 = \omega^2(\epsilon_0 \epsilon \mu_0 \mu) = \frac{\omega^2}{c^2}(\epsilon\mu) = \frac{\omega^2}{v^2}\,.$$

The Laplacian in rectangular coordinates was separated in a straightforward manner; however, we did constrain the components k_m of the wave vector **k** to be the same for each projection of the field amplitude, say $F_n(\mathbf{r})$. This assumption allowed a plane wave solution. Without this constraint the resulting amplitude would have been complicated and difficult to interpret.

The plane wave solutions do form a complete orthogonal set of solutions, and in principle a special solution for a given curvilinear coordinate system could be expanded in terms of these waves; however, the resulting expansion would prove to be extremely tedious and in addition could not illustrate the basic properties of the radiation.

On the other hand, when one attempts a separation of variables for the equation $(\nabla^2 + k^2)\mathbf{F} = 0$ in spherical coordinates, the three components of **F**; F_r, F_θ, and F_ϕ, are each complicated functions of r, θ, and ϕ. Unless we begin with some symmetry requirement as was done in the case of plane waves in cartesian coordinates the solution may not be useful.

Radiation from a source at the origin can be expanded in terms of the multipoles of the time-varying electric and the time-varying magnetic fields. These multipole fields have a nonzero net angular momentum relative to the origin.

The quantity $(1/c^2)\mathbf{S} = (1/c^2)(\mathscr{E} \times \mathbf{H})$ has the properties of linear momentum. At any point designated by **r** the quantity

$$\mathbf{r} \times \left(\frac{1}{c^2}\mathbf{S}\right)$$

represents the instantaneous angular momentum of the field at the point.

There are two ways of viewing the radiation problem. We can either conceive of the energy of the radiation field as propagating in a radial direction with at least one nontransverse amplitude, or the propagation can be viewed as nonradial with both $\mathscr{E}$ and **H** transverse.

We shall take the following point of view. One field amplitude will be expanded in vector functions (of θ and ϕ only) tangent to the surface of a sphere of radius r. The final amplitude will consist of a radial function R(r) which is a function of r only, times the vector harmonic.

The other field amplitude will then be generated by taking the curl of the first function. This second amplitude will have a radial component associated with it. Whether the radiation is characteristic of an electric multipole or a magnetic multipole is determined by the initial field amplitude which is taken tangent to the sphere.

That is, if the magnetic amplitude **M(r)** is tangent to the surface of the sphere, the electric amplitude **F(r)** has a radial component and

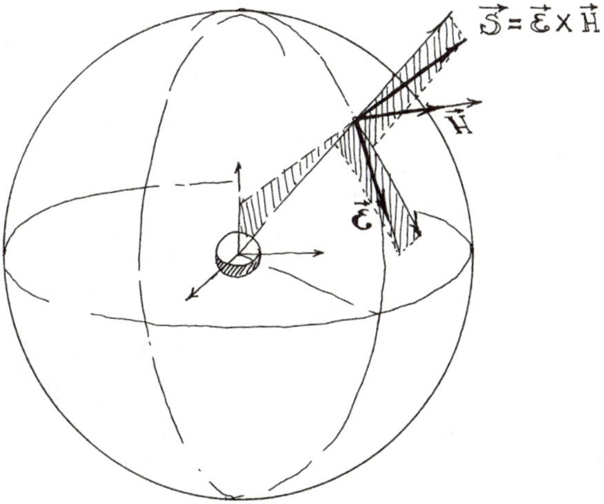

ELECTRIC MULTIPOLE RADIATION

$\vec{\mathcal{E}}$ has a radial component

the radiation is that of an *electric multipole*. Remember that $\mathscr{E} = \mathbf{F}(\mathbf{r})e^{-i\omega t}$, and $\mathbf{H} = -i\left[\frac{\epsilon_0\epsilon}{\mu_0\mu}\right]^{1/2}\mathbf{M}(\mathbf{r})e^{-i\omega t}$. Thus **F** is in direction of $\mathscr{E}$, and **M** is in the direction of $-\mathbf{H}$.

When we start with the electric vector **F** tangent, the magnetic vector or **M** has a radial component and the radiation is characteristic of a *magnetic multipole*.

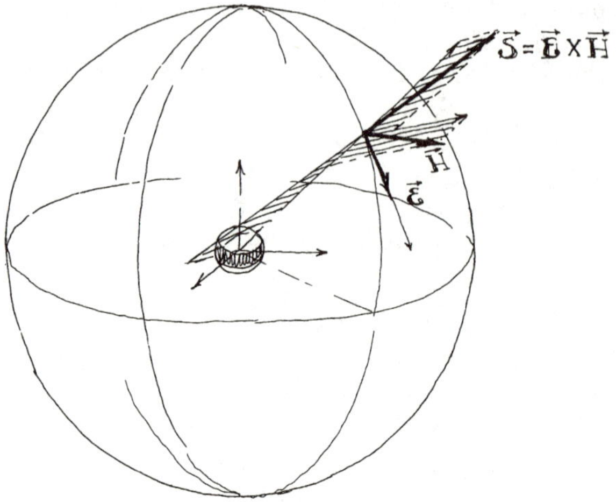

The *vector harmonics* will be symbolized by $\mathbf{X}_{LM}(\theta,\phi)$. They are tangent to the surface of a sphere centered at the origin (i.e., no ϵ_r component) and are functions of θ and ϕ only. L is the separation constant for the P_L^M equation and is a positive integer. M is the separation constant for the equation in the variable ϕ. Remember that $|M| \leqslant L$.

We shall use an electric multipole as an illustrative example to develop these functions. The total solution to the reduced wave equation $(\nabla^2 + k^2)\mathbf{M} = 0$ is written as the product of a scalar radial function $R_L(r)$ and a vector harmonic $\mathbf{X}_{LM}(\theta,\phi)$ where $\mathbf{X}_{LM}$ has no radial component:

$$\mathbf{M}_{LM}^{(E)} = R_L(r)\mathbf{X}_{LM}(\theta,\phi).$$

Our first task is to show that the particular form chosen for $\mathbf{X}_{LM}$ does indeed satisfy the wave equation. To show that $\mathbf{M}_{LM}^{(E)}$ permits a separation of variables we can expand $\nabla^2 R_L \mathbf{X}_{LM}$. In Appendix A the expansion ∇^2 in terms of (curl curl) and (grad div) is shown, and in addition we shall utilize the expansion of curl $(R\mathbf{X})$.

$$\nabla^2\{R_L(r)\mathbf{X}_{LM}(\theta,\phi)\} = -\text{curl curl}\,(R_L\mathbf{X}_{LM}) + \text{grad div}\,(R_L\mathbf{X}_{LM}).$$

In free space div $\mathbf{M}_{LM}^{(E)}$ must vanish. However, because $\mathbf{X}_{LM} = y_\theta \boldsymbol{\epsilon}_\theta + y_\phi \boldsymbol{\epsilon}_\phi$, we find that

$$\text{div}\,(R_L \mathbf{X}_{LM}) = \frac{R_L}{r^2 \sin\theta}\left\{\frac{\partial}{\partial\theta}(y_\theta r \sin\theta) + \frac{\partial}{\partial\phi}(y_\phi r)\right\} = 0,$$

and

$$\text{div}\,\mathbf{X}_{LM}(\theta,\phi) = 0.$$

This indicates that $\mathbf{X}_{LM}(\theta,\phi)$ can be defined by the curl of another vector, say $\mathbf{W}$. Since the curl involves terms of the order $1/r$, this intermediate vector $\mathbf{W}_{LM}$ must go as r in the numerator.

A reasonable choice for $\mathbf{W}$ is

$$\begin{aligned}\mathbf{W}_{LM}(r,\theta,\phi) &= \mathbf{r} N_L^M P_L^M(\theta) e^{iM\phi} \\ &= \mathbf{r} Y_L^M(\theta,\phi).\end{aligned}$$

Here $\mathbf{r} = r\boldsymbol{\epsilon}_r$,

$Y_L^M(\theta,\phi)$ = the normalized spherical harmonics (see Appendix D);

and $P_L^M(\theta)$ = the associated Legendre polynomials.

Anticipating the customary normalization of the vector harmonics we can write*

$$\mathbf{X}_{LM}(\theta,\phi) = \frac{i}{\sqrt{L(L+1)}}\,\text{curl}\,(\mathbf{r}Y_L^M).$$

The reader can check this to show that

$$\begin{aligned}\mathbf{X}_{LM}(\theta,\phi) &= \frac{-i}{\sqrt{L(L+1)}}\{\mathbf{r} \times \text{grad}\, Y_L^M\} \\ &= \frac{i}{\sqrt{L(L+1)}}\left\{\frac{\boldsymbol{\epsilon}_\theta}{\sin\theta}\frac{\partial Y_L^M}{\partial\phi} - \boldsymbol{\epsilon}_\phi \frac{\partial Y_L^M}{\partial\theta}\right\}.\end{aligned}$$

To perform a separation of variables, write r^2 times the Laplacian as

$$r^2\nabla^2 = \frac{\partial}{\partial r}\left(r^2 \frac{\partial}{\partial r}\right) + \mathscr{L}^2(\theta,\phi),$$

* The expansion of curl $(\mathbf{r}Y_L^M)$ is

$$\text{curl}\,(\mathbf{r}Y_L^M) = -\mathbf{r} \times \text{grad}\, Y_L^M + Y_L^M\,\text{curl}\,\mathbf{r} = -\mathbf{r} \times \text{grad}\, Y_L^M.$$

Because Y_L^M is a function of θ and ϕ only, the resulting cross product is independent of both r and $\boldsymbol{\epsilon}_r$.

where $\mathscr{L}^2(\theta,\phi) = \frac{1}{\sin\theta}\frac{\partial}{\partial\theta}\left(\sin\theta\frac{\partial}{\partial\theta}\right) + \frac{1}{\sin^2\theta}\frac{\partial^2}{\partial\phi^2}.$

We have assumed a solution of the form $\mathbf{M}_{LM}^{(E)} = R_L(r)\mathbf{X}_{LM}(\theta,\phi)$. Substituting into

$$\{r^2\nabla^2 + r^2k^2\}\mathbf{M}_{LM}^{(E)} = 0,$$

we obtain

$$\mathbf{X}_{LM}\frac{d}{dr}\left(r^2\frac{dR_L}{dr}\right) + R_L\mathbf{X}_{LM}r^2k^2 = -R_L\mathscr{L}^2(\theta,\phi)\mathbf{X}_{LM}.$$

The components of $\mathbf{X}_{LM}$ have been designated as y_θ and y_ϕ. The vector equation above can be written in terms of the components of $\mathbf{X}_{LM}$. Dividing by y_m in the scalar equations we obtain

$$\frac{1}{R_L}\frac{d}{dr}\left(r^2\frac{dR_L}{dr}\right) = -\frac{1}{y_\theta}\mathscr{L}^2y_\theta = L(L+1);$$

also $$\frac{1}{R_L}\frac{d}{dr}\left(r^2\frac{dR_L}{dr}\right) = -\frac{1}{y_\phi}\mathscr{L}^2y_\phi = L(L+1).$$

Because the two scalar equations are each equal to the same radial equation, when separated both are equal to the same constant.* Now multiply the first equation by $\boldsymbol{\epsilon}_\theta$ and the second by $\boldsymbol{\epsilon}_\phi$. The resulting separated set can be written as

$$\frac{1}{r^2}\frac{d}{dr}\left\{r^2\frac{dR_L}{dr}\right\} + \left(k^2 - \frac{L(L+1)}{r^2}\right)R_L = 0$$

and $$\mathscr{L}^2\mathbf{X}_{LM} + L(L+1)\mathbf{X}_{LM} = 0.$$

The radial equation is the defining equation for the spherical Bessel functions $j_L(kr)$ and the spherical Neumann functions $n_L(kr)$, or a linear combination of these, the spherical Hankel functions of the first and second kind, $h_L^{(1)}(kr)$ and $h_L^{(2)}(kr)$. The choice of representation depends on the physical conditions of the problem. In problems involving spherically diverging electromagnetic waves we are concerned with the spherical Hankel functions of the first kind $h_L^{(1)}(kr)$. The "second kind" of functions correspond to spherically converging (or collapsing) waves. The Bessel, Neumann,

* The parameter L is a positive integer ranging from zero to N where N is arbitrarily large. This particular choice of separation constant has been taken because we finally will define the $\mathbf{X}_{LM}$ in terms of the associated Legendre polynomials P_L^M.

and Hankel functions are not independent. Their interrelationships and asymptotic forms are noted below.

$$h_L^{(1)}(\rho) = j_L(\rho) + in_L(\rho)$$

$$h_L^{(2)}(\rho) = j_L(\rho) - in_L(\rho)$$

$$j_L(\rho) \xrightarrow[\text{large } \rho]{} \frac{\cos[\rho - (L+1)\pi/2]}{\rho}$$

$$n_L(\rho) \xrightarrow[\text{large } \rho]{} \frac{\sin[\rho - (L+1)\pi/2]}{\rho}$$

$$h_L^{(1)}(\rho) \xrightarrow[\text{large } \rho]{} \frac{e^{i\rho}}{\rho} e^{-i(L+1)\pi/2}$$

$$h_L^{(2)}(\rho) \xrightarrow[\text{large } \rho]{} \frac{e^{-i\rho}}{\rho} e^{i(L+1)\pi/2}.$$

From these relations it can be seen that the $h_L^{(1)}$ solutions are the only ones suited to the physical conditions of our problem. Thus up to a normalization term,

$$R_L(r) = h_L^{(1)}(kr).$$

The following proof demonstrates that our assumed form of

$$\mathbf{X}_{LM} = \frac{i}{\sqrt{L(L+1)}} \operatorname{curl}(\mathbf{r}Y_L^M(\theta,\phi))$$

is in fact a solution to

$$\{\mathscr{L}^2 + L(L+1)\}\mathbf{X}_{LM} = 0.$$

Because $\mathbf{X}_{LM}(\theta,\phi)$ is a function of the variables θ and ϕ only, we can write

$$\mathscr{L}^2\mathbf{X}_{LM} = r^2\nabla^2\mathbf{X}_{LM}.$$

Now remembering that div $\mathbf{X} = 0$,

$$\nabla^2\mathbf{X}_{LM} = -\operatorname{curl}\operatorname{curl}\mathbf{X}_{LM} + \operatorname{grad}(\operatorname{div}\mathbf{X}_{LM}) = -\operatorname{curl}\operatorname{curl}\mathbf{X}_{LM}.$$

Up to a constant then

$$\begin{aligned}
\nabla^2\{-\mathbf{r}\times\operatorname{grad}Y_L^M\} &= -\operatorname{curl}\operatorname{curl}\{-\mathbf{r}\times\operatorname{grad}Y_L^M\} \\
&= \operatorname{curl}\{\mathbf{r}\operatorname{div}(\operatorname{grad}Y_L^M) - (\operatorname{div}\mathbf{r})\operatorname{grad}Y_L^M \\
&\qquad + (\operatorname{grad}Y_L^M\cdot\nabla)\mathbf{r} - (\mathbf{r}\cdot\nabla)\operatorname{grad}Y_L^M\} \\
&= \operatorname{curl}\{\mathbf{r}\nabla^2Y_L^M - \operatorname{grad}Y_L^M\} \\
&= \{\nabla^2Y_L^M\}\operatorname{curl}\mathbf{r} - \mathbf{r}\times\operatorname{grad}(\nabla^2Y_L^M) \\
&= -\mathbf{r}\times\operatorname{grad}\{\nabla^2Y_L^M\}.
\end{aligned}$$

In the somewhat tedious expansion above we have utilized the facts that curl grad $Y_L^M = 0$ and that curl $\mathbf{r} = 0$.

Finally,

$$r^2\nabla^2\mathbf{X}_{LM} = -r^2\{\mathbf{r} \times \text{grad}\,(\nabla^2 Y_L^M)\} = -\mathbf{r} \times \text{grad}\,(r^2\nabla^2 Y_L^M).$$

This operation is permissible since ($\mathbf{r} \times$ grad) is independent of r.

Last of all

$$\{\mathscr{L}^2 + L(L+1)\}\mathbf{X}_{LM} = -\mathbf{r} \times \text{grad}\,\{[\mathscr{L}^2 + L(L+1)]Y_L^M\} = 0.$$

The defining equation for the spherical harmonics is (see Appendix D)

$$\{\mathscr{L}^2(\theta,\phi) + L(L+1)\}Y_L^M = 0;$$

Therefore $\mathbf{X}_{LM}(\theta,\phi)$ is indeed a solution of

$$\{\mathscr{L}^2(\theta,\phi) + L(L+1)\}\mathbf{X}_{LM}(\theta,\phi) = 0.$$

As stipulated, the vector harmonics are related to the spherical harmonics by

$$\mathbf{X}_{LM} = \frac{-i}{\sqrt{L(L+1)}}\{\mathbf{r} \times \text{grad}\, Y_L^M(\theta,\phi)\},$$

or

$$\mathbf{X}_{LM} = \frac{-i}{\sqrt{L(L+1)}}\left\{\frac{\boldsymbol{\epsilon}_\theta}{\sin\theta}\frac{\partial Y_L^M}{\partial\phi} - \boldsymbol{\epsilon}_\phi\frac{\partial Y_L^M}{\partial\theta}\right\}.$$

To list a few sample functions,

$$\mathbf{X}_{0,0} = 0,$$

$$\mathbf{X}_{1,1} = \frac{-e^{i\phi}}{\sqrt{2}}\{\boldsymbol{\epsilon}_\theta + i\cos\theta\,\boldsymbol{\epsilon}_\phi\},$$

$$\mathbf{X}_{1,0} = \frac{+i}{\sqrt{2}}\sin\theta\,\boldsymbol{\epsilon}_\phi,$$

and

$$\mathbf{X}_{1,-1} = \frac{e^{-i\phi}}{\sqrt{2}}\{\boldsymbol{\epsilon}_\theta - i\cos\theta\,\boldsymbol{\epsilon}_\phi\}.$$

A particular angular momentum or multipolarity is associated with fields characterized by the parameter L (a positive integer).

To construct *magnetic multipole fields* of order L we begin with

$$\mathbf{F}_{LM}^{(H)} = R_L(r)\mathbf{X}_{LM}$$

and

$$\mathbf{M}_{LM}^{(H)} = \frac{1}{k}\,\text{curl}\,(R_L\mathbf{X}_{LM}),$$

$$R_L(r) = h_L^{(1)}(kr).$$

The fields $\mathbf{M}_{LM}^{(E)} = R_L(r)\mathbf{X}_{LM}(\theta,\phi)$ do not form a complete set of vector functions for the radiation field. In the case of the radiation field (excluding longitudinal fields) the functions $\mathbf{M}_{LM}^{(E)} = R_L\mathbf{X}_{LM}(\theta,\phi)$ along with the functions $\mathbf{M}_{LM}^{(H)} = (1/k)\,\text{curl}\,(R_L\mathbf{X}_{LM})$ form a complete set, and any arbitrary magnetic radiation field $\mathbf{M}$ can be Fourier analyzed in terms of the combination of $\mathbf{M}_{LM}^{(E)}$ and $\mathbf{M}_{LM}^{(H)}$.

It is sufficient to compute either $\mathbf{M}$ or $\mathbf{F}$ because one vector completely determines the other.

If $\mathbf{M}(\mathbf{r})$ is a function well behaved for large r and continuous in the function and in its first derivative, then

$$\mathbf{M}(\mathbf{r}) = \sum_{L=0}^{\infty}\sum_{M=-L}^{M=L}\{a_{LM}^{(E)}\mathbf{M}_{LM}^{(E)}(\mathbf{r}) + a_{LM}^{(H)}\mathbf{M}_{LM}^{(H)}(\mathbf{r})\};$$

where
$$\mathbf{M}_{LM}^{(E)} = R_L(r)\mathbf{X}_{LM}(\theta,\phi)$$

and
$$\mathbf{M}_{LM}^{(H)} = \frac{1}{k}\,\text{curl}\,(R_L(r)\mathbf{X}_{LM}(\theta,\phi)).$$

The a_{LM} are the Fourier expansion coefficients.

Because $\mathbf{M}(\mathbf{r})$ and $\mathbf{F}(\mathbf{r})$ are not independent, we could work in an electric field representation where

$$\mathbf{F}_{LM}^{(E)} = \frac{1}{k}\,\text{curl}\,(R_L(r)\mathbf{X}_{LM}(\theta,\phi))$$

and
$$\mathbf{F}_{LM}^{(H)} = R_L(r)\mathbf{X}_{LM}(\theta,\phi);$$

with an arbitrary $\mathbf{F}(\mathbf{r})$ represented by

$$\mathbf{F}(\mathbf{r}) = \sum_{L=0}^{\infty}\sum_{M=-L}^{M=L}\{b_{LM}^{(E)}\mathbf{F}_{LM}^{(E)} + b_{LM}^{(H)}\mathbf{F}_{LM}^{(H)}\}.$$

Here the b_{LM} are the Fourier coefficients.

The great advantage of this representation is that it emphasizes fields of unique multipolarity and parity. The functions $\mathbf{M}_{LM}^{(E)}$ have multipolarity L and parity $(-1)^L$; the parity $(-1)^L$ is associated with ELECTRIC MULTIPOLES. The functions $\mathbf{M}_{LM}^{(H)}$ have parity $-(-1)^L$ and multipolarity L, and are associated with MAGNETIC MULTIPOLES. Thus the function

$$\mathbf{M}_{10}^{(E)} = R_1(r)\mathbf{X}_{10}$$

has parity -1 and corresponds to the magnetic field of an electric multipole of order 1, called electric dipole radiation. The function

$$\mathbf{M}_{10}^{(H)} = \frac{1}{k}\,\text{curl}\,(\mathbf{R}_1(r)\mathbf{X}_{10})$$

has parity $-(-1) = +1$ and is the magnetic radiation vector associated with a magnetic moment (or dipole).

The parity refers to the sign of the vector function when there is reflection through the origin. If

$$\mathbf{M}_{LM}(r,\theta,\phi) = +\mathbf{M}_{LM}(r, \pi - \theta, \pi + \phi)$$

the parity is even, i.e., equal to $+1$. On the other hand, if

$$\mathbf{M}_{LM}(r,\theta,\phi) = -\mathbf{M}_{LM}(r, \pi - \theta, \pi + \phi)$$

the parity is odd, equaling -1. Equivalent statements hold for the $\mathbf{F}_{LM}$.

In order to illustrate a typical field calculation let us compute the field of an electric dipole. $\mathbf{M}_{10}^{(E)}$ defines the electric multipole of order 1 and projection (M) of zero.

$$\mathbf{M}_{10}^{(E)} = h_1^{(1)}(kr)\mathbf{X}_{10} = \frac{i}{\sqrt{2}}\left(1 + \frac{i}{kr}\right)\frac{e^{ikr}}{kr}\sin\theta\,\boldsymbol{\epsilon}_\phi.$$

Then $$\mathbf{F}_{10}^{(E)} = \frac{1}{k}\,\text{curl}\,\mathbf{M}_{10}^{(E)}$$

$$= \frac{i}{\sqrt{2}}\left\{2\left(1 + \frac{i}{kr}\right)\frac{e^{ikr}}{(kr)^2}\cos\theta\,\boldsymbol{\epsilon}_r - \left(i - \frac{1}{kr} - \frac{i}{(kr)^2}\right)\frac{e^{ikr}}{kr}\sin\theta\,\boldsymbol{\epsilon}_\theta\right\}.$$

To obtain a qualitative understanding of the sources of such a field we can compute the time variation of charge associated with this field on the surface of a conducting sphere of radius a at the origin. The first Maxwell equation relates the charge density at any point to the electric displacement vector. Assume that the dielectric constant at the surface of the sphere is 1; then

$$\{\epsilon_0\,\text{div}\,\mathscr{E}_{10}^{(E)}(\mathbf{r},t)\}_{r=a} = -\rho(a,t).$$

Since $$\mathscr{E}^{(E)}_{(\mathbf{r},t)} = \mathscr{E}_0\mathbf{F}_{10}^{(E)}e^{-i\omega t}$$

($\mathscr{E}_0$ is a constant carrying the field dimensions), using the form above for $\mathbf{F}_{10}^{(E)}$ we find

$$\rho(a,t) = -\mathscr{E}_0\epsilon_0\frac{\sqrt{2}}{a}\left\{\frac{1}{(ka)^2} - \frac{i}{(ka)^3}\right\}e^{i(ka-\omega t)}\cos\theta.$$

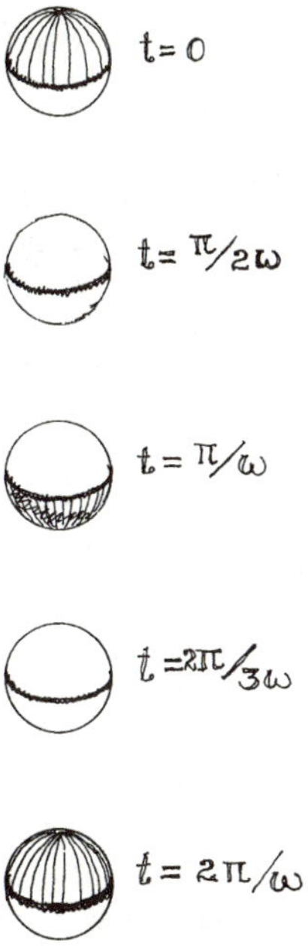

ELECTRIC DIPOLE RADIATION

FLOW OF SOURCE CHARGES

The real part of this expression represents the physical solution. It is clear that the charge density is a maximum at the poles and that the sign at one pole is opposite to the sign of the charge at the other. The time variation of the charge is sinesoidal.

E. SOURCES OF THE RADIATION FIELD

1. THE RETARDED POTENTIALS

The solution to the inhomogeneous wave equations is quite complicated. As we discovered in electrostatics, the inhomogeneous

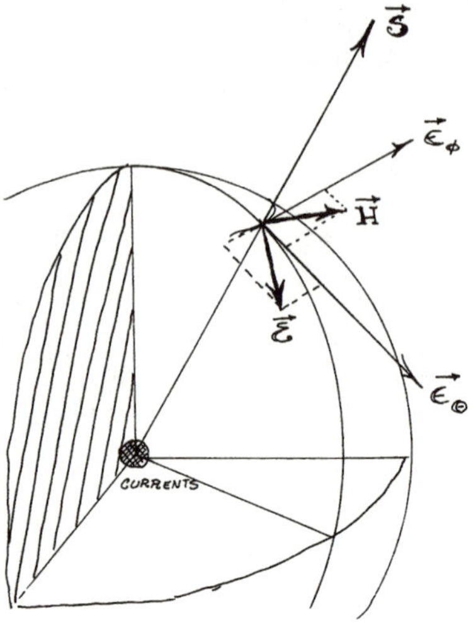

differential equation can in some instances be converted to an equivalent integral equation by means of a Green's function. The Green's function for the Laplacian differential operator is

$$\left\{\frac{1}{4\pi}\frac{1}{|\mathbf{r}-\mathbf{r}'|}\right\}$$

and when under an integral sign corresponds in a sense to the inverse of the ∇^2 operator.

The most convenient method for handling problems which contain the sources of the field is to work with the scalar and vector potentials which define the field.

Suppose that the conductivity of all regions of space is zero ($\beta = 0$). The Lorentz condition relating the potential functions is then

$$\operatorname{div} \mathbf{A} + \frac{1}{v^2}\frac{\partial V}{\partial t} = 0$$

where $$\frac{1}{v^2} = \epsilon_0 \epsilon \mu_0 \mu.$$

The defining equations for V and **A** are then

$$\Box V = -\rho/\epsilon_0\epsilon$$

and $$\Box \mathbf{A} = -\mu_0\mu \mathbf{J}.$$

Here ρ and $\mathbf{J}$ represent the sources of the electromagnetic field. The symbol $\square$ is the d'Alembertian operator,

$$\square = \nabla^2 - \frac{1}{v^2}\frac{\partial^2}{\partial t^2}\,.$$

Some of the physical features of the solution can be seen by a simple discussion. Not all solutions to the inhomogeneous wave equation will be acceptable because some solutions violate causality. In other words if radiation is emitted from a source charge

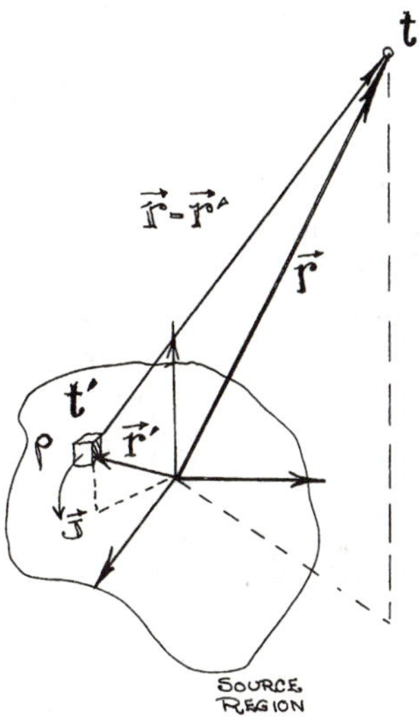

at a position $\mathbf{r}'$ and at a source time t′, and if this radiation is picked up by a receiver at a position $\mathbf{r}$ and at a receiving time t, we can immediately write down the relations between the time interval and the spatial separation when the velocity of propagation is v:

$$(t - t') = \frac{|\mathbf{r} - \mathbf{r}'|}{v}\,.$$

In general the maximum possible velocity is c; therefore

$$(t - t') \geqslant \frac{|\mathbf{r} - \mathbf{r}'|}{c}\,.$$

The mathematics employed in finding the potential functions actually provides solutions satisfying such a constraint plus an additional solution in which the sign of $|\mathbf{r} - \mathbf{r}'|/v$ is negative. This latter solution is *nonphysical*, having an associated time difference which corresponds to a signal being received before it is sent. These nonphysical solutions are known as "advanced potentials." The advanced potentials are excluded by a requirement that

$$(t - t') \geqslant \frac{|\mathbf{r} - \mathbf{r}'|}{c}.$$

This condition is similar to a boundary condition when we demand that a function be zero everywhere outside of a specified interval.

The mathematical problem to be solved is actually *beyond the stated limits of this course.* The developmet of the Green's function $G((\mathbf{r} - \mathbf{r}');(t - t'))$ for the d'Alembertian $\square$ *has been presented* in Appendix E *for completeness.* Thus the reader if interested can familiarize himself with a method of solution for the problem.

Because the operator $\square$ contains the time, the Green's function will contain terms of the form $(t - t')$ and will require an integration over t' from $-\infty$ to $+\infty$. This is not surprising, for the source terms $\rho(r',t')$ and $\mathbf{J}(r',t')$ are functions of t' and must in some manner be related to the reception time t. Relating t' to t forms perhaps the most formidable part of the problem at hand.

From the arguments concerning "causality" the reader might surmise that the time-dependent form of $G((\mathbf{r} - \mathbf{r}');(t - t'))$ will do two things:

(a) It will require that $t' = t - \dfrac{|\mathbf{r} - \mathbf{r}'|}{v}$, and

(b) It will require that the solution be zero for $(t - t') < 0$.

Consider the inhomogeneous equation for the scalar potential $V(\mathbf{r},t)$ in a vacuum $\epsilon = 1$:

$$\square V = -\rho(r,t)/\epsilon_0.$$

The Green's function for $\square$ is

$$G((r - r');(t - t')) = \frac{1}{4\pi\,|\mathbf{r} - \mathbf{r}'|}\,\Gamma(t - t')\,\delta\left(t' - \left(t - \frac{|\mathbf{r} - \mathbf{r}'|}{c}\right)\right).$$

The function $\Gamma(y)$ is a step function:

$$\Gamma(y) = +1 \qquad 0 < y,$$
$$\Gamma(y) = 0 \qquad y < 0.$$

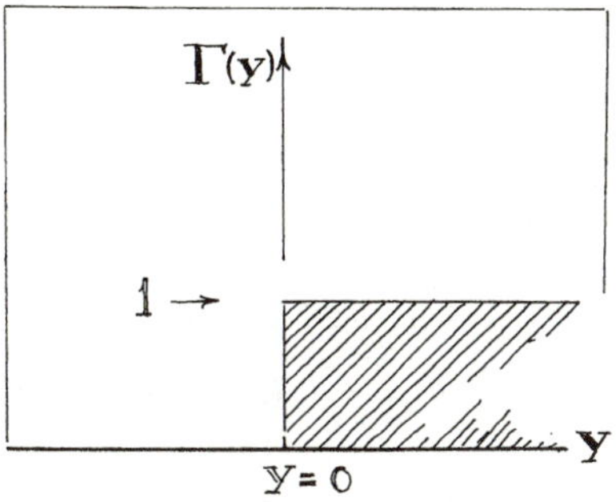

Thus the reception time can never be less than $|\mathbf{r} - \mathbf{r}'|/c$, and $(t - t') \geqslant 0$.

The function

$$\delta\left(t' - \left(t - \frac{|\mathbf{r} - \mathbf{r}'|}{c}\right)\right)$$

is the Dirac delta function which fixes t′ as

$$\left(t - \frac{|\mathbf{r} - \mathbf{r}'|}{c}\right).$$

The integral solution for $V(\mathbf{r},t)$ is

$$V(\mathbf{r},t) = \frac{1}{4\pi\epsilon_0} \iiint_\tau \int_{-\infty}^{\infty} G((\mathbf{r} - \mathbf{r}'),(t - t'))\rho(\mathbf{r}',t')\, d\tau'\, dt'$$

Integrating over t′ we obtain the integral for the retarded potentials

$$V(\mathbf{r},t) = \frac{1}{4\pi\epsilon_0} \iiint_\tau \frac{\rho\left(\mathbf{r}', \left(t - \frac{|\mathbf{r} - \mathbf{r}'|}{c}\right)\right)}{|\mathbf{r} - \mathbf{r}'|} \Gamma(t - t')\, d\tau'.$$

In the same manner

$$\mathbf{A}(\mathbf{r},t) = \frac{\mu_0}{4\pi} \iiint_\tau \frac{\mathbf{J}\left(\mathbf{r}', \left(t - \frac{|\mathbf{r} - \mathbf{r}'|}{c}\right)\right)}{|\mathbf{r} - \mathbf{r}'|} \Gamma(t - t')\, d\tau'.$$

2. THE LIENARD-WIECHERT POTENTIALS

The scalar and vector potentials of a moving point charge q can also be derived using the Green's function of the operator $\square$.

In electrostatics the charge density of the point charge was written as

$$\rho(\mathbf{r}') = q\delta(\mathbf{r}' - \mathbf{R}),$$

where the charge is located by a position vector **R**.

If the point charge is moving along a path such that at any source time t′ the position of q is given as $\mathbf{R}(t')$, then

$$\rho(\mathbf{r}',t') = q\delta(\mathbf{r}' - \mathbf{R}(t')).$$

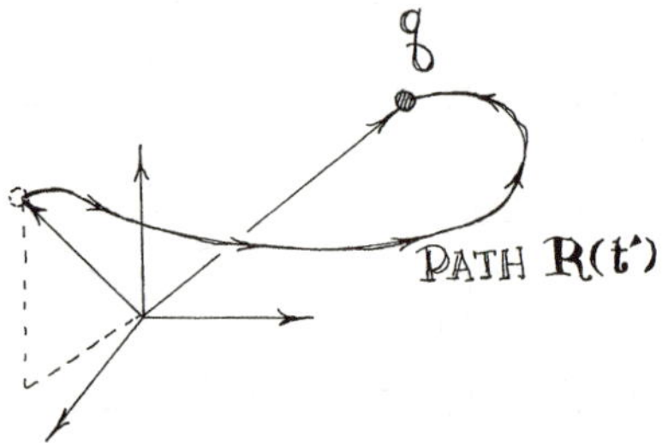

Assume that $\epsilon = 1$ and $\mu = 1$; then $v = c$.

The potential $V(\mathbf{r},t)$ is then

$$V(\mathbf{r},t) = \frac{1}{4\pi\epsilon_0} \times \iiint_{\text{All Space}} \int_{-\infty}^{\infty} \frac{\Gamma(t - t')q\delta(\mathbf{r}' - \mathbf{R}(t'))\,\delta\left(t' - \left(t - \frac{|\mathbf{r} - \mathbf{r}'|}{c}\right)\right)}{|\mathbf{r} - \mathbf{r}'|}\, d\tau'\, dt'.$$

We now integrate over $d\tau'$, obtaining

$$V(\mathbf{r},t) = \frac{q}{4\pi\epsilon_0}\int_{-\infty}^{\infty} \frac{\Gamma(t - t')\,\delta\left(t' - \left(t - \frac{|\mathbf{r} - \mathbf{R}(t')|}{c}\right)\right)}{|\mathbf{r} - \mathbf{R}(t')|}\, dt'.$$

To integrate this expression we use a theorem which states that

$$\int_{-\infty}^{\infty} g(t')\,\delta[h(t')]\,dt' = \frac{g(t_0)}{\left|\left(\frac{dh}{dt'}\right)_{t_0}\right|},$$

where t_0 is the solution of

$$h(t_0) = 0.$$

In our problem

$$h(t') = -\left(t - \frac{|\mathbf{r} - \mathbf{R}(t')|}{c}\right) + t'$$

and

$$g(t') = \frac{\Gamma(t - t')}{|\mathbf{r} - \mathbf{R}(t')|}.$$

Now

$$\frac{dh}{dt'} = \left\{\frac{[\mathbf{r} - \mathbf{R}(t')] \cdot \dfrac{d\mathbf{R}}{dt}}{|\mathbf{r} - \mathbf{R}(t')|} - 1\right\}.$$

Setting $h(t_0') = 0$

we find that $t_0' = f(t,\mathbf{r})$.

For instance, suppose that the charge q accelerates along the z axis with a constant acceleration a; then

$$\mathbf{R}(t') = \tfrac{1}{2}at'^2\boldsymbol{\epsilon}_3,$$

and

$$t - \frac{1}{v}\sqrt{\left[x^2 + y^2 + \left(z - \frac{a}{2}t_0'^2\right)^2\right]} - t_0' = 0.$$

Solving for t_0' we obtain

$$t_0' = f(x,y,z;t).$$

Then

$$V(\mathbf{r},t) = \frac{q\Gamma(t - t_0')}{4\pi\epsilon_0} \frac{1}{\left(R' - \dfrac{\mathbf{v}_0 \cdot \mathbf{R}'}{c}\right)}$$

where

$$\mathbf{R}' = \mathbf{r} - \mathbf{R}(t_0')$$

and

$$\mathbf{v}_0 = \left[\frac{d\mathbf{R}}{dt'}\right]_{t_0}.$$

The same approach gives $\mathbf{A}(\mathbf{r},t)$; remembering that

$$\mathbf{J} \to q\mathbf{v}\delta(\mathbf{r}' - \mathbf{R}(t')),$$

$$\mathbf{A}(\mathbf{r},t) = \frac{\mu_0}{4\pi}\mathbf{v}_0 \frac{\Gamma(t - t_0')}{\left(R' - \dfrac{\mathbf{v}_0 \cdot \mathbf{R}'}{c}\right)}.$$

To compute the fields from these equations some care must be exercised since t_0' is a function of $\mathbf{r}$ and t.

$$\mathscr{E} = -\text{grad}_{\mathbf{r}}\, V - \frac{\partial \mathbf{A}}{\partial t},$$

and

$$\mathbf{H} = \frac{1}{\mu_0}\text{curl}\, \mathbf{A}.$$

The derivatives of $\Gamma(t - t'_0)$ will not contribute since $\dfrac{d\Gamma}{dt} = \delta(t - t'_0)$, which evaluates the field only at $t = t'_0$.

As an example,

$$\frac{\partial A}{\partial t} = \frac{\mu_0}{4\pi} q\Gamma(t - t'_0)\left\{\frac{\dfrac{d\mathbf{v}_0}{dt}}{\left(R' - \dfrac{\mathbf{v}_0 \cdot \mathbf{R}'}{c}\right)} - \mathbf{v}_0 \frac{\dfrac{dR'}{dt} - \dfrac{d}{dt}\left(\dfrac{\mathbf{v}_0 \cdot \mathbf{R}'}{c}\right)}{\left(R' - \dfrac{\mathbf{v}_0 \cdot \mathbf{R}'}{c}\right)^2}\right\}.$$

The retardation requirement is that

$$f(\mathbf{r},t) - t'_0 = 0$$

indicating that $\dfrac{\partial t}{\partial t'} = \dfrac{1}{\partial t'/\partial t}$ for fixed $\mathbf{r}$.

The gradient and curl operators must be applied both to d/dt terms and to R' terms, because $\partial/\partial x_j$ operates both on $\mathbf{r}$ and on t'_0. For instance

$$\operatorname{curl}\left\{\frac{\mathbf{v}_0}{\left(R' - \dfrac{\mathbf{v}_0 \cdot \mathbf{R}'}{c}\right)}\right\} = \frac{1}{\left(R' - \dfrac{\mathbf{v}_0 \cdot \mathbf{R}'}{c}\right)} \operatorname{curl} \mathbf{v}_0 - \mathbf{v}_0 \times \operatorname{grad}\left\{\frac{1}{\left(R' - \dfrac{\mathbf{v}_0 \cdot \mathbf{R}'}{c}\right)}\right\}.$$

Now $$\operatorname{curl} \mathbf{v}_0 = \{\operatorname{grad}(t'_0)\} \times \left[\frac{d\mathbf{v}_0}{dt_0}\right] = \{\nabla t_0\} \times \mathbf{a}_0$$

where $\mathbf{a}_0$ is the acceleration of the charge at the retarded time t'_0.

$$\text{If } \frac{|\mathbf{v}_0|}{c} \ll 1, \quad \text{then} \quad R' - (\mathbf{v}_0 \cdot \mathbf{R}'/c) \simeq R'.$$

If we look in the far field $|\mathbf{R}| \ll |\mathbf{r}|$, then $t_0 \simeq t - |\mathbf{r}|/c$, and

$$\mathscr{E}(\mathbf{r},t) \simeq \frac{\mu_0 q}{4\pi}\left\{\frac{\mathbf{r} \times (\mathbf{r} \times \mathbf{a}_0)}{r^3}\right\}$$

with $$\mathbf{B}(\mathbf{r},t) \simeq \frac{\mu_0 q}{4\pi c}\left\{\frac{\mathbf{a}_0 \times \mathbf{r}}{r^2}\right\}.$$

The far field Poynting vector then reduces to

$$\mathbf{S} = \frac{\mu_0 q^2}{16\pi^2 c}\left\{\frac{(\mathbf{r} \times \mathbf{a}_0)^2 \mathbf{r}}{r^5}\right\}.$$

The total power radiated across the surface of a sphere of radius r by a slowly moving (low velocity) accelerated charge located near the origin is

$$-\frac{\partial U}{\partial t} = \int_{\text{Sphere}} \mathbf{S} \cdot \mathbf{n}\, dS = \frac{2}{3}\frac{\mu_0 q^2}{4\pi c} a_0{}^2.$$

XII. Time-Varying Current Circuits

A. INTRODUCTION

In the preceding development of direct current circuit theory, we have seen that the whole system depends upon Ohm's law and the current and voltage laws of Kirchhoff. That there do exist substances which have the properties satisfying Ohm's law is simply an experimental fact, and given this fact, the laws of Kirchhoff follow as a consequence of the general field theory of electromagnetism. The theory is greatly simplified in this case because the field quantities, or in terms of circuits the voltages and currents, are independent of time.

Alternating current circuits by definition are those in which the voltages and currents do vary with time, which means in terms of field theory that the electromagnetic field is time-dependent. This means that the development of an appropriate "circuit theory" requires certain restrictions.

We will have to satisfy two requirements:

(1) That there be no appreciable loss of energy by electromagnetic radiation, and

(2) That Kirchhoff's voltage and current laws be true for the instantaneous voltages and currents.

Obviously, Kirchhoff's laws cannot be satisfied if the wavelength of the current waves, as defined by

$$\lambda = c/f$$

or

$$kc = \omega$$

(where c = velocity of light, f = frequency of oscillation = $\omega/2\pi$, and λ = wavelength = $2\pi/k$), is of the same order of magnitude as the physical dimensions of the circuit or circuit components. The same criterion roughly holds for the possibility of electromagnetic radiation. However, a simple calculation will show that for anything apt to be found on a laboratory bench, a very high frequency, of the order of megacycles per second, is required for λ to become small enough to worry about. It suffices to say that in the circuits to be discussed, to fall into a situation where electromagnetic radiation is a serious factor requires some ingenuity.

In particular, we shall develop the theory of alternating current circuits in terms of *lumped parameters*. That is, we shall speak of resistances, capacitances, and inductances as circuit parameters without special dimensions. In short, we shall deal with voltages, currents, and these lumped parameters, all of which are the integrated results of field theory. We shall ignore the fact that a condenser, for example, is not a pure capacitance; rather we shall proceed as if these parameters had a pristine existence apart from the vicissitudes of the material world. This is in no wise a process of approximation; capacitance and inductance, for example, have rigorous definitions in terms of permittivity, permeability, and geometric factors. But it is true that it is impossible to realize physically these quantities in a pure form, and this makes experiments a little more difficult than drawing circuit diagrams. It is worth pointing out, however, that because we do know exactly what these circuit parameters mean, we can always completely describe, say, a condenser (which we would like to be a capacitance) in terms of capacitance, resistance, and inductance.

B. ELEMENTARY VOLTAGE AND CURRENT RELATIONS

Kirchhoff's current and voltage laws hold for the instantaneous quantities in a mesh. Besides the voltage drops V_R across resistors R we now include the induced emf V_L across self-inductances L_{jj} and the mutual inductances L_{jk} ($j \neq k$) plus the voltage drops V_c across capacitors C.

In an elementary circuit carrying a time-varying current I(t), the voltage drop across an inductance opposes the change in current while the voltage across the condenser is given by the charge Q(t) residing upon C at the time in question:

$$V_R = I(t)R; \quad R = \text{the resistance in ohms}$$

$$V_L = L\frac{dI}{dt}; \quad L = \text{the self-inductance in henries}$$

$$V_C = \frac{Q(t)}{C}; \quad C = \text{the capacitance in farads.}$$

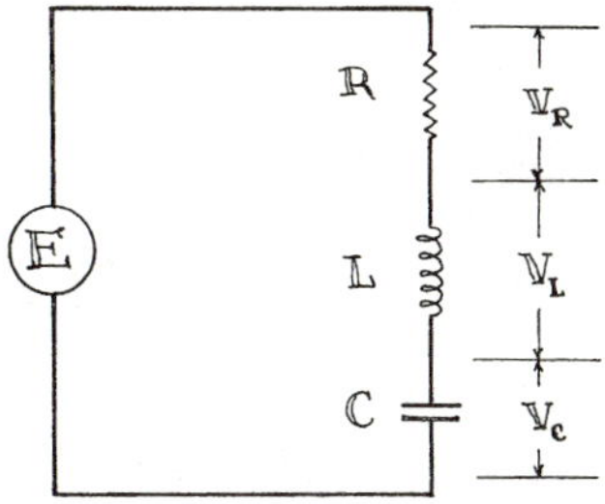

If at t = 0 Q(0) = 0, then because I(t) = dQ/dt,

$$V_C = \int_0^t \frac{I(t)}{C}\,dt.$$

For more complicated circuits which contain mutual inductances, the emf V_j induced in the jth branch by the kth current $I_k(t)$ is

$$V_j = L_{jk}\frac{dI_k(t)}{dt}.$$

C. TRANSIENTS IN ELEMENTARY CIRCUITS

Single or double element single loop circuits can be solved in a straightforward manner. Because these examples are usually treated in the standard texts on linear differential equations we shall merely note the problems and their solutions.

1. THE LR CIRCUIT

We consider the LR circuit initially carrying zero current which at t = 0 is suddenly connected in series with a battery of emf E.

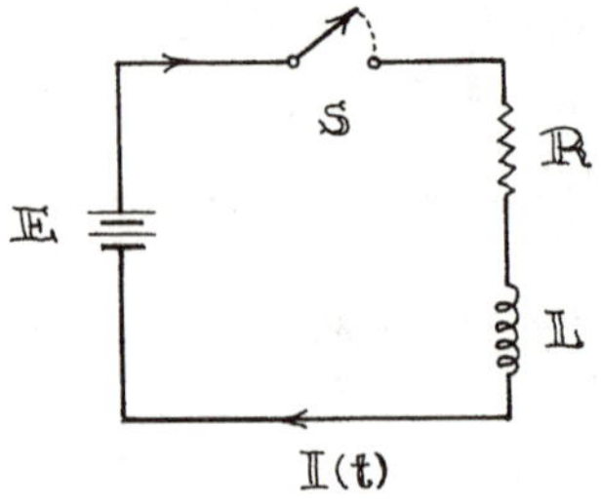

At any later time $0 \leqslant t$ the voltage relations are

$$E = V_R(t) + V_L(t),$$

or

$$E = I(t)R + L\frac{dI}{dt}.$$

The complementary solution is $I_c = I_0e^{-Rt/L}$ while the particular solution is $I_p = E/R$. The complete solution is a linear combination of these with the initial condition that $I(0) = 0$. Therefore

$$I(t) = \frac{E}{R}\{1 - e^{-Rt/L}\}.$$

In this circuit the current reaches a value $(E/R)(1 - e^{-1})$ in a time equal to its time constant $t_L = L/R$. The voltage drop across L drops to $e^{-1}E$ in this time.

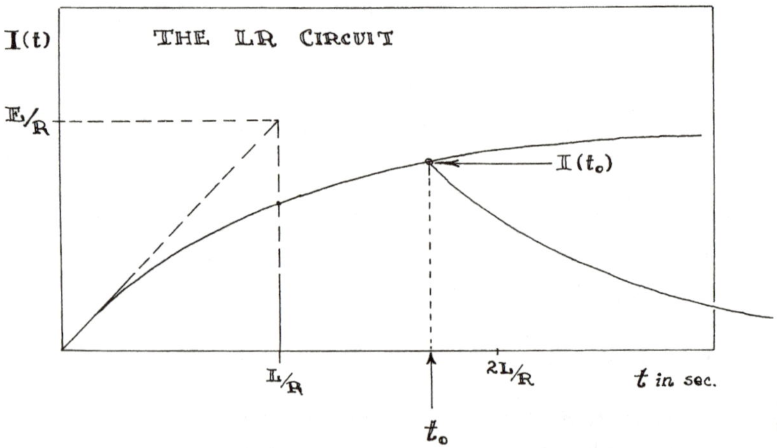

If the battery is now shorted out at $t = t_0$, the current drops to zero as $t \to \infty$. The reader should show that when the circuit is shorted out at t_0 then

$$I(t) = I(t_0)e^{-(R/L)(t-t_0)}, \quad t_0 \leq t.$$

2. THE RC CIRCUIT

A resistor R and a condenser C in series are connected across a

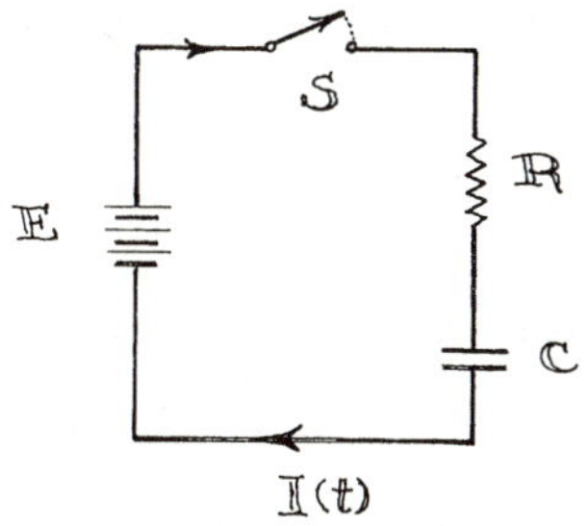

battery emf at $t = 0$. The voltage equation is

$$E = V_R(t) + V_C(t),$$

or

$$E = R\frac{dQ(t)}{dt} + \frac{Q(t)}{C}.$$

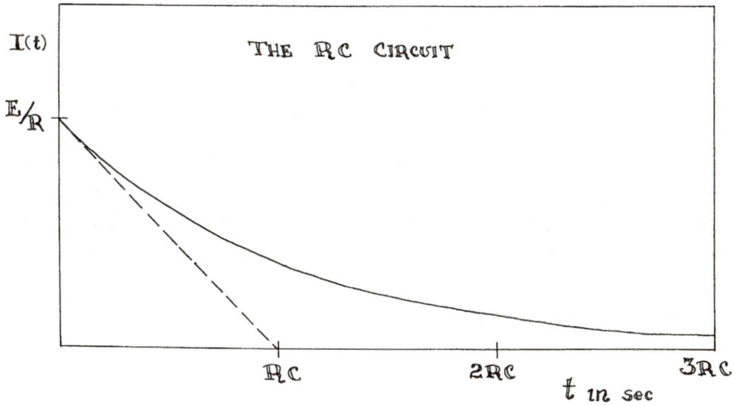

Solving for Q(t) with the initial condition that the charge on the condenser is zero at $t = 0$, that is, $Q(0) = 0$,

$$Q(t) = EC\{1 - e^{-t/RC}\},$$

and

$$I(t) = \frac{dQ}{dt} = \left(\frac{E}{R}\right)e^{-t/RC}.$$

The current drops to $e^{-1}[E/R]$ in a time equal to the time constant $t_c = RC$. If at $t = t_0$ the battery is shorted out, the current reverses and

$$I(t) = \frac{Q(t_0)}{RC}\,e^{-(t-t_0)/RC}, \quad t_0 \leqslant t.$$

3. THE LRC CIRCUIT

For a series circuit the sum of the voltages is

$$E = V_C(t) + V_R(t) + V_L(t),$$

or

$$E = \frac{Q}{C} + R\,\frac{dQ}{dt} + L\,\frac{d^2Q}{dt^2}.$$

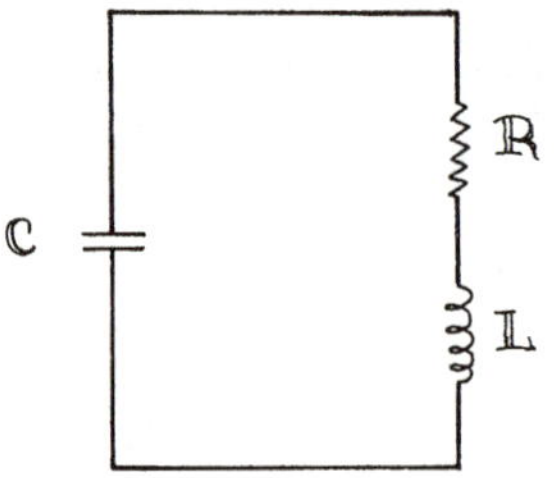

To illustrate a typical solution we let $E = 0$ (we use the complementary solution only) and assume that $Q(0) = Q_0$ (i.e., the condenser charged at $t = 0$). At a later time $0 < t$ the charge variations in the circuit may behave according to one of three characteristic solutions:

The circuit is said to be OVERDAMPED when

$$\left\{\frac{R^2}{4L^2} - \frac{1}{LC}\right\} > 0.$$

Defining the quantities α, β, and ω_0 as

$$\alpha = R/2L,$$
$$\omega_0^2 = 1/LC,$$
$$\beta = \{\alpha^2 - \omega_0^2\}^{\frac{1}{2}},$$

we can write the solution Q(t) as

$$Q(t) = \frac{EC}{\beta} e^{-\alpha t}\{\beta \cosh \beta t + \alpha \sinh \beta t\}.$$

The corresponding current I(t) is

$$I(t) = \frac{dQ}{dt} = \frac{-EC}{\beta} \omega_0^2 e^{-\alpha t} \sinh \beta t.$$

If $\{\alpha^2 - \omega_0^2\} = \beta^2 = 0$, the solution of the previous section does not apply. This is the condition for CRITICAL DAMPING. Assuming a solution of the form $(A + Bt)\, e^{mt}$, we find that

$$Q(t) = Q_{(t=0)}(1 + \alpha t)e^{-\alpha t},$$

and
$$I(t) = -Q_{(t=0)}\alpha^2 t e^{-\alpha t}.$$

The current reaches an extreme value $-Q(0)\alpha e^{-1}$ in a time $t_{max} = 1/\alpha$.

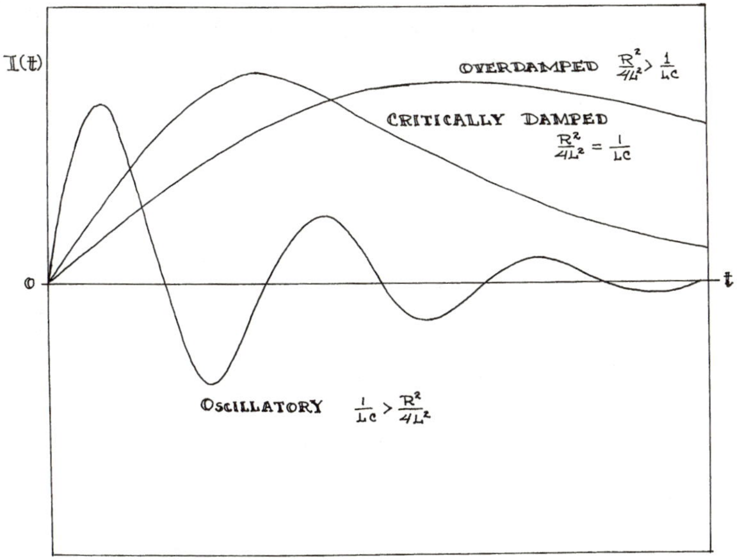

When

$$\left\{\frac{1}{LC} - \frac{R^2}{4L^2}\right\} = \{\omega_0^2 - \alpha^2\} > 0,$$

the circuit is said to be OSCILLATORY. Let

$$\omega^2 = \omega_0^2 - \alpha^2;$$

then $$Q(t) = Q_{(t=0)}\left(\frac{\omega_0}{\omega}\right)e^{-\alpha t} \cos(\omega t - \phi_0).$$

The phase angle ϕ_0 for the initial conditions we have chosen is

$$\phi_0 = \tan^{-1}(\alpha/\omega).$$

The instantaneous current is

$$I(t) = \dot{Q}(t) = -Q(0)(\omega_0^2/\omega)e^{-\alpha t} \sin \omega t.$$

The zeros of I(t) are regular and occur for

$$t = n\pi/\omega$$

where n is an integer. The maxima and minima are also regular. Thus the wave has a period of

$$2\pi/\omega.$$

The logarithmic decrement of a damped oscillating function is defined as the log of the ratio of the nth positive maximum to the $(n + 1)^{th}$ positive maximum. The ratio of successive maxima is

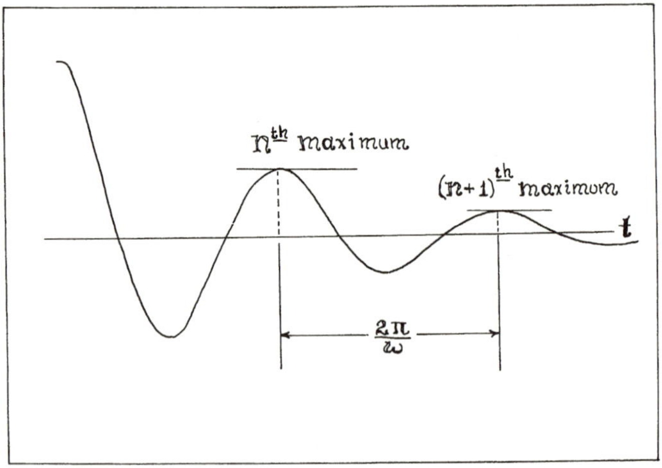

$e^{-\pi R/L\omega}$; thus the logarithmic decrement is

$$\delta = \log_e \left\{ \frac{i_{max}^{(n)}}{i_{max}^{(n+1)}} \right\} = \frac{\pi R}{L\omega}.$$

D. FORCED OSCILLATIONS OF AN LRC CIRCUIT

In the accompanying figure is shown a series LRC circuit in which a source of time-varying emf

$$\begin{aligned} E(t) &= \mathscr{R}(E_0 e^{i\omega t}) \\ &= E_0 \cos \omega t \end{aligned}$$

has been inserted in series with the passive elements L, R, and C. The instantaneous voltage drops must equal the instantaneous emf; therefore

$$V_L + V_R + V_C = E(t),$$

or allowing for complex currents I'

$$L\frac{dI'}{dt} + RI' + \frac{Q'(t)}{C} = E_0 e^{i\omega t}.$$

The complementary solution (to the homogeneous equation) was completed in the preceding section on transients. If we consider solutions at a time which is long compared to $1/\alpha$, the *particular* solution will predominate. In other words, after the circuit is turned on the transient solution damps out as $e^{-\alpha t}$, and for $t \gg 1/\alpha$ the transient solution for our purposes can be neglected.

On differentiation with respect to t, the equation becomes

$$L\frac{d^2I'}{dt^2} + R\frac{dI'}{dt} + \frac{I'}{C} = i\omega E_0 e^{i\omega t}.$$

Assuming a solution of the form $I'_0 e^{i\omega t}$ we find that

$$I'_0 = \frac{E_0}{R + i\left(\omega L - \dfrac{1}{\omega C}\right)},$$

or

$$I'(t) = \frac{E_0 e^{i\omega t}}{R + i\left(\omega L - \dfrac{1}{\omega C}\right)}.$$

I(t) is complex, and we define the complex impedance Z as

$$Z = R + i\left(\omega L - \frac{1}{\omega C}\right) = R + iX(\omega).$$

$X(\omega)$ is the reactance of the circuit and is a function of the driving frequency $\omega/2\pi$ of the source of emf. Thus

$$I'(t) = \frac{E(t)}{Z(\omega)}.$$

Because we are interested only in the real part of $I'(t)$, which is the real current $I(t)$, we write

$$Z(\omega) = \{R^2 + X^2(\omega)\}^{1/2}\, e^{i\phi}$$

where $$\phi = \text{the phase} = \tan^{-1}\left\{\frac{\omega L - \dfrac{1}{\omega C}}{R}\right\}.$$

Then $$\mathscr{R}I'(t) = \mathscr{R}\left\{\frac{E_0 e^{i(\omega t-\phi)}}{\sqrt{R^2 + X^2}}\right\},$$

and $$I(t) = \frac{E_0 \cos(\omega t - \phi)}{\sqrt{R^2 + X^2}}.$$

The reactance $X(\omega)$ is composed of two parts, the inductive reactance $= X_L = \omega L$ and the capacitive reactance $= X_C = 1/\omega C$.

The total reactance is $X = X_L - X_C$, while the magnitude of the impedance is

$$|Z| = \sqrt{R^2 + X^2}.$$

The voltage drop across the inductance V_L is $L\,dI/dt$, and using our solution for $I(t)$

$$V_L = \frac{-\omega L E_0 \sin(\omega t - \phi)}{|Z|}.$$

V_L leads the current $I(t)$ by 90°.

The phase relations between V_L and the magnitude of $I(t)$ are carried by the complex representation of Z. If we represent the relations between E_0 and I_0 as

$$E_0 = ZI_0,$$

then $$E_0 = |Z|\, I_0 e^{i\phi}.$$

In the complex notation

$$V_L = +iX_L I_0 = \omega L I_0 e^{i\pi/2}.$$

Here the factor $e^{i(\omega t-\phi)}$ has been suppressed.

Across the resistance the voltage drop is

$$V_R = RI(t)$$

with V_R in phase with I(t). Utilizing the complex representation,

$$V_R = RI_0.$$

The drop across the capacitive reactance is

$$V_C = +\frac{E_0}{\omega C\,|Z|}\sin(\omega t - \phi),$$

or

$$V_C = -iX_C I_0 = \frac{I_0}{\omega C}e^{-i\pi/2}.$$

The voltage drops can be represented in the complex plane. We let the phase of the current be zero; thus V_R lies along the real axis. The emf E_0 is the complex vector sum of V_L, V_R, and V_C, while the phase angle is measured as the angle between E_0 and the real axis.

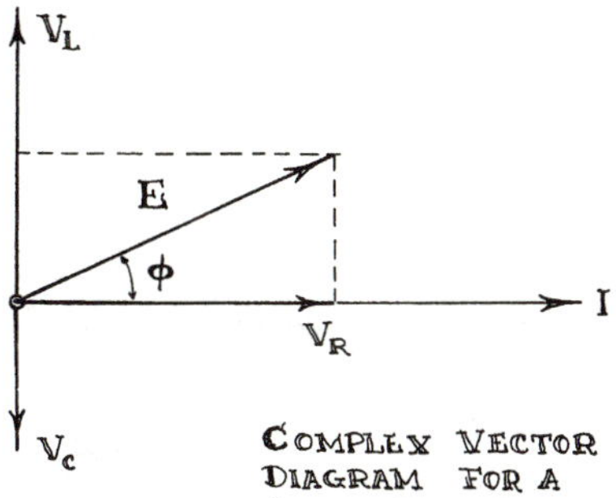

COMPLEX VECTOR DIAGRAM FOR A SERIES LRC CIRCUIT

The instantaneous power dissipated in this circuit is the real part of E times the real part of I,

$$P(t) = \mathscr{R}E(t)\mathscr{R}I(t) = E_0I_0\cos\omega t\cos(\omega t - \phi).$$

Over one period $2\pi/\omega$ the average power is

$$\langle P\rangle = \frac{\omega}{2\pi}\int_0^{2\pi/\omega} E_0I_0\{\cos^2\omega t\cos\phi + \cos\omega t\sin\omega t\sin\phi\}\,dt.$$

Because the term ($\cos\omega t \sin\omega t$) integrates to zero,

$$\langle P\rangle = \frac{E_0I_0}{2}\cos\phi.$$

If we represent these quantities in terms of the complex functions

$$E_0 e^{i\omega t} \quad \text{and} \quad I_0 e^{i(\omega t - \phi)},$$

then

$$\langle P \rangle = \tfrac{1}{2}\mathscr{R}(E^* I) = \frac{E_0 I_0}{2} \cos \phi.$$

The average power dissipated in L and C is zero, and the average power dissipated in the resistor R represents the total average power $\langle P \rangle$. The factor $\cos \phi$ is called the POWER FACTOR.

It is convenient in a.c. circuits to suppress the time dependence; then the current and voltage are represented in terms of their root mean square values I and E:

$$I = \left\{\frac{\omega}{2\pi} \int_0^{2\pi/\omega} I^2(t)\, dt\right\}^{\frac{1}{2}} = \frac{I_0}{\sqrt{2}},$$

and in the same manner

$$E = \frac{E_0}{\sqrt{2}}.$$

These quantities prove convenient in that

$$P = EI \cos \phi.$$

E. SERIES AND PARALLEL CIRCUITS

Series and parallel impedances are treated in the same manner as the pure resistance combinations.

1. SERIES CIRCUITS

The current is common to all elements in a series combination. For N lumped complex impedances Z_j, the total voltage across the combination is

$$E = \sum_{j=1}^{N} V_j = \sum_{j=1}^{N} Z_j I = I\left\{\sum_{j=1}^{N} Z_j\right\}.$$

The total impedance of the series combination is

$$Z_T = \sum_{j=1}^{N} Z_j = \sum_{j=1}^{N} \{R_j + iX_j\} = R_T + iX_T$$

where R_j and X_j are the resistance and reactance associated with Z_j:

$$R_T = \sum_{j=1}^{N} R_j,$$

$$X_T = \sum_{j=1}^{N} X_j.$$

The current in a series circuit can exhibit resonance structure as the frequency $\omega/2\pi$ is varied. Since

$$I = \frac{E}{Z} = \frac{E\{R - i(X_L - X_C)\}}{R^2 + (X_L - X_C)^2},$$

the current I passes through a maximum value E/R when $X_L = X_C$ or when

$$\omega^2 = {\omega_0}^2 = \frac{1}{LC}.$$

As the frequency is varied from zero to values above ω_0, the phase angle between E and I varies from $-\pi/2$ at $\omega = 0$ to values approaching $+\pi/2$ as $\omega \to \infty$.

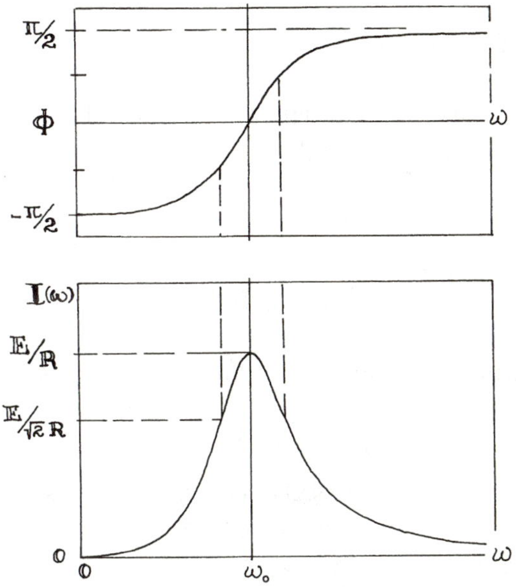

PHASE AND CURRENT CHARACTERISTICS FOR A SERIES RLC CIRCUIT

The phase angle ϕ is $\tan^{-1}\{(X_L - X_C)/R\}$. At low frequencies (below ω_0), $X_L < X_C$ and the reactance is capacitive. Then at high frequencies (above ω_0), $X_C < X_L$ and the reactance is inductive.

The series resonant circuit has a figure of merit associated with it called the Q value, where

$$Q = \frac{\text{Circulating energy}}{\text{Dissipated energy}} = \frac{\omega_0 L}{R} = \sqrt{\frac{L}{C}}\left(\frac{1}{R}\right).$$

The average power I^2R reaches a maximum E^2/R at resonance. The two frequencies ω_1 and ω_2 correspond to the situation in which the power has a value of one half the maximum, $E^2/2R$. Solving for ω_1 and ω_2 we find that

$$\omega_2 - \omega_1 = R/L,$$

and

$$Q = \frac{\omega_0 L}{R} = \frac{\omega_0}{\omega_2 - \omega_1}.$$

The narrowness of the resonance is a measure of the Q value of the circuit.

2. PARALLEL COMBINATIONS

In a parallel combination of N impedances Z_j, the voltage V is the common element. The current through any one of the parallel branches is

$$I_j = \frac{V}{Z_j}.$$

The total current is

$$I_T = \sum_{j=1}^{N} I_j = V\left\{\sum_{j=1}^{N} \frac{1}{Z_j}\right\}.$$

The equivalent impedance Z_T of the parallel combination is

$$\frac{1}{Z_T} = \sum_{j=1}^{N} \frac{1}{Z_j}.$$

The ADMITTANCE of a parallel branch is designated by Y_j:

$$Y_j = \frac{1}{Z_j},$$

with

$$Y_T = \frac{1}{Z_T} = \sum_{j=1}^{N} Y_j = \sum_{j=1}^{N}\left\{\frac{R_j}{(R_j^2 + X_j^2)} - \frac{iX_j}{(R_j^2 + X_j^2)}\right\}.$$

As an example consider a characteristic two-element network which has a capacitance C in parallel with a resistive inductance. Then

$$Y_C = +i/X_C,$$

and

$$Y_L = \frac{R}{(R^2 + X_L^2)} - \frac{iX_L}{(R^2 + X_L^2)}.$$

Solving for the input impedance,

$$Z_T = \frac{1}{Y_T} = \frac{1}{Y_C + Y_L} = \frac{R}{R^2\omega^2C^2 + (\omega^2LC - 1)^2} + \frac{i\omega\{L - C(R^2 + X_L^2)\}}{R^2\omega^2C^2 + (\omega^2LC - 1)^2}.$$

The impedance passes through a maximum at a specified frequency. The resonance frequency of this simple tank circuit can be defined when the reactance of the parallel combination is zero. Setting the reactance to zero and solving for ω_r

$$\omega_r = \left\{\frac{1}{LC} - \frac{R^2}{L^2}\right\}^{1/2} = \omega_0\left\{1 - \frac{1}{Q^2}\right\}^{1/2}$$

where

$$\omega_0^2 = \frac{1}{LC},$$

and

$$Q = \frac{\omega_0 L}{R}.$$

If the Q is large, $\omega_r \simeq \omega_0$. The resonant characteristic of this circuit is quite different from that of the series circuit. When $\omega \simeq \omega_0$ the reactance approaches zero and the real part of the impedance approaches a maximum and the input current becomes a minimum. In case Q is not very large, the resonant frequency is *not* given by ω_0 nor is the maximum of $|Z_T|$ given by ω_r.

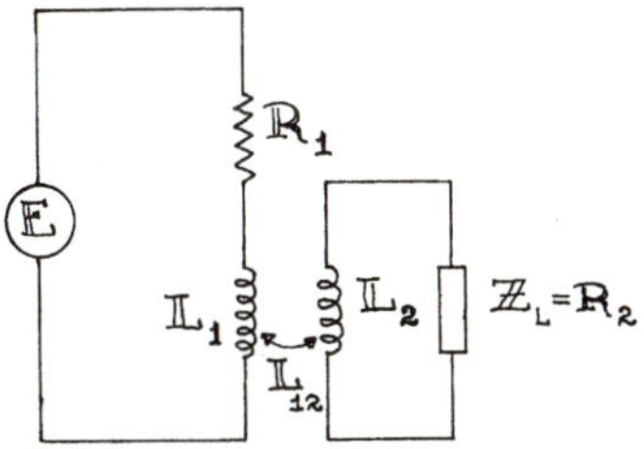

3. COUPLED CIRCUITS

When mutual inductances are present, the currents in the separate loops are coupled. In the simple example shown, the equations representing the voltage drops are

$$i\omega L_{11}I_1 + R_1I_1 + i\omega L_{12}I_2 = E_1$$
$$i\omega L_{22}I_2 + R_2I_2 + i\omega L_{21}I_1 = 0.$$

Solving for I_2 in the second equation and substituting in the first, we find that

$$I_1 = \frac{(R_2 + i\omega L_{22})E_1}{R_1R_2 + \omega^2(L_{12}^2 - L_{11}L_{22}) + i\omega(R_1L_{22} + R_2L_{11})},$$

and

$$I_2 = \frac{-i\omega L_{21}E_1}{R_1R_2 + \omega^2(L_{12}^2 - L_{11}L_{22}) + i\omega(R_1L_{22} + R_2L_{11})}.$$

This simple example serves to illustrate the method of treatment. Use has been made of the properties of the L_{jk}'s derived in Chapter VIII, i.e., $L_{jk} = L_{kj}$.

F. NETWORKS

1. MATRIX REPRESENTATION

We shall treat the a.c. networks by the same methods as used before to discuss the resistance networks. The major difference which we shall notice is that the matrices are complex and a separate matrix must be computed for each frequency present. As before, the power of the matrix development will be emphasized by the use of the inverse matrix. The form of the inverse of the complex matrix remains unchanged. If $\mathbf{Z}$ is a complex matrix, then the elements Y_{jk} of the inverse of $\mathbf{Z}$ are

$$Y_{jk} = \frac{(-1)^{j+k}\text{minor } Z_{kj}}{|\mathbf{Z}|},$$

where $\mathbf{Z}^{-1} = \mathbf{Y}$ and $\mathbf{Z} \cdot \mathbf{Y} = \mathbf{I}$.

To initiate the general derivations, a special problem can be considered. Consider two loops with complex loop currents I_1 and I_2. The branch common to 1 and 2 carries a current which is the vector difference of I_1 and I_2,

$$i_{12} = I_1 - I_2.$$

The negative sign occurs because we have taken both loop currents in a counterclockwise direction.

The algebraic relations specifying the sum of the voltage drops in the two loops are

$$(z_{11} + z_{21})I_1 - z_{12}I_2 = E_{11} - E_{21},$$
$$-z_{21}I_1 + (z_{12} + z_{22})I_2 = E_{12} - E_{22}.$$

The individual branch impedances have been written z_{mn}. Writing the individual loop currents as the components of a two-dimensional complex column vector **I**,

$$\mathbf{I} = \begin{bmatrix} I_1 \\ I_2 \end{bmatrix}.$$

In the same manner, the column vector **E** specifies the sum of emfs about each loop. These rms values must correspond to the same frequency,*

$$\mathbf{E} = \begin{bmatrix} E_{11} - E_{21} \\ E_{12} - E_{22} \end{bmatrix} = \begin{bmatrix} F_1 \\ F_2 \end{bmatrix};$$

$$F_m = \sum_{k=1}^{2} E_{km}, \quad \text{with} \quad E_{mk} = -E_{km}.$$

The reader should notice that the phasing of the various sources of emf E_{mn} must be carefully specified.

In our special problem $E_{21} = E_{22} = 0$; therefore the only non-vanishing component of **E** is $F_1 = E_{11}$.

The impedance matrix is $\mathbf{Z}$, where

$$\mathbf{Z} = \begin{pmatrix} Z_{11} & Z_{12} \\ Z_{21} & Z_{22} \end{pmatrix} = \begin{pmatrix} (z_{11} + z_{21}) & -z_{12} \\ -z_{21} & (z_{12} + z_{22}) \end{pmatrix}.$$

$\mathbf{Z}$ is a symmetric matrix with $Z_{jk} = Z_{kj}$. In this representation our algebraic equations become

$$\mathbf{Z} \cdot \mathbf{I} = \mathbf{E}.$$

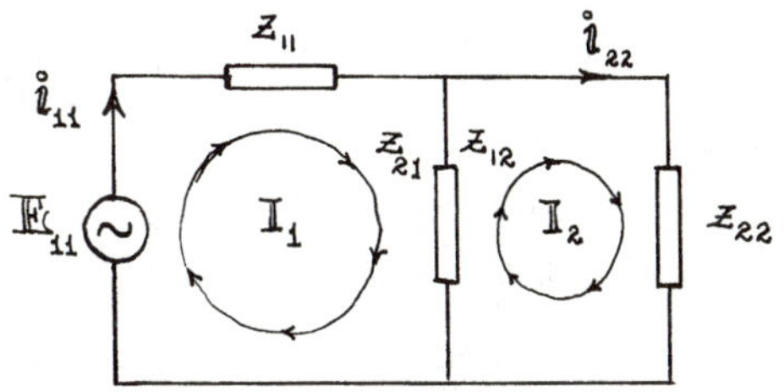

* A separate calculation must be performed for each frequency present in the network. Later by including the time dependence these results can be superposed.

The solutions for the loop currents are obtained by multiplying this equation by $\mathbf{Z}^{-1} = \mathbf{Y}$:

$$\mathbf{I} = \mathbf{Y} \cdot \mathbf{E}.$$

From the details of the matrix, the inverse can be computed in terms of the individual impedances Z_{mn},

$$\mathbf{Y} = \frac{1}{|\mathbf{Z}|}\begin{pmatrix} Z_{22} & -Z_{12} \\ -Z_{21} & Z_{11} \end{pmatrix} = \begin{pmatrix} Y_{11} & Y_{12} \\ Y_{21} & Y_{22} \end{pmatrix},$$

where $|\mathbf{Z}| = z_{11}z_{22} + z_{22}z_{21} + z_{11}z_{12}$.

Then $$\mathbf{I} = \begin{bmatrix} I_1 \\ I_2 \end{bmatrix} = \frac{1}{|\mathbf{Z}|}\begin{pmatrix} (z_{22} + z_{12}) & -z_{12} \\ -z_{21} & (z_{11} + z_{21}) \end{pmatrix}\begin{bmatrix} E_{11} \\ 0 \end{bmatrix}.$$

Finally, $$I_1 = \frac{(z_{22} + z_{12})E_{11}}{|\mathbf{Z}|}$$

and $$I_2 = \frac{-z_{21}E_{11}}{|\mathbf{Z}|}.$$

The reader should check the method by solving the simultaneous algebraic equations directly. The magnitude of the matrix $\mathbf{Z}$ is a complex number because it is the determinant of $\mathbf{Z}$.

The extension of this formalism to a flat network of N loops is straightforward. We list all individual impedances adjacent to the exterior region as z_{mm}, and we list impedances mutual to the jth and k^{th} loops as $z_{jk} = z_{kj}$. Keeping track of all of the relative phases of the sources of emf in each branch,

$$F_m = \sum_{j=1}^{N} E_{jm}$$

where the sum extends over all branches (jm) adjacent to the mth loop. With these rules the column vector listing the loop currents is

$$\mathbf{I} = \begin{bmatrix} I_1 \\ I_2 \\ I_3 \\ \cdot \\ \cdot \\ \cdot \\ I_N \end{bmatrix} = \mathbf{Y} \cdot \mathbf{E} = \mathbf{Y} \cdot \begin{bmatrix} F_1 \\ F_2 \\ F_3 \\ \cdot \\ \cdot \\ \cdot \\ F_N \end{bmatrix}$$

in which $$\mathbb{Y} = \mathbb{Z}^{-1}$$

with
$$\mathbb{Z} = \begin{pmatrix} Z_{11} & Z_{12} & \cdots & Z_{1N} \\ Z_{21} & & & \cdot \\ \cdot & & & \cdot \\ \cdot & & & \cdot \\ \cdot & & & \cdot \\ Z_{N1} & \cdots & \cdots & Z_{NN} \end{pmatrix}.$$

The components Z_{mn} are given by the equations

$$Z_{mm} = \sum_{j=1}^{N} z_{jm}$$

$$Z_{mn} = -z_{m_n} \quad (m \neq n).$$

2. THÉVENIN'S THEOREM

To demonstrate the power of the matrix description we shall derive the theorem of Thévenin. This was stated previously in Section E-4 Chapter V. A flat network, say of $N-1$ loops, with the last branch (the Nth) constructed as a two-terminal connection to the load impedance Z_L, can be

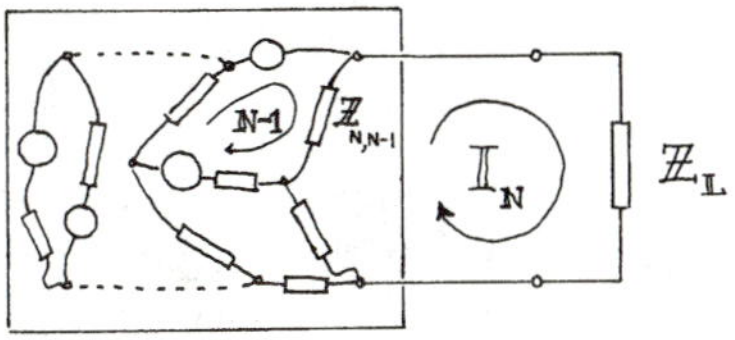

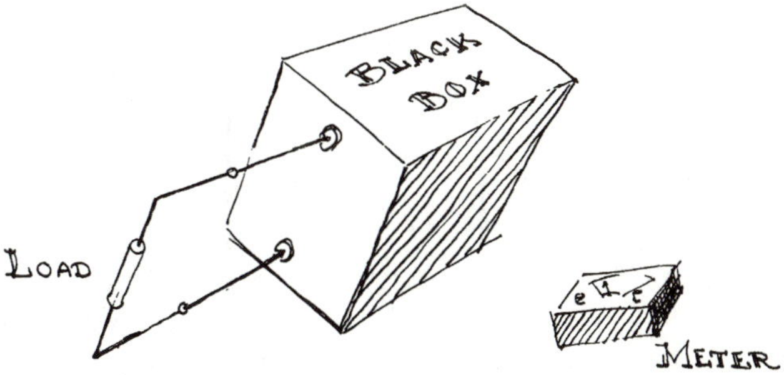

replaced by a single equivalent impedance Z_G in series with a single equivalent source of emf E_G. We further state that a measurement of the complex open circuit voltage E_{oc} corresponding to a single frequency and the complex short circuit current I_{sc} provide Z_G and E_G. The internal series impedance is given by

$$Z_G = \frac{E_{oc}}{I_{sc}}$$

and

$$E_G = E_{oc}.$$

In other words, if a load impedance Z_L is connected across the two output terminals, the load current I_N is given by*

$$\left\{\frac{E_{oc}}{I_{sc}} + Z_L\right\} I_N = E_{oc}.$$

First we examine the general form of the impedance matrix separating the load impedance from the NN element:

$$\mathbf{Z} \cdot \mathbf{I} = \begin{pmatrix} Z_{11} & Z_{12} & \cdots & Z_{1N} \\ Z_{21} & Z_{22} & \cdots & Z_{2N} \\ \cdot & & & \cdot \\ \cdot & & & \cdot \\ \cdot & & & \cdot \\ Z_{N1} \cdots & & & \cdots (Z_{NN} + Z_L) \end{pmatrix} \begin{bmatrix} I_1 \\ I_2 \\ \cdot \\ \cdot \\ \cdot \\ I_N \end{bmatrix} = \begin{bmatrix} F_1 \\ F_2 \\ \cdot \\ \cdot \\ \cdot \\ F_N \end{bmatrix}.$$

The load impedance can be separated from the remaining impedance by use of the special matrix N in which all elements of N are zero except the NN element which is 1:

$$N_{mn} = \delta_{mn}\delta_{nN},$$

or

$$N = \begin{pmatrix} 0 & 0 & 0 & \cdots & 0 \\ 0 & 0 & 0 & & 0 \\ 0 & 0 & 0 & & \cdot \\ \cdot & & & \cdot & \cdot \\ \cdot & & & \cdot & \cdot \\ \cdot & & & \cdot & \cdot \\ 0 & \cdots & & \cdots & 1 \end{pmatrix}.$$

* When many frequencies are present the principle of superposition allows us to perform this measurement for each individual frequency. Then the results can be superposed by including the time dependence.

The remaining impedance matrix is designated as $\mathbb{Z}^{(s)}$ where

$$\mathbb{Z}^{(s)} = \mathbb{Z} - Z_L\mathbb{N} = \begin{pmatrix} Z_{11} & Z_{12} & \dots & Z_{1N} \\ Z_{21} & Z_{22} & \dots & Z_{2N} \\ \cdot & & & \cdot \\ \cdot & & & \cdot \\ \cdot & & & \cdot \\ Z_{N1} & \dots\dots & & Z_{NN} \end{pmatrix}$$

Then $$\{\mathbb{Z}^{(s)} + Z_L\mathbb{N}\} \cdot \mathbf{I} = \mathbf{E}.$$

Because we want the solution for the load current I_N, we multiply both sides from the left by $(\mathbb{Z}^{(s)})^{-1} = \mathbb{Y}^{(s)}$,

$$\{\mathbb{I} + Z_L\mathbb{Y}^{(s)} \cdot \mathbb{N}\} \cdot \mathbf{I} = \mathbb{Y}^{(s)} \cdot \mathbf{E}.$$

Define $\mathbf{\Lambda}$ as

$$\mathbf{\Lambda} = \{\mathbb{I} + Z_L\mathbb{Y}^{(s)} \cdot \mathbb{N}\} = \begin{pmatrix} 1 & 0 & 0 & \dots & Z_LY_{1N} \\ 0 & 1 & 0 & \dots & Z_LY_{2N} \\ 0 & 0 & 1 & \dots & Z_LY_{3N} \\ \cdot & & \cdot & & \cdot \\ \cdot & & \cdot & & \cdot \\ \cdot & & \cdot & & \cdot \\ 0 & 0 & 0 & \dots & (1 + Z_LY_{NN}) \end{pmatrix}.$$

$\mathbf{\Lambda}$ is a unique matrix; the only nonvanishing elements are the diagonal elements Λ_{jj} and the elements in the last column $\Lambda_{jN} = Z_LY_{jN}$ $(j \neq N)$ and $\Lambda_{NN} = (1 + Z_LY_{NN})$.

Because of this unique form, the solution for I_N, the current in the load, has a particularly simple form. Solving for **I** by multiplying by $\mathbf{\Lambda}^{-1}$ from the left,

$$\mathbf{I} = \mathbf{\Lambda}^{-1} \cdot \mathbb{Y}^{(s)} \cdot \mathbf{E}.$$

The mth component of the vector $\mathbb{Y}^{(s)} \cdot \mathbf{E}$ is $\sum_{j=1}^{N} Y_{mj}F_j$.

The load current is given by I_N where

$$I_N = \sum_{k=1}^{N} \Lambda'_{Nk} \sum_{j=1}^{N} Y_{kj}F_j.$$

Here the Λ'_{Nk} represent the elements in the LAST ROW of Λ^{-1}. We can demonstrate, however, that all of these elements in the last row are zero EXCEPT for Λ'_{NN}, that is,

$$\Lambda'_{Nk} = \Lambda'_{Nk}\delta_{Nk}.$$

The proof is quite simple:

$$\Lambda'_{Nk} = \frac{(-1)^{N+k} \text{ minor } \Lambda_{kN}}{|\mathbf{\Lambda}|}.$$

The minors of all $\mathbf{\Lambda}_{kN}$ for $k \neq N$ are zero because the last row of each of these minors is zero. To see this examine $\mathbf{\Lambda}$.

The element

$$\Lambda'_{NN} = \frac{1}{|\mathbf{\Lambda}|} = \frac{1}{(1 + Z_L Y_{NN})}.$$

Thus the load current I_N is

$$I_N = \frac{\sum_{j=1}^{N} Y_{Nj}F_j}{(1 + Z_L Y_{NN})}.$$

Rearranging this result we find

$$\left\{\frac{1}{Y_{NN}} + Z_L\right\} I_N = \frac{\sum_{j=1}^{N} Y_{Nj}F_j}{Y_{NN}}.$$

The short circuit current I_{sc} is obtained when $Z_L \to 0$,

$$I_{sc} = \sum_{j=1}^{N} Y_{Nj}F_j.$$

The open circuit current becomes zero as $Z_L \to \infty$. However, the product $Z_L I_N \to E_{oc}$, the open circuit voltage. Assuming that $Y_{NN} \neq 0$ (this has been implicit throughout) as $Z_L \to \infty$, we can neglect Y_{NN}^{-1}, and

$$E_{oc} = \lim_{Z_L \to \infty} (Z_L I_N) = \frac{\sum_{j=1}^{N} Y_{Nj}F_j}{Y_{NN}}.$$

Thus our solution for I_N for a single frequency becomes

$$\left\{\frac{E_{oc}}{I_{sc}} + Z_L\right\} I_N = E_{oc}.$$

This statement proves the theorem.

Appendix

A. Vectors and Geometry

1. GENERAL CONSIDERATIONS

In many standard texts on advanced calculus or vector analysis one can find a thorough treatment of vector analysis. It serves our purpose here merely to mention some important aspects of this subject. To make our review the more useful we will survey this material in terms of generalized coordinates.

The geometric vector is a mathematical quantity having both direction and magnitude. Vectors of equal length having the same direction are said to be equal. If two vectors have equal length but have opposite sense (or direction) one is called the negative of the other. A scalar m times a vector produces a vector m times as long, in the same direction if m is positive or in the opposite direction if m is negative.

A vector is defined at a point in space. Therefore six numbers are actually required to specify it: three numbers to specify the point and three numbers to specify the components of the vector at the point.

Vectors can be combined by addition or by multiplication. To be combined, two vectors must be defined at the same point in space. It does not make sense to attempt the combination of two vectors defined at different points. We shall observe later that this constraint is quite important when handling vector combinations in curvilinear coordinates.

Addition obeys the Commutative Law,

$$\mathbf{A} + \mathbf{B} = \mathbf{B} + \mathbf{A}.$$

Addition obeys the Associative Law,

$$(\mathbf{A} + \mathbf{B}) + \mathbf{C} = \mathbf{A} + (\mathbf{B} + \mathbf{C}).$$

Addition obeys the Distributive Law,

$$m(\mathbf{A} + \mathbf{B}) = m\mathbf{A} + m\mathbf{B}.$$

There are two products defined, the dot or scalar product and the cross or vector product. The dot product of two vectors **A** and **B** is defined as

$$\mathbf{A} \cdot \mathbf{B} = |\mathbf{A}|\,|\mathbf{B}| \cos \sphericalangle {}^{\mathbf{B}}_{\mathbf{A}},$$

where $\cos \sphericalangle {}^{\mathbf{B}}_{\mathbf{A}}$ is the cosine of the smallest angle between **A** and **B**. This product is a scalar quantity.

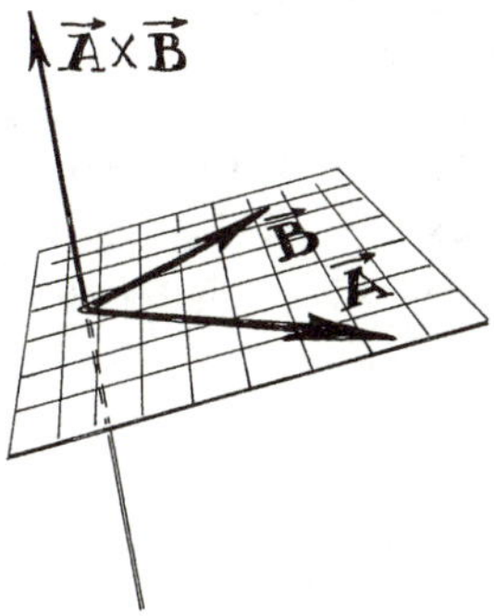

The vector product of two vectors **A** and **B** is a vector quantity, and it is written

$$\mathbf{A} \times \mathbf{B} = |\mathbf{A}|\,|\mathbf{B}| \sin \sphericalangle {}^{\mathbf{B}}_{\mathbf{A}}\, \mathbf{n},$$

where **n** is a unit vector directed perpendicular to the plane of **A** and **B**. The direction of **n** is obtained by the right-hand rule. If the fingers of the right hand are initially aligned along the first vector of the product, and if then they are moved toward the second vector in an operation of closure, the thumb then indicates the direction of the vector product.

These few introductory remarks suffice to allow the introduction of base vectors. It is possible to expand any vector along a complete set of base vectors. These bases need not be orthogonal but merely need to be a linearly independent set of vectors. If **A**, **B**, and **C** form a linearly independent set, then the relation

$$a\mathbf{A} + b\mathbf{B} + c\mathbf{C} = 0$$

can be satisfied only if a = b = c = 0.

2. CARTESIAN BASES

A set of bases is complete if any arbitrary vector can be expanded in terms of its projections along the bases. For instance, two linearly independent (nonparallel) vectors which lie in the xy plane do not form a complete set of bases for a vector in three dimensions; however, these two would form a complete set for all problems pertaining to two-dimensional vectors which lie in the xy plane.

Throughout our discussion we will be interested in three-dimensional orthogonal spaces. In other words, the specification of orthogonal unit base vectors will suit our needs. To start we shall use the cartesian bases **i**, **j**, and **k**. These are unit vectors aligned along the x, y, and z axes of a cartesian space.

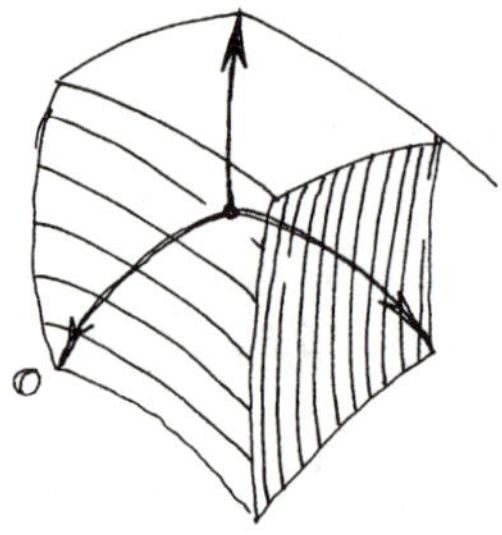

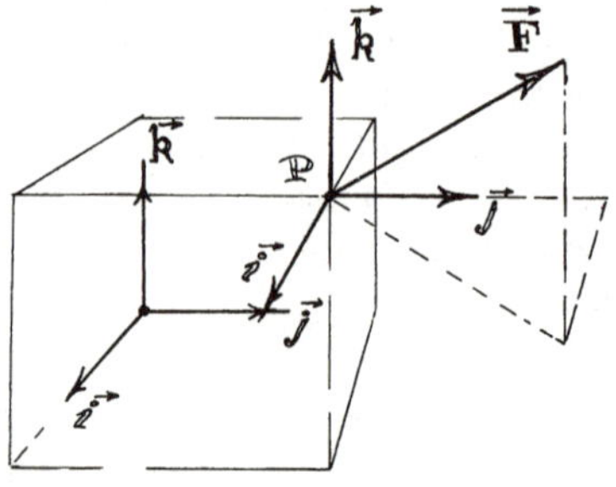

At every point in space a set of cartesian bases are defined. An arbitrary vector **F** can be projected on the base vectors by taking the inner product of **F** with the bases.

The expansion of **F** is written

$$\mathbf{F} = F_x\mathbf{i} + F_y\mathbf{j} + F_z\mathbf{k}$$

where

$$F_x = \mathbf{F}\cdot\mathbf{i}, \quad F_y = \mathbf{F}\cdot\mathbf{j}, \quad F_z = \mathbf{F}\cdot\mathbf{k}.$$

To make our notation more compact we can write

$$F_x = F_1, \quad \mathbf{i} = \boldsymbol{\epsilon}_1, \quad x = x_1,$$

$$F_y = F_2, \quad \mathbf{j} = \boldsymbol{\epsilon}_2, \quad y = x_2,$$

$$F_z = F_3, \quad \mathbf{k} = \boldsymbol{\epsilon}_3, \quad z = x_3.$$

Then

$$\mathbf{F} = \sum_{m=1}^{3} F_m\boldsymbol{\epsilon}_m = F_m\boldsymbol{\epsilon}_m.$$

In this last equation on the extreme right the repeated index signifies that this index should be summed over all of its possible values.

3. GENERALIZED ORTHOGONAL COORDINATE SYSTEMS

Instead of considering the various curvilinear coordinate systems independently, it is most instructive to work out the general vector properties of an abstract system, which can readily be applied to any specific coordinate system.

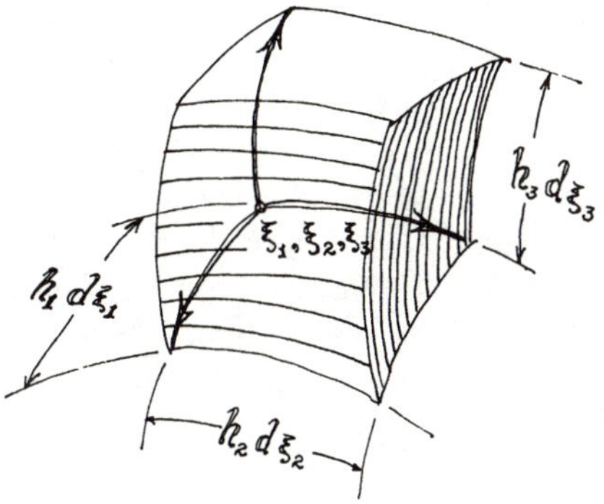

Let us consider three independent coordinate variables ξ_1, ξ_2, and ξ_3. At any point in this space a set of bases $\boldsymbol{\epsilon}_1$, $\boldsymbol{\epsilon}_2$, and $\boldsymbol{\epsilon}_3$ will be defined. We are interested in three-dimensional spaces which can be developed from a cartesian space.

Important examples of the spaces in which we are interested are

(1) Cylindrical coordinates with variables a, ϕ, z, and

(2) Spherical coordinates with variables r, θ, ϕ.

To perform the transformation from the cartesian variables, the x_m are expressed as scalar functions of the generalized variables ξ_1, ξ_2, ξ_3,

$$x_m = x_m(\xi_1,\xi_2,\xi_3).$$

For instance, in the transformation from cartesian variables to spherical coordinates, $x_1 = r \sin \phi \cos \phi$, etc. The magnitude (or magnitude squared) of a length element **dr** is invariant under the transformation:

$$(dr)^2 = \mathbf{dr} \cdot \mathbf{dr} = (dx_1)^2 + (dx_2)^2 + (dx_3)^2 = \sum_{m=1}^{3} (dx_m)^2.$$

The element dx_j expressed in terms of ξ_k is

$$dx_j = \sum_{k=1}^{3} \frac{\partial x_j}{\partial \xi_k} d\xi_k.$$

Then

$$(dr)^2 = \sum_{i=1}^{3} \sum_{j=1}^{3} \sum_{k=1}^{3} \frac{\partial x_i}{\partial \xi_j} \frac{\partial x_i}{\partial \xi_k} d\xi_j \, d\xi_k$$

$$= \sum_{j=1}^{3} \sum_{k=1}^{3} g_{jk} \, d\xi_j \, d\xi_k.$$

The components g_{jk} are elements of the metric tensor and are called the metrical coefficients. From the relations above,

$$g_{jk} = \sum_{i=1}^{3} \frac{\partial x_i}{\partial \xi_j} \frac{\partial x_i}{\partial \xi_k}.$$

We can regard our spaces as being defined by the specification of the metric. Orthogonal spaces have diagonal metrics. In other words, terms of the type $d\xi_m \, d\xi_n$ with $m \neq n$ do not appear in the metric. This discussion will be confined to orthogonal metrics. Define a weight factor h_k by

$$g_{mn} = 0, \quad m \neq n$$

$$g_{kk} = h_k^2.$$

Therefore for an orthogonal space

$$(dr)^2 = \sum_{k=1}^{3} h_k^2\,(d\xi_k)^2, \quad \text{where} \quad h_k = +\left\{\sum_{i=1}^{3}\left(\frac{\partial x_i}{\partial \xi_k}\right)^2\right\}^{1/2}.$$

We see immediately that the weight factors h_k are functions of ξ_1, ξ_2, and ξ_3; further, $h_k\, d\xi_k$ has the dimensions of a length dl_k. For instance, in spherical coordinates $\xi_2 = \theta$ while $h_2 = r$; thus $h_2\, d\xi_2 = r\, d\theta$ which has the proper length dimension.

In general,

$$d\mathbf{r} = \sum_{m=1}^{3} dl_m\, \boldsymbol{\epsilon}_m = \sum_{m=1}^{3} h_m\, d\xi_m\, \boldsymbol{\epsilon}_m.$$

Thus the base vectors are defined by

$$\boldsymbol{\epsilon}_k = \frac{d\mathbf{r}}{dl_k} = \frac{1}{h_k}\frac{d\mathbf{r}}{\partial \xi_k} = \frac{1}{\sqrt{g_{kk}}}\frac{\partial \mathbf{r}}{\partial \xi_k}.$$

The metric tensor is a square array of 9 terms (for a three-dimensional space). It is convenient to specify the three spaces encountered most often, in terms of the associated metrics:

$$\text{Cartesian } \mathbf{g} = \begin{pmatrix} 1 & 0 & 0 \\ 0 & 1 & 0 \\ 0 & 0 & 1 \end{pmatrix} \qquad \begin{aligned} \xi_1 &= x_1 \\ \xi_2 &= x_2 \\ \xi_3 &= x_3; \end{aligned}$$

$$\text{Cylindrical } \mathbf{g} = \begin{pmatrix} 1 & 0 & 0 \\ 0 & a^2 & 0 \\ 0 & 0 & 1 \end{pmatrix} \qquad \begin{aligned} \xi_1 &= a \\ \xi_2 &= \phi \\ \xi_3 &= z; \end{aligned}$$

$$\text{Spherical } \mathbf{g} = \begin{pmatrix} 1 & 0 & 0 \\ 0 & r^2 & 0 \\ 0 & 0 & r^2 \sin^2\theta \end{pmatrix} \qquad \begin{aligned} \xi_1 &= r \\ \xi_2 &= \theta \\ \xi_3 &= \phi. \end{aligned}$$

4. MULTIPLICATION OF VECTORS

a. THE SCALAR PRODUCT OF VECTORS EXPANDED IN ORTHOGONAL BASES $\boldsymbol{\epsilon}_m$. This is

$$\mathbf{F}\cdot\mathbf{G} = \left\{\sum_{m=1}^{3} F_m \boldsymbol{\epsilon}_m\right\}\cdot\left\{\sum_{n=1}^{3} G_n \boldsymbol{\epsilon}_n\right\}$$

$$= \sum_{m=1}^{3}\sum_{n=1}^{3} F_m G_n (\boldsymbol{\epsilon}_m \cdot \boldsymbol{\epsilon}_n).$$

Since
$$\boldsymbol{\epsilon}_m \cdot \boldsymbol{\epsilon}_n = \delta_{mn} = 1 \quad \text{if } m = n$$
$$= 0 \quad \text{if } m \neq n,$$
we then have
$$\mathbf{F} \cdot \mathbf{G} = \sum_{m=1}^{3} \sum_{n=1}^{3} F_m G_n \delta_{mn} = \sum_{m=1}^{3} F_m G_m.$$

It should be remembered that this operation implies that **F** and **G** are defined at the same point.

b. THE VECTOR PRODUCT. This product is defined for a space of three dimensions. The cross or vector product of orthonormal bases is defined by a density function e_{mkn}:
$$\boldsymbol{\epsilon}_m \times \boldsymbol{\epsilon}_k = e_{mkn} \boldsymbol{\epsilon}_n.$$
The quantity

$e_{mkn} = 1$ when $m \neq k \neq n \neq m$ and when the order of mkn is 123, 312, or 231, i.e., cyclic;

$e_{mkn} = -1$ when $m \neq k \neq n \neq m$ and the order is anticyclic, i.e., 213, 321, or 132;

and $e_{mkn} = 0$ if two or more of the indices are equal.

With these definitions
$$\mathbf{F} \times \mathbf{G} = \sum_{m=1}^{3} \sum_{n=1}^{3} \sum_{k=1}^{3} F_m G_n \boldsymbol{\epsilon}_k e_{mnk}.$$

This is the expression for a 3×3 determinant,
$$\begin{vmatrix} \boldsymbol{\epsilon}_1 & \boldsymbol{\epsilon}_2 & \boldsymbol{\epsilon}_3 \\ F_1 & F_2 & F_3 \\ G_1 & G_2 & G_3 \end{vmatrix}.$$

c. THE TRIPLE SCALAR PRODUCT
$$\mathbf{A} \cdot (\mathbf{B} \times \mathbf{C}) = \sum_{i=1}^{3} A_i \boldsymbol{\epsilon}_i \cdot \left\{ \sum_{m,n,k} B_m C_n \boldsymbol{\epsilon}_k e_{mnk} \right\}.$$
$$= \sum_{i,m,n,k} A_i B_m C_n (\boldsymbol{\epsilon}_i \cdot \boldsymbol{\epsilon}_k) e_{mnk}$$
$$= \sum_{k=1}^{3} \sum_{m=1}^{3} \sum_{n=1}^{3} A_k B_m C_n e_{kmn}.$$

This is the expression for the expansion of a 3×3 determinant; thus

$$\mathbf{A} \cdot (\mathbf{B} \times \mathbf{C}) = \begin{pmatrix} A_1 & A_2 & A_3 \\ B_1 & B_2 & B_3 \\ C_1 & C_2 & C_3 \end{pmatrix}.$$

d. THE TRIPLE VECTOR PRODUCT. The expansion of the triple vector product can be rendered in a convenient manner if we notice the characteristics of the product $e_{mnk}e_{jks}$ when summed over say k. Using the density notation, the triple vector product is

$$\mathbf{A} \times (\mathbf{B} \times \mathbf{C}) = \sum_{s=1}^{3} \sum_{j=1}^{3} \sum_{k=1}^{3} A_j(\mathbf{B} \times \mathbf{C})_k \epsilon_s e_{jks}$$

where $(\mathbf{B} \times \mathbf{C})_k$ is the kth component of the product $\mathbf{B} \times \mathbf{C}$; i.e.,

$$(\mathbf{B} \times \mathbf{C})_k = \sum_{m=1}^{3} \sum_{n=1}^{3} B_m C_n e_{mnk}.$$

Substituting we obtain

$$\mathbf{A} \times (\mathbf{B} \times \mathbf{C}) = \sum_{s,j,k} \sum_{m,n} A_j B_m C_n \epsilon_s e_{mnk} e_{jks}.$$

The sum over k of $e_{mnk}e_{jks}$ takes on a very simple form:

$$\sum_{k=1}^{3} e_{mnk}e_{jks} = -\sum_{k=1}^{3} e_{mnk}e_{jsk}$$
$$= -\delta_{mj}\delta_{ns} + \delta_{nj}\delta_{ms}.$$

The reader can demonstrate this result by tabulating the nonzero values of e_{mnk} and then listing the corresponding nonzero forms of e_{jsk}.

Summing over k and inserting the resulting Kronecker deltas into the equation for $\mathbf{A} \times (\mathbf{B} \times \mathbf{C})$, we find

$$\mathbf{A} \times (\mathbf{B} \times \mathbf{C}) = \sum_{s,j} \sum_{m,n} \{A_j B_m C_n \epsilon_s [\delta_{nj}\delta_{ms} - \delta_{mj}\delta_{ns}]\}.$$

Summing over j and m in the first term and over j and n in the second,

$$\mathbf{A} \times (\mathbf{B} \times \mathbf{C}) = \sum_{s=1}^{3} \sum_{n=1}^{3} A_n B_s C_n \epsilon_s - \sum_{s=1}^{3} \sum_{m=1}^{3} A_m B_m C_s \epsilon_s$$
$$= \mathbf{B}(\mathbf{A} \cdot \mathbf{C}) - \mathbf{C}(\mathbf{A} \cdot \mathbf{B}).$$

5. DIFFERENTIATION AND INTEGRATION OF VECTORS

Because the sections to follow treat the differentiation and integration of vector point functions in detail, we shall merely list here a few simple cases.

(a) Let
$$\mathbf{F} = \sum_{m=1}^{3} F_m(\xi_1\xi_2\xi_3)\boldsymbol{\epsilon}_m(\xi_1\xi_2\xi_3);$$

then
$$\frac{\partial \mathbf{F}}{\partial \xi_j} = \sum_{m=1}^{3}\left\{\frac{\partial F_m}{\partial \xi_j}\boldsymbol{\epsilon}_m + F_m\frac{\partial \boldsymbol{\epsilon}_m}{\partial \xi_j}\right\}.$$

The reason that the vector **F** has been written in such detail is that in most cases other than the special example of a cartesian space, a variation in the variable ξ_j will give rise to a variation in $\boldsymbol{\epsilon}_m$. For instance, in cylindrical coordinates

$$\frac{\partial \boldsymbol{\epsilon}_a}{\partial \phi} = \boldsymbol{\epsilon}_\phi.$$

When the geometry is pictorially simple a result such as the one given above is visually obvious. In a more complicated geometry this type of result may be more difficult to develop.

Time derivatives are quite important and the total time derivative of a vector **F** is obtained by applying the standard expansion of the total derivative:

$$\frac{d\mathbf{F}}{dt} = \frac{d}{dt}\left\{\sum_m F_m\boldsymbol{\epsilon}_m\right\} = \sum_{m=1}^{3}\left\{\frac{dF_m}{dt}\boldsymbol{\epsilon}_m + F_m\frac{d\boldsymbol{\epsilon}_m}{dt}\right\},$$

where $\frac{dF_m}{dt} = \sum_{j=1}^{3}\frac{\partial F_m}{\partial \xi_j}\frac{d\xi_j}{dt} + \frac{\partial F_m}{\partial t}$ and $\frac{d\boldsymbol{\epsilon}_m}{dt} = \sum_{j=1}^{3}\frac{\partial \boldsymbol{\epsilon}_m}{\partial \xi_j}\frac{d\xi_j}{dt}$.

(b) $$\frac{d(\mathbf{A}\cdot\mathbf{B})}{dl_j} = \frac{1}{h_j}\frac{d}{d\xi_j}(\mathbf{A}\cdot\mathbf{B}) = \frac{1}{h_j}\left\{\frac{d\mathbf{A}}{d\xi_j}\cdot\mathbf{B} + \mathbf{A}\cdot\frac{d\mathbf{B}}{d\xi_j}\right\}.$$

(c) Line Integrals

$$\int_C \mathbf{F}\cdot d\mathbf{r} = \sum_{k=1}^{3}\int_C F_k(\xi_1\xi_2\xi_3)h_k\, d\xi_k.$$

6. THE GRADIENT

The directional derivative of a scalar point function $V(\xi_1,\xi_2,\xi_3) = K$ is merely the rate of change of the function in a given direction. The gradient is the rate of change of the function in a direction NORMAL to the surface represented by the function at the point of evaluation.

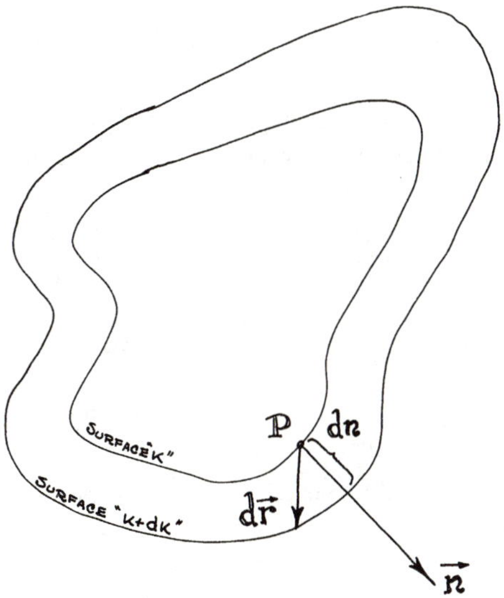

It can be shown that the gradient is the maximum rate of change of the function V at a point.

The form of the gradient in various coordinate systems can be developed from the definition of the total derivative.

At a point P construct a unit normal **n**. The variation of V between the surface K and the surface defined by K + dK is

$$dV\,\mathbf{n} = (\text{grad } V)\,dn,$$

or

$$\text{grad } V = \frac{dV}{dn}\mathbf{n}.$$

The maximum change in V along **n** is

$$dV = \frac{dV}{dn}\,dn = \frac{dV}{dn}\,\mathbf{n}\cdot d\mathbf{r} = (\text{grad } V)\cdot d\mathbf{r}.$$

The total derivative in general coordinates is

$$dV = \sum_{k=1}^{3}\frac{\partial V}{\partial l_k}\,dl_k = \sum_{k=1}^{3}\frac{\partial V}{\partial l_k}\boldsymbol{\epsilon}_k\cdot\left\{\sum_{j=1}^{3}dl_j\,\boldsymbol{\epsilon}_j\right\},$$

or $$dV = \left\{\sum_{k=1}^{3} \frac{\partial V}{\partial l_k} \boldsymbol{\epsilon}_k\right\} \cdot d\mathbf{r} = (\text{grad } V) \cdot d\mathbf{r}.$$

Equating the coefficients of $\{d\mathbf{r}\}_j$ we find

$$\text{grad } V = \sum_{k=1}^{3} \frac{\partial V}{\partial l_k} \boldsymbol{\epsilon}_k = \sum_{k=1}^{3} \frac{1}{h_k} \frac{\partial V}{\partial \xi_k} \boldsymbol{\epsilon}_k.$$

In the previous development, the h_k's and ξ_k's corresponding to the three most familiar spaces were specified. The gradient in each of these spaces is

Cartesian: $$\text{grad } V = \frac{\partial V}{\partial x_1} \mathbf{i} + \frac{\partial V}{\partial x_2} \mathbf{j} + \frac{\partial V}{\partial x_3} \mathbf{k} = \sum_{m=1}^{3} \frac{\partial V}{\partial x_m} \boldsymbol{\epsilon}_m;$$

Cylindrical: $$\text{grad } V = \frac{\partial V}{\partial a} \boldsymbol{\epsilon}_a + \frac{1}{a} \frac{\partial V}{\partial \phi} \boldsymbol{\epsilon}_\phi + \frac{\partial V}{\partial z} \boldsymbol{\epsilon}_3;$$

Spherical: $$\text{grad } V = \frac{\partial V}{\partial r} \boldsymbol{\epsilon}_r + \frac{1}{r} \frac{\partial V}{\partial \theta} \boldsymbol{\epsilon}_\theta + \frac{1}{r \sin \theta} \frac{\partial V}{\partial \phi} \boldsymbol{\epsilon}_\phi.$$

The symbol ∇ should be used for cartesian coordinates only, and is defined as

$$\nabla = \sum_{k=1}^{3} \boldsymbol{\epsilon}_k \frac{\partial}{\partial x_k}.$$

Many arguments result from differences concerning the implied operation of $\nabla \cdot \mathbf{A}$ or $\nabla \times \mathbf{A}$. We shall observe directly that if the operation ∇V means grad V, and if the operations $\nabla \cdot \mathbf{F}$ and $\nabla \times \mathbf{F}$ mean divergence **F** and curl **F** respectively, there is no resulting confusion when these symbols are utilized with the curvilinear coordinates. In such instances one must merely keep in mind that ∇ is not a separate mathematical object in curvilinear spaces. The use of $(\nabla \cdot)$ or $(\nabla \times)$ in these spaces must designate the full operation of divergence or curl. In spaces other than cartesian these operations cannot be separated in a multiplicative fashion.

7. THE DIVERGENCE

If we consider the vector field **F** as a flux density (i.e., a measure of directed lines per unit area), then the integral

$$\iint_{S(\text{closed})} \mathbf{F} \cdot \mathbf{n} \, dS$$

(where capital S represents a surface) is a measure of the net number of lines issuing from or terminating within the closed surface S.

The divergence theorem relates the divergence of the field in the volume τ enclosed by S, to the closed surface integral shown above.

$$\iint\limits_{S(\tau)} \mathbf{F}\cdot\mathbf{n}\,dS = \iiint\limits_{\tau} (\text{div}\,\mathbf{F})\,d\tau.$$

The volume τ can be taken very small. Let us designate this small volume element as $\Delta\tau$; then

$$\text{div}\,\mathbf{F} = \lim_{\Delta\tau\to 0} \frac{\displaystyle\iint\limits_{S(\Delta\tau)} \mathbf{F}\cdot\mathbf{n}\,dS}{\Delta\tau} = \frac{\text{Net flux in or out}}{\text{Unit volume}}.$$

To compute the form of div **F** in a general system of coordinates we merely take a volume element having edges $dl_k = h_k\,d\xi_k$ and

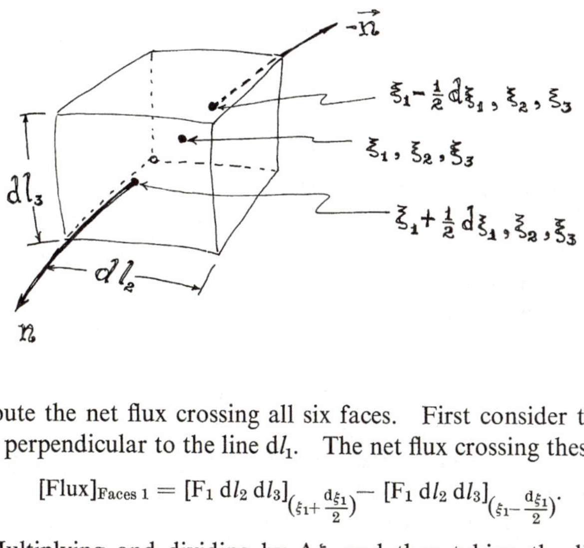

compute the net flux crossing all six faces. First consider the two faces perpendicular to the line dl_1. The net flux crossing these faces is

$$[\text{Flux}]_{\text{Faces 1}} = [F_1\,dl_2\,dl_3]_{\left(\xi_1+\frac{d\xi_1}{2}\right)} - [F_1\,dl_2\,dl_3]_{\left(\xi_1-\frac{d\xi_1}{2}\right)}.$$

Multiplying and dividing by $\Delta\xi_1$ and then taking the limit as $\Delta\xi_1 \to 0$,

$$[\text{Flux}]_1 = \frac{\partial}{\partial\xi_1}\{F_1\,dl_2\,dl_3\} = \frac{\partial}{\partial\xi_1}\{F_1 h_2 h_3\}\,d\xi_1\,d\xi_2\,d\xi_3$$

$$= \frac{1}{h_1 h_2 h_3}\frac{\partial}{\partial\xi_1}\{F_1 h_2 h_3\}\,d\tau.$$

The fact that $d\tau = dl_1\, dl_2\, dl_3 = h_1h_2h_3\, d\xi_1\, d\xi_2\, d_3$ has been used.

Carrying out similar developments for the other two sets of faces, we find that

$$\operatorname{div} \mathbf{F} = \frac{1}{h_1h_2h_3} \sum_{j=1}^{3} \frac{\partial}{\partial \xi_j} \{F_j h_k h_m\}, \quad j \neq k \neq m \neq j.$$

This general equation then allows us to compute div **F** in the three familiar systems:

Cartesian:

$$\operatorname{div} \mathbf{F} = \frac{\partial F_1}{\partial x_1} + \frac{\partial F_2}{\partial x_2} + \frac{\partial F_3}{\partial x_3};$$

Cylindrical:

$$\operatorname{div} \mathbf{F} = \frac{1}{a} \left\{ \frac{\partial}{\partial a} (F_a a) + \frac{\partial}{\partial \phi} (F_\phi) + \frac{\partial}{\partial x_3} (F_3 a) \right\};$$

Spherical:

$$\operatorname{div} \mathbf{F} = \frac{1}{r^2 \sin \theta} \left\{ \frac{\partial}{\partial r} (F_r r^2 \sin \theta) + \frac{\partial}{\partial \theta} (F_\theta r \sin \theta) + \frac{\partial}{\partial \phi} (F_\phi r) \right\}.$$

8. THE CURL

Vector fields fall into two main classes. Those fields which have a zero curl are called irrotational or conservative. If the curl of a vector field is nonzero, the field is known as a rotational field.

The curl is defined from Stokes' law which relates the integral of the curl over a cap surface to the line integral about the contour bounding the cap:

$$\iint_{S(\text{cap})} [\operatorname{curl} \mathbf{F}] \cdot \mathbf{n}\, dS = \oint_{C(\text{cap})} \mathbf{F} \cdot d\mathbf{r}.$$

The curl of **F** is a measure of the circulation of **F**. If one considers a small surface element **n** ΔS(cap) bounded by a closed contour ΔC, then the *component* of the curl parallel to the surface normal **n** is

$$[\operatorname{curl} F]_n = \lim_{\Delta S \to 0} \frac{\oint_{\Delta C} \mathbf{F} \cdot d\mathbf{r}}{\Delta S}.$$

Because the curl has three components it behaves in many ways like a vector (although under certain transformations it behaves like a second-order tensor). To calculate the components of curl **F** we consider a set of three surface elements each perpendicular to the coordinate corresponding to the component of curl **F**. For example, to evaluate the third component of curl **F** we take the square bounded by dl_1 and dl_2. We compute $\oint_{\Delta C} \mathbf{F} \cdot d\mathbf{r}$ for this square and assign to it

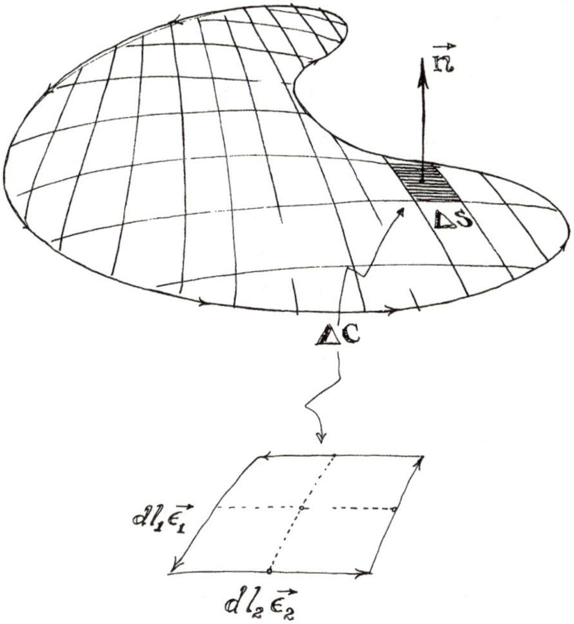

a direction given by a base vector $\boldsymbol{\epsilon}_3$. **F** is expanded about the midpoint of the square; thus for the third component

$$[\text{curl F}]_3 \, \Delta l_1 \, \Delta l_2$$

$$= \sum_k \mathbf{F} \cdot \{\Delta l_k \boldsymbol{\epsilon}_k\}$$

$$= F_1\left(\xi_1, \left[\xi_2 - \frac{\Delta\xi_2}{2}\right], \xi_3\right)\Delta l_1 + F_2\left(\left[\xi_1 + \frac{\Delta\xi_1}{2}\right], \xi_2, \xi_3\right)\Delta l_2$$

$$- F_1\left(\xi_1, \left[\xi_2 + \frac{\Delta\xi_2}{2}\right], \xi_3\right)\Delta l_1 - F_2\left(\left[\xi_1 - \frac{\Delta\xi_1}{2}\right], \xi_2, \xi_3\right)\Delta l_2.$$

Multiply and divide the 1st and 3rd terms by $\Delta\xi_2$; multiply and divide the 2nd and 4th terms by $\Delta\xi_1$. Taking the limit as $\Delta\xi_1 \to 0$ and as $\Delta\xi_2 \to 0$, we find

$$[\text{curl } F]_3 = \frac{1}{h_1h_2}\left\{\frac{\partial}{\partial\xi_1}(h_2F_2) - \frac{\partial}{\partial\xi_2}(h_1F_1)\right\}.$$

The same type of calculation can now be performed on surface elements perpendicular to $\boldsymbol{\epsilon}_1$ and $\boldsymbol{\epsilon}_2$ respectively. The resulting form for the curl of **F** is

$$\text{curl } \mathbf{F} = \sum_{k=1}^{3}\sum_{j=1}^{3}\sum_{m=1}^{3}\frac{1}{h_jh_m}\left\{\frac{\partial}{\partial\xi_j}(h_mF_m)\right\}\boldsymbol{\epsilon}_k\, e_{jmk}.$$

The expansion of this equation in terms of cartesian, cylindrical, and spherical coordinates is

Cartesian:

$$\text{curl } \mathbf{F} = \begin{vmatrix} \mathbf{i} & \mathbf{j} & \mathbf{k} \\ \dfrac{\partial}{\partial x_1} & \dfrac{\partial}{\partial x_2} & \dfrac{\partial}{\partial x_3} \\ F_1 & F_2 & F_3 \end{vmatrix};$$

Cylindrical:

$$\begin{aligned}\text{curl } \mathbf{F} &= \frac{1}{a}\left\{\frac{\partial}{\partial\phi}(F_3) - \frac{\partial}{\partial x_3}(aF_\phi)\right\}\boldsymbol{\epsilon}_a \\ &+ \frac{1}{1}\left\{\frac{\partial}{\partial x_3}(F_a) - \frac{\partial}{\partial a}(F_3)\right\}\boldsymbol{\epsilon}_\phi \\ &+ \frac{1}{a}\left\{\frac{\partial}{\partial a}(aF_\phi) - \frac{\partial}{\partial\phi}(F_a)\right\}\boldsymbol{\epsilon}_3;\end{aligned}$$

Spherical:

$$\begin{aligned}\text{curl } \mathbf{F} &= \frac{1}{r^2\sin\theta}\left\{\frac{\partial}{\partial\theta}(r\sin\theta\, F_\phi) - \frac{\partial}{\partial\phi}(rF_\theta)\right\}\boldsymbol{\epsilon}_r \\ &+ \frac{1}{r\sin\theta}\left\{\frac{\partial}{\partial\phi}(F_r) - \frac{\partial}{\partial r}(r\sin\theta\, F_\phi)\right\}\boldsymbol{\epsilon}_\theta \\ &+ \frac{1}{r}\left\{\frac{\partial}{\partial r}(rF_\theta) - \frac{\partial}{\partial\theta}(F_r)\right\}\boldsymbol{\epsilon}_\phi.\end{aligned}$$

9. MISCELLANEOUS OPERATOR IDENTITIES

In the following identities, U, V (scalar point functions), **A**, and **B** (vector point functions) are arbitrary. They do not refer to any specific physical quantities. The relations shown below hold only for points at which the functions are defined.

(a) $\text{grad}\,[UV] = U\,\text{grad}\,V + V\,\text{grad}\,U$

(b) $\text{grad}\,[\mathbf{A}\cdot\mathbf{B}] = (\mathbf{A}\cdot\nabla)\mathbf{B} + (\mathbf{B}\cdot\nabla)\mathbf{A} + \mathbf{A}\times\text{curl}\,\mathbf{B} + \mathbf{B}\times\text{curl}\,\mathbf{A}$

(c) $\text{curl grad}\,U = 0$

(d) $\text{div curl}\,\mathbf{B} = 0$

(e) $\text{curl curl}\,\mathbf{A} = \text{grad div}\,\mathbf{A} - \nabla^2\mathbf{A}$

(f) $\nabla^2 = \text{div grad}$

$$= \frac{1}{h_1h_2h_3}\sum_{\substack{j=1\\ j\neq k\neq m\neq j}}^{3}\frac{\partial}{\partial\xi_j}\left\{h_kh_m\frac{1}{h_j}\frac{\partial}{\partial\xi_j}\right\}$$

(g) $\text{div}\,(V\mathbf{A}) = V\,\text{div}\,\mathbf{A} + (\mathbf{A}\cdot\nabla)V$

(h) $\text{div}\,(\mathbf{A}\times\mathbf{B}) = \mathbf{B}\cdot\text{curl}\,\mathbf{A} - \mathbf{A}\cdot\text{curl}\,\mathbf{B}$

(i) $\text{curl}\,(U\mathbf{A}) = U\,\text{curl}\,\mathbf{A} + (\text{grad}\,U)\times\mathbf{A}$

(j) $\text{curl}\,(\mathbf{A}\times\mathbf{B}) = \mathbf{A}\,\text{div}\,\mathbf{B} - \mathbf{B}\,\text{div}\,\mathbf{A} + (\mathbf{B}\cdot\nabla)\mathbf{A} - (\mathbf{A}\cdot\nabla)\mathbf{B}$

(k) $\displaystyle\iint_{s(\tau)} U\mathbf{n}\,dS = \iiint_{\tau}\text{grad}\,U\,d\tau$

(l) $\displaystyle\iint_{s(\tau)} \mathbf{n}\times\mathbf{F}\,dS = \iiint_{\tau}\text{curl}\,\mathbf{F}\,d\tau$

(m) $\displaystyle\oint_{c}(\mathbf{A}\cdot\mathbf{B})\,d\mathbf{r} = \iint_{s(\text{cap})}\{\text{grad}\,(\mathbf{A}\cdot\mathbf{B})\}\times\mathbf{n}\,dS$

10. TRANSFORMATION OF THE TIME DERIVATIVE

If we consider a set of orthonormal base vectors $\boldsymbol{\epsilon}_1'$, $\boldsymbol{\epsilon}_2'$, $\boldsymbol{\epsilon}_3'$ rotating with an instantaneous angular velocity $\boldsymbol{\omega}$ about a common origin O relative to a stationary set of bases $\boldsymbol{\epsilon}_1$, $\boldsymbol{\epsilon}_2$, and $\boldsymbol{\epsilon}_3$, the representations of the time derivative of a vector $\mathbf{f}$ in terms of these two sets of bases have different forms.

To understand that a problem exists involving the transformation of the time derivative, one need merely regard a simple example.

Consider a time-varying vector **f**(t) expressed in terms of the stationary bases $\boldsymbol{\epsilon}_k$. Assume that in an interval Δt the vector **f** changes by $\Delta\mathbf{f}$. Here $\Delta\mathbf{f}$ is measured by an observer in the stationary frame. In the rotating reference system $\boldsymbol{\epsilon}_m'$, the observations are somewhat different. An observer in the moving frame sees **f** change by $\Delta\mathbf{f}'$. Because in Δt the bases $\boldsymbol{\epsilon}_m'$ move relative to the bases $\boldsymbol{\epsilon}_k$, it is apparent that $\Delta\mathbf{f}$ and $\Delta\mathbf{f}'$ will not in general be equal.

The relations between $\Delta\mathbf{f}$ and $\Delta\mathbf{f}'$ can be obtained by developing the time derivatives of the moving bases relative to the stationary bases.

Expanding the vector **f** in the moving bases, we obtain at any instant

$$\mathrm{f} = \sum_{m=1}^{3} \mathrm{f}_m' \boldsymbol{\epsilon}_m'.$$

In view of the fact that the $\boldsymbol{\epsilon}_m'$ are time-dependent, the time derivative of **f** in the primed (or moving) system is

$$\frac{d\mathbf{f}}{dt} = \sum_{m=1}^{3} \left\{ \left(\frac{d\mathrm{f}_m'}{dt} \right) \boldsymbol{\epsilon}_m' + \mathrm{f}_m' \frac{d\boldsymbol{\epsilon}_m'}{dt} \right\}.$$

This equation should be contrasted with that presenting the expansion of **f** in the stationary bases,

$$\frac{d\mathbf{f}}{dt} = \sum_{k=1}^{3} \left\{ \frac{d\mathrm{f}_k}{dt} \right\} \boldsymbol{\epsilon}_k,$$

where

$$\frac{d\boldsymbol{\epsilon}_k}{dt} = 0.$$

The variation of any one of the moving bases in a time Δt is

$$\Delta\boldsymbol{\epsilon}_m' = \{\boldsymbol{\omega} \times \boldsymbol{\epsilon}_m'\} \Delta t,$$

where $\boldsymbol{\omega}$ is the instantaneous angular velocity at the beginning of the interval. In the limit as Δt goes to zero,

$$\frac{d\boldsymbol{\epsilon}_m'}{dt} = \boldsymbol{\omega} \times \boldsymbol{\epsilon}_m'.$$

The time derivative of **f** expressed in terms of the moving bases is therefore

$$\frac{d\mathbf{f}}{dt} = \sum_{m=1}^{3} \left\{ \left(\frac{d\mathrm{f}_m'}{dt} \right) \boldsymbol{\epsilon}_m' + \mathrm{f}_m'(\boldsymbol{\omega} \times \boldsymbol{\epsilon}_m') \right\},$$

or

$$\frac{d\mathbf{f}}{dt} = \left[\frac{d\mathbf{f}'}{dt} \right]_{\mathrm{rot}} + \boldsymbol{\omega} \times \mathbf{f}'.$$

The symbol $\left[\dfrac{d\mathbf{f}'}{dt}\right]_{rot}$ merely implies that the time derivative of the coefficients has been taken and not the derivatives of the bases:

$$\left[\frac{d\mathbf{f}'}{dt}\right]_{rot} = \sum_{m=1}^{3}\left\{\frac{df_m'}{dt}\right\}\boldsymbol{\epsilon}_m'.$$

The change $\Delta\mathbf{f}'$ measured by an observer in the rotating system is related only to the variations of the components of f_m'. In other words, in an interval Δt

$$\Delta\mathbf{f}' = \left[\frac{d\mathbf{f}'}{dt}\right]_{rot}\Delta t = \sum_{m=1}^{3}\left\{\frac{df_m'}{dt}\Delta t\right\}\boldsymbol{\epsilon}_m'.$$

The change $\Delta\mathbf{f}$ is related to $\Delta\mathbf{f}'$ by means of the complete transformation equation

$$\Delta\mathbf{f} = \frac{d\mathbf{f}}{dt}\Delta t = \left[\frac{d\mathbf{f}'}{dt}\right]_{rot}\Delta t + \{\boldsymbol{\omega}\times\mathbf{f}'\}\,\Delta t$$
$$= \Delta\mathbf{f}' + \{\boldsymbol{\omega}\times\mathbf{f}'\}\,\Delta t.$$

Rotating systems are noninertial systems. The time derivative transformation provides a technique by which velocities and accelerations as observed in rotating systems can be related to the corresponding velocities and accelerations as observed in a stationary system.

If $\mathbf{r}$ is the position vector of a point, the velocity of the point in a stationary system (unprimed) is related to the velocity in a rotating (primed) system by

$$\frac{d\mathbf{r}}{dt} = \left[\frac{d\mathbf{r}'}{dt}\right]_{rot} + \boldsymbol{\omega}\times\mathbf{r}'.$$

The transformation of the acceleration requires that the time derivatives of all base vectors on the right-hand side of the equation above be included:

$$\frac{d^2\mathbf{r}}{dt^2} = \frac{d}{dt}\left\{\sum_{k=1}^{3}(\dot{x}_k'\boldsymbol{\epsilon}_k' + \boldsymbol{\omega}\times(x_k'\boldsymbol{\epsilon}_k'))\right\},$$

$$= \sum_{k=1}^{3}\{\ddot{x}_k'\boldsymbol{\epsilon}_k' + 2\boldsymbol{\omega}\times(\dot{x}_k'\boldsymbol{\epsilon}_k') + \boldsymbol{\omega}\times[\boldsymbol{\omega}\times(x_k'\boldsymbol{\epsilon}_k')]\},$$

or
$$\frac{d^2\mathbf{r}}{dt^2} = \left[\frac{d^2\mathbf{r}'}{dt^2}\right]_{rot} + 2\boldsymbol{\omega}\times\left[\frac{d\mathbf{r}'}{dt}\right]_{rot} + \boldsymbol{\omega}\times(\boldsymbol{\omega}\times\mathbf{r}').$$

The reader will recognize the last two terms. The term $2\boldsymbol{\omega} \times \dot{\mathbf{r}}$ is the Coriolis acceleration, while $\boldsymbol{\omega} \times (\boldsymbol{\omega} \times \mathbf{r}')$ is the centripetal acceleration.

This last relationship for the acceleration is quite general. It does not, however, completely demonstrate the details necessary to obtain acceleration components in a specific curvilinear coordinate system in which $\boldsymbol{\omega}$ is described in terms of the coordinates.

A method of virtual displacements permits a general development of the components of the acceleration vector in any orthonormal system without specific analysis of the time variation of the bases of the system in question.

Employing the notation set down earlier in this appendix, the variation of the position vector of a point is

$$\delta\mathbf{r} = \sum_{j=1}^{3}\left\{\frac{\partial\mathbf{r}}{\partial\xi_j}\right\}\delta\xi_j = \sum_{m=1}^{3}\{\delta l_m\}\boldsymbol{\epsilon}_m = \sum_{m=1}^{3}\{h_m\delta\xi_m\}\boldsymbol{\epsilon}_m.$$

The acceleration vector resolved along specific bases and expanded in terms of the generalized coordinates ξ_k with metrical coefficients h_k^2 will be designated as $\ddot{\mathbf{r}}$. The symbol for acceleration with its associated components is

$$\mathbf{a} = \sum_{i=1}^{3} a_i\boldsymbol{\epsilon}_i.$$

Starting with

$$\mathbf{a} = \ddot{\mathbf{r}},$$

we can remove the base vectors by using the form

$$\mathbf{a}\cdot\delta\mathbf{r} = \ddot{\mathbf{r}}\cdot\delta\mathbf{r},$$

or

$$\sum_{=1}^{3}\left\{a_j h_j - \ddot{\mathbf{r}}\cdot\frac{\partial\mathbf{r}}{\partial\xi_j}\right\}\delta\xi_j = 0.$$

The variations in ξ_j are arbitrary; therefore

$$a_j h_j = \ddot{\mathbf{r}}\cdot\frac{\partial\mathbf{r}}{\partial\xi_j}.$$

The right-hand side of this equation can be expanded in terms of the coordinates:

$$\ddot{\mathbf{r}}\cdot\frac{\partial\mathbf{r}}{\partial\xi_j} = \frac{d}{dt}\left\{\dot{\mathbf{r}}\cdot\frac{\partial\mathbf{r}}{\partial\xi_j}\right\} - \dot{\mathbf{r}}\cdot\frac{d}{dt}\left\{\frac{\partial\mathbf{r}}{\partial\xi_j}\right\}.$$

Using the equality

$$\dot{\mathbf{r}} = \frac{d\mathbf{r}}{dt} = \sum_{K=1}^{3}\left\{\frac{\partial\mathbf{r}}{\partial\xi_K}\dot{\xi}_K + \frac{\partial\mathbf{r}}{\partial t}\right\},$$

we find that $$\frac{\partial \dot{\mathbf{r}}}{\partial \dot{\xi}_j} = \frac{\partial \mathbf{r}}{\partial \xi_j}.$$

Substituting into the forms containing the acceleration,

$$\ddot{\mathbf{r}} \cdot \frac{\partial \mathbf{r}}{\partial \xi_j} = \frac{d}{dt}\left\{\dot{\mathbf{r}} \cdot \frac{\partial \dot{\mathbf{r}}}{\partial \dot{\xi}_j}\right\} - \dot{\mathbf{r}} \cdot \frac{\partial \dot{\mathbf{r}}}{\partial \xi_j}$$
$$= \frac{d}{dt}\left\{\frac{\partial}{\partial \dot{\xi}_j}(\tfrac{1}{2}\dot{\mathbf{r}} \cdot \dot{\mathbf{r}})\right\} - \frac{\partial}{\partial \xi_j}(\tfrac{1}{2}\dot{\mathbf{r}} \cdot \dot{\mathbf{r}}).$$

This result completes our derivation. The mth component of the acceleration is therefore

$$a_m = \frac{1}{h_m}\left[\frac{d}{dt}\left\{\frac{\partial}{\partial \dot{\xi}_m}(\tfrac{1}{2}\dot{\mathbf{r}}^2)\right\} - \frac{\partial}{\partial \xi_m}(\tfrac{1}{2}\dot{\mathbf{r}}^2)\right].$$

The reader familiar with classical mechanics will recognize the basic elements of the derivation of the Euler-Lagrange equations.

In the case of an orthonormal system,

$$\tfrac{1}{2}\dot{\mathbf{r}}^2 = \sum_{k=1}^{3} \tfrac{1}{2} h_k{}^2 \dot{\xi}_k{}^2.$$

Employing this result in the expression for a_m, we obtain

$$a_m = h_m \ddot{\xi}_m + \sum_{k=1}^{3}\left\{2\dot{\xi}_m \dot{\xi}_k \frac{\partial h_m}{\partial \xi_k} - \frac{h_k}{h_m}\dot{\xi}_k{}^2 \frac{\partial h_k}{\partial \xi_m}\right\}.$$

This general expression can be checked for instance by computing the radial acceleration component in spherical coordinates. For these specific coordinates,

$$\xi_1 = r, \quad \xi_2 = \theta, \quad \xi_3 = \phi$$
$$h_1 = 1, \quad h_2 = r, \quad h_3 = r \sin \theta,$$

and $$a_r = \ddot{r} - r\dot{\theta}^2 - r \sin^2\theta\, \dot{\phi}^2.$$

The last two terms in this expression are the radial components of the centripetal acceleration.

B. Matrices

1. GENERAL PROPERTIES

Any ordered rectangular array of mathematical quantities (such as numbers, functions, etc.) can be considered as a matrix. While this section is in no manner intended as a complete discussion of matrices, it will suffice as an introduction to enable the reader to understand clearly the matrix calculations performed in the text.

We shall confine our discussion to square arrays and to row or column arrays.

A typical square array $\mathbb{A}$ of 3×3 elements is

$$\mathbb{A} = \begin{pmatrix} A_{11} & A_{12} & A_{13} \\ A_{21} & A_{22} & A_{23} \\ A_{31} & A_{32} & A_{33} \end{pmatrix}.$$

It is quite important to keep the notation firmly in mind. We shall be mainly concerned with linear algebraic equations. The advantage of the matrix notation to us will lie in our ability to imply a lengthy calculation very concisely.

An $N \times N$ square array will be designated by an open faced letter such as $\mathbb{A}$ above. If written in terms of the elements A_{jk} the square matrix will be enclosed in parentheses, as is demonstrated on the right-hand side of the equality.

Notice that the elements A_{jk} of the square array have double subscripts. The first subscript designates the row and the second specifies the column in which the element lies. Thus A_{23} is the element belonging in the second row and placed in the third column.

As an example of a numerical matrix, consider the array

$$\mathbb{A} = \begin{pmatrix} 5 & 1 & 3 \\ 0 & 9 & 2 \\ 7 & 5 & 7 \end{pmatrix}.$$

This particular example has components which have been assigned specific numerical values. In this array

$$\begin{array}{lll} A_{11} = 5 & A_{12} = 1 & A_{13} = 3 \\ A_{21} = 0 & A_{22} = 9 & A_{23} = 2 \\ A_{31} = 7 & A_{32} = 5 & A_{33} = 7. \end{array}$$

Two matrices can be added by summing like elements. Consider the addition

$$\mathbb{A} + \mathbb{B} = \mathbb{C}.$$

Then $$C_{jk} = A_{jk} + B_{jk}.$$

Matrix addition obeys:

The Commutative Law,

$$\mathbb{A} + \mathbb{B} = \mathbb{B} + \mathbb{A};$$

The Associative Law,

$$(\mathbb{A} + \mathbb{B}) + \mathbb{D} = \mathbb{A} + (\mathbb{B} + \mathbb{D});$$

The Distributive Law,

$$n(\mathbb{A} + \mathbb{B}) = n\mathbb{A} + n\mathbb{B},$$

when n is a scalar.

The other type of matrix in which we shall be quite interested is the row matrix or the column matrix. This particular array will in many instances be referred to as a *vector* since it transforms like a vector. We shall designate the column matrix by a bold-faced symbol in the same manner as the vector was specified in Appendix A. The column matrix is encompassed by square brackets. For example the position vector **r** can be written

$$\mathbf{r} = \begin{bmatrix} x_1 \\ x_2 \\ x_3 \end{bmatrix}.$$

In this representation the bases are suppressed.

The row vector is a transposed column matrix. This matrix is designated by a bold-faced symbol with a $\sim$ over it,

$$\tilde{\mathbf{r}} = [x_1, x_2, x_3].$$

Later it will be seen that column vectors are used in multiplying from the right and row vectors when multiplying from the left. The inner product of a vector with itself is the product of the row matrix times the column matrix:

$$r^2 = \tilde{\mathbf{r}} \cdot \mathbf{r} = [x_1, x_2, x_3] \cdot \begin{bmatrix} x_1 \\ x_2 \\ x_3 \end{bmatrix}$$
$$= x_1^2 + x_2^2 + x_3^2.$$

The technique of multiplying a row times a column will turn out to be general, in that elements of the products of square arrays are obtained in this manner.

To introduce matrix multiplication let us consider two simultaneous algebraic equations in two unknowns x_1 and x_2,

$$\sum_{k=1}^{2} A_{jk}x_k = y_j,$$

where j takes on each of the values 1 and 2 to provide the equations. In the matrix notation this is written

$$\mathbb{A} \cdot \mathbf{r} = \mathbf{y},$$

or

$$\begin{pmatrix} A_{11} & A_{12} \\ A_{21} & A_{22} \end{pmatrix} \begin{bmatrix} x_1 \\ x_2 \end{bmatrix} = \begin{bmatrix} A_{11}x_1 + A_{12}x_2 \\ A_{21}x_1 + A_{22}x_2 \end{bmatrix} = \begin{bmatrix} y_1 \\ y_2 \end{bmatrix},$$

where the arrows indicate the first two steps in obtaining the product.

If two matrices are equal their elements are equal; thus in the column matrices at the right,

$$y_1 = A_{11}x_1 + A_{12}x_2, \quad \text{etc.}$$

We see that the elements of the product between $\mathbb{A}$ and $\mathbf{r}$ are obtained by multiplying a row in $\mathbb{A}$ times the column $\mathbf{r}$ to obtain one element of the product $\mathbf{y}$. This same procedure of course applies to $N \times N$ matrices and N-dimensional vectors.

As an exercise, regard the equations

$$3x_1 + 2x_2 = 8$$
$$x_1 - 4x_2 = -2.$$

In matrix notation this is written

$$\begin{pmatrix} 3 & 2 \\ 1 & -4 \end{pmatrix} \begin{bmatrix} x_1 \\ x_2 \end{bmatrix} = \begin{bmatrix} 8 \\ -2 \end{bmatrix}.$$

The product of two square arrays $\mathbb{A}$ and $\mathbb{B}$ can be taken as if the array on the right is composed of an ordered set of column matrices. Consider the product $\mathbb{A} \cdot \mathbb{B}$,

$$\mathbb{A} \cdot \mathbb{B} = \begin{pmatrix} A_{11} & A_{12} \\ A_{21} & A_{22} \end{pmatrix} \begin{pmatrix} B_{11} & B_{12} \\ B_{21} & B_{22} \end{pmatrix} \to \begin{pmatrix} A_{11} & A_{12} \\ A_{21} & A_{22} \end{pmatrix} \begin{bmatrix} B_{11} \\ B_{21} \end{bmatrix} \begin{bmatrix} B_{12} \\ B_{22} \end{bmatrix}$$

giving

$$\to \begin{bmatrix} A_{11}B_{11} + A_{12}B_{21} \\ \hline A_{21}B_{11} + A_{22}B_{21} \end{bmatrix} \begin{bmatrix} A_{11}B_{12} + A_{12}B_{22} \\ \hline A_{21}B_{12} + A_{22}B_{22} \end{bmatrix}$$

$$\to \left(\begin{array}{c|c} A_{11}B_{11} + A_{12}B_{21} & A_{11}B_{12} + A_{12}B_{22} \\ \hline A_{21}B_{11} + A_{22}B_{21} & A_{21}B_{12} + A_{22}B_{22} \end{array}\right).$$

In the example above a set of 2×2 matrices has been utilized. Needless to say the rules as applied here hold for arrays of any dimensions. If the dimensionality of the space is N, then the jk^{th} element of the product matrix is given by the relation

$$C_{jk} = \sum_{n=1}^{N} A_{jn}B_{nk},$$

where $$\mathbb{C} = \mathbb{A} \cdot \mathbb{B}.$$

To illustrate the procedure we will again utilize a set of 2×2 matrices. This time we will make the computation with specific values assigned to the elements. For a special case let

$$\mathbb{A} = \begin{pmatrix} 5 & 3 \\ 9 & 7 \end{pmatrix}; \quad \mathbb{B} = \begin{pmatrix} 1 & 0 \\ -2 & 3 \end{pmatrix}.$$

Then $$\mathbb{C} = \begin{pmatrix} 5 & 3 \\ 9 & 7 \end{pmatrix} \begin{pmatrix} 1 & 0 \\ -2 & 3 \end{pmatrix} = \left(\begin{array}{c|c} 5 - 6 & 0 + 9 \\ \hline 9 - 14 & 0 + 21 \end{array}\right)$$

$$= \begin{pmatrix} -1 & 9 \\ -5 & 21 \end{pmatrix}.$$

The reader can check this result as an exercise in the multiplication of matrices.

2. SPECIAL MATRICES

a. THE TRANSPOSE MATRIX. The transpose matrix $\tilde{\mathbb{A}}$ of a matrix $\mathbb{A}$ is obtained by interchanging the rows and columns. In other words the A_{mn} element of $\mathbb{A}$ becomes the $\tilde{A}_{nm}$ element of the transpose $\tilde{\mathbb{A}}$,

$$\tilde{A}_{nm} = A_{mn}.$$

If
$$\tilde{\mathbb{A}} = \begin{pmatrix} 5 & 1 & 3 \\ 0 & 9 & 2 \\ 7 & 5 & 7 \end{pmatrix},$$

then
$$\tilde{\mathbb{A}} = \begin{pmatrix} 5 & 0 & 7 \\ 1 & 9 & 5 \\ 3 & 2 & 7 \end{pmatrix}.$$

b. THE ADJOINT MATRIX. The adjoint matrix of $\mathbb{A}$ is designated by the symbol $\mathbb{A}^\dagger$, and is the complex conjugate of the transpose $\mathbb{A}$. Thus

$$\mathbb{A}^\dagger = (\tilde{\mathbb{A}})^*,$$

and the elements of $\mathbb{A}^\dagger$ are $A^\dagger_{nm}$,

$$A^\dagger_{nm} = (\tilde{A}_{nm})^* = (A_{mn})^*.$$

Obviously this matrix is applied only in cases in which the elements of $\mathbb{A}$ are complex.

Consider as an example

$$\mathbb{A} = \begin{pmatrix} (1 + i3) & 4 \\ (7 + i5) & i6 \end{pmatrix};$$

then
$$\mathbb{A}^\dagger = \begin{pmatrix} (1 - i3) & (7 - i5) \\ 4 & -i6 \end{pmatrix}$$

where
$$i = \sqrt{-1}.$$

c. THE UNIT MATRIX. The unit or identity matrix is defined as the matrix which transforms another matrix into itself, i.e.,

$$\mathbb{I} \cdot \mathbf{r} = \mathbf{r};$$

also
$$\mathbb{I} \cdot \mathbb{A} = \mathbb{A}.$$

The elements of $\mathbb{I}$ are δ_{nm} where δ_{nm} is the Kronecker delta

$$\delta_{nm} = 1, \quad n = m$$
$$= 0, \quad n \neq m.$$

The 3 × 3 unit matrix is

$$\mathbb{I} = \begin{pmatrix} 1 & 0 & 0 \\ 0 & 1 & 0 \\ 0 & 0 & 1 \end{pmatrix}.$$

d. THE MAGNITUDE OF A MATRIX. The magnitude of a square array is simply the determinant of the matrix:

$$|\mathbb{A}| = \text{Det } \mathbb{A}.$$

e. THE INVERSE MATRIX. *If the magnitude of a matrix is nonzero,* it is possible to define an inverse such that

$$\mathbb{A} \cdot \mathbb{A}^{-1} = \mathbb{I};$$

or

$$A_{jn}A'_{nk} = \delta_{jk}$$

where A'_{nk} is an element of $\mathbb{A}^{-1}$. The derivation of the relation between $\mathbb{A}^{-1}$ and $\mathbb{A}$ is obtained from the definition of the determinant,

$$\text{Det } \mathbb{A} = |\mathbb{A}| = \sum_{n=1}^{N} A_{jn}(-1)^{j+n} \text{minor } A_{jn}$$

or

$$\delta_{jk} = \sum_{n=1}^{N} A_{jn}\left\{\frac{(-1)^{k+n} \text{minor } A_{kn}}{|\mathbb{A}|}\right\}.$$

Thus the elements of $\mathbb{A}^{-1}$ are

$$A'_{nk} = \frac{(-1)^{k+n} \text{minor } A_{kn}}{|\mathbb{A}|}.$$

To illustrate the technique, let us compute $\mathbb{A}^{-1}$ if

$$\mathbb{A} = \begin{pmatrix} 1 & 0 & 3 \\ 7 & 5 & 2 \\ 1 & 4 & 6 \end{pmatrix}.$$

$$|\mathbb{A}| = \begin{vmatrix} 1 & 0 & 3 \\ 7 & 5 & 2 \\ 1 & 4 & 6 \end{vmatrix} = 1\begin{vmatrix} 5 & 2 \\ 4 & 6 \end{vmatrix} - 0\begin{vmatrix} 7 & 2 \\ 1 & 6 \end{vmatrix} + 3\begin{vmatrix} 7 & 5 \\ 1 & 4 \end{vmatrix}$$

$$= 1(30 - 8) + 3(28 - 5) = 22 + 69 = 91.$$

Now in this example

$$A'_{11} = \frac{(-1)^2}{91}\begin{vmatrix} 5 & 2 \\ 4 & 6 \end{vmatrix} = \frac{22}{91}$$

$$A'_{21} = \frac{(-1)^3}{91}\begin{vmatrix} 7 & 2 \\ 1 & 6 \end{vmatrix} = -\frac{40}{91}$$

$$A'_{31} = \frac{(-1)^4}{91}\begin{vmatrix} 7 & 5 \\ 1 & 4 \end{vmatrix} = \frac{23}{91}$$

$$A'_{12} = \frac{(-1)^3}{91}\begin{vmatrix} 0 & 3 \\ 4 & 6 \end{vmatrix} = +\frac{12}{91}.$$

The reader should now as an exercise compute the remaining elements and show that

$$\mathbb{A}^{-1} = \frac{1}{91}\begin{pmatrix} 22 & 12 & -15 \\ -40 & 3 & 19 \\ 23 & -4 & 5 \end{pmatrix};$$

and further, by direct multiplication, demonstrate that in fact

$$\mathbb{A} \cdot \mathbb{A}^{-1} = \mathbb{I}.$$

It should be remarked that although this initial computation is tedious it is systematic, and it can be checked quite readily. The advantage of this approach lies in the fact that once $\mathbb{A}^{-1}$ has been evaluated the remaining details of a solution of a set of simultaneous linear algebraic equations are reduced to a trivial operation. In the section on circuits (Chapter V) we find that the currents in each loop are listed in a column vector **I** where

$$\mathbb{R} \cdot \mathbf{I} = \mathbf{E}.$$

When R and E are given, the solution, I, is obtained by multiplying from the left by $\mathbb{R}^{-1}$(using the fact that $\mathbb{R}^{-1} \cdot \mathbb{R} = \mathbb{I}$); thus

$$\mathbb{R}^{-1} \cdot \mathbb{R} \cdot \mathbf{I} = \mathbb{I} \cdot \mathbf{I} = \mathbf{I} = \mathbb{R}^{-1} \cdot \mathbf{E}.$$

Evidently, carrying out the multiplication $\mathbb{R}^{-1} \cdot \mathbf{E}$ renders all the components of **I** in one operation. It is of course true that this manipulation is completely equivalent to the standard techniques for solving simultaneous algebraic equations. However, the standard techniques are quite subject to mistakes, and are difficult to check.

f. SYMMETRIX MATRICES. A matrix is symmetric if it is equal to its transpose

$$A_{nm} = A_{mn}.$$

In other words, $\mathbb{A} = \tilde{\mathbb{A}}$, if the matrix is symmetric.

g. HERMITIAN CONJUGATE MATRICES. A matrix is Hermitian conjugate if it equals its adjoint:

$$A_{nm} = (A_{mn})^*,$$

or

$$\mathbb{A} = \mathbb{A}^\dagger.$$

Symmetric matrices are in the class known as Hermitian conjugate.

h. ORTHOGONAL TRANSFORMATIONS. The application of a square array to a vector **r** produces a new vector which we can label **y**:

$$\mathbb{A} \cdot \mathbf{r} = \mathbf{y}.$$

A special class of transformations known as orthogonal leaves the magnitude of a real vector unchanged. Rotations are geometric examples of this.

Let $\mathbb{S}$ be an orthogonal matrix. The application of $\mathbb{S}$ to a position vector **r** rotates the vector to **r**′,

$$\mathbf{r}' = \mathbb{S} \cdot \mathbf{r},$$

and

$$\tilde{\mathbf{r}}' = \tilde{\mathbf{r}} \cdot \tilde{\mathbb{S}}.$$

The magnitude r'^2 is then

$$r'^2 = \tilde{\mathbf{r}}' \cdot \mathbf{r}' = \tilde{\mathbf{r}} \cdot \tilde{\mathbb{S}} \cdot \mathbb{S} \cdot \mathbf{r} = \tilde{\mathbf{r}} \cdot \mathbb{I} \cdot \mathbf{r} = \tilde{\mathbf{r}} \cdot \mathbf{r} = r^2.$$

Thus

$$\tilde{\mathbb{S}} \cdot \mathbb{S} = \mathbb{I},$$

and since

$$\mathbb{S}^{-1} \cdot \mathbb{S} = \mathbb{I},$$

$$\tilde{\mathbb{S}} = \mathbb{S}^{-1}.$$

The equality of $\tilde{\mathbb{S}}$ and $\mathbb{S}^{-1}$ in the case of the orthogonal matrices is useful in computations of $\mathbb{S}^{-1}$.

Rotations and space reflections in two and three dimensions are familiar examples of orthogonal transformations. A reflection of the coordinates is sometimes called an "improper rotation" and has a determinant equal to -1, while a rotation matrix has a determinant of $+1$.

A counterclockwise rotation of the coordinates in a two-dimensional space through an angle α transforms a vector **f** to **f′** by

$$\mathbf{f}' = \mathbb{S} \cdot \mathbf{f}$$

or

$$\begin{bmatrix} f_1' \\ f_2' \end{bmatrix} = \begin{pmatrix} \cos\alpha & \sin\alpha \\ -\sin\alpha & \cos\alpha \end{pmatrix} \begin{bmatrix} f_1 \\ f_2 \end{bmatrix}.$$

The most general rotation in three dimensions can be reduced to three two-dimensional rotations. First there is a rotation about the **k** (or $\boldsymbol{\epsilon}_3$) axis through an angle ϕ. Call this matrix $\mathbb{C}(\phi)$ where

$$\mathbb{C}(\phi) = \begin{pmatrix} \cos\phi & \sin\phi & 0 \\ -\sin\phi & \cos\phi & 0 \\ 0 & 0 & 1 \end{pmatrix}.$$

After the first rotation the new x axis is **i′**. The second rotation is counterclockwise about **i′** through an angle θ, bringing **k** to **k′** and **j′** to **j″**. This rotation is labeled $\mathbb{B}(\theta)$,

$$\mathbb{B}(\theta) = \begin{pmatrix} 1 & 0 & 0 \\ 0 & \cos\theta & \sin\theta \\ 0 & -\sin\theta & \cos\theta \end{pmatrix}.$$

The last rotation is about **k′** counterclockwise through an angle ψ. The matrix is $\mathbb{A}(\psi)$ where

$$\mathbb{A}(\psi) = \begin{pmatrix} \cos\psi & \sin\psi & 0 \\ -\sin\psi & \cos\psi & 0 \\ 0 & 0 & 1 \end{pmatrix}.$$

This last rotation takes **i′** to **i″** and **j″** to **j‴**. The total rotation matrix $\mathbb{S}$ is the application of $\mathbb{C}(\phi)$, $\mathbb{B}(\theta)$, and $\mathbb{A}(\psi)$ in the appropriate order,

$$\mathbb{S} = \mathbb{A}(\psi)\mathbb{B}(\theta)\mathbb{C}(\phi)$$

or

$$S_{mn} = \sum_{p=1}^{3} \sum_{q=1}^{3} A_{mp}B_{pq}C_{qn}.$$

The reader should compute the 9 elements of $\mathbb{S}$ in terms of ϕ, θ and ψ. Most standard works on analytical dynamics provide the expressions for the S_{mn}.

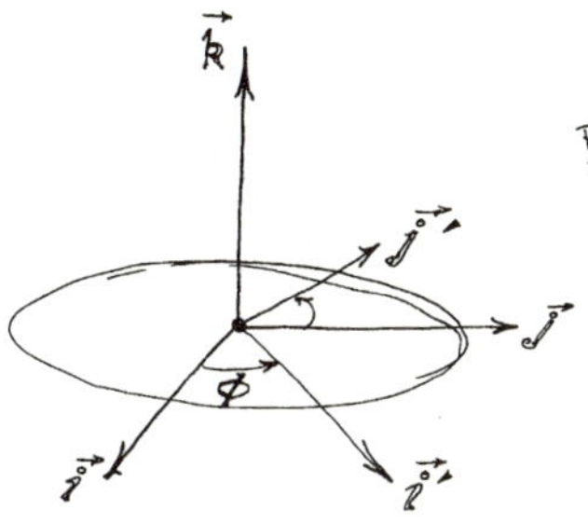

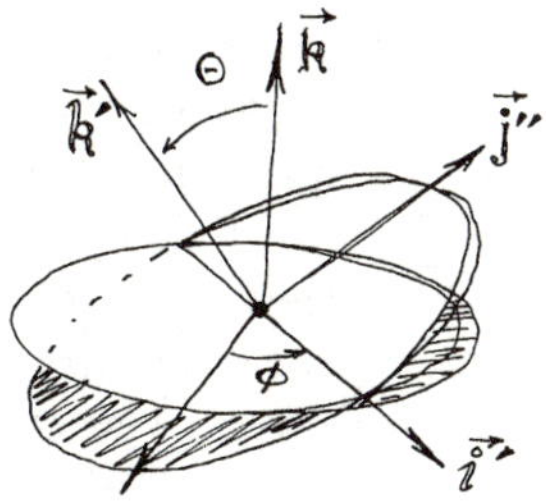

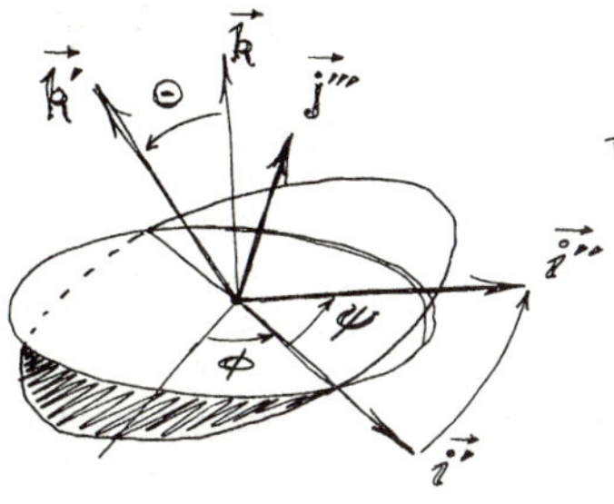

i. UNITARY TRANSFORMATIONS. These transformations include orthogonal transformations as a subgroup. The magnitude of a complex vector $\mathbf{v}$ is

$$\mathbf{v}^{\dagger} \cdot \mathbf{v} = |\mathbf{v}|^2.$$

A unitary transformation also leaves the magnitude of the complex vector invariant. Then

$$\mathbf{v}' = \mathbb{S} \cdot \mathbf{v},$$

and

$$\mathbf{v}'^{\dagger} = \mathbf{v}^{\dagger} \cdot \mathbb{S}^{\dagger}.$$

Under such circumstances

$$|\mathbf{v}'|^2 = \mathbf{v}'^{\dagger} \cdot \mathbf{v}' = \mathbf{v}^{\dagger} \cdot \mathbb{S}^{\dagger} \cdot \mathbb{S} \cdot \mathbf{v} = \mathbf{v}^{\dagger} \cdot \mathbb{I} \cdot \mathbf{v} = \mathbf{v}^{\dagger} \cdot \mathbf{v} = |\mathbf{v}|^2.$$

Therefore, because

$$\mathbb{S}^{-1} \cdot \mathbb{S} = \mathbb{I},$$

$$\mathbb{S}^{\dagger} = \mathbb{S}^{-1}$$

if the transformation is unitary.

3. QUADRATIC FORMS AND THE EIGENVALUE PROBLEM

A quadratic form in most cases of physical interest can be represented in terms of a symmetric matrix or an Hermitian conjugate matrix. Because the Hermitian conjugates and unitary transformations apply equally well to problems involving real spaces in which we have transposed matrices, symmetric matrices, and orthogonal transformations, the discussion of quadratics and the eigenvalue problem will be written in terms of the more comprehensive quantities. In other words, the real matrices are a subclass of the more general complex matrices.

The general quadratic form (translated to the symmetry point) can be written

$$\tilde{\mathbf{r}}^{\dagger} \cdot \mathbb{A} \cdot \mathbf{r} = K \quad \text{(a scalar)}.$$

Writing this equation out in detail,

$$\sum_{k=1}^{N} \sum_{j=1}^{N} x_k^* A_{kj} x_j = K.$$

This form can be diagonalized by a *unitary transformation.*

If $\qquad \mathbf{r}' = \mathbb{S} \cdot \mathbf{r},$

then $\qquad \mathbf{r} = \mathbb{S}^{-1} \cdot \mathbf{r}';$

and $\qquad \mathbf{r}^{\dagger} = \mathbf{r}'^{\dagger} \cdot (\mathbb{S}^{-1})^{\dagger} = \mathbf{r}'^{\dagger} \cdot \mathbb{S}.$

Substituting for $\mathbf{r}$ and $\mathbf{r}^{\dagger}$,

$$\mathbf{r}'^{\dagger} \cdot (\mathbb{S} \cdot \mathbb{A} \cdot \mathbb{S}^{-1}) \cdot \mathbf{r}' = \mathbf{r}'^{\dagger} \cdot \mathbb{\Gamma} \cdot \mathbf{r}' = K.$$

Using the fact that $\mathbb{S}$ is unitary (or orthogonal in the case of real elements) one can demonstrate that if $\mathbb{A}$ is Hermitian conjugate (or symmetric) then an $\mathbb{S}$ can be found which produces* a diagonal $\mathbb{\Gamma}$,

$$\mathbb{S} \cdot \mathbb{A} \cdot \mathbb{S}^{-1} = \mathbb{\Gamma},$$

and

$$\Gamma_{jk} = \lambda_j \delta_{jk}.$$

* The fact that $\mathbb{A}$ can be diagonalized to $\mathbb{\Gamma}$ is demonstrated by assuming that $\mathbb{\Gamma}$ is diagonal. Then employing the Hermitian conjugate condition on $\mathbb{A}$ and the unitary

If $\mathbb{A}$ is Hermitian conjugate (or if real, symmetric), the λ_j can be obtained as follows: Utilize the invariance of $|\mathbf{r}|^2$; subtract $\lambda|\mathbf{r}|^2 = K$ from $\mathbf{r}^\dagger \cdot \mathbb{A} \cdot \mathbf{r} = K$, giving

$$\mathbf{r}^\dagger \cdot \{\mathbb{A} - \lambda\mathbb{I}\} \cdot \mathbf{r} = \mathbf{r}'^\dagger\{\mathbf{\Gamma} - \lambda\mathbb{I}\} \cdot \mathbf{r}' = 0.$$

If this be the case, then the magnitude $|\mathbb{A} - \lambda\mathbb{I}|$ is negative definite. Taking the maximum value

$$|\mathbb{A} - \lambda\mathbb{I}| = |\mathbf{\Gamma} - \lambda\mathbb{I}| = 0.$$

Because $\mathbf{\Gamma}$ is diagonal,

$$|\mathbb{A} - \lambda\mathbb{I}| = \prod_{j=1}^{N} (\lambda_j - \lambda) = (\lambda_1 - \lambda)(\lambda_2 - \lambda) \cdots (\lambda_N - \lambda) = 0,$$

and the roots of $\qquad |\mathbb{A} - \lambda\mathbb{I}| = 0$

must be the $\qquad \lambda_j = \Gamma_{jj}.$

In most problems of this type, we are interested in the linear form associated with the maximum value of $|\mathbb{A} - \lambda\mathbb{I}|$ in the case of a real space. A sufficient condition for the vanishing of the quadratic $\mathbf{r}^\dagger \cdot \{\mathbb{A} - \lambda\mathbb{I}\} \cdot \mathbf{r} = 0$ is that $\mathbf{r}$ becomes a vector $\mathbf{R}$ such that

$$\{\mathbb{A} - \lambda\mathbb{I}\} \cdot \mathbf{R} = 0.$$

This is the *eigenvalue problem.*

In order that the solutions to this equation be nontrivial (i.e., $R^2 \neq 0$) it is necessary that

$$|\mathbb{A} - \lambda\mathbb{I}| = 0.$$

---

condition on $\mathbb{S}$, the proof can be demonstrated. Multiply

$$\mathbb{S} \cdot \mathbb{A} \cdot \mathbb{S}^{-1} = \mathbf{\Gamma}$$

by $\mathbb{S}^{-1}$, giving

$$\mathbb{A} \cdot \mathbb{S}^{-1} = \mathbb{S}^{-1} \cdot \mathbf{\Gamma}$$

or

$$\sum_{j=1}^{N} A_{mj} S'_{jk} = \sum_{j=1}^{N} S'_{mj} \Gamma_{jk}.$$

Assume that $\mathbf{\Gamma}$ is diagonal with $\Gamma_{km} = \lambda_k \delta_{km}$; then

$$\sum_j A_{mj} S'_{jk} = S'_{mk} \Gamma_{kk} = S'_{mk} \lambda_k,$$

or

$$\sum_{j=1}^{N} \{A_{mj} - \lambda_k \delta_{kj}) S'_{jk} = 0.$$

If we are to have a nontrivial solution for S'_{jk} then $|\mathbb{A} - \lambda\mathbb{I}| = 0$. The reader should then demonstrate that if the columns of $\mathbb{S}$ form a set of mutually orthogonal vectors, $\mathbb{A}$ must be Hermitian conjugate.

Again this equation has N roots λ_j called the EIGENVALUES of $\mathbb{A}$.

If we label the components of the associated eigenvector R_j as R_{mj}, the R_{mj}'s are found by substituting λ_j into the eigenvalue problem and solving for the components of R_j:

$$\{\mathbb{A} - \lambda_j \mathbb{I}\} \cdot \mathbf{R}_j = 0,$$

or

$$A_{mk} R_{kj} = X_{mj} \lambda_j.$$

We should include the fact that the elements of $\mathbb{S}^{-1}$ are related to the components of the normalized **R**'s ($|\mathbf{R}| = 1$) by

$$S'_{mj} = R_{mj},$$

assuming that the R_j's have been normalized, i.e.,

$$\mathbf{R}_j{}^\dagger \cdot \mathbf{R}_j = 1.$$

Also, most books on this subject demonstrate that when $\mathbb{A}$ is Hermitian conjugate, the eigenvalues are *real* and the eigenvectors are orthogonal:

$$\lambda_j = \lambda_j^*,$$

and

$$\mathbf{R}_k{}^\dagger \cdot \mathbf{R}_j = \delta_{kj} \quad \text{(if normalized)}.$$

Proofs of these latter assertions are left to the reader. Try them.

C. The Complete Elliptic Functions

Many practical problems involve circular geometry. In electrostatics and electromagnetics the experimenter will continually encounter sets of apparatus utilizing circular configurations. The elliptic functions under such circumstances become an important mathematical tool in developing the problems encountered.

Like the trigonometric functions the elliptic functions must be developed in a series and tabulated. In conjunction with the use of the tables, however, the student must be familiar with the general manipulation of the functions.

The elliptic function involves a general function which we shall label as $\Delta(k,\alpha)$:

$$\Delta(k,\alpha) = \sqrt{1 - k^2 \sin^2 \alpha}.$$

$\Delta(k,\alpha)$ is a two-parameter function.

The complete elliptic function of the first kind is denoted by $K(k)$ and is defined by the integral

$$K(k) = \int_0^{\pi/2} \Delta^{-1}(k,\alpha)\, d\alpha = \int_0^{\pi/2} \frac{d\alpha}{\sqrt{1 - k^2 \sin^2 \alpha}}.$$

The complete elliptic function of the second kind $E(k)$ is defined as

$$E(k) = \int_0^{\pi/2} \Delta(k,\alpha)\, d\alpha = \int_0^{\pi/2} \sqrt{1 - k^2 \sin^2 \alpha}\, d\alpha.$$

A third kind function is also available, although we shall not find much use for it in the problems developed in this text. The complete elliptic function of the third kind is

$$\Pi(n,k) = \int_0^{\pi/2} \frac{\Delta^{-1}(k,\alpha)}{(1 + n \sin^2 \alpha)}\, d\alpha = \int_0^{\pi/2} \frac{d\alpha}{(1 + n \sin^2 \alpha)\sqrt{1 - k^2 \sin^2 \alpha}}.$$

The electrostatic potential of a circular ring of charge is given by the first kind of function K(k), and the vector magnetic potential arising from a circular current loop is also a first-kind function. To obtain electrostatic and magnetic fields of these configurations one must take the gradient or the curl of the respective potential functions. These operations involve derivatives of the elliptic functions. The derivatives which must be used are

$$\frac{dE(k)}{dk} = \frac{1}{k}\{E(k) - K(k)\},$$

and

$$\frac{dK(k)}{dk} = \frac{1}{k}\left\{\frac{E(k)}{(1 - k^2)} - K(k)\right\}.$$

The derivative of E with respect to k is readily obtained by differentiation under the integral sign. In the case of $\partial K/\partial k$ the proof of the relation stated above is not obvious. The trick is to differentiate $(1 - x)K(x)$ with respect to x where $x = k^2$. After this is done one can solve for $\partial K/\partial k$, obtaining the result quoted above.

In solving problems, the derivatives are taken with respect to x_1, x_2, and x_3. Because k is a function of the three coordinate variables, the partial derivative with respect to the jth variable of say K(k) is merely

$$\frac{\partial K}{\partial x_j} = \frac{\partial k}{\partial x_j}\frac{dK(k)}{dk}.$$

The values of K(k) and E(k) for a given k are tabulated in most mathematical tables. Thus the reader is referred to these tables for the numerical solutions to problems.

D. Partial Differential Equations

1. SEPARATION OF VARIABLES

Many of the problems which are of interest to us in this text involve the Laplacian of a function. A typical problem might appear as follows:

$$\nabla^2\psi(x_1,x_2,x_3) = f(x_1,x_2,x_3)$$

where f is specified. Solutions to the homogeneous equation

$$\nabla^2\psi(x_1,x_2,x_3) = 0$$

can be obtained by the method of separation of variables. The final separated forms are quite different for the different coordinate systems. Thus it will suit our purpose to indicate briefly the manner in which the variables separate in the three main coordinate systems.

a. CARTESIAN COORDINATES

$$\nabla^2 = \sum_{j=1}^{3} \frac{\partial^2}{\partial x_j^2},$$

and

$$\nabla^2\psi = \sum_{m=1}^{3} \frac{\partial^2\psi}{\partial x_m^2} = 0.$$

A solution of the form $\psi = X_1(x_1)X_2(x_2)X_3(x_3)$ is assumed. Substituting and dividing by ψ we obtain

$$\frac{1}{X_3}\frac{\partial^2 X_3}{\partial x_3^2} = -\frac{1}{X_1}\left\{\frac{\partial^2 X_1}{\partial x_1^2} + \frac{1}{X_2}\frac{\partial^2 X_2}{\partial x_2^2}\right\}.$$

Because the functions $X_j(x_j)$ are independent, this equality requires that both sides of this equation be equal to a constant, say $-n_3^2$. Then

$$X_3 = A_3 e^{in_3x_3} + B_3 e^{-in_3x_3}.$$

The equations in X_1 and X_2 can be separated in the same manner:

$$\frac{1}{X_2}\frac{\partial^2 X_2}{\partial x_2^2} = -\frac{1}{X_1}\frac{\partial^2 X_1}{\partial x_1^2} + n_3^2 = -n_2^2.$$

Finally we obtain solutions of the type

$$X_2 = A_2 e^{in_2x_2} + B_2 e^{-in_2x_2},$$

and

$$X_1 = A_1 e^{n_1x_1} + B_1 e^{-n_1x_1}.$$

No reality restrictions have been placed upon the separation constants; they can be real or complex. However, we have required that

$$n_1^2 = -n_2^2 - n_3^2,$$

which requires some real exponentials.

Thus

$$\psi(x_1,x_2,x_3) = \prod_{j=1}^{3} X_j(x_j)$$

and the parameters A_j, B_j, and n_j are fixed by the boundary conditions of the problem.

The reduced wave equation in Chapter XI is of the form

$$\{\nabla^2 + k^2\}\psi = 0.$$

Again solutions of the form $\psi = X_1X_2X_3$ can be set up. After substituting and performing separation of variables as before,

$$X_j(x_j) = \{A_j e^{ik_jx_j} + B_j e^{-ik_jx_j}\}$$

where

$$k^2 = k_1^2 + k_2^2 + k_3^2.$$

b. CYLINDRICAL COORDINATES. Laplace's equation is

$$\nabla^2\psi = \frac{1}{a}\frac{\partial}{\partial a}\left(a\frac{\partial\psi}{\partial a}\right) + \frac{1}{a^2}\frac{\partial^2\psi}{\partial\phi^2} + \frac{\partial^2\psi}{\partial x_3^2} = 0.$$

For this problem we assume a solution

$$\psi(a,\phi,x_3) = \mathscr{A}(a)\Phi(\phi)X_3(x_3).$$

Dividing by ψ we first separate the x_3 equation,

$$\frac{1}{X_3}\frac{\partial^2 X_3}{\partial x_3^2} = -\left\{\frac{1}{a\mathscr{A}}\frac{\partial}{\partial a}\left(a\frac{\partial\mathscr{A}}{\partial a}\right) + \frac{1}{a^2\Phi}\frac{\partial^2\Phi}{\partial\phi^2}\right\} = +n_3^2;$$

$$X_3 = A_3 e^{n_3x_3} + B_3 e^{-n_3x_3}.$$

Here n_3 can be real or complex. Then

$$\frac{1}{\Phi}\frac{\partial^2\Phi}{\partial\phi^2} = \frac{a}{\mathscr{A}}\frac{d}{da}\left(a\frac{d\mathscr{A}}{da}\right) - n_3{}^2a^2 = -m^2,$$

giving $$\Phi = A_2e^{im\phi}.$$

Because Φ must be single valued, m is required to assume only integral real values (positive or negative). Finally,

$$\frac{1}{a}\frac{d}{da}\left(a\frac{d\mathscr{A}}{da}\right) + \left\{an_3{}^2 - \frac{m^2}{a^2}\right\}\mathscr{A} = 0.$$

Particularly simple solutions are obtained when the problem is not x_3-dependent. In this case

$$\frac{1}{a}\frac{d}{da}\left(a\frac{d\mathscr{A}}{da}\right) - \frac{m^2}{a^2}\mathscr{A} = 0,$$

with solutions $$\mathscr{A} = a^{\pm m}, \quad m \neq 0$$

and $$\mathscr{A} = \log a, \quad m = 0.$$

A simplified reduced wave equation in two variables can be constructed:

$$\frac{1}{a}\frac{\partial}{\partial a}\left(a\frac{\psi}{\partial a}\right) + \frac{1}{a^2}\frac{\partial^2\psi}{\partial\phi^2} + k^2\psi = 0.$$

The Φ solutions are still of the form $e^{im\phi}$, while the radial solutions are cylindrical Bessel and Neumann functions of order m,

$$\mathscr{A}_m(a) = J_m(ka), \text{ and } N_m(ka).$$

A general two-dimensional solution is formed by an expansion of the type

$$\psi(a,\phi) = \sum_{m=\infty}^{\infty}\{A_mJ_m(ka) + B_mN_m(ka)\}\, e^{im\phi}.$$

The A_j and B_k are the expansion coefficients.

C. SPHERICAL COORDINATES. A number of the examples in the text involve the Laplacian in spherical coordinates,

$$\nabla^2 = \frac{1}{r^2}\frac{\partial}{\partial r}\left(r^2\frac{\partial\psi}{\partial r}\right) + \frac{1}{r^2\sin\theta}\frac{\partial}{\partial\theta}\left(\sin\theta\frac{\partial\psi}{\partial\theta}\right) + \frac{1}{r^2\sin^2\theta}\frac{\partial^2\psi}{\partial\phi^2}.$$

The methods by which each separated differential equation is solved in series will be discussed briefly in a later section of this Appendix. The separation of variables, though performed in the same manner as before, must however be carried out with care. A solution is assumed of the form

$$\psi(r,\theta,\phi) = R(r)P(\theta)\Phi(\phi).$$

Divide by ψ, multiply by $r^2 \sin^2 \theta$, and separate the Φ equation:

$$\frac{1}{\Phi}\frac{d^2\Phi}{d\phi^2} = -\left\{\frac{\sin^2\theta}{R}\frac{d}{dr}\left(r^2\frac{dR}{dr}\right) + \frac{\sin\theta}{P}\frac{d}{d\theta}\left(\sin\theta\frac{dP}{d\theta}\right)\right\}$$
$$= -M^2.$$

Then $$\Phi(\phi) = A_3 e^{iM\phi}$$

again with the restriction that M be an integer (positive or negative).

Performing the next separation,

$$P\frac{1}{\sin\theta}\frac{d}{d\theta}\left\{\sin\theta\frac{dP}{d\theta}\right\} - \frac{M^2}{\sin^2\theta} = -\frac{1}{R}\frac{d}{dr}\left\{r^2\frac{dR}{dr}\right\} = -L(L+1).$$

Here it can be demonstrated either by factorization or by solution in series, that the only well-behaved solutions from $\theta = 0$ to $\theta = \pi$ for P exist when the separation constant is $L(L + 1)$, where L is a *positive integer* and $|M| \leqslant L$.

The solutions of

$$\frac{1}{\sin\theta}\frac{d}{d\theta}\left\{\sin\theta\frac{dP}{d\theta}\right\} + L(L+1)P - \frac{M^2}{\sin^2\theta}P = 0$$

are the associated Legendre polynomials $P_L^M(\cos\theta)$.

Some of the lowest order polynomials are listed below. The differential equation defines P_L^M only in terms of M^2; therefore, as one would expect, $P_L^M = P_L^{-M}$.

L	|M|	P_L^M
0	0	1
1	1	$\sin\theta$
1	0	$\cos\theta$
2	2	$\sin^2\theta$
2	1	$\sin\theta\cos\theta$
2	0	$(3\cos^2\theta - 1)$
3	3	$\sin^3\theta$
3	2	$\sin^2\theta\cos\theta$
3	1	$\sin\theta\,(5\cos^2\theta - 1)$
3	0	$\frac{1}{3}\cos\theta\,(5\cos^2\theta - 3)$

If the problem is independent of the azimuthal angle ϕ, in other words if $M = 0$, the resulting solutions $P_L^0(\cos\theta)$ are the Legendre polynomials. The P_L^0's, as they are written, are the solutions of

$$\frac{1}{\sin\theta}\frac{d}{d\theta}\left\{\sin\theta\frac{dP_L^0}{d\theta}\right\} + L(L+1)P_L^0 = 0.$$

These functions are shown in the table in the rows corresponding to $M = 0$.

Finally, the solutions of the radial equation are obtained from

$$\frac{d}{dr}\left\{r^2\frac{dR_L}{dr}\right\} - L(L+1)R_L = 0.$$

There are two independent solutions to this equation,

$$R_L = r^L$$

and

$$R_L = r^{-(L+1)}.$$

The choice between the two depends upon whether the problem excludes the origin or excludes regions in which r is arbitrarily large.

The most general solution is then a linear combination of all of these solutions:

$$\psi(r,\theta,\phi) = \sum_{L=0}^{\infty}\sum_{M=-L}^{M=+L}\{A_{LM}r^L + B_{LM}r^{-(L+1)}\}P_L^M(\cos\theta)e^{iM\phi}.$$

The product $P_L^M(\cos\theta)e^{iM\phi}$ is usually combined and normalized to give the *spherical harmonics* Y_L^M:

$$Y_L^M(\theta,\phi) = \left\{\frac{(2L+1)[(L-|M|)!]}{4\pi[(L+|M|)!]}\right\}^{1/2} P_L^M(\cos\theta)e^{iM\phi}.$$

Some of the spherical harmonics are:

$$Y_0^0 = \frac{1}{2\sqrt{\pi}}$$

$$Y_1^{\pm1} = \frac{1}{2}\sqrt{\frac{3}{2\pi}}\sin\theta\, e^{\pm i\phi}$$

$$Y_1^0 = \frac{1}{2}\sqrt{\frac{3}{\pi}}\cos\theta$$

$$Y_2^{\pm2} = \frac{1}{4}\sqrt{\frac{15}{2\pi}}\sin^2\theta\, e^{\pm i2\phi}$$

$$Y_2^{\pm1} = \frac{1}{2}\sqrt{\frac{15}{2\pi}}\sin\theta\cos\theta\, e^{\pm i\phi}$$

$$Y_2^0 = \frac{1}{4}\sqrt{\frac{5}{\pi}}(3\cos^2\theta - 1)$$

$$Y_3^{\pm 3} = \frac{1}{8}\sqrt{\frac{35}{\pi}}\sin^3\theta\, e^{\pm 3i\phi}$$

$$Y_3^{\pm 2} = \frac{1}{4}\sqrt{\frac{105}{2\pi}}\sin^2\theta\cos\theta\, e^{\pm 2i\phi}$$

$$Y_3^{\pm 1} = \frac{1}{8}\sqrt{\frac{21}{\pi}}\sin\theta\,(5\cos^2\theta - 1)e^{\pm i\phi}$$

$$Y_3^0 = \frac{3}{4}\sqrt{\frac{7}{\pi}}\left(\frac{5}{3}\cos^2\theta - 1\right)\cos\theta.$$

The orthogonality and normalization satisfy the relation

$$\int_0^{\pi}\int_0^{2\pi}(Y_{L'}^{M'})^* Y_L^M \sin\theta\, d\theta\, d\phi = \delta_{LL'}\delta_{MM'}$$

These functions are more convenient for expansions in spherical coordinates:

$$\psi(r,\theta,\phi) = \sum_{L=0}^{\infty}\sum_{M=-L}^{M=+L}\{a_{LM}r^L + b_{LM}r^{-(L+1)}\}Y_L^M(\theta,\phi).$$

An oft used radial solution is obtained for the equation

$$\{\nabla^2 + k^2\}\psi = 0.$$

Separation of variables produces the same solution as before for Φ and P_L^M; however, the radial equation becomes

$$\frac{1}{r^2}\frac{d}{dr}\left\{r^2\frac{dR_L}{dr}\right\} + k^2R_L - \frac{L(L+1)}{r^2}R_L = 0$$

and has solutions which are the spherical Bessel and Neumann functions $j_L(kr)$ and $n_L(kr)$. If we define

$$\rho = kr$$

then our equation becomes

$$\frac{1}{\rho^2}\frac{d}{d\rho}\left\{\rho^2\frac{dR_L}{d\rho}\right\} + \left\{k^2 - \frac{L(L+1)}{\rho^2}\right\}R_L = 0.$$

A few of the spherical radial solutions are shown below

Bessel Functions:	*Neumann Functions:*
$j_0(\rho) = \frac{\sin\rho}{\rho}$	$n_0(\rho) = -\frac{\cos\rho}{\rho}$
$j_1(\rho) = \frac{\sin\rho}{\rho^2} - \frac{\cos\rho}{\rho}$	$n_1(\rho) = -\frac{\cos\rho}{\rho^2} - \frac{\sin\rho}{\rho}$
$j_2(\rho) = \left\{\frac{3}{\rho^3} - \frac{1}{\rho S}\right\}\sin\rho - \frac{3}{\rho_2}\cos\rho$;	$n_2(\rho) = -\left\{\frac{3}{\rho^3} - \frac{1}{\rho}\right\}\cos\rho - \frac{3}{\rho^2}\sin\rho.$

Hankel Functions:

$$h_L^{(1)}(\rho) = j_L(\rho) + in_L(\rho)$$

$$h_L^{(2)}(\rho) = j_L(\rho) - in_L(\rho);$$

$$h_0^{(1)}(\rho) = -\frac{i}{\rho}\, e^{i\rho}$$

$$h_1^{(1)}(\rho) = -\left\{\frac{i}{\rho^2} + \frac{1}{\rho}\right\} e^{i\rho}$$

$$h_2^{(1)}(\rho) = \left\{\frac{i}{\rho} - \frac{3}{\rho^2} - \frac{3i}{\rho^3}\right\} e^{i\rho}.$$

The Neumann functions n_L are not defined at the origin; therefore the use of such a function in a general solution again depends upon whether or not the origin is excluded from the problem.

The spherical Bessel and Neumann functions are related to two other functions, the Hankel functions of the first and second kinds $h_L^{(1)}(\rho)$ and $h_L^{(2)}(\rho)$. These relations are analogous to the relation between the exponential and trigonometric functions.

Expanding in the functions of this set, a function $F(\mathbf{r})$ becomes

$$F(r,\theta,\phi) = \sum_{L=0}^{\infty} \sum_{M=-L}^{M=+L} \{a_{LM} j_L(kr) + b_{LM} n_L(kr)\} Y_L^M(\theta,\phi).$$

The method of evaluation of the expansion coefficients (or Fourier coefficients) is described in Appendix E.

2. THE SOLUTION OF LINEAR DIFFERENTIAL EQUATIONS

This subject encompasses a large amount of material. Here an attempt will be made to discuss the problems involved in a series solution to a given differential equation. Finally, a very qualitative discussion will be presented to indicate why the various angular and radial solutions shown in the previous section have the forms they have.

The differential equations with which we shall be concerned in this text are *linear* and of *second order*. We consider a function u(z) defined by the linear differential equation

$$\frac{d^2u}{dz^2} + p(z)\frac{du}{dz} + q(z)u = 0;$$

certain characteristics of the coefficients p(z) and q(z) serve as a guide in solving the differential equation.

The singularities of p(z) and q(z) determine the behavior of u(z), that is, its convergence properties at a point, etc. Under certain conditions the solutions can be expanded in a power series about a singularity of p(z) and or q(z) to provide two independent solutions $u_1(z)$ and $u_2(z)$ which are everywhere well behaved. In many instances we can find but one well-behaved solution, and in others it may not be possible to find an infinite series which converges at all points in the interval of interest.

The Legendre equation is an example of an equation which has two different singular points in the interval of interest $0 \leqslant \theta \leqslant \pi$, and it is not possible to expand about one singular point and at the same time find the series converging at the other. Such a situation is met by terminating the series after a finite number of terms and forming polynomial solutions. Besides multiple singularities there are other reasons which require that a series be treated in a special fashion. The purpose of this discussion is to present an informal survey of some of the methods of solution. The reader is referred to an excellent treatment of this subject by Morse and Feshbach.*

In some cases in which a singular point of p(z) and q(z) is *regular*, the solutions can be expanded in a power series about the regular point to give a well-behaved solution.

A point c is said to be regular if

$$(z - c)p(z) = P(z) = \text{a function analytic at } c,$$

and/or

$$(z - c)^2 q(z) = Q(z) = \text{a function analytic at } c.$$

If P(z) and Q(z) are analytic, they in turn can be expanded in a Taylor's series about c,

$$P(z) = P_0 + P_1(z - c) + P_2(z - c)^2 + \dots$$

$$Q(z) = Q_0 + Q_1(z - c) + Q_2(z - c)^2 + \dots.$$

The function u(z) is now expanded about c in the form

$$u(z) = (z - c)^\alpha \sum_{n=0}^{\infty} a_n (z - c)^n.$$

* Morse and Feshbach, *Methods of Theoretical Physics*, Vol. I, Chaps. 4 and 5. New York; McGraw-Hill Book Co., Inc., 1953.

Substituting this series into the differential equation written in the form

$$(z - c)^2 \frac{d^2u}{dz^2} + (z - c)\{P_0 + P_1(z - c) + \ldots\} \frac{du}{dz}$$
$$+ \{Q_0 + Q_1(z - c) + \ldots\}u = 0;$$

the coefficients of each term $(z - c)^m$ must be zero. The coefficient of $(z - c)^\alpha$ (the lowest term) is set equal to zero and is called the *indicial equation*,

$$\alpha^2 + (P_0 - 1)\alpha + Q_0 = 0.$$

The roots α_1 and α_2 of this equation determine the two solutions to the differential equation. If the roots differ by an integer or zero there will be at most one independent solution.

Often there is more than one singular point, and in addition the singular points may *not* be regular. In other words, multiplication by $(z - c)^2$ will not remove the singularity. The procedure outlined is still useful. In general one removes all regular singularities and attempts to solve the remainder by a cut-off technique which provides a polynomial solution.

To illustrate the method we shall examine a few of the more familiar differential equations occurring in the study of electromagnetic theory. These are developed from separations of Laplace's equation in various coordinate systems and from the reduced wave equation.

a. LAPLACE'S EQUATION IN CYLINDRICAL COORDINATES. The separation of variables was demonstrated in Section 1 of this appendix. The solutions had the form

$$\psi(a,\phi,z) = \mathscr{A}(a)\Phi(\phi)X_3(x_3).$$

$X_3(x_3)$ and $\Phi(\phi)$ were obtained in terms of exponential functions,

$$X(x_3) = A_3 e^{n_3 x_3} + B_3 e^{-n_3 x_3}$$

and

$$\Phi(\phi) = A_2 \cos m\phi + B_2 \sin m\phi$$

with

$$m = \text{an integer}.$$

When the x_3 dependence was omitted, the defining equation for $\mathscr{A}(a)$ was

$$\frac{1}{a}\frac{d}{da}\left(a\frac{d\mathscr{A}}{da}\right) - \frac{m^2\mathscr{A}}{a^2} = \frac{d^2\mathscr{A}}{da^2} + \frac{1}{a}\frac{d\mathscr{A}}{da} - \frac{m^2\mathscr{A}}{a^2} = 0.$$

This equation has a regular point at a = 0 and an irregular point at a → ∞. At most we can expect but one well-behaved solution at a = 0. A solution is assumed of the form

$$\mathscr{A} = a^{\alpha} \sum_{n=0}^{\infty} a_n a^n.$$

The indicial equation for

$$a^2 \frac{d^2\mathscr{A}}{da^2} + a \frac{d\mathscr{A}}{da} - m^2\mathscr{A} = 0$$

is

$$\alpha^2 + (1 - 1)\alpha - m^2 = 0$$

or

$$\alpha = \pm m.$$

Then

$$\mathscr{A}_m(a) = a^{\pm m}, \quad \text{for } m \neq 0.$$

In other words the solution a^{α} satisfies the differential equation except when m = 0.

When m = 0 a straightforward integration gives

$$\mathscr{A}_0(a) = \log a, \quad (m = 0).$$

b. LAPLACE'S EQUATION IN SPHERICAL COORDINATES. In Section 1 a solution of the form

$$\psi(r,\theta,\phi) = R_L(r)P_L^M(\theta)\Phi_M(\phi)$$

gave

$$\Phi_M(\phi) = e^{iM\phi},$$

where

$$M = \text{an integer}.$$

The function P(θ) is defined by

$$\frac{1}{\sin\theta}\frac{d}{d\theta}\left\{\sin\theta\frac{dP}{d\theta}\right\} + \lambda P - \frac{M^2}{\sin^2\theta}P = 0$$

where λ is a separation constant to be determined.

We let μ = cos θ. The differential equation then becomes

$$(1 - \mu^2)\frac{d^2P}{d\mu^2} - 2\mu\frac{dP}{d\mu} + \left\{\lambda - \frac{M^2}{(1 - \mu^2)}\right\}P = 0.$$

This equation has regular points at μ = +1 and μ = −1. Since the interval of interest is $0 \leqslant \theta \leqslant \pi$ or $-1 \leqslant \mu \leqslant +1$, the two singular points are included. We can at best attempt to remove the singularities and then to expand about an intermediate point (μ = 0) hoping to be able to terminate the expansion.

Regarding the indicial equation for each regular point, we find that a partial removal of the singularity is achieved by the expansion

$$P = (1-\mu)^{\frac{|M|}{2}}(1+\mu)^{\frac{|M|}{2}} \sum_{n=0}^{\infty} g_n \mu^n.$$

The series $\sum_{n=0}^{\infty} g_n \mu^n$ represents an expansion about the intermediate point $\mu = 0$.

Substituting the series for P into the differential equation, we find a recursion relation between g_{n+2} and g_n,

$$g_{n+2} = \left\{\frac{(n+|M|)(n+|M|+1)-\lambda}{(n+1)(n+2)}\right\} g_n.$$

This series diverges at $\mu = \pm 1$ if $n \to \infty$; therefore we cut the series off by setting $\lambda = L(L+1)$ where L is a positive integer and equals the maximum value of $n + |M|$ in a given series.

Polynomial solutions constructed in this manner are known as associated Legendre polynomials, $P_L^M(\theta)$. The polynomials corresponding to $M = 0$ are known as Legendre polynomials. Some of the polynomials are listed in Section 1 of this appendix.

The radial equation defining $R_L(r)$ is, after multiplying by r^2,

$$r^2 \frac{d^2 R_L}{dr^2} + 2r \frac{dR_L}{dr} - L(L+1)R_L = 0,$$

where L must be a positive integer.

The point $r = 0$ is regular but $r \to \infty$ is irregular; we find solutions either well behaved at $r = 0$ (but not at $r \to \infty$) or well behaved at infinity (but not at $r = 0$).

Removal of the regular point at $r = 0$ is achieved by the expansion

$$R_L(r) = r^{\alpha} \sum_{n=0}^{\infty} a_n r^n.$$

Since $P_0 = 2$ and $Q_0 = -L(L+1)$, the indicial equation is

$$\alpha^2 + \alpha - L(L+1) = 0$$

with roots $\alpha = L \quad \text{or} \quad -(L+1),$

differing by an integer.

Again our radial differential equation is satisfied by

$$R(r) = r^L \qquad \text{(not well behaved as } r \to \infty\text{)},$$

or

$$R(r) = r^{-(L+1)} \qquad \text{(not well behaved at } r = 0\text{)}.$$

C. THE REDUCED WAVE EQUATION IN SPHERICAL COORDINATES

$$\{\nabla^2 + k^2\}\psi(r,\theta,\phi) = 0.$$

Previously we have shown by a separation of variables that

$$\psi = R_L(r)Y_L^M(\theta,\phi)$$

where Y_L^M is the spherical harmonic. $R_L(r)$ is defined by

$$\frac{1}{\rho^2}\frac{d^2R_L}{d\rho^2} + \frac{2}{\rho}\frac{dR_L}{d\rho} + \left\{1 - \frac{L(L+1)}{\rho^2}\right\}R_L = 0$$

where $\rho = kr$.

This equation has a regular point at $r = 0$; the roots of the indicial equation differ by an integer $(2L + 1)$, and only one well-behaved solution $j_L(kr)$ can be expected.

The other solution $n_L(kr)$ is singular at $r = 0$. For our purposes the spherical Hankel function of the first kind $h_L^{(1)}(\rho)$ is the more useful. The Hankel functions are linear combinations of $j_L(kr)$ and $n_L(kr)$. As a result both $h_L^{(1)}(\rho)$ and $h_L^{(2)}(\rho)$ are singular at $r = 0$.

For large ρ

$$\frac{d^2R_L}{d\rho^2} + R_L \simeq 0 \quad \text{for all L.}$$

Extracting a factor $e^{i\rho}$, from $R_L(\rho)$

$$R_L(\rho) = e^{i\rho}G_L(\rho).$$

Substituting and solving the indicial equation corresponding to G_L, we find

$$G_L \to (-i\rho)^\alpha \sum_{n=0}^{\infty} g_n^{(L)}(-i\rho)^n,$$

and the indicial equation gives

$$\alpha = L \quad \text{and} \quad -(L+1).$$

Because we are looking for solutions which are well behaved as $r \to \infty$, we take

$$G_L \to (-i\rho)^{-(L+1)} \sum_{n=0}^{\infty} g_n^{(L)}(-i\rho)^n = (-i\rho)^{-(L+1)}\Gamma_L(\rho).$$

Expanding $\Gamma_L(\rho)$ in a power series about $r = 0$ we obtain

$$g_n^{(L)} = \frac{2^n L!\,(2L-n)!}{n!\,(L-n)!\,(2L)!}\,g_0^{(L)}.$$

The series cuts off at

$$n = L.$$

Then the spherical Hankel function of the first kind is

$$h_L^{(1)}(\rho) = e^{i\rho}(-i\rho)^{-(L+1)}g_0^{(L)} \sum_{n=0}^{L} \left\{\frac{2^n L!\,(2L-n)!}{n!\,(L-n)!\,(2L)!}\right\}(-i\rho)^n$$

The second kind of function $h_L^{(2)}(\rho)$ is obtained by extracting $e^{-i\rho}$ from $R_L(\rho)$.

A few of the $h_L^{(1)}(\rho)$ functions (with $g_0^{(L)} = -1$) are

$$h_0^{(1)} = -\frac{i}{\rho}e^{i\rho}$$

$$h_1^{(1)}(\rho) = -\left\{\frac{1}{\rho} + \frac{i}{\rho^2}\right\}e^{i\rho}$$

$$h_2^{(1)}(\rho) = \left\{\frac{i}{\rho} - \frac{3}{\rho^2} - \frac{3i}{\rho^3}\right\}e^{i\rho}.$$

E. Fourier Series

1. INTRODUCTION

In Appendix B the eigenvalue problem in a finite vector space was touched upon. The eigenvectors of a symmetric (or Hermitian) matrix were stated to form a complete set of orthogonal base vectors. Any arbitrary vector in the space defined by these base vectors can be expanded in terms of these bases.

If the $\mathbf{R}_j$'s in the eigenvalue problem form a complete orthonormal set of bases, then a vector $\mathbf{F}$ in the space (of N dimensions) defined by the $\mathbf{R}_j$'s can be written as

$$\mathbf{F} = \sum_{n=1}^{N} a_n \mathbf{R}_n.$$

The separation of partial differential equations subject to particular boundary conditions provides a differential eigenvalue problem. The eigensolutions (or eigenfunctions) are analogous to the eigenvectors of the finite vector problem. The main difference lies in the fact that the differential equation may produce an infinite number of eigenfunctions or bases.

Regard the one-dimensional vibrating string. The problem pertains to a string of length L and mass per unit length μ stretched to a tension T between rigid supports at $x = 0$ and $x = L$. The wave equation for the string is

$$\frac{\partial^2 y(x,t)}{\partial x^2} - \frac{\mu}{T}\frac{\partial^2 y(x,t)}{\partial t^2} = 0$$

where $y(x,t)$ is the displacement of the string at any time t at a point x $(0 \leqslant x \leqslant L)$.

Separation of variables gives

$$y(x,t) = X(x)T(t)$$

where $$T(t) = e^{\pm ikvt}, \quad v = \sqrt{T/\mu},$$

and
$$\frac{d^2X_k}{dx^2} + k^2X_k = 0.$$

The boundary conditions limit k^2 to *special values* or *eigenvalues.* The solutions X_k corresponding to a given k represent an eigenfunction or base function for the problem.

The boundary conditions on the string are that

$$X(0) = X(L) = 0;$$

in other words, the string is fixed at the ends. These boundary conditions restrict the k's to the values

$$k_n = n\pi/L,$$

and
$$X_k(x) = A_k \sin k_n x.$$

The most general solution for the problem is a linear combination of *all* of the acceptable solutions $X_k(x)T_k(t)$,

$$y(x,t) = \sum_{\text{all } k_n} a_n \sin(k_n x)\cos(k_n vt - \phi_n),$$

or
$$y(x,t) = \sum_{n=0}^{\infty} a_n \sin\left(\frac{n\pi}{L}x\right)\cos\left(\frac{n\pi v}{L}t - \phi_n\right).$$

This is an expansion of the general function y(x,t) in terms of the unnormalized bases X_kT_k. The constants a_n and ϕ_n are established from the initial conditions of the problem. Actually the X_k's can be normalized by using as a base

$$\frac{1}{\sqrt{L}}\sin\left(\frac{n\pi}{L}x\right).$$

The similarity between the Fourier series above and the expansion in a finite vector space seems apparent.

2. HERMITIAN OPERATORS AND THEIR EIGENFUNCTIONS

The linear differential equations which are of importance in our studies consist of a linear operator $\mathcal{O}$ in an equation of the type

$$\{\mathcal{O}(x) - \lambda w(x)\}u_\lambda(x) = 0.$$

Here λ is the eigenvalue and w(x) is the weight function. $w(x) = 1$ for cartesian spaces and $w(x) = x^2$ for the radial equation in a spherical coordinate system. $\mathcal{O}$ is an operator of the form

$$\mathcal{O}(x) = \frac{d}{dx}\left\{p(x)\frac{d}{dx}\right\} + q(x).$$

The function $u_\lambda(x)$ from the previous equation is an eigenfunction of the operator $\mathcal{O}$, and it is associated with an eigenvalue which is determined from the boundary conditions of the problem.

As an example, owing to the requirement that the associated Legendre polynomials be finite at $\theta = 0$ and π, the separation constant of the differential equation had to be $L(L + 1)$ where L is a positive integer.

Assume that the functions $u_\lambda(x)$ of the equation

$$\{\mathcal{O}(x) - \lambda w(x)\}u_\lambda(x) = 0$$

are defined in the closed interval $a \leqslant x \leqslant b$. The most general boundary condition is that

$$\left[p(x)u_\lambda \frac{du_\lambda^*}{dx}\right]_{x=a} = \left[p(x)u_\lambda \frac{du_\lambda^*}{dx}\right]_{x=b} = 0.$$

In most problems a milder boundary condition can be employed. If u_λ and $u_{\lambda'}$ are two different solutions to the differential equation, the most convenient boundary condition is that

$$\left[pu_\lambda \frac{du_{\lambda'}}{dx}\right]_a = \left[pu_\lambda \frac{du_{\lambda'}}{dx}\right]_b = 0$$

$$= \left[pu_{\lambda'}^* \frac{du_\lambda}{dx}\right]_a = \left[pu_{\lambda'}^* \frac{du_\lambda}{dx}\right]_b = 0.$$

With these boundary conditions we can show that $\mathcal{O}$ is Hermitian, which means that

$$\int_a^b u_{\lambda'}^* \mathcal{O} u_\lambda \, dx = \int_a^b \{\mathcal{O} u_{\lambda'}^*\} u_\lambda \, dx.$$

This proof is quite straightforward. One substitutes the form of $\mathcal{O}$ and integrates by parts. The terms evaluated at a and b outside the final integral vanish because of our boundary condition. The result is the equality shown above.

Because $\mathcal{O}$ is Hermitian, the eigenvalues λ are real and the eigenfunctions u_λ are orthogonal. Consider

$$\{\mathcal{O} - \lambda w(x)\}u_\lambda(x) = 0$$

and $$\{\mathcal{O} - \lambda'^* w(x)\}u_{\lambda'}^*(x) = 0.$$

Multiply the first equation from the left by $u_{\lambda'}^*$ and the second from the right by u_λ. Now integrate both over x from a to b. When the equations are subtracted we find that

$$\int_a^b u_{\lambda'}^* \mathcal{O} u_\lambda \, dx - \int_a^b \{\mathcal{O} u_{\lambda'}^*\} u_\lambda \, dx - (\lambda - \lambda'^*) \int_a^b u_{\lambda'}^* u_\lambda w(x) \, dx = 0.$$

The first two integrals cancel, giving

$$(\lambda - \lambda'^*) \int_a^b u_{\lambda'}^* u_\lambda w(x) \, dx = 0.$$

If $\lambda \neq \lambda'$ then $\int_a^b u_{\lambda'}^* u_\lambda w \, dx = 0$, indicating that $u_{\lambda'}$ and u_λ are orthogonal. If $\lambda = \lambda'$ then the integral should not be zero, showing that

$$\lambda = \lambda^*,$$

or that the eigenvalues are real.

The reader should be aware from the previous formalism that

$$\int_a^b u_{\lambda'}^* u_\lambda w(x) \, dx$$

is analogous to the inner product of the two base vectors

$$\mathbf{R}_j^\dagger \cdot \mathbf{R}_k.$$

In fact we define the inner product of a function f(x) and a function g(x) in the interval as

$$\int_a^b f^*(x) g(x) w(x) \, dx = (f(x), g(x)).$$

3. THE FOURIER SERIES

The more familiar sets of functions which we use are complete orthogonal sets. The idea of completeness is much the same for functions as it is for the base vectors of a finite vector space, although the completeness is much more difficult to prove for an infinite series.

A system of orthogonal functions is said to be complete if an arbitrary function f(x) defined in the interval of the set, with a finite number of discontinuities and quadratically integrable, i.e.,

$$\int_a^b |f|^2 w(x) \, dx = \text{finite value},$$

can be approximated in the *mean* with the boundary conditions of the set.

Let $u_j(x)$ be the eigenfunction of the jth eigenvalue λ_j; then

$$\{\mathcal{O} - \lambda_j w(x)\}u_j(x) = 0.$$

We approximate f(x) with a sum over N of the functions $u_j(x)$:

$$f_N(x) \simeq \sum_{n=1}^{N} a_n u_n(x).$$

The function f(x) is approximated in the mean if

$$\lim_{N\to\infty} \int_a^b \left\{f^*(x) - \sum_{n=1}^{N} a_n^* u_n^*(x)\right\}\left\{f(x) - \sum_{m=1}^{N} a_m u_m(x)\right\} w(x)\,dx \equiv 0.$$

The proof that a given set $u_n(x)$ is complete will not be given. However, we can indicate the conditions which must be satisfied if f(x) is to be approximated as shown. Define the integral above before the limit is taken, as $\mathcal{J}_N$. Then

$$\mathcal{J}_N = \int_a^b |f|^2\, w\,dx - \sum_{n=1}^{N} a_n^* \int_a^b u_n^* f w\,dx - \sum_{m=1}^{N} a_m \int_a^b f^* u_m w\,dx$$
$$+ \sum_{n=1}^{N}\sum_{m=1}^{N} a_n^* a_m \int_a^b u_n^* u_m w\,dx.$$

We minimize $\mathcal{J}_N$ with respect to variations in the Fourier coefficients a_n; i.e., set

$$\frac{\partial \mathcal{J}_N}{\partial a_n^*} = 0 \quad \text{and} \quad \frac{\partial \mathcal{J}_N}{\partial a_m} = 0.$$

Carrying out the differentiations, we find that these minimum conditions require that

$$a_k = \int_a^b u_k^*(x) f(x) w(x)\,dx,$$

and

$$a_j^* = \int_a^b f^* u_j w\,dx.$$

Substituting these values into $\mathcal{J}_N$,

$$\mathcal{J}_N = \int_a^b f^* f w\,dx - \sum_{n=1}^{N}\sum_{m=1}^{N} a_n^* a_m \int_a^b u_n^* u_m w\,dx.$$

The u_j's are assumed to be normalized; thus we can write

$$\int_a^b u_n^* u_m w\,dx = \delta_{nm};$$

and

$$\mathcal{J}_N = \int_a^b |f(x)|^2\, w(x)\,dx - \sum_{n=1}^{N} |a_n|^2.$$

If the set is complete, $\mathscr{I}_N \to 0$ in the limit as $N \to \infty$, and

$$\int_a^b |f(x)|^2 w(x)\, dx = \sum_{n=1}^{\infty} |a_n|^2.$$

The infinite series

$$f(x) = \sum_{\text{all } n} a_n u_n(x)$$

is called a Fourier series, with the projection of f(x) on $u_n(x)$ (the Fourier component) given by

$$a_n = \int_a^b u_n^*(x) f(x) w(x)\, dx.$$

The standard Fourier series in one dimension, periodic in the interval $-L \leqslant x \leqslant L$, is

$$f(x) = \sum_{n=-\infty}^{\infty} a_n e^{i(n\pi/Lx)}$$

where a_n is a complex number. This is also written as

$$f(x) = \frac{A_0}{2} + \sum_{n=1}^{\infty} \left\{ A_n \cos\left(\frac{n\pi x}{L}\right) + B_n \sin\left(\frac{n\pi x}{L}\right) \right\}.$$

The boundary condition here is less restrictive than in the special case of the vibrating string.

4. THE FOURIER INTEGRAL OR TRANSFORM

The Fourier series expansion is applicable for any function f(x) which is periodic in an interval $-L$ to L. The Fourier integral may be considered as the limit of the Fourier series as the period tends to infinity.

The Fourier series periodic in $-L$ to L is written

$$f(x) = \sum_{n=-\infty}^{\infty} a_n e^{i(n\pi/L)x}$$

with

$$a_n = \frac{1}{2L} \int_{-L}^{L} e^{-i(n\pi/L)x} f(x)\, dx.$$

Then

$$a_n(k) = \frac{1}{2L} \int_{-L}^{L} e^{-ik_n x} f(x)\, dx.$$

In the limit as L becomes arbitrarily large,

$$k_n = n\pi/L,$$

$$\Delta k = k_{n+1} - k_n = \pi/L,$$

and

$$\lim_{L\to\infty} \pi/L = dk.$$

Thus to define f(x) in the interval $-L \leqslant x \leqslant L$ as $L \to \infty$,

$$f(x) = \sum_{n=-\infty}^{\infty} e^{ik_n x} \lim_{L\to\infty} \frac{\Delta k_n}{2\pi} \int_{-L}^{L} e^{-ik_n x'} f(x')\, dx'.$$

As $L \to \infty$ the sum over n goes to an integral. The limits of this integral over k are set at $-K$ and $+K$ with $K \to \infty$. We now write

$$f(x) = \lim_{K\to\infty} \lim_{L\to\infty} \frac{1}{2\pi} \int_{-K}^{K} e^{ikx}\, dk \int_{-L}^{L} e^{-ikx'} f(x')\, dx'.$$

Define

$$a(k) = \frac{1}{\sqrt{2\pi}} \int_{-\infty}^{\infty} e^{-ikx'} f(x')\, dx'$$

and then allow $K \to \infty$; this gives

$$f(x) = \frac{1}{\sqrt{2\pi}} \int_{-\infty}^{\infty} a(k)\, e^{ikx}\, dk.$$

The function a(k) is the Fourier transform of f(x).

In three dimensions (by the same arguments),

$$f(\mathbf{r}) = \frac{1}{(2\pi)^{3/2}} \underset{\text{All k space}}{\iiint} a(\mathbf{k}) e^{i\mathbf{k}\cdot\mathbf{r}}\, d\mathbf{k},$$

and

$$a(\mathbf{k}) = \frac{1}{(2\pi)^{3/2}} \underset{\text{All space}}{\iiint} e^{-i\mathbf{k}\cdot\mathbf{r}} f(\mathbf{r})\, d\mathbf{r}$$

where $d\mathbf{k} = k^2\, dk \sin\theta\, d\theta\, d\phi$, or $d\mathbf{k} = dk_1\, dk_2\, dk_3$;

$d\mathbf{r} = r^2\, dr \sin\theta\, d\theta\, d\phi$, or $d\mathbf{r} = dx_1\, dx_2\, dx_3$.

5. THE DIRAC DELTA FUNCTION

The Dirac delta function $\delta(x)$, introduced in Chapter II, has the property that

$$\int_{-\infty}^{\infty} \delta(x) f(x)\, dx = f(0).$$

In a bounded interval $a \leqslant x \leqslant b$ we were able to normalize the complete set of functions u_j according to

$$\int_a^b u_m^*(x) u_n(x)\, dx = \delta_{mn}.$$

(The weight function w(x) has been deleted for convenience.)

The functions f(x) defined over the infinite interval must be quadratically integrable,

$$\int_{-\infty}^{\infty} f^*f\,dx = \text{a finite value.}$$

Using a general set $u_m(x)$ (such as $(1/\sqrt{2\pi})e^{ikx}$) we can write

$$f(x) = \int_{-\infty}^{\infty} a(k)u_k(x)\,dk,$$

or
$$a(k) = \int_{-\infty}^{\infty} u_k^*(x')f(x')\,dx'.$$

Substituting for a(k) in the first expression and changing the order of integration*

$$f(x) = \int_{-\infty}^{\infty} \left\{\int_{-\infty}^{\infty} u_k(x)u_k^*(x')\,dk\right\} f(x')\,dx',$$

it is apparent that

$$\int_{-\infty}^{\infty} u_k(x)u_k{}^*(x')\,dk = \delta(x' - x).$$

In terms of the trigonometric series,†

$$\delta(x' - x) = \frac{1}{2\pi}\int_{-\infty}^{\infty} e^{ik(x-x')}dk.$$

This form is made up of a continuum of k's (i.e., for the infinite interval). By similar methods we can show that for the complete orthonormal set $v_m(x)$ defined in the finite interval $a \leqslant x \leqslant b$,

$$\delta(x' - x) = \sum_{\text{all } m} v_m(x)v_m^*(x').$$

The proof of this last statement is obtained by expanding f(x) in a Fourier series and substituting the integral form of the Fourier coefficients into the series.

The preceding sections do not pretend to thoroughness, but are intended to provide the reader with some feeling for the mathematics discussed.

* Because we assume that the delta function is always used under an integral sign, these changes of order are permissible.

† The Dirac delta function is even, thus $\delta(x' - x) = \delta(x - x')$. Care must be exercised with the odd derivatives of the delta function. They change sign upon reflection.

6. THE GREEN'S FUNCTION

The differential operators used in electromagnetic theory have equivalent integral representations which prove to be more useful when an inhomogeneous equation is encountered.

Two of the equations most important to us are

$$\nabla^2 V = -\rho(\mathbf{r})/\epsilon_0 \quad \text{(Poisson's equation)},$$

and

$$\Box\psi = -f(\mathbf{r},t) \quad \text{(the inhomogeneous wave equation)}.$$

Consider the simplest equation first,

$$\nabla^2 V = -\rho(\mathbf{r})/\epsilon_0.$$

The operator ∇^2 has a source function or Green's function $G(\mathbf{r},\mathbf{r}')$ associated with it. In Chapter II we demonstrated that G was associated with the spatial dependence of the potential of a point charge.

The source function $G(\mathbf{r},\mathbf{r}')$ is defined from an inhomogeneous equation in which the inhomogeneity is a delta function;

$$\nabla_r^2 G(\mathbf{r},\mathbf{r}') = -\delta(\mathbf{r}' - \mathbf{r}).$$

We assert that under these conditions a solution of Poisson's equation is

$$V(\mathbf{r}) = \frac{1}{\epsilon_0} \iiint_{\text{All Space}} G(\mathbf{r},\mathbf{r}')\rho(\mathbf{r}')\, d\tau'.$$

To demonstrate this result, take ∇_r^2 of both sides of this equation. Then

$$\nabla^2 V(r) = \frac{1}{\epsilon_0} \iiint \nabla_r^2 G(\mathbf{r},\mathbf{r}')\rho(\mathbf{r}')\, d\tau' = \frac{-1}{\epsilon_0} \iiint \delta(\mathbf{r}' - \mathbf{r})\rho(\mathbf{r}')\, d\tau'$$

$$= -\frac{\rho(\mathbf{r})}{\epsilon_0}.$$

There are several methods by which $G(\mathbf{r},\mathbf{r}')$ can be developed in a given coordinate representation. One of the simplest is to utilize the Fourier transform of $G(\mathbf{r},\mathbf{r}')$ and of $\delta(\mathbf{r}' - \mathbf{r})$. Write $\Delta(\mathbf{k})$ as the transform of $G(\mathbf{r},\mathbf{r}')$, then

$$G(\mathbf{r},\mathbf{r}') = \frac{1}{(2\pi)^{3/2}} \iiint_{\text{All k Space}} \Delta(\mathbf{k}) e^{i\mathbf{k}\cdot(\mathbf{r}-\mathbf{r}')}\, d\mathbf{k}$$

and

$$\nabla_r^2 G = \frac{1}{(2\pi)^{3/2}} \iiint \Delta(\mathbf{k})(-k^2) e^{i\mathbf{k}\cdot(\mathbf{r}-\mathbf{r}')}\, d\mathbf{k}.$$

On the right side of the defining equation for G we note that

$$\delta(\mathbf{r}' - \mathbf{r}) = \frac{1}{(2\pi)^3} \iiint\limits_{\text{All k Space}} \{1\} e^{i\mathbf{k}\cdot(\mathbf{r}-\mathbf{r}')} d\mathbf{k}.$$

Then

$$\nabla^2 G + \delta(\mathbf{r}' - \mathbf{r}) = \frac{1}{(2\pi)^{3/2}} \iiint \left\{-k^2 \Delta(\mathbf{k}) + \frac{1}{(2\pi)^{3/2}}\right\} e^{i\mathbf{k}\cdot(\mathbf{r}-\mathbf{r}')} d\mathbf{k} = 0;$$

or

$$\Delta(k) = \frac{1}{(2\pi)^{3/2} k^2}.$$

We can now find the $G(\mathbf{r},\mathbf{r}')$ for the operator ∇_r^2.

$$G(\mathbf{r},\mathbf{r}') = \lim_{\eta\to 0} \frac{1}{(2\pi)^3} \int_0^\infty \int_0^\pi \frac{e^{ik|\mathbf{r}-\mathbf{r}'|\cos\theta}\, 2\pi \sin\theta\, d\theta\, k^2\, dk}{(k + i\eta)(k - i\eta)}.$$

The small imaginary component (η is a positive number) is inserted in the definition of k to place the singularities of the integrand off the real axis. Perform the integration over θ. The integral of $-\int_0^\infty e^{-ik|\mathbf{r}-\mathbf{r}'|}$ is equivalent to the integral $+\int_{-\infty}^0 e^{ik|\mathbf{r}-\mathbf{r}'|}$, so

$$G(\mathbf{r},\mathbf{r}') = \lim_{\eta\to 0} \frac{1}{(2\pi)^2} \int_{-\infty}^\infty \frac{e^{ik|\mathbf{r}-\mathbf{r}'|}}{i|\mathbf{r}-\mathbf{r}'|\,(k + i\eta)(k - i\eta)}\, k\, dk.$$

To obtain convergence we integrate in the upper half plane. This integral can be obtained by using the *theorem of residues*, giving

$$G(\mathbf{r},\mathbf{r}') = \lim_{\eta\to 0(+)} \left\{\frac{e^{-\eta|\mathbf{r}-\mathbf{r}'|}}{4\pi\,|\mathbf{r}-\mathbf{r}'|}\right\} = \frac{1}{4\pi\,|\mathbf{r}-\mathbf{r}'|}.$$

Thus

$$V(\mathbf{r}) = \frac{1}{4\pi\epsilon_0} \iiint\limits_{\text{All Space}} \frac{\rho(\mathbf{r}')\, d\tau'}{|\mathbf{r}-\mathbf{r}'|}.$$

As expected, $G(\mathbf{r},\mathbf{r}')$ represents the spatial dependence of the potential of a point source. The defining equation for the electrostatic $G(\mathbf{r},\mathbf{r}')$ is Poisson's equation for a point distribution.

The source function for the d'Alembertian $\Box$ is obtained in much the same manner. Care must be exercised because the differential equation is hyperbolic. The equation to be solved is

$$\left\{\nabla^2 - \frac{1}{c^2}\frac{\partial^2}{\partial t^2}\right\} \psi(\mathbf{r},t) = \Box\psi(\mathbf{r},t) = -f(\mathbf{r},t).$$

With a corresponding source function $G(\mathbf{r},\mathbf{r}';t,t')$,

$$\Box_{r,t} G(\mathbf{r},\mathbf{r}';t,t') = -\delta(\mathbf{r}' - \mathbf{r})\,\delta(t' - t);$$

then $$\psi(\mathbf{r},t) = \iiint_{\text{All space}} d\tau' \int_{-\infty}^{\infty} dt'\, G(\mathbf{r},\mathbf{r}';t,t')f(\mathbf{r}',t').$$

The four-dimensional Fourier transform is used, where ω is the variable conjugate to t:

$$G(\mathbf{r},\mathbf{r}';t,t') = \frac{1}{(2\pi)^2}\iiint_{\tau}\int_{-\infty}^{\infty} \Delta(\mathbf{k},\omega)e^{i\mathbf{k}[\cdot(\mathbf{r}-\mathbf{r}')-\omega(t-t')]}\,d\mathbf{k}\,d\omega,$$

$$\delta(\mathbf{r}'-\mathbf{r})\,\delta(t'-t) = \frac{1}{(2\pi)^4}\iiint_{\tau}\int_{-\infty}^{\infty}\{1\}e^{i[\mathbf{k}\cdot(\mathbf{r}-\mathbf{r}')-\omega(t-t')]}\,d\mathbf{k}\,d\omega.$$

Applying $\square_{r,t}$ to G we obtain

$$G+\delta = \frac{1}{(2\pi)^2}\iiint_{\tau}\int_{-\infty}^{\infty}\left[\left\{-k^2+\frac{\omega^2}{c^2}\right\}\Delta+\frac{1}{(2\pi)^2}\right]e^{i[\mathbf{k}\cdot(\mathbf{r}-\mathbf{r}')-\omega(t-t')]}\,d\mathbf{k}\,d\omega$$
$$= 0.$$

Therefore

$$\Delta(\mathbf{k},\omega) = \frac{1}{(2\pi)^2\left(|k|+\dfrac{\omega}{c}\right)\left(|k|-\dfrac{\omega}{c}\right)}.$$

The Green's function is then

$$G(\mathbf{r},\mathbf{r}';t,t') = \frac{1}{(2\pi)^4}\iiint_{\tau}\int_{-\infty}^{\infty}\frac{e^{i(\mathbf{k}\cdot\boldsymbol{\xi}-\omega\gamma)}}{\left(|k|+\dfrac{\omega}{c}\right)\left(|k|-\dfrac{\omega}{c}\right)}\,d\mathbf{k}\,d\omega$$

where $$\boldsymbol{\xi} = \mathbf{r}-\mathbf{r}'$$
and $$\gamma = t-t'.$$

The integral over ω can be performed first by writing $k = \lim_{\eta\to 0}(k - i\eta)$ and closing the contour in the lower half plane. This choice eliminates the advanced potential terms which represent nonphysical situations.

$$\lim_{\eta\to 0}\int_{-\infty}^{\infty}\frac{e^{-i\omega\gamma}}{\left(k-i\eta+\dfrac{\omega}{c}\right)\left(k-i\eta-\dfrac{\omega}{c}\right)}\,d\omega = \lim_{\eta\to 0}\left\{\frac{2\pi i\, e^{-i(|k|-i\eta)\gamma c}}{2(k-i\eta)}\right\}$$
$$= \pi i\,\frac{e^{-i\gamma kc}}{|k|},\quad (0 \leqslant \gamma).$$

$\gamma = (t - t')$ and must be positive; otherwise the real part of the exponential in the lower half plane will cause the integral to diverge on the contour. $\gamma = (t - t')$ is specified positive by combining the result of the integration over ω with a step function, $\Gamma(\gamma)$, where

$$\Gamma(\gamma) = 1, \quad 0 < \gamma$$
$$= 0, \quad \gamma < 0.$$

The source function is now

$$G = \frac{\pi i \Gamma(\gamma)}{(2\pi)^4} \int_0^\infty \int_0^\pi \frac{e^{i|\mathbf{k}||\boldsymbol{\xi}| \cos\theta} \, e^{-ikc\gamma}}{|\mathbf{k}|} 2\pi \sin\theta \, d\theta \, k^2 \, dk.$$

Integrating over θ we find that

$$G(\mathbf{r},\mathbf{r}';t,t') = \frac{\Gamma(\gamma)}{2(2\pi)^2} \int_0^\infty \frac{e^{ikc\left(\frac{\xi}{c} - \gamma\right)}}{|\boldsymbol{\xi}|} dk.$$

The integral portion contains the one-dimensional delta function,

$$\delta\left(\gamma - \frac{\xi}{c}\right) = \frac{1}{2\pi} \int_{-\infty}^\infty e^{-ikc\left(\gamma - \frac{\xi}{c}\right)} dk.$$

Therefore substituting into the integral for G we find that

$$G(\mathbf{r},\mathbf{r}';t,t') = \frac{\Gamma(t - t')}{4\pi |\mathbf{r} - \mathbf{r}'|} \delta\left[t' - \left(t - \frac{|\mathbf{r} - \mathbf{r}'|}{c}\right)\right].$$

This is the form used in Chapter XI to obtain the retarded potentials and the Lienard-Wiechert potentials.

F. Conformal Transformations

1. INTRODUCTION

Many two-dimensional electrostatic problems can be worked out in terms of a complex representation. The particular topics of conjugate functions and conformal mapping will be sketched in this appendix.

Initially, it may be found useful to review some important properties of complex quantities. A point in the complex z plane is designated by the variables x and y, where

$$z = x + iy$$

and $$i = \sqrt{-1}.$$

In polar coordinates,

$$z = a \cos \phi + ia \sin \phi = ae^{i\phi},$$

with $$a = (x^2 + y^2)^{1/2}$$

and $$\phi = \tan^{-1}(y/x).$$

The magnitude of z is the real positive radius a, known as the modulus of z,

$$|z| = a,$$

while the polar angle ϕ is called the argument of z,

$$\arg z = \phi.$$

The complex conjugate of z is written z*, and

$$z^* = x - iy = ae^{-i\phi}.$$

Thus $$|z|^2 = z^*z = x^2 + y^2 = a^2.$$

A function of z, say w = f(z), is also a complex function. If we let the real and imaginary parts of w be u and v respectively, then u and v are also functions of x and y:

$$w = f(z) = u(x,y) + iv(x,y).$$

2. ANALYTIC FUNCTIONS

A single-valued function of z, f(z), which is analytic at a point z_0 has the property that its derivative with respect to z is unique at z_0. This uniqueness implies that $(df/dz)_{z_0}$ is the same for *all paths* approaching the point z_0 along a straight line segment Δz, where

$$\Delta z = (z - z_0) = \Delta x + i\Delta y = (x - x_0) + i(y - y_0).$$

Then the condition of uniqueness implies that $\Delta f/\Delta z$ will be independent of $\Delta y/\Delta x$. Let

$$w = f(z) = u(x,y) + iv(x,y);$$

then
$$\frac{\Delta w}{\Delta z} = \frac{\Delta f}{\Delta z} = \frac{\left\{\frac{\partial u}{\partial x} + i\frac{\partial v}{\partial x}\right\}\Delta x + \left\{\frac{\partial u}{\partial y} + i\frac{\partial v}{\partial y}\right\}\Delta y}{\Delta x + i\,\Delta y}.$$

Rearranging,

$$\frac{\Delta f}{\Delta z} = \frac{\left\{\frac{\partial u}{\partial x} + i\frac{\partial v}{\partial x}\right\}}{\left\{1 + i\frac{\Delta y}{\Delta x}\right\}}\left\{1 + i\frac{\Delta y}{\Delta x}\left[\frac{\frac{\partial v}{\partial y} - i\frac{\partial u}{\partial y}}{\frac{\partial u}{\partial x} + i\frac{\partial v}{\partial x}}\right]\right\}.$$

If $\Delta f/\Delta z$ is not a function of $\Delta y/\Delta x$ in the vicinity of z_0, the derivative will be independent of the direction of the path taken to the point z_0. Under the special circumstance that

$$\frac{\partial u}{\partial x} + i\frac{\partial v}{\partial x} = \frac{\partial v}{\partial y} - i\frac{\partial u}{\partial y},$$

$\Delta f/\Delta z$ is *not* a function of $\Delta y/\Delta x$.

Equating real and imaginary terms we obtain two equations which must be satisfied if f(z) is to be analytic. These are

$$\frac{\partial u}{\partial x} = \frac{\partial v}{\partial y} \quad \text{and} \quad \frac{\partial u}{\partial y} = -\frac{\partial v}{\partial x}.$$

These two equations are known as the Cauchy-Riemann conditions, and they provide a necessary condition that f(z) be analytic at z_0. A sufficient condition for analyticity at z_0 is that the derivatives of u and v be continuous at z_0.

3. CONJUGATE FUNCTIONS

In electrostatics we are concerned with functions which are analytic everywhere except at certain specified points. These points represent the sources and sinks of the fields. The potential surface of a *line of charge* in three dimensions can be represented as a potential curve in the two-dimensional complex plane.

One can demonstrate that the two conjugate sets of functions u and v simultaneously satisfy the two-dimensional Laplace equation and form two families of mutually orthogonal curves. As a result, one set of functions can be employed as a family of equipotentials and the other to represent the corresponding field lines for the potentials.

Differentiating the Cauchy-Riemann conditions with respect to x and y, we find

$$\frac{\partial^2 u}{\partial x^2} + \frac{\partial^2 u}{\partial y^2} = 0,$$

$$\frac{\partial^2 v}{\partial x^2} + \frac{\partial^2 v}{\partial y^2} = 0.$$

In polar coordinates,

$$\left\{\frac{1}{a}\frac{\partial}{\partial a}\left(a\frac{\partial}{\partial a}\right) + \frac{1}{a^2}\frac{\partial^2}{\partial \phi^2}\right\}\begin{bmatrix} u \\ v \end{bmatrix} = 0.$$

To show that u(x,y) and v(x,y) curves are mutually orthogonal, we use vector notation and demonstrate that the inner product of the gradient of u by the gradient of v is zero. Because the gradients are perpendicular to the curves, the curves must be orthogonal at every point.

$$[\text{grad } u] \cdot [\text{grad } v] = \frac{\partial u}{\partial x}\frac{\partial v}{\partial x} + \frac{\partial u}{\partial y}\frac{\partial v}{\partial y}$$

$$= -\frac{\partial u}{\partial x}\frac{\partial u}{\partial y} + \frac{\partial u}{\partial y}\frac{\partial u}{\partial x} = 0.$$

4. CONFORMAL TRANSFORMATIONS

The relation $w = f(z)$ can be viewed as a transformation of the two-dimensional z plane to the two-dimensional w plane. For every point (x,y) in the z plane there is at least one corresponding point (u,v) in the w plane.

The transformations represented by analytic functions have the property that the angles between corresponding line segments in the two planes are invariant under the transformation. A line segment $(z_1 - z_0)$ extending from the point z_0 can be designated as

$$\Delta z_1 = |z_1 - z_0|\, e^{i\gamma_1} = |\Delta z_1|\, e^{i\gamma_1}.$$

A second segment extending from z_0 is

$$\Delta z_2 = |z_2 - z_0|\, e^{i\gamma_2} = |\Delta z_2|\, e^{i\gamma_2}.$$

The angle between Δz_1 and Δz_2 is $(\gamma_2 - \gamma_1)$.

Transform Δz_j to the w plane; then

$$\Delta f_j = \Delta z_j \left(\frac{df}{dz}\right)_{z_0}.$$

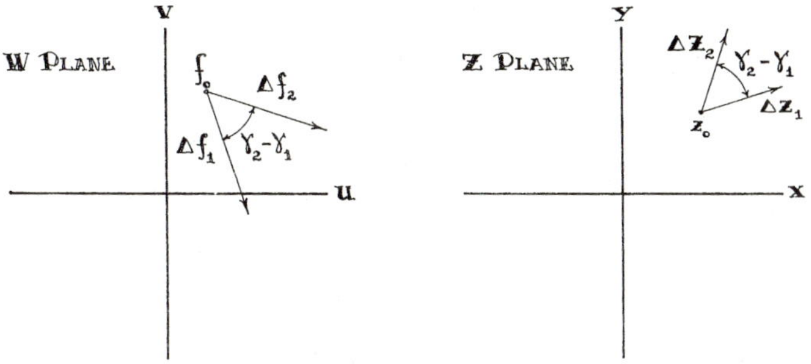

If f(z) is analytic, then

$$\left(\frac{df}{dz}\right)_{z_0} = \left|\frac{df}{dz}\right|_{z_0} e^{i\delta}$$

where the derivative is independent of the direction of approach to z_0 and therefore independent of γ_1 and γ_2. Thus

$$\Delta f_j = |\Delta z_j| \left|\frac{df}{dz}\right|_{z_0} e^{i(\gamma_j + \delta)},$$

and the angle between Δf_2 and Δf_1 is $(\gamma_2 - \gamma_1)$.

The transformation scales the line segments by the factor $|df/dz|_{z_0}$. The presence of a scale factor indicates that to be conformal at a point the transformation must have the property that the derivative df/dz is nonzero at that point. If df/dz vanishes at z_0, the inverse transformation from f to z(f) at z_0 will be singular.

The conditions for a conformal transformation are therefore that f(z) be analytic in a sufficiently small region about z_0 and that z(f) be analytic. The inverse analyticity implies that df/dz is not zero at z_0.

When $df/dz \to 0$, a large region in the z plane can be compressed into a very small region in the w plane.

Successful use of conformal mapping depends a great deal upon insight and experience. The game has a relatively simple aim: one attempts to transform a complicated geometrical figure in the z plane to a simpler geometrical figure in the w plane. One transformation may not suffice. A series of conformal mappings may be necessary to achieve simplicity. The ideal is to resolve a z-plane geometrical figure into a series of coordinate lines in the w plane.

As an example, regard a line of charge of λ coulombs/m. In two dimensions this becomes a point charge of strength λ. The curves of constant potential are represented by the real part of

$$w = -\frac{\lambda}{2\pi\epsilon_0}\log z,$$

or

$$w = u + iv = -\frac{\lambda}{2\pi\epsilon_0}\{\log|z| + i\arg z\}.$$

The curves of constant v are straight lines representing the field lines which are perpendicular to the equipotential circles of constant u.

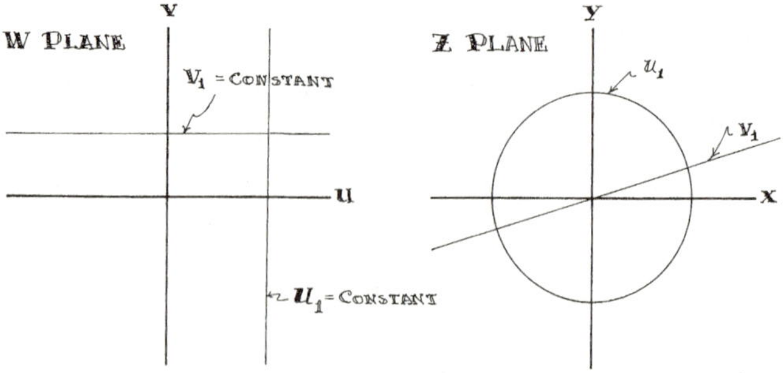

To solve the equation above we write

$$u = -\frac{\lambda}{2\pi\epsilon_0}\log|z| = -\frac{\lambda}{2\pi\epsilon_0}\log a,$$

or

$$x^2 + y^2 = e^{-4\pi\epsilon_0 u/\lambda};$$

and

$$v = -\frac{\lambda}{2\pi\epsilon_0}\arg z = -\frac{\lambda}{2\pi\epsilon_0}\phi,$$

or

$$y/x = \tan\{-2\pi\epsilon_0 v/\lambda\}.$$

To introduce the method we will solve a number of transformations and then attempt to fit them to physical problems. The transformation

$$w = \log\left\{\frac{z + ib}{z - ib}\right\}$$

can be viewed as the complex representation of two lines of charge, one negative at $z = -ib$ and one positive at $z = +ib$.

Any pair of the resulting equipotential curves can be viewed also as a pair of charged cylinders, and then the exterior region will correspond to the equipotentials of a system of two cylinders. To see this let us extend the development:

$$w = \log\left\{\frac{1 + i(b/z)}{1 - i(b/z)}\right\} = 2i\cot^{-1}(z/b),$$

or

$$z = b\cot\left\{\frac{u + iv}{2i}\right\} = x + iy.$$

By equating real and imaginary parts,

$$x = \frac{b\sin v}{\cosh u - \cos v},$$

$$y = \frac{b\sinh u}{\cosh u - \cos v}.$$

Eliminating v,

$$x^2 + (y - b\coth u)^2 = b^2\operatorname{csch}^2 u.$$

This is the equation of a circle centered at $(b\coth u)$ with a radius $(b\cosh u)$. Another family of circles (or in our problem it could be field lines) can be obtained by eliminating u.

The result shown above can be used to obtain the capacitance between two cylinders of radii R_1 and R_2, their centers separated by a

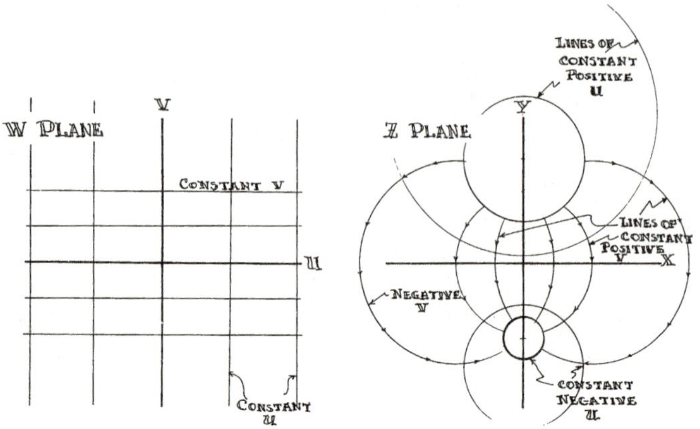

distance d. The potential of one is determined by u = constant = u_1 while the potential of the other is u = constant = u_2. u_1, u_2 and the parameter b will be set by the geometry R_1, R_2, and d where $(R_1 + R_2) < d$. Let

$$R_1 = b\,|\text{csch}\, u_1|, \quad R_2 = b\,|\text{csch}\, u_2|,$$

and

$$d = b\,|\coth u_1| + b\,|\coth u_2|.$$

Solving for $\cosh(u_2 - u_1)$,

$$\cosh(u_2 - u_1) = \frac{d^2 - R_1^{\ 2} - R_2^{\ 2}}{2R_1R_2}.$$

To select the appropriate units we set $u_j = 2\pi U_j/\lambda$, where U_j is the potential in volts on the jth cylinder and λ is the magnitude of the charge per unit length on each cylinder. The capacitance per unit length between the two cylinders is therefore

$$C = \frac{\lambda}{U_2 - U_1} = 2\pi\left\{\cosh^{-1}\left(\frac{d^2 - R_1^{\ 2} - R_2^{\ 2}}{2R_1R_2}\right)\right\}^{-1}.$$

5. THE SCHWARZ-CHRISTOFFEL TRANSFORMATION

This very useful transformation maps a polygon in the z plane into the upper half of the w plane by mapping the sides of the polygon onto the real axis of the w plane. In unfolding the sides of the polygon onto the real w axis it is necessary that the transformation of the angles at the vertices of the polygon be nonconformal. Because the vertices are not mapped conformally, the transformation function or its inverse are not analytic at the vertex points. The *nonconformal* transformation is set up by intentionally placing singular points or zeros in the derivative of z with respect to w.

Before deriving the formal transformation it will prove instructive to examine a few special cases which illustrate the general method.

a. SEMI-INFINITE CHARGED CONDUCTING PLANE. Consider a charged semi-infinite conducting plane which appears as the positive real axis in the z plane. The transformation to obtain the equipotential curves and field lines consists of mapping the path from $x \to +\infty$ to the origin and back to $x \to +\infty$, onto the real axis of the w plane. This transformation will merely be given at this stage of the discussion. It is

$$z = w^2, \quad \text{or} \quad w = +z^{1/2}.$$

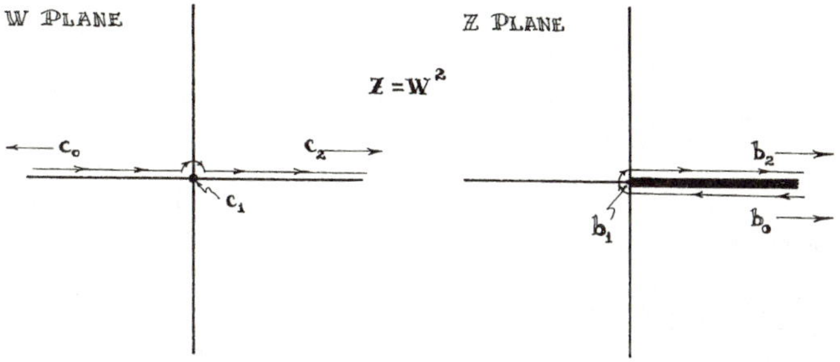

The positive square root is taken in order that the positive real axis of the z plane be described. The negative root describes the negative real axis. The vertex of the conducting plane is at the point b_1 $(z = 0)$, which in this problem is the origin. The corresponding vertex on the real axis of the w plane is c_1. We shall take c_1 at $w = 0$. Later the reader will find that there is some freedom of choice in the selection of the w-plane vertices. The choice will not affect the result when the number of vertices is three or less, as long as the subsequent development remains consistent.

Let the polar form of the complex variable in the w plane be written

$$w = u + iv = \rho e^{i\zeta}$$

where $\rho = (u^2 + v^2)^{1/2}$

and $\zeta = \tan^{-1}(v/u)$.

Returning to the transformation $z = w^2$, we find

$$z = ae^{i\phi} = \rho^2 e^{i2\zeta}.$$

We can now relate the points in the two planes, as shown in the table.

Point	u axis	ζ	ϕ	a
b_0	$u < 0$	π	2π	ρ^2
$b_1(-)$	$u \to 0(-)$	π	2π	$a \to 0$
$b_1(+)$	$u \to 0(+)$	0	0	$a \to 0$
b_2	$0 < u$	0	0	ρ^2

The curves of constant potential correspond to lines of constant v, since these lines are parallel to the real w axis which we had previously identified with the semi-infinite plane at a constant potential.

The curves of constant u are perpendicular to the equipotentials; therefore, after the conformal transformation, u(x,y) represents the field lines. Solving for u and v we obtain

$$w = u + iv = z^{1/2} = (x^2 + y^2)^{1/4} e^{i\phi/2};$$

with $u^2 = a \cos^2 \tfrac{1}{2}\phi = \tfrac{1}{2}(a + x)$,

or $y^2 = 4u^2(u^2 - x)$.

As indicated, this family of parabolas represents the field lines.

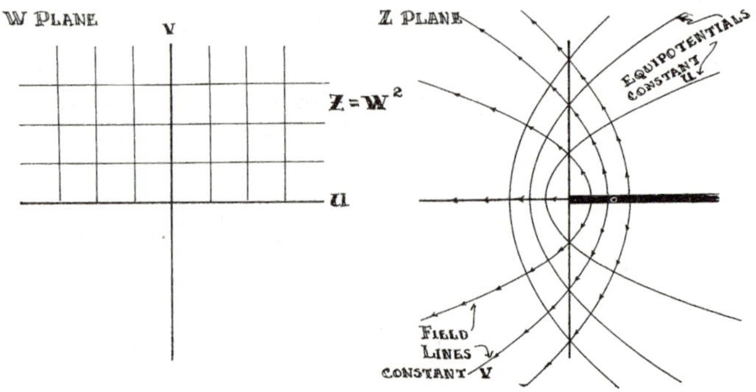

Eliminating the variable u gives

$$v^2 = a \sin^2 \tfrac{1}{2}\phi = \tfrac{1}{2}(a - x),$$

or

$$y^2 = 4v^2(v^2 + x).$$

This family of parabolas represents the curves of equipotential. These curves are orthogonal at every point to the curves for constant u.

The edge of the plane can be shifted by using

$$(z - b) = w^2.$$

The derivative of z with respect to w will be the starting point in establishing these particular transformations. Therefore it will prove instructive to note the features of the special cases which we are examining now. In the present problem,

$$dz/dw = 2(w - 0)^{(2-1)}.$$

The zero in the quantity (w − 0) corresponds to the vertex c_1 in the w plane. The power of the element (w − 0) is the negative of the phase change divided by π. In this specific problem the path in the z plane is initially directed to the left. At the vertex Δz is rotated through an angle of $(-\pi)$, and the path is then directed to the right. The negative of the change of phase divided by π is then equal to plus one or (2 − 1).

b. THE CONDUCTING RIGHT ANGLE. The transformation which we must consider is

$$z = w^{1/2} = ae^{i\phi} = \rho^{1/2} e^{i\zeta/2},$$

or

$$w = z^2.$$

This maps the first quadrant of the z plane onto the upper half of the w plane. Again we place the singular points or vertices b_0 and c_0 at $z = 0$ and $w = 0$ respectively.

Point	u axis	ζ	ϕ	a
b_0	$u < 0$	π	$\pi/2$	$\rho^{1/2}$
$b_1(-)$	$u \to 0(-)$	π	$\pi/2$	$a \to 0$
$b_1(+)$	$u \to 0(+)$	0	0	$a \to 0$
b_2	$0 < u$	0	0	$\rho^{1/2}$

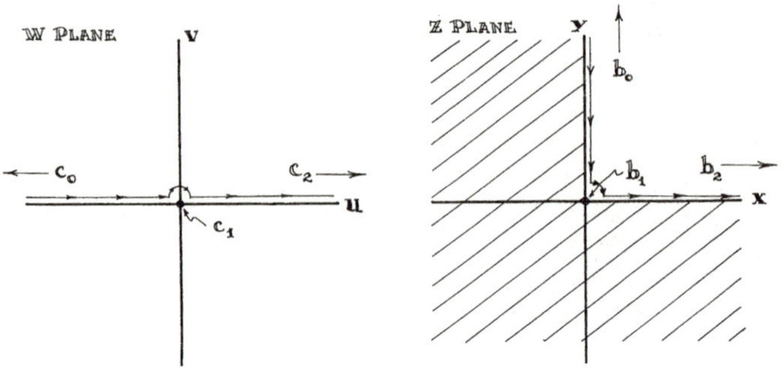

Solving for the lines of u and v we obtain

$$w = u + iv = z^2 = (x^2 - y^2) + i2xy$$

with the field lines $\qquad u = (x^2 - y)^2,$

and the lines of equipotential

$$v = 2xy.$$

In this problem

$$\frac{dz}{dw} = \tfrac{1}{2}(w - 0)^{(\frac{1}{2}-1)},$$

indicating a phase change at the vertex b_1 of $+\pi/2$.

c. A CONDUCTING ANGLE $\beta\pi$ ($\beta < 1$). From the previous example we deduce that the transformation

$$z = w^\beta = ae^{i\phi} = \rho^\beta e^{i\beta\zeta}$$

provides the field lines and potential curves for two conducting planes which have been joined to subtend an angle $\beta\pi$. The equipotentials for this problem are given by

$$v = (x^2 + y^2)^{1/\beta} \sin\left\{\frac{1}{\beta}\tan^{-1}\frac{y}{x}\right\}.$$

The phase change at b_1 is $(1 - \beta)\pi$.

d. THE TRANSFORMATION OF A POLYGON. Consider now a portion of a polygon as shown in the accompanying figure. A vertex is placed at b_1 and the phase change is α_1. Here the real axis of the z plane up to the vertex b_1 is mapped onto the real w axis between $-\infty$ and c_1. There is a vertex b_c, and correspondingly c_0, at $x \to -\infty$

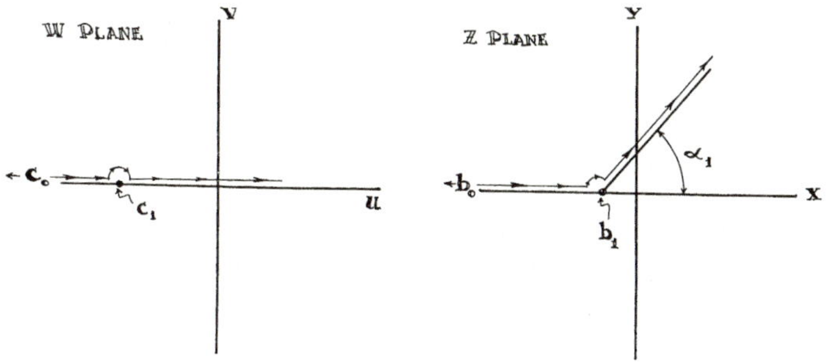

and $u \to -\infty$; however, the associated phase change is zero, which drops this vertex from the problem.

At the vertex c_1, $(w - c_1)$ or Δw changes phase by $-\pi$ while $z = (z - b_1)$ changes phase by α_1. Thus the equation

$$\Delta z \xrightarrow[w \to c_1]{} A(w - c_1)^{\beta-1} \Delta w$$

describes the relation between Δz and Δw in the region near the vertex b_1. The term $(w - c_1)^{(\beta-1)}$ either is singular at c_1 or its inverse is singular. This factor thus causes the transformation to be *nonconformal* at c_1. β can be related to the phase changes by examining the argument of Δz:

$$\arg \Delta z = \arg A + (\beta - 1) \arg (w - c_1) + \arg \Delta w.$$

At $u \to c_1(-)$,

$$\arg \Delta z = \arg A + (\beta - 1)\pi + 0,$$

and at $u \to c_1(+)$,

$$\arg \Delta z' = \arg A + 0.$$

As w passes c_1, the argument of Δz changes by $(-\beta + 1)\pi$:

$$\text{change in arg } \Delta z = -(\beta - 1)\pi = \alpha_1;$$

thus

$$\beta = 1 - (\alpha_1/\pi).$$

Our relation between Δz and Δw at c_1 can now be written

$$\Delta z \xrightarrow[w \to c_1]{} A(w - c_1)^{-\alpha_1/\pi} \Delta w,$$

and

$$\frac{dz}{dw} \xrightarrow[w \to c_1]{} A(w - c_1)^{-\alpha_1/\pi}.$$

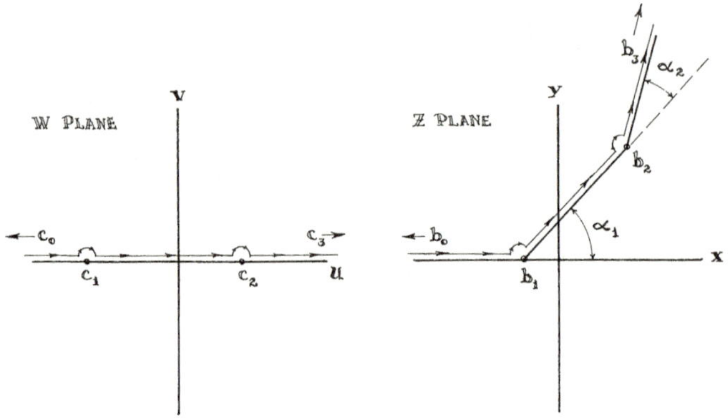

We can now insert another vertex following b_1 at b_2. At this vertex the phase change in Δz is α_2. Therefore

$$\frac{dz}{dw} \xrightarrow[w \to c_2]{} A(w - c_2)^{-\alpha_2/\pi}.$$

Suppose we now have a polygon of N sides and N vertices. The first vertex b_0 is no longer at $x \to -\infty$ and therefore must be included. The transformation is obtained from

$$\frac{dz}{dw} = A(w - c_0)^{-\alpha_0/\pi}(w - c_1)^{-\alpha_1/\pi} \cdots (w - c_N)^{-\alpha_N/\pi}.$$

Because the polygon is closed,

$$\alpha_0 + \alpha_1 + \alpha_2 + \ldots \alpha_N = \pi.$$

Finally, the transformation equation is

$$z = z_0 + A\int (w - c_0)^{-\alpha_0/\pi}(w - c_1)^{-\alpha_1/\pi} \cdots (w - c_N)^{-\alpha_N/\pi}\, dw.$$

The constants z_0, $|A|$, and arg A must be adjusted to provide the correct position, orientation, and scale for the polygon. The c_j must be taken in some consistent manner to correspond to the b_j. The arbitrary nature of z_0, $|A|$, and arg A allows us the freedom to choose any three of the c_j points as we wish as long as the ordering is maintained.

In many problems one of the vertices is at infinity with an associated phase angle of zero. As stated previously, the corresponding $(w - c_j)^0$ enters the calculation as a multiplication by one.

The preceding special cases $z = w^2$, $z = w^{1/2}$, and $z = w^{\beta}$ are now obvious examples of the general expression derived above.

The exterior of a polygon in the z plane can be mapped onto the real w axis by traversing the sides of the polygon in a clockwise direction. The corresponding interior angles at each vertex are now negatives of the phase changes α_j which are associated with the mapping of the interior.

In mapping the region exterior to a polygon the region of the z plane at infinity must be described by a pole in the w plane at a point p not on the real axis. A multiplicative term of the form $(w - p)^{-1}$ is not sufficient to take this into account, because the imaginary part of p will cause the phase to change as points are taken along the real axis of the w plane. To maintain a real phase when w is specified to be real we describe the region at $z \to \infty$ by

$$(w - p)^{-1}(w - p^*)^{-1}.$$

The reader can observe that if

$$p = r + is,$$

then
$$(w - p)^{-1}(w - p^*)^{-1} = \{(u - r)^2 + s^2\}^{-1},$$

and the required multiplicative factor is real. Thus, to map the region outside of a polygon,

$$z = z_0 + A\int (w - p)^{-1}(w - p^*)^{-1}(w - c_0)^{+\alpha_0/\pi} \times (w - c_1)^{+\alpha_1/\pi} \cdots (w - c_N)^{+\alpha_N/\pi}\, dw.$$

To demonstrate the derivation of transformations, we shall consider a few special cases. As yet nothing has been said about assigning specific potentials to the various sides of a polygon. This is done by labeling the segments on the real w axis between vertices with the appropriate potentials. A second conformal transformation then labels the sides in the z plane. The technique for accomplishing this second transformation will be demonstrated in the illustrative examples to follow.

e. TWO INFINITE PARALLEL CONDUCTING PLANES. In this problem we shall map the region $0 < y < d$ into the upper half of the w plane. The vertex b_0 is taken at $x \to -\infty$, $y = 0$, while the second vertex b_1 occurs at $x \to +\infty$. At b_1 there is a phase change of $+\pi$ and the path returns along the line $z = (x + id)$ to $x \to -\infty$ at b_0 again.

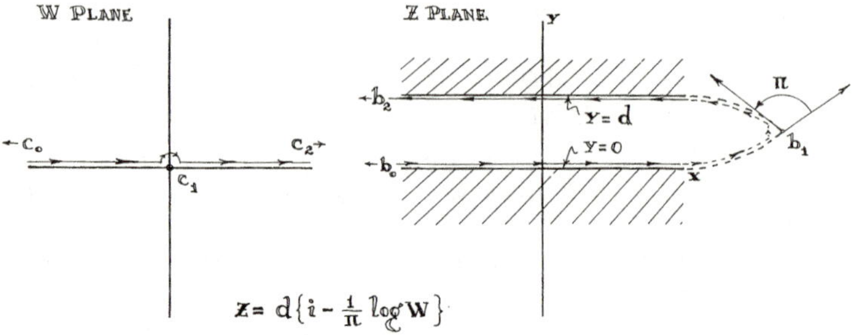

The corresponding vertex c_1 in the w plane will be taken for convenience at $w = 0$.

Remembering that $\alpha_1 = +\pi$, we can write

$$z = z_0 + A\int (w - 0)^{-1}\,dw = z_0 + A \log w.$$

For $u < 0$, the real part of z must be

$$x = z_0 + A \log(-\rho) = z_0 + A \log \rho + Ai\pi.$$

For instance, we can scale the problem so that at $w = -1$, $z = 0$. Then

$$0 = z_0 + Ai\pi, \quad \text{or} \quad A = -z_0/i\pi.$$

As $x \to +\infty$, $u \to 0(-)$, and A must be negative. In the region $0 < u$, $z = x + id$. We can choose $u = +1$ to correspond to the point $z = id$; then

$$id = z_0 + 0,$$

and

$$A = -d/\pi.$$

Finally,

$$z = d\left(i - \frac{1}{\pi}\log w\right),$$

or

$$w = \exp\left\{-\frac{\pi}{d}(z - id)\right\}.$$

This particular transformation is quite important. It has the form necessary for labeling the segments on the real w axis with specified potentials.

We establish a t plane (a third complex plane with a standard connection to the w plane), analogous to the z plane, in which the variables are U and V;

$$t = U + iV.$$

Using the same procedure as above for the z plane we can relate the $V = 0$ coordinate to the interval $u < 0$, and the $V = V_1$ coordinate to the interval $0 < u$. In other words, in the previous example we substitute t for z, U for x, V for y, and V_1 for d. Then

$$w = \exp\left\{-\frac{\pi}{V_1}(t - iV_1)\right\}.$$

f. A SEMI-INFINITE CONDUCTING PLANE ABOVE AND PARALLEL TO AN INFINITE GROUNDED CONDUCTING PLANE. To describe the z and

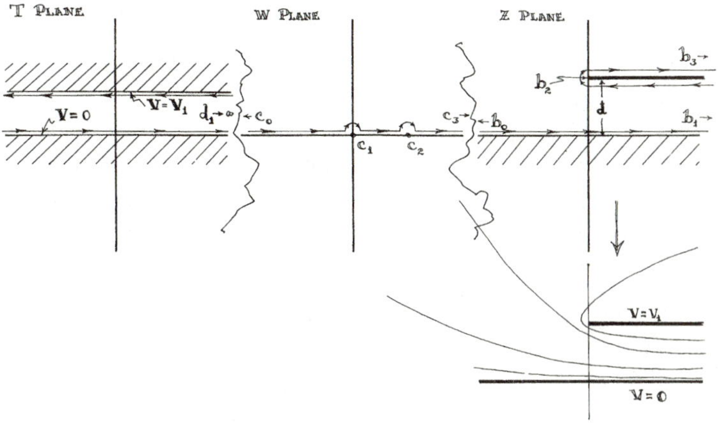

w planes as shown in the accompanying figure we note the mapping given in the table.

Vertex in the z plane	Change of phase	Vertex in w plane
$b \to -\infty$	0	$c_0 \to -\infty$
$b_1 \to +\infty$	$+\pi$	$c_1 = 0$
$b_2 = id$	$-\pi$	$c_2 = +1$
$b_3 \to +\infty$	0	$c_3 \to +\infty$

The transformation is

$$z = z_0 + A\int (w - 0)^{-1}(w - 1)^1\, dw = z_0 + A(w - \log w).$$

When $u < 0$, z is real; thus

$$x = x_0 + iy_0 + A(-\rho - \log \rho - i\pi),$$

giving

$$A = y_0/\pi.$$

At $u = 0$, $x \to +\infty$, showing that A is positive. When $u = 1$, $z = id = z_0 + A$, which gives

$$y_0 = d, \quad x_0 = -d/\pi,$$

and

$$A = d/\pi.$$

Finally,

$$z = \frac{d}{\pi}(-1 + i\pi + w - \log w).$$

This equation establishes the geometry of the field lines and equipotentials if both planes are at the same potential.

By use of the t plane transformation of w with vertices corresponding to the z plane transformation, the planes can be placed at different potentials. Set the infinite plane ($y = 0$) at the potential $V = 0$ and the semi-infinite plane ($y = d$) at $V = V_1$. The vertex d_1 in the t plane is at $U \to +\infty$ and corresponds to c_1 at $u = 0$ in the w plane.

The transformation assigning $V = 0$ to the interval $u < 0$ and $V = V_1$ to the interval $0 < u$ is

$$w = \exp\left\{-\frac{\pi}{V_1}(t - iV_1)\right\}.$$

For convenience we define new variables

$$\eta = \pi U/V_1, \qquad x' = \pi x/d,$$

$$\xi = \pi V/V_1, \qquad y' = \pi y/d,$$

and

$$\tau = \pi t/V_1, \qquad z' = \pi z/d;$$

thus

$$w = -e^{-\tau}.$$

By eliminating w between the equations z(w) and t(w), we obtain

$$z' = \{\tau - e^{-\tau} - 1\},$$

$$(x' - 1) = \{\eta - e^{-\eta} \cos \xi\},$$

and

$$y' = \{\xi + e^{-\eta} \sin \xi\}.$$

Eliminating U (or η) we find that

$$(x' - 1) = \log \left\{ \frac{\sin \xi}{y' - \xi} \right\} - (y' - \xi) \cot \xi.$$

The behavior of the equipotentials can be gauged by examining some special values. When $V = V_1/2$, $\xi = \pi/2$ and

$$y' - (\pi/2) = e^{-(x'-1)}.$$

This curve approaches a constant value $y' = \pi/2$, or $y = d/2$, for large positive values of x. For large negative values of x the equipotential goes as $e^{|x|}$, which seems reasonable.

When we examine the potential near the real axis, $y = 0$, we expect to find an almost parallel curve (i.e. parallel to first order). Let $V = \delta V_1$, where $\delta \ll 1$; then

$$y' \xrightarrow[\delta \ll 1]{} \delta\pi, \quad \text{or} \quad y \xrightarrow[\delta \ll 1]{} \delta d.$$

This equipotential is essentially parallel to the $y = 0$ plane. The symmetry of the fringing field of a parallel plate condenser allows us to apply the solution given above to any quadrant of the condenser. There is an approximation involved which requires that the plate separation be much smaller than the edge dimensions of the faces.

G. Physical Constants

TABLE I. CONVERSION OF UNITS AND PHYSICAL CONSTANTS

a. CONVERSION OF ELECTRICAL UNITS. A formula given in cgs emu or cgs esu units may be expressed in rationalized mks units by replacing each symbol with its value in the emu or esu column, respectively. For more precise work, the factor 3 should be replaced by $c/10^8 = 2.998$ wherever it appears in the table.

Quantity	Rationalized	Emu	Esu
Capacitance	C farad	10^{-9} C	9×10^{11} C
Charge, quantity	q coulomb	10^{-1} q	3×10^{9} q
Current	I ampere	10^{-1} I	3×10^{9} I
Field intensity, electric	$\mathscr{E}$ volt/meter	$10^{6}\ \mathscr{E}$	$3^{-1}\ 10^{-4}\ \mathscr{E}$
Field intensity, magnetic	H amp-turn/m	$4\pi\ 10^{-3}$ H	$12\pi\ 10^{7}$ H
Flux, magnetic	Φ weber	$10^{8}\ \Phi$	$(300)^{-1}\ \Phi$
Impedance	Z ohm	10^{9} Z	$9^{-1}\ 10^{-11}$ Z
Inductance	L henry	10^{9} L	$9^{-1}\ 10^{-11}$ L
Induction, magnetic	B weber/meter^2	10^{4} B	$3^{-1}\ 10^{-6}$ B
Magnetic moment (dipole)	M′ weber-meter	$(4\pi)^{-1}\ 10^{10}$ M′	$(12\pi)^{-1}$ M′
Pole strength	m weber	$(4\pi)^{-1}\ 10^{8}$ m	$(1200\pi)^{-1}$ m
Potential, electric scalar	V volt	10^{8} V	$(300)^{-1}$ V
Reactance	X ohm	10^{9} X	$9^{-1}\ 10^{-11}$ X
Resistance	R ohm	10^{9} R	$9^{-1}\ 10^{-11}$ R

b. VALUES OF THE PHYSICAL CONSTANTS*

Avogadro's number	$N_0 = 6.0247 \times 10^{23}$ molecules/mole
Velocity of light	$c = 2.9979 \times 10^8$ meters/sec
Electronic charge	$e = \begin{cases} 4.802 \times 10^{-10} \text{ esu (statcoulombs)} \\ 1.602 \times 10^{-20} \text{ emu (abcoulombs)} \\ 1.602 \times 10^{-19} \text{ coulomb} \end{cases}$
Bohr magneton	$\mu_0 = \dfrac{e\hbar}{2m_0} = \begin{cases} 9.273\,(10)^{-21} \text{ erg/gauss} \\ 9.273\,(10)^{-24} \text{ joule-m}^2\text{/weber} \end{cases}$
Charge to mass ratio of the electron	$e/m_0 = \begin{cases} 5.273 \times 10^{17} \text{ esu/gram} \\ 1.759 \times 10^7 \text{ emu/gram} \\ 1.759 \times 10^8 \text{ coulomb/gram} \end{cases}$
Faraday's constant	$F = 96{,}520$ coulombs/mole (physical scale)
Planck's constant	$h = \begin{cases} 6.627 \times 10^{-27} \text{ erg-second} \\ 6.627 \times 10^{-34} \text{ watt-second} \end{cases}$
Neutron rest mass	$M_n = \begin{cases} 1.008982 \text{ atomic mass units} \\ 1.67476 \times 10^{-24} \text{ gram} \end{cases}$
Proton rest mass	$M_p = \begin{cases} 1.007593 \text{ atomic mass units} \\ 1.67245 \times 10^{-24} \text{ gram} \end{cases}$
Electron rest mass	$m_0 = \begin{cases} 5.4876 \times 10^{-4} \text{ atomic mass unit} \\ 9.1086 \times 10^{-28} \text{ gram} \end{cases}$

1 Atomic mass unit $= 931$ Mev $= 9.31 \times 10^8$ electron volts $= 1841\, m_0$

1 Electron volt $= 1.602 \times 10^{-12}$ erg $= 1.602 \times 10^{-19}$ joule

First Bohr radius $a_0 = \dfrac{\hbar^2 4\pi\epsilon_0}{m_0 e^2} = 5.29 \times 10^{-9}$ centimeter

Classical electron radius $r_0 = \dfrac{e^2}{4\pi\epsilon_0 m_0 c^2} = 2.818\,(10)^{-15}$ meter.

* J. A. Bearden and J. S. Thomsen, Nuovo Cimento, 5, 267 (1957). In many of the expressions above the numbers are only given to three or four places for convenience.

TABLE II. DIELECTRIC CONSTANTS*

Substance		ϵ
Air at 0°C		1.000590
SPT		
Helium		$1 + 6.8\,(10)^{-5}$
Oil, silicone		
Dow Corning D.C. 120	at 25°C	>2.2–2.76
Carbon tetrachloride	at 20°C	2.24
Water (distilled)	0°C	88
	20°C	80
	100°C	48
Calcite (1 optic axis)		8.5
Calcite (11 optic axis)		8.0
Tourmaline (1 optic axis)		7.10
Tourmaline (11 optic axis)		6.3
Rutile (1 optic axis)		86
(11 optic axis)		170
Nylon		3.5
Polyethylene		2.3
Polytetrafluorethylene		2.0
Alumina (ceramic)		4.5–8.4
Titanium dioxide		14–110
Corning 0120 glass		6.65
Fused quartz		3.75–4.1

TABLE III. RESISTIVITY†

Substance	β in mhos per meter
Aluminum	$3.53\,(10)^7$
Copper	$5.92\,(10)^7$
Gold	$4.10\,(10)^7$
Iron (0°C)	$1.13\,(10)^7$
Magnesium	$2.30\,(10)^7$
Nickel	$1.38\,(10)^7$
Silver	$6.81\,(10)^7$
Tungsten	$1.81\,(10)^7$
Tin	$0.870\,(10)^7$
German silver	$0.30\,(10)^7$
Carbon (graphite)	$5(10)^{-5}$

* *Handbook of Chemistry and Physics*, 38th edition.

† *Handbook of Chemistry and Physics*, 38th edition, and *American Institute of Physics Handbook*, McGraw-Hill, New York, 1957.

TABLE III.—*continued*

Substance	β in mhos per meter
Germanium (pure)	0.455
(10^{-6}% As)	0.200
Silicon (pure)	0.637
(10^{-6}% As)	0.1
Ga Sb	1
Ceramic: AlSi Mag	$<10^{-16}$
Zircon porcelain	$143(10)^{-17}$
Glass	10^{-14}–10^{-19}
Bakelite micarta	$2\,(10)^{-13}$
Hard rubber	10^{-15}

TABLE IV. MAGNETIC SUSCEPTIBILITY AT ROOM TEMPERATURE*

Substance	χ
Aluminum	$8.3(10)^{-6}$
Bismuth	$-1.70(10)^{-5}$
Carbon (graphite)	$-4.41(10)^{-5}$
Copper	$-0.107(10)^{-6}$
Gadolinium oxide (Gd_2O_3)	$164(10)^{-5}$
Gold	$-1.89(10)^{-6}$
Magnesium	$+6.93(10)^{-6}$
Manganese	$12.5(10)^{-5}$
Silver	$-2.6(10)^{-5}$
Tungsten	$+6.8(10)^{-5}$
Hydrogen (1 atm)	$-0.21(10)^{-8}$
Nitrogen (1 atm)	$-0.51(10)^{-8}$
Oxygen (1 atm)	$+209.0(10)^{-8}$

* *American Institute of Physics Handbook*, McGraw-Hill, New York, 1957.

Problems

CHAPTER I

1. Is the vector point function $\mathbf{F} = x^2\,\mathbf{i} + xy\,\mathbf{j}$ conservative?
2. If $V = xy + 2z^2 + 3zyx^3$, find the associated force field.
3. Given that $\mathbf{F} = xy^3\,\mathbf{i} + x^2\,\mathbf{j}$ and $d\mathbf{r} = dx\,\mathbf{i} + dy\,\mathbf{j}$, evaluate the line integral $\int_a^b \mathbf{F}\cdot d\mathbf{r}$ along the path $y = \frac{1}{2}x^2$ from $a = (0,0)$ to $b = (2,2)$.
4. Consider a force $\mathbf{F} = \dfrac{a(2x^2 - y^2)\,\mathbf{i}}{(x^2 + y^2)^{5/2}} + \dfrac{3axy\,\mathbf{j}}{(x^2 + y^2)^{5/2}}$. Compute the potential energy at any point P relative to $r \to \infty$ $\left(\text{i.e., } \int_\infty^{r_p} \mathbf{F}\cdot d\mathbf{r}\right)$.
5. Given a vector function

$$\mathbf{A} = \frac{\mu_0}{4\pi}\frac{\mathbf{M}\times\mathbf{r}}{r^3}$$

and a scalar function

$$V = \frac{\mu_0}{4\pi}\frac{\mathbf{M}\cdot\mathbf{r}}{r^3};$$

where **M** is a constant vector; show that curl **A** = −grad V. Explain this result. (*Hint:* examine at $r \to 0$).

6. Find the scalar point function which generates the vector function

$$\mathbf{F} = \frac{(y^2 - x^2)}{(x^2 + y^2)}\{-y\,\mathbf{i} + x\,\mathbf{j}\}.$$

7. Compute the work required to move a point particle in a circle of radius 1 meter when the force field is given by $\mathbf{F} = C(y\,\mathbf{i} - x\,\mathbf{j})$. Does the direction of integration influence the result?
8. A force field

$$\mathbf{F} = C\left[\frac{x\,\mathbf{i} + y\,\mathbf{j} + z\,\mathbf{k}}{\{x^2 + y^2 + z^2\}^{3/2}}\right]$$

is defined at all points except at $r = 0$ (the units of x, y, and z will be meters, c is a constant).

(a) Compute the work to move a point particle from ∞ to the point (0,1,4). Choose the path to consist of the two segments, $x = 0$, $z = 0$ from $y = \infty$ to $y = 1$, and $y = 1$, $x = 0$ from $z = 0$ to $z = 4$.

(b) Compute the work to move a point particle in a circle of radius 5 meters about the origin. Let the circle lie in the xy plane.

9. Given a force field

$$\mathbf{F} = \frac{x\,\mathbf{i} + y\,\mathbf{j}}{(x^2 + y^2)},$$

find the work necessary to go from point (5,0) to point (5,4) along the path $x = 5$; then from (5,4) to $(-3,4)$ along the path $y = 4$.

10. The function

$$f(x,y,z) = \frac{(\mathbf{k} \times \mathbf{a})}{a^2}$$

where

$$\mathbf{a} = x\,\mathbf{i} + y\,\mathbf{j}$$

has the interesting property that curl $\mathbf{f} = 0$, while the line integral on a circle about the origin does not vanish.
(a) Show that curl $\mathbf{f} = 0$. Does this relation hold everywhere in space?
(b) Using a circle of radius R in the xy plane and centered at the origin, show that $\oint_{\text{circle}} \mathbf{f} \cdot d\mathbf{r} = 2\pi$.
(c) Employ Stokes' theorem to show that the surface integral of curl $\mathbf{f}$ over the cap bounded by the circle of part (b) is 2π. Explain this in view of the apparent vanishing of the curl.
(d) Is there any connection between this problem and Cauchy's Principle Value?

CHAPTER II

1. Three equal positive point charges q are located at (0,0,1m), (0,0,0), and (0,2m,0). What is the value of the integral of the normal component of $\mathscr{E}$, the electric field intensity, over the surface of a sphere having a radius of 1.5 meters and centered at (0,0,0)?
2. Four equal positive charges q are placed at points (1,0,0), (0,1,0), (0,0,1), and (0,0,0) respectively. Find the electric field vector and the potential V at the point (1,1,1).
3. An electric field is set up by two charges, $q_1 = 10^{-10}$ coulombs at (1,0,0) and $q_2 = 2(10)^{-10}$ coulombs at (0,1,0). Find the work necessary to move a charge of magnitude $q_t = 10^{-12}$ coulomb from (64,0,4) to (0,0,0) along the path $x = 4z^2$.
4. An external electric field is given by $\mathscr{E} = 10^4(\mathbf{i} + 3\,\mathbf{j} + 3\,\mathbf{k})$ volts/m. Show that no work is done in moving a test charge from the origin to $\mathbf{r} = (0.06\,\mathbf{i} - 0.04\,\mathbf{j} + 0.03\,\mathbf{k})$ meters.
5. Eight equal charges q are placed one on each corner of a cube having edges of length l. Compute the total force on any one of the charges.
6. Consider four equal positive charges placed upon the corners of a square of side l. A charge +q placed at the geometric center of the square is in unstable equilibrium. Compute the force on the charge at the center for small displacements δ ($\delta \ll l$) from the center.
7. Three small pith balls of the same mass m are suspended by equal weightless threads of length l from a common point. A charge q is divided equally between the balls and they come to equilibrium at the corners of a horizontal

equilateral triangle whose sides are d. Show that

$$q^2 = 36\pi\epsilon_0 \, mgd \sqrt{l^2 - (d^2/3)}$$

where g = the acceleration of gravity.

8. Coulomb's law is found to hold for the electrostatic repulsion between nuclei at very short distances. Compute the electrostatic force between an alpha particle of atomic number 2 and a gold nucleus of atomic number 79 at a separation of $3(10)^{-12}$ cm.
9. Two electrons are separated by a distance r. What is the ratio of the gravitational attractive force to the force of electrostatic repulsion?
10. A point charge $q = 10^{-6}$ coulombs is fixed at the origin. An electric dipole $\mathbf{p} = 10^{-10}\,\mathbf{k}$ coulomb-meter is at the point (0,1,0).
    (a) What is the torque on the dipole about its geometrical center?
    (b) What is the total torque on the dipole?
    (c) What is the potential energy of the dipole?
    (d) What is the translational force on the dipole (a vector)?
11. An electrostatic dipole of moment $|\mathbf{p}| = 10^{-8}$ coulomb-meter is placed in a constant electric field $\mathscr{E} = 10^{-6}\,\mathbf{k}$ volt/meter. Compute the frequency of the oscillations which occur when $\mathbf{p}$ is allowed to oscillate about its center, being initially parallel to $\mathscr{E}$ and then displaced by a small angle from equilibrium and allowed to move under the action of the field. Assume that the moment of inertia of the dipole is 10^{-14} kg-m^2.
12. Demonstrate that the zero equipotential of two point charges $+q_1$ and $-q_2$ separated a distance l is a sphere. Obtain the radius of the sphere in terms of q_1, q_2, and l.
13. An electric dipole moment $\mathbf{p} = p_0\,\mathbf{k}$ coulomb-m is placed at the origin in the presence of an electric field $\mathscr{E} = -\text{grad}\,[\mathscr{E}_0 z e^{-(y^2/\alpha^2)}]$ at that point. What is the vector torque and translational force on $\mathbf{p}$?
14. Three charges are positioned along the z axis in the following manner: $+q$ at $(0,0,-l)$, $-3q$ at $(0,0,0)$ and $4q$ at $(0,0,2l)$. Compute the quadrupole tensor. Does this distribution have moments below that of the quadrupole?
15. A point dipole $\mathbf{p}$ lies displaced from the origin of coordinates and is located by a position vector $\mathbf{R}$. Make a multipole expansion of the dipole potential, and obtain the relation describing the geometric quadrupole moment which is observed as a result of the displacement.
16. A dipole is defined by the parameters $|q| = 10^{-9}$ coulomb and $\boldsymbol{l} = 10^{-4}\,\mathbf{k}$ meter. What is the work done in moving a point charge $Q = 10^{-10}$ coulomb from a point $\mathbf{r}_1 = 10\,\mathbf{j}$ meters to a point $\mathbf{r}_2 = 10\,\mathbf{k}$ meters?
17. Two positive point charges $q = 10^{-6}$ coulomb are located one at $\mathbf{r}_1 = 1\text{m}\,\mathbf{j}$ and the other at $\mathbf{r}_2 = -1\text{m}\,\mathbf{j}$. A point dipole $\mathbf{p} = 10^{-14}\,\mathbf{k}$ coulomb-m is placed at the origin (r = 0). What is the magnitude and direction of the translational force on the dipole?
18. An electron initially traveling along the z axis with a $\mathbf{k}$ velocity of 10^5 m/sec enters at z = 0 an electric field cylindrically symmetric about the z axis. If the field is $\mathscr{E}_{(a=0)} = 10e^{-\alpha z}\,\mathbf{k}$ volts/m (where $\alpha = 100$ cm^{-1}) and the electron leaves the field at z = 10 cm, find the final velocity (see diagram).

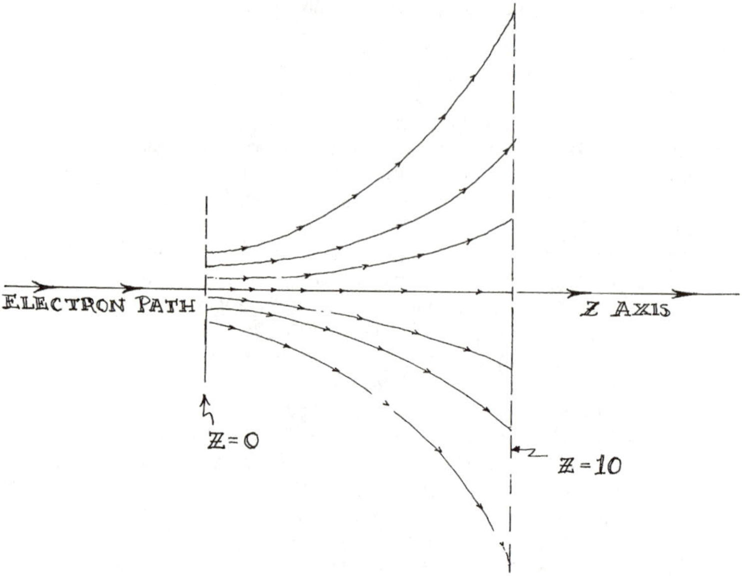

19. If the particle in Problem 18 is an electric dipole of moment $\mathbf{p} = 10^{-10}\,\mathbf{k}$ coulomb-m, what is the final velocity of the dipole?
20. A point charge of 10^{-6} coulomb on the $\mathbf{k}$ axis is 0.01 meter from a fixed dipole $\mathbf{p} = 10^{-10}\,\mathbf{k}$ coulomb-m. Compute the magnitude and direction of the force which the dipole exerts upon the point charge.
21. A square quadrupole is formed by placing point charges of alternating sign and of magnitude 10^{-9} coulomb at the corners of a square 10^{-3} m on a side. What is the total energy stored in this system?
22. Perform a multipole expansion of the following charge configuration,

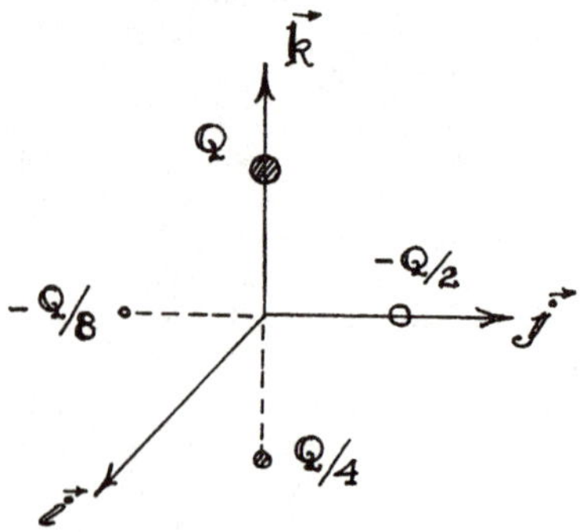

obtaining all terms up to and *including* the quadrupole terms: charge Q at $l\,\mathbf{k}$; $-Q/2$ at $l\,\mathbf{j}$; $Q/4$ at $-l\,\mathbf{k}$, and $-Q/8$ at $-l\,\mathbf{j}$.

23. A square quadrupole ql^2 centered about the origin with its corners positioned on the y and z axes exerts a translational force on a point charge Q located at (0,L,0). Find the force on Q.
24. A rectangular point quadrupole is characterized by two dipoles $\mathbf{p}$ and $-\mathbf{p}$ separated by a distance $\mathbf{d}$, where $\mathbf{d} \cdot \mathbf{p} = 0$. The quadrupole is located at $\mathbf{r} = y\,\mathbf{j}$. Compute the force and torque on the quadrupole when a point charge q is located at the origin.

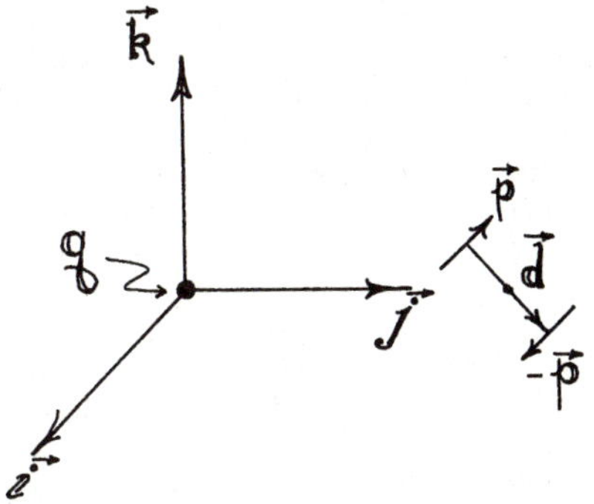

25. Compute the first three multipole contributions to the potential function of the following charge configuration: there are six equal positive point charges $+q$ at (1,0,0), $(-1,0,0)$, (0,1,0), $(0,-1,0)$, (0,0,1), and $(0,0,-1)$, combined with one point charge $-6q$ at the origin.
26. A charge $+q$ is placed on a small pith ball of mass m hanging by a thread vertically along the z axis in the earth's gravitational field. The pith ball is located at the origin. Three other charges $+q$ are placed on the corners of an equilateral triangle of side l. If the plane of the triangle is parallel to the xy plane, find the distance between the pith ball at the origin and the plane of the triangle when the total force on the pith ball is zero.

CHAPTER III

1. A plane circular disk of radius R carries a uniform surface charge density σ. Develop the electric field on the symmetry axis of the disk.
2. A flat circular disk of radius R carries a uniform surface charge density σ. A circular hole of radius b is cut from the center. If the center of the disk is at the origin and if the disk lies in the xy plane, show that the axial field is given by

$$\mathscr{E}_{\text{axial}} = \frac{\sigma z_p\,\mathbf{k}}{2\epsilon_0}\left\{\frac{1}{(b^2 + z_p^2)^{1/2}} - \frac{1}{(R^2 + z_p^2)^{1/2}}\right\},$$

3. Find the electric field along the symmetry axis of a hemispherical volume containing a constant volume distribution of charge ρ. Assume that the radius of the hemisphere is R.
4. Find the electric field of a line of charge extending along the z axis from $-L/2$ to L. The charge per unit length varies linearly as $\lambda = \lambda_0 z/L$.
5. A charged circular ring carries a constant charge per unit length of λ. Show that the electric field evaluated at the plane of the loop has no component perpendicular to that plane.
6. Compute the axial field and potential of a charged ring if the charge per unit length varies as $\lambda\phi/2\pi$. This function is discontinuous at $\phi = 0$.
7. Assume that the total electrostatic potential of the electron (with respect to infinity) is given by that potential which is present if the charge of the electron is uniformly distributed on a spherical surface of radius R. If the electric potential times the electron charge equals the rest mass energy of the electron (m_0c^2), find a "classical radius" of the electron.
8. A spherical volume of radius R carries a constant volume charge density ρ_0. Find the fields and potentials outside and inside the sphere. Plot as a function of r.
9. A cylindrical volume of radius R contains a constant volume density of charge ρ_0. Compute the electric fields and potentials inside and outside the volume. Assume that the axis of the cylinder extends from $-\infty$ to $+\infty$ along the z axis.
10. Find the electric field on the axis at one end of a cylindrical distribution of charge of constant density ρ. The length of the cylinder is L and the radius is R. Also, what is the field at the center?
11. Given a spherical distribution of charge with radius R and a volume density $\rho = kr'^2$ (where k = const. and r′ is the distance from the center, $r' \leqslant R$). Find the fields and potentials as a function of r.
12. An electric field is created by a uniformly charged rod of zero thickness and length $2l$ bearing total charge q. A spherical shell of radius a bearing a uniform surface charge density of total magnitude Q is moved from a great distance to a point near the rod. The center of the sphere is placed at a distance R from the center of the rod ($R > a$). The rod is perpendicular to the line of centers. Find the work required to move the sphere into this position.
13. By direct integration show that an electric field of uniform charge density σ which is distributed upon the surface of a spherical shell of radius R, is the same, for all points outside the sphere, as the field of a point charge of magnitude $4\pi R^2\sigma$ located at the center. (*Hint:* Line up P, the point of examination, on the z axis. Construct a circular filament of charge $\sigma 2\pi R^2 \sin\theta\, d\theta$ which has its axis of symmetry along the z axis.)
14. Find the dipole moment of the charge distribution given by

$$\rho(a,z) = \rho_0 \alpha z e^{-\alpha^2(a^2+z^2)}$$

where $$a^2 = x^2 + y^2.$$

15. View the electron as a spherical volume of radius R cm containing a total charge equal to $1.6(10)^{-19}$ coulomb distributed in a constant volume density

of charge. If the stored energy is equal to the rest mass energy $m_0c^2 = 0.511$ Mev, find the radius required to account for this if equal to the electrostatic stored energy. Compare this result with the result of Problem 7.

16. Consider a volume distribution of quadrupoles contained in a cube with edges a and

$$Q_v = Q_0 \sin\left[\frac{\pi x}{2a}\right] \sin\left[\frac{\pi y}{2a}\right]\begin{pmatrix} 0 & 0 & 0 \\ 0 & 0 & 1 \\ 0 & 1 & 0 \end{pmatrix}.$$

Find the electric field in the medium.

17. A cylindrical cavity of radius R′ is constructed inside the volume of charge described above. The axis of the cavity is parallel to the axis of the cylinder of charge and displaced a distance l, where $(l + r') \leqslant R$.
   (a) Find the fields and potentials in the three regions

$$0 \leqslant r \leqslant R', \quad R' \leqslant r \leqslant R, \quad \text{and} \quad R \leqslant r$$

   when $l = 0$.
   (b) Obtain the fields and potentials outside the charge cylinder when $R' = l = \frac{1}{2}R$.

18. An infinite, thin cylindrical shell of radius R carries a uniform surface density of charge σ. Compute the fields and potentials inside and outside the shell.

19. A volume distribution of charge $\rho = \rho_0 r/R$ is enclosed in a sphere of radius R. Find the electric fields and potential functions inside and outside the sphere.

20. From Laplace's equation obtain the expression for the potential in the space between the plates of a parallel plate condenser. Assume that the plates are square, and that they are parallel to the xy plane and equidistant from the origin. Assume the potential of the lower plate to be zero and the potential of the upper plate to be constant and equal to V_0.

21. Find the electric field and potential of a volume distribution of charge

$$\rho = \rho_0 e^{-r^2/\alpha^2}.$$

Sketch the potential function.

22. A constant density of charge ρ_0 is contained in a sphere of radius R, and a

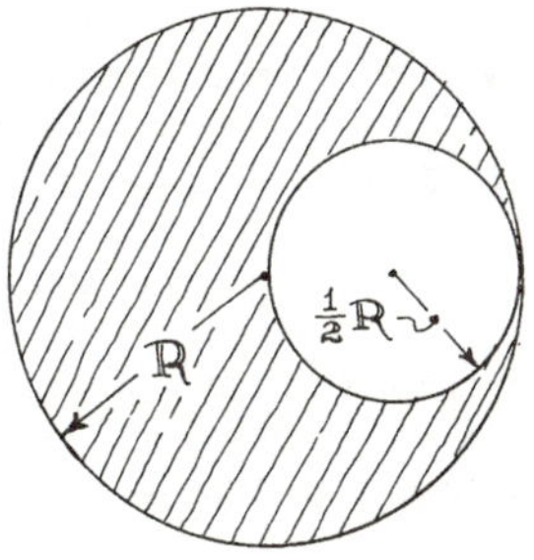

spherical cavity of radius $\frac{1}{2}R$ is constructed inside the sphere. The center of the cavity is at a distance $\frac{1}{2}R$ from the center of the sphere. Compute the dipole moment of the resulting distribution relative to the center.

23. A spherically symmetric charge distribution varies as $\rho = \rho_0 e^{-\alpha r}$. Find the expressions for the field and potential of this distribution.
24. A positive point charge q is situated in an infinite sea of positive charge of constant density ρ_0. The point charge is located at a point $\mathbf{r}_q = r_0\,\mathbf{j}$. A

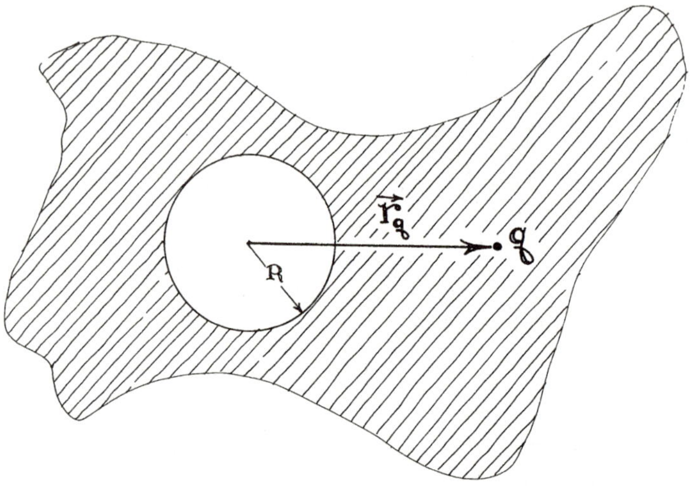

spherical cavity is formed in the sea; the center of the cavity is at the origin, and the radius is R.
(a) Compute the force between the point charge and the cavity.
(b) What is the energy stored in the system composed of q and the cavity?

25. For many purposes an atom may be considered as a point nucleus bearing a positive charge Ze, surrounded by a cloud of electrons whose average charge density decreases rapidly with the distance from the center. Assuming that the atom is electrically neutral, and that the electronic charge distribution is spherically symmetric with density proportional to $e^{-r/a}$, find the potential at all points within the atom. Next, assuming that the electronic charge distribution is unchanged in shape but now contains only $Z - 1$ electrons, i.e., that the atom is singly ionized, again find the potential at all points. Make a small sketch showing the behavior of the potential in the two cases, and find simplified expressions for the potentials at very small and large distances.
26. The volume density of negative charge in an atom having a nuclear charge $+Ze$ varies as $\rho_0 e^{-\alpha r}$, where $\alpha = 10^6\ \mathrm{m}^{-1}$. Compute the total energy stored in this system. Evaluate ρ_0 by requiring that the total amount of negative charge be $-Ze$.

27. A cylindrical distribution of charge having a constant volume density of charge $+\rho$ is symmetric along the z axis, extending along the z axis from $-\infty$ to $+\infty$. The radius of the cylinder is R. A spherical cavity of radius R centered at the origin is cut from the cylinder. What is the electric field at any point (three regions must be considered)?
28. An electric field is set up by a spherical distribution of negative charge of constant density,

$$\rho(r) = -\rho_0, \quad r \leqslant R$$
$$\rho(r) = 0, \qquad R < r.$$

A point positive charge $+Q$ of mass m is released at a point $\mathbf{r} = l\mathbf{k}$ on the z axis inside the distribution ($l < R$). The point charge moves freely under the influence of the field set up by $\rho(r)$, and thus its motion is simple harmonic. What is the frequency of the oscillation in terms of Q, R, ρ_0, and m?
29. A spherical charge distribution $\rho(r)$ is described by

$$\rho(r) = +\rho_0, \quad 0 < r < \frac{R}{2^{1/3}};$$
$$\rho(r) = -\rho_0, \quad \frac{R}{2^{1/3}} < r < R;$$
$$\rho(r) = 0, \qquad R < r.$$

Find the fields and potentials of this distribution.
30. Charge ρ is distributed uniformly in a spherical shell of inner radius a and outer radius b. Obtain the total energy stored in this system.
31. A constant volume density of charge ρ_0 is contained in a spherical volume of radius R. Show that the total energy stored in this system is

$$U = 4\pi\rho_0{}^2R^5/15\epsilon_0.$$

32. A semi-infinite slab of thickness 2L carries a constant volume density of positive charge. The plane of the slab is parallel to the xy plane.
(a) By Gauss's theorem, obtain the expression for the electric field intensity everywhere along the z axis.
(b) If the potential at $z = 0$ is zero, what are the expressions for the potential in the three regions (above, below, and inside)? Assume that $z = 0$ occurs in the central plane of the slab.

CHAPTER IV

1. A weightless spring with spring constant k suspends a small conducting sphere of mass m and radius r (where $r \ll d$) above a grounded conducting plane.
(a) If when uncharged the sphere is d cm above the plane, what is the distance between the center of the sphere and the plane when a charge q is placed upon it?

(b) If the natural period of oscillation of the spring-mass system is $2\pi\sqrt{m/k}$, show that the period of oscillation for small displacements when the charge q is placed upon the sphere is approximately

$$T \simeq 2\pi \left\{\frac{m}{k - (q^2/8\pi\epsilon_0 d^3)}\right\}^{1/2}.$$

2. An electric field is set up between two long concentric conducting cylindrical shells of radii a and b, with $b = 2a$. The charge per unit length is $\pm\lambda$. The inside of the outer cylinder carries the negative charge. What is the translational force on a dipole **p** which is oriented to lie along a radius at a point midway between a and b (i.e., at $r = \frac{3}{2}a$)?
3. A charged conducting sphere of 1 meter radius has an electric field intensity at the surface of 100 volts/m. What is the surface charge density?
4. Prove that in the problem of the point charge q above an infinite grounded conducting plane (discussed in this chapter), the total charge induced upon the grounded conducting plane is equal to q, i.e., show that

$$\iint_{\text{plane}} \sigma\, dA = q.$$

5. A long wire of radius 1 cm is supported parallel to and 10 m above the earth. (Assume that the surface of the earth is a conducting plane.) If the potential difference between the wire and the earth is 10,000 volts,
(a) Find the charge per unit length on the wire.
(b) Find the electric field intensity at the surface of the earth directly below the wire.
6. A flat disk is charged with a constant surface charge density σ. The disk has a concentric circular hole cut into it so that the inner radius is R and the outer radius is $\sqrt{6}\,R$. A dipole of moment $\mathbf{p} = p_0\,\mathbf{k}$ is placed at a distance z above the plane of the disk on the **k** axis (the symmetry axis of the disk).
(a) What is the axial field of the charged disk at any point z on the axis?
(b) Show that the energy of the dipole in the axial field of the charged disk at any point on the z axis is

$$U = \frac{-p_0\sigma z}{2\epsilon_0}\left\{\frac{1}{(R^2 + z^2)^{1/2}} - \frac{1}{(6R^2 + z^2)^{1/2}}\right\}.$$

(c) Obtain the translational force on the dipole and show that when the dipole is $\sqrt{3}\,R$ meters above the plane of the disk, the magnitude of the force is

$$F = \frac{11}{144}\frac{p_0\sigma}{\epsilon_0 R}.$$

What is the direction of **F**?
7. An electric dipole is formed from two point charges of $\pm 10^{-7}$ coulombs. Their separation is 10^{-4} m. What is the force of attraction on the dipole when it is 10^{-2} m from a grounded conducting sphere of radius 10^{-3} m, if the direction of the dipole moment is along a radius vector from the center of the sphere?

8. A vacuum diode consists of a cylindrical cathode 0.05 cm in radius mounted coaxially within a cylindrical anode 0.45 cm in radius. The potential of the anode is 300 volts above that of the cathode. An electron leaves the surface of the cathode with zero initial velocity. What is the velocity of the electron when it strikes the anode?
9. A spherical conductor of radius R (located *in vacuo*) carries a charge q. If the sphere is made up of two separable hemispheres, find the force required to hold the two halves together.
10. A conducting balloon of radius $R = 1$ m carries a total charge $q_T = 1$ microcoulomb. At $t = 0$ the gas filling the balloon begins to leak out at the rate of 10^{-6} m^3/sec. If the surface tension of the balloon is constant at all radii and equal to 1 newton/m^2, at what radius will the electrostatic pressure overcome the surface tension, so that the balloon will cease deflating?
11. In Problem 10, a point charge of 10^{-9} coulomb is suspended at the end of a spring lying on an axis passing through the center of the balloon. At $t = 0$ the point charge is in equilibrium and is 10 cm from the surface of the balloon.
    (a) If the spring constant is 0.1 newton/m, what is the initial deflection of the spring?
    (b) Describe the motion of the charge on the spring as the balloon deflates.
12. Consider an infinite line of charge of λ coulombs/m lying parallel to and a distance d from the axis of a grounded conducting cylinder of infinite length. Assume the radius of the cylinder to be R and that $R < d$. By the method of images find an expression for the potential and field in the exterior region.
13. In the problem in the text concerning the point charge q outside of a grounded conducting sphere of radius R, compute the geometric dipole and geometric quadrupole contributions which arise because of the displacement of the singlet and doublet configurations (use the charge q and the image charge in the analysis).
14. Show that $\left[\frac{1}{r}\frac{\partial V}{\partial \theta}\right]_{r=R}$ is zero for the problem of the point charge outside a grounded conducting sphere of radius R.
15. By direct integration of the pressure on the surface, show that, in the problem of the point charge outside a grounded sphere, the force on the grounded sphere is minus the force exerted by the image on the charge $+q$.
16. An infinite conducting cylinder of radius R is placed in an external uniform electric field $\mathscr{E}$, with the cylinder axis perpendicular to the field.
    (a) Find the expressions for the potential and the field in region $R \leqslant r$.
    (b) Obtain the expression for the induced surface charge density.
17. A dipole **p** is located a distance d above a semi-infinite grounded conducting plane. Obtain the expression for the force on the dipole. (A semi-infinite plane extends to $\pm\infty$ in the coordinates x and y.)
18. In Problem 17, assume that the dipole **p** is oriented perpendicular to the plane. Find the expression for the induced surface charge.

19. Two equal positive point charges are located at $(0,l,0)$ and $(0,-l,0)$. Construct a grounded conducting sphere at the origin which reduces the repulsive force on either charge to zero. What is the radius of the sphere?
20. Compute the dipole and quadrupole moments of the system described in the preceding problem.
21. Two conducting spheres of radius R are connected by a thin wire. If the separation of the centers is l, show that the capacitance is

$$2R \sinh\left(\frac{l}{2R}\right) \sum_{j=1}^{\infty} (-1)^{j+1} \operatorname{csch}\left(\frac{jl}{2R}\right)$$

22. Obtain the potential of a hemispherical sheet of constant surface charge density σ.
23. A grounded conducting spherical shell of radius b surrounds and is insulated from a conducting sphere of radius a. The centers of the two spheres are a distance δ apart, where $\delta \ll a$. If the total charge on the inner sphere is Q, find the potential function between the conductors.
24. (a) Find the charge distribution on the surface of the inner sphere in the preceding problem.
    (b) Find the capacitance between the two spheres.
25. A grounded conducting disk of radius R is influenced by a point charge q on the axis of rotational symmetry and a distance from the disk. Find the potential function of the system.
26. Two semi-infinite grounded conducting planes intersect in a right angle. If a point charge q is placed in the angle such that its perpendicular distance to one plane is a and to the other plane is b, find the distribution of charge on the two planes. What is the total induced charge?
27. Two semi-infinite parallel grounded conducting planes are separated by a distance l. If a point charge q is placed a quarter of the distance between the planes from one plane, compute the first three (singlet through quadrupole) multipoles of this system.
28. An infinite straight line of charge (λ coulombs per unit length) is parallel to and a distance δ from the axis of an infinite grounded cylindrical conducting cavity. Find the potential function in the cavity and the charge per unit length induced on the wall.
29. An infinite conducting straight wire of radius δ is parallel to and a distance l above a semi-infinite grounded conducting plane. If $l \gg \delta$, compute the approximate capacitance per unit length between the plane and the wire.
30. Two conducting spheres of radii a and b lie a large distance R apart, $R \gg a$, $R \gg b$. Using the method of images, find expressions for the potential and capacitance coefficients of the system they form, correct to terms of order $1/R^3$.
31. A sphere of radius a contains a uniform volume density of charge and total charge Q, and lies with its center a distance d from an infinite grounded conducting plane ($d > a$). Assuming the dielectric properties of the sphere to be negligible ($\epsilon = 1$), find the total electrostatic energy of the configuration.

32. A pair of semi-infinite conducting sheets lying side by side in a horizontal plane have parallel edges separated by a distance 2c. An infinitely long straight wire bearing charge λ coulombs per meter lies in the same plane midway between the edges of the sheets. The sheets are grounded. Find their surface charge densities.
33. An uncharged conducting sphere is placed in an external electric field

$$\mathscr{E} = C\left\{\frac{x\,\mathbf{i} + y\,\mathbf{j} + z\,\mathbf{k}}{(x^2 + y^2 + z^2)^{3/2}}\right\} \text{ volts/m.}$$

The center of the sphere is on the z axis 1 meter from the origin. Find the translational force exerted on the sphere by the external field $\mathscr{E}$.
34. Regard the configuration composed of a conducting sphere at a potential of 9 volts and a point charge q of 10^{-8} coulomb a distance of 5 m from the center of the sphere. What is the magnitude and direction of the force on q?
35. A point charge of 10^{-7} coulomb is in equilibrium at a point midway between two semi-infinite horizontal parallel grounded conducting planes.
(a) If the separation between the planes is 10^{-2} m, what is the total induced charge on each plane?
(b) Assume that q is displaced a distance δ downward from the midpoint; using the two closest images to q, find the force on q to this order of approximation.
36. An infinite grounded conducting plane has a hemispherical conducting boss (or bump) of radius R. A point charge q is situated a distance d > R

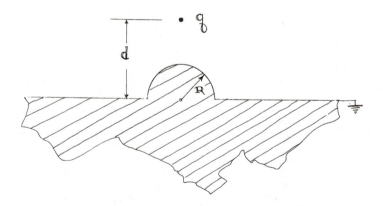

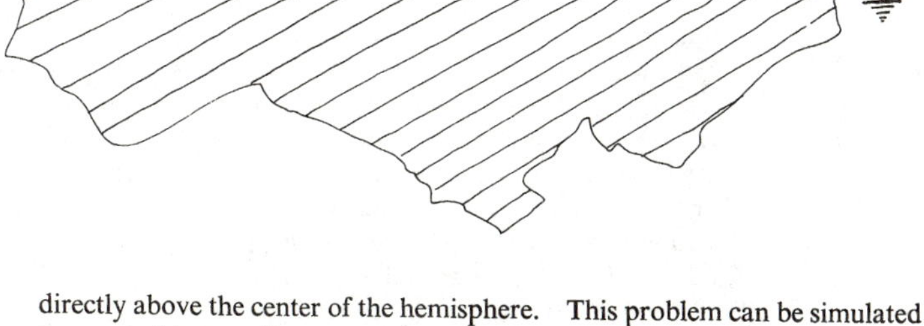

directly above the center of the hemisphere. This problem can be simulated by a set of image charges.
(a) Sketch the image charge configuration.
(b) What is the effective dipole moment relative to the origin O, the origin being at the center of the hemisphere?

(c) What is the effective quadrupole moment relative to O?
(d) What is the force on the point charge?

37. Two positive charges q are at equal distances d from the center of a grounded conducting sphere of radius R, where $R < d$. What is the total energy stored in the system if $d = 2R$ and if the angle between the position vectors of the two charges is 90°?

38. If a positive charge q is inside of an uncharged hollow conducting spherical shell of radius R at a distance l from the center, what is the force on q? Consider the sphere to be isolated and not grounded.

39. Two equal point charges q are placed a distance d from the center of a grounded conducting sphere of radius R. The point charges lie on opposite sides and on a straight line through the center.
(a) Find the total charge induced on the sphere. What is the sign of this charge?
(b) In the far field the system acts as a point charge at the origin. What is the magnitude and sign of this charge?

40. A long straight wire of radius δ is located parallel to and a distance d from the axis of a long grounded conducting cylinder of radius R ($d > R$). Assuming a voltage V impressed between the wire and the grounded cylinder, derive the equation for the field between the wire and cylinder.

41. A point charge of q coulombs is placed outside of an *uncharged* insulated metal sphere of radius R. The charge q is a distance d from the center of the sphere.
(a) Find the magnitudes and signs of the "*two*" image charges.
(b) Show that the force between the charge q and the sphere is

$$F = -\frac{q^2R^3}{4\pi\epsilon_0 d^3}\left\{\frac{2d^2 - R^2}{(d^2 - R^2)^2}\right\}$$

(*Hint:* One image charge can be placed at the center of the sphere and the spherical potential at $r = R$ can be maintained.)

42. A conducting sphere of radius R and carrying total charge Q is placed in an external field which initially was constant. Derive the final potential of this system.

43. An isolated conducting sphere of radius R is a distance d ($R < d$) from a semi-infinite grounded conducting plane. Using the method of images, find the capacitance of this system to an accuracy of the order $(R/d)^4$.

44. An isolated set of concentric conducting spheres of radii 10 cm and 5 cm respectively are insulated from one another. The region exterior to and between the spheres is a vacuum. A charge of $q_1 = +10^{-6}$ coulomb is placed on the outer sphere while the inner sphere carries a charge of $q_2 = -\frac{1}{2}(10)^{-6}$ coulomb. Under these conditions,
(a) Find the $\mathbb{C}$ and $\mathbb{P}$ matrices for this system.
(b) Find the potentials of both spheres relative to infinity.

45. A spherical conducting shell of radius R is split into two hemispheres. The hemispheres are separated by an insulating strip of negligible width. If one hemisphere is held at a potential $+V_0$ and the other at $-V_0$, using

these boundary conditions show that

$$V(r,\theta) = V_0\left\{\frac{3}{2}\left(\frac{r}{R}\right)P_1 - \frac{7}{8}\left(\frac{r}{R}\right)^3 P_3 + \frac{11}{16}\left(\frac{r}{R}\right)^5 P_5 + \dots\right\}, \quad 0 \leqslant r < R;$$

$$V(r,\theta) = V_0\left\{\frac{3}{2}\left(\frac{R}{r}\right)^2 P_1 - \frac{7}{8}\left(\frac{R}{r}\right)^4 P_3 + \frac{11}{16}\left(\frac{R}{r}\right)^6 P_5 + \dots\right\}, \quad R < r,$$

where P_l is the lth Legendre polynomial.

46. Three condensers of capacity 2, 4, and 6 μf respectively are connected in series. A voltage of 200 volts is impressed on the system.
    (a) Compute the charge on each.
    (b) Compute the voltage across each condenser.
    (c) Compute the energy stored in each condenser.
47. The three condensers of Problem 46 are disconnected and reconnected in parallel without the battery (+ charges together, − charges together).
    (a) Compute the charge on each.
    (b) Compute the total voltage across the parallel combination.
    (c) Compute the energy stored in each condenser.
    (d) Is the total energy larger or smaller than that determined in part (c) of Problem 46?
48. The plates of a parallel plate condenser are square, each side being w meters

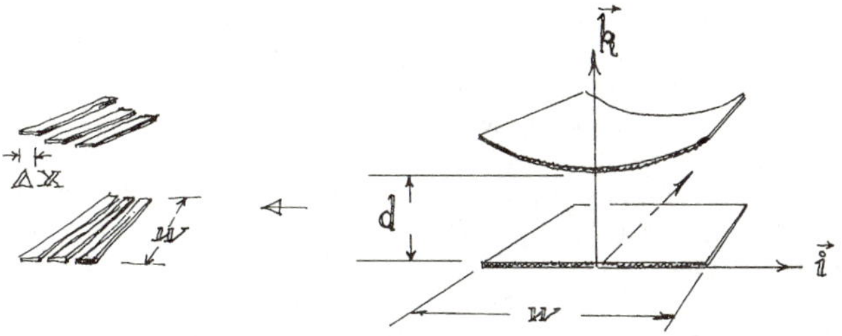

long. Initially the plate separation is d meters, but then the upper plate is warped in one dimension such that the separation of the plates z is given by

$$100(z - d) = x^2.$$

The separation z at any x is the same for all y. Compute the capacitance of the deformed condenser. (*Hint:* Assume a set of infinitesimal parallel plate condensers connected appropriately.)

49. The plates of a condenser are not quite parallel. The separation at one edge is $d + \delta$ and that at the opposite edge is $d - \delta$ ($\delta \ll d$). Neglecting edge effects, show that the capacity is approximately

$$C = \frac{\epsilon_0 A}{d}\left\{1 + \frac{1}{3}\frac{\delta^2}{d^2}\right\}$$

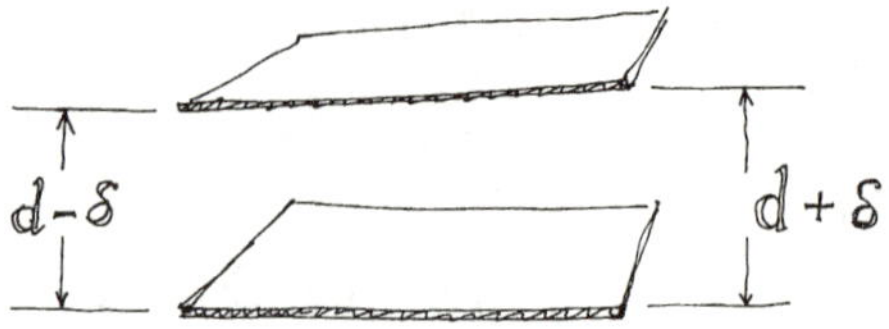

where A is the area of each plate, and use is made of the expansion

$$\log\left\{\frac{1+x}{1-x}\right\} = 2\left\{x + \frac{x^3}{3} + \ldots\right\}.$$

50. A slab of dielectric of dielectric constant ϵ fills the gap in a parallel plate condenser. A small spherical cavity of radius R is then cut into the dielectric near its center. A voltage V is impressed between the plates. The plate separation is d. Assuming that the dielectric remains uniformly polarized to first order, find the magnitude of the electric field intensity at the center of the spherical cavity in terms of the impressed voltage.
51. A conducting sphere of radius a carries a constant surface charge density σ. The sphere is surrounded by a shell of dielectric of inner radius a and outer radius b. If the dielectric constant is ϵ, find the total energy stored by this system.
52. A parallel plate condenser has plate areas A and plate separation d. The condenser is charged to a potential V by a battery of zero internal impedance.
    (a) Compute the force between the plates.
    (b) The battery remains connected and the plates are allowed to come together until their separation is d/2. How much mechanical work is done?
    (c) In part (b), how much energy is supplied by the battery? Account for the difference ($W_{battery} - W_{mech}$).
53. A parallel plate condenser of face area A and plate separation d is connected to a battery V. An uncharged conducting plate of face area A and thickness δ is then suspended by insulating supports between the plates of the condenser.
    (a) What is the capacitance without the inserted plate?
    (b) What is the capacitance after the plate has been inserted?
    (c) What is the change in the charge on the plates of the condenser?
    (d) Does the force between the condenser plates increase or decrease?
    (e) What is the change in stored energy? Account for the difference.
54. A parallel plate condenser is charged to a voltage V. The plate area is A and the separation between the plates is d. A molten dielectric of dielectric constant ϵ is then poured between the plates, completely filling the gap volume. The dielectric has the property that when it cools and solidifies, the dipole moment per unit volume, $\mathbf{p}_v$, is frozen, maintaining the alignment it had while cooling. The dielectric is allowed to solidify between the plates of the condenser with the voltage difference V impressed between the plates, and is then removed from the condenser. Derive the expression for

the magnitude and direction of the resultant field inside the isolated frozen dielectric, in terms of A, d, V, ϵ_0, and ϵ (or χ).

55. A parallel plate condenser has plate area A and plate separation d_1. A voltage V is established between the plates by a battery. A dielectric slab

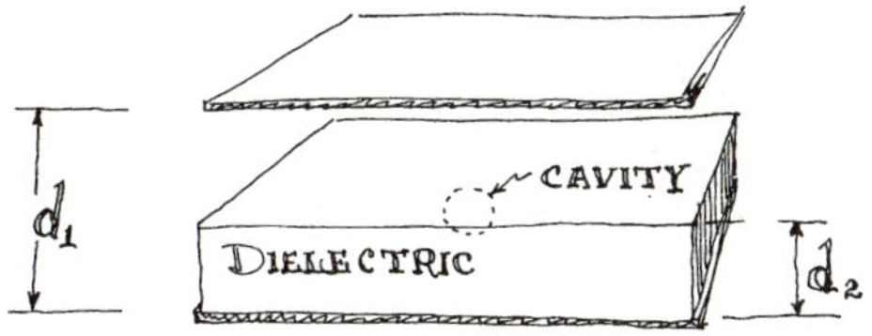

of dielectric constant $\epsilon = 1 + \chi$, face area A, and thickness d_2 ($d_2 < d_1$), is placed on one plate of the condenser. The dielectric slab has a spherical cavity of radius R ($R \ll d_2$) cut out of it. Derive the expression for the field at the center of the cavity in terms of V, d_1, d_2, A, R, and χ.

56. A student is given two boxes, each having two leads connected to the electrical equipment inside. He is told that one box contains a parallel plate condenser and the other a concentric spherical condenser. Given a battery, a ground, and a ballistic galvanometer, what simple measurements will specify which box contains the spherical condenser and which contains the parallel plate condenser? Explain. (Assume that the boxes do not affect the electrical part of the problem, i.e., that the boxes only hide the condensers.)

57. A parallel plate condenser has circular plates each of radius 20 cm and separated by a distance of 1.0 cm. The condenser is charged to a potential difference of 1000 volts and then insulated so that no charge escapes. An uncharged circular metal plate of radius 20 cm and thickness 0.5 cm is inserted between and parallel to the original plates without making contact. Neglect edge effects.

(a) Find the initial and the final stored energies.

(b) Find the final potential difference between the outside plates of the condenser.

(c) How much mechanical work has been done, and by what agent?

58. A parallel plate condenser has a plate area of 0.01 m^2 and a plate separation of 0.01 m. The condenser is charged to 1000 volts and the battery is removed (without dissipation of Q). With the plates insulated, a dielectric slab of face area 0.005 meter^2 and thickness 0.01 m is inserted between the plates, the slab being completely contained in the gap. The dielectric constant is $\epsilon = 5$.

(a) What was the initial total force between the plates before the dielectric was inserted?

(b) What is the final total force between the plates?

59. A parallel plate condenser is formed by two square conducting plates l meters on a side. Their separation is d. A wedge of dielectric is inserted

between the plates, the wedge thickness being d cm at one end and (d − δ) cm at the other (the face of the dielectric is square and l meters on a side). Show that the capacitance of the condenser with the wedge of dielectric between the plates is approximately

$$C = \frac{\epsilon_0 \epsilon l^2}{x\delta} \log \left\{1 + \frac{x\delta}{d}\right\}$$

where $\delta \ll d$.

60. A dielectric slab having a dielectric constant of $\epsilon = 5$ is placed in a homogeneous constant electric field of $+10^4$ **k** volts/m. A unit normal **n** perpendicular to the face of the slab makes an angle of 30° with the field. What is

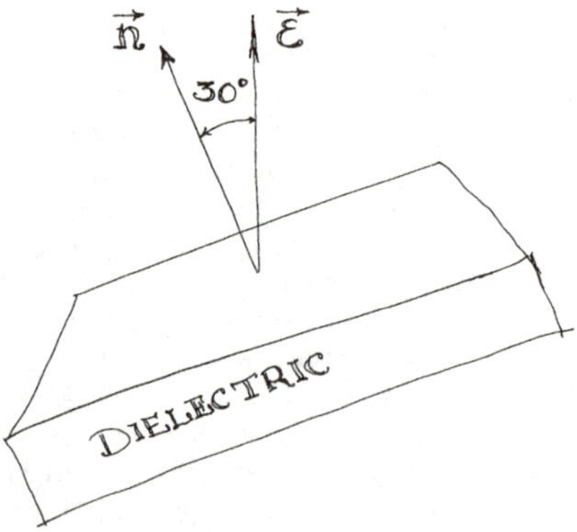

the magnitude and direction of the electric field inside of the dielectric slab?

61. A parallel plate condenser has a face area of 1 cm² and a gap width of 0.5 cm. A voltage of 1000 volts is impressed across the plate.

(a) Compute the force between the plates.

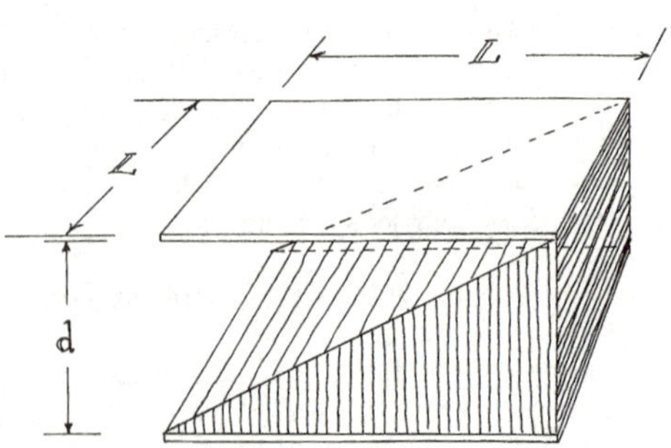

(b) A cube 0.5 cm on a side and having $\epsilon = 5$ is slid between the plates. What is the force between the plates under these conditions?

62. A parallel plate condenser has a face area of 100 cm^2 and a gap width of 1 cm. The condenser is charged and isolated. The magnitude of the charge is 10^{-6} coulomb. One plate surface is now coated 0.1 cm thick with a dielectric paint. The dielectric constant of the paint is 10. What is the change in voltage on the condenser?

63. In Problem 62, compute:
(a) The energy stored initially in the condenser gap.
(b) The energy stored after the dielectric has been inserted.
(c) The polarization energy stored in the dielectric.

64. Compute the field at the center of a dielectric sphere placed in a constant field set up between the plates of a parallel plate condenser. Assume that the radius of the sphere R is small compared to the plate separation, and that therefore the exciting field and the induced polarization can be taken as constant.

65. A parallel plate condenser has a gap partially filled with dielectric as shown.

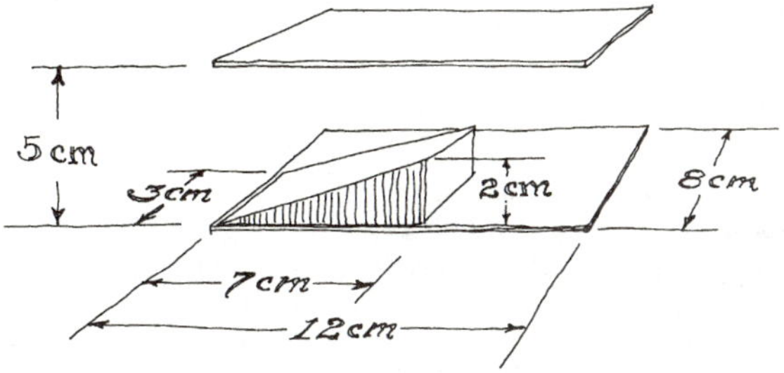

Compute the capacitance of this system. The dielectric constant is $\epsilon =$ 3.57.

66. A point charge q is imbedded in a semi-infinite sea of dielectric of dielectric constant ϵ. The charge q is a distance d from a grounded conducting plane extending to infinity at its boundaries. The plane and dielectric make contact. Assume the region above the xy plane at $z = 0$ contains the dielectric.
(a) Is the magnitude of the total charge induced on the plane greater than, equal to, or less than q? Explain.
(b) Develop the expression for the electric field in the dielectric.
(c) What is the force on q?

67. A point charge $q = 10^{-6}$ coulomb has a position vector $\mathbf{r}_q = 1\mathbf{k}$ m above a grounded conducting xy plane. An external agent moves the charge to a position $\mathbf{r}_q' = (-\mathbf{j} + 2\mathbf{k})$ m. How much work is done by the external system?

68. An insulated metal sphere having an outer radius of 3 cm carries a positive

charge of 10^{-9} coulomb. This sphere is surrounded concentrically by a hollow insulated conducting sphere which carries a charge of $-0.5(10)^{-9}$ coulomb. The outer sphere has an inner radius of 6 cm and an outer radius of 9 cm.

(a) Compute the charge on the inner surface and on the outer surface of the outer hollow sphere.

(b) Find the potential difference between the two spheres.

(c) Find the potential of the system with respect to $r = \infty$.

(d) Compute the electric field intensity at $r = 4.5$, 7, and 12 cm.

69. For a monatomic gas of atomic number Z assume that the electronic charge density is $\rho = \rho_0 e^{-\alpha r}$ where $\alpha \simeq 10^8\ \text{cm}^{-1}$.

(a) Show that $\rho_0 = -(Ze/8\pi)\alpha^3$.

(b) Using Gauss's law, show that the electric field of the electronic charge at a distance r from the center is

$$\mathscr{E} = \frac{-Ze}{4\pi\epsilon_0}\left\{(1 - e^{-\alpha r}) - (\alpha r)\left(1 + \frac{\alpha r}{2}\right)e^{-\alpha r}\right\}\frac{\mathbf{r}}{r^3}.$$

(c) Place the atom in an external constant electric field $\mathscr{E}_0$. If the displacement of the positive nucleus from the center of the electronic cloud is small compared to α^{-1}, that is, $\alpha r \ll 1$, show that the electric susceptibility is

$$\chi \simeq \left\{\frac{6\pi n_v}{\alpha^3}\right\}$$

where n_v = no. atoms/unit volume.

70. A dipole of moment **p** is a distance d above a grounded conducting plane. Obtain the expression for the force which the plane exerts upon the dipole.

CHAPTER V

1. The current density in a gas at S.T.P. is $10^{-8}\ \text{amp/m}^2$. Assuming that this current is set up by positive and negative heavy ions of the same mass, compute the mean ion velocity if the mean charge of either ion is $\pm 450e$, e being the electronic charge.)
2. Compute the power dissipated per meter by No. 14 copper wire carrying a current of 10 amperes. (The reader should look up a standard table of specifications for various wire sizes.)
3. Find the resistance of a copper tube of 1 mm outer diameter and 0.98 mm inner diameter.
4. A spherical copper shell has a diameter of 10 cm and a thickness of 0.1 mm. Find the resistance to electrical current when two contacts are placed at opposite poles (i.e., at the extremities of a diameter).
5. Twelve 10 ohm resistors are connected to form the edges and corners of a cube. Compute the total resistance between contacts placed at two diagonally opposite corners.

6. A resistance variable from zero to R is connected in parallel with a fixed resistor R. Plot the total resistance of the combination as a function of the resistance of the variable resistor.
7. In the circuit shown, determine the voltage between a and b and the magnitude and direction of the current from point a to point b.

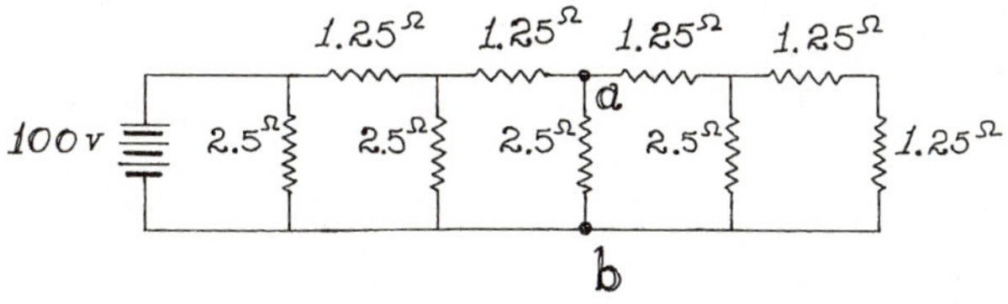

8. What is the voltage across the 1 microfarad condenser in the circuit shown?

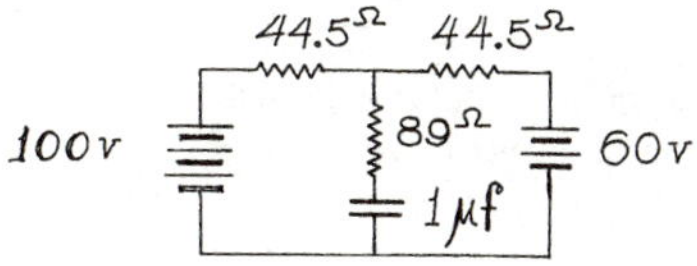

9. Find the value of the resistor R′ shown in the diagram such that the maximum possible power is dissipated in R.

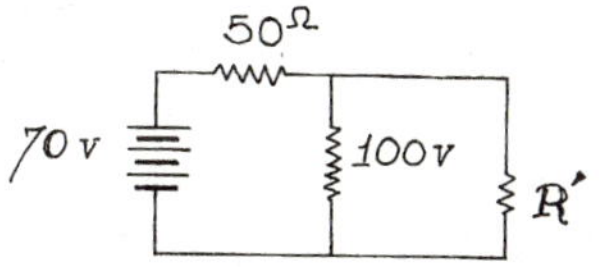

10. In the circuit shown,

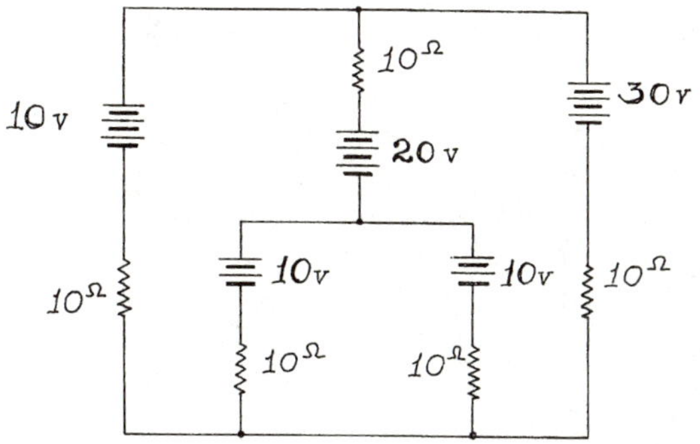

(a) Find the resistance matrix and the $\mathbb{R}^{-1}$ matrix. Check your answer by demonstrating that $\mathbb{R} \cdot \mathbb{R}^{-1} = \mathbb{I}$.

11. (a) Prove that the equivalence of the Y and Δ networks shown requires

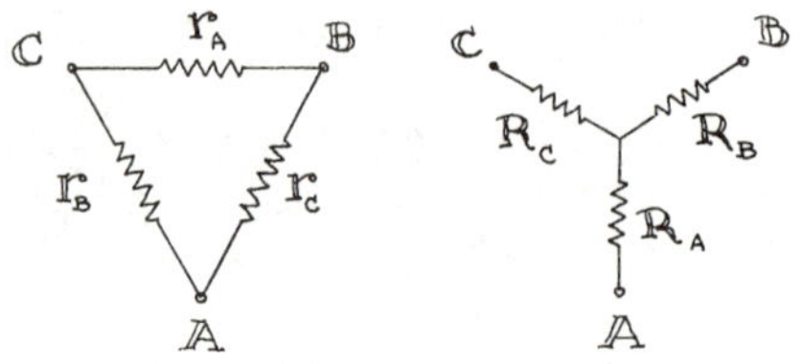

that the resistances be related in the following way:

$$R_A = \frac{r_B r_C}{r_A + r_B + r_C}\,; \quad R_B = \frac{r_C r_A}{r_A + r_B + r_C}\,; \quad R_C = \frac{r_A r_B}{r_A + r_B + r_C}\,.$$

The small r_j's are components of the delta network.

(b) Derive the components of the delta (the r_j's) in terms of the R_k's.

12. (a) Find the currents through A_1 and A_2.

(b) Find the potential difference across CD.

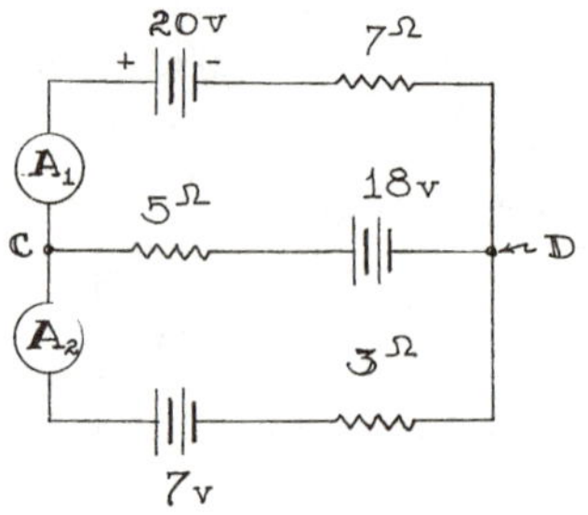

13. (a) Find the potential difference between A and B with S open.

(b) Find the current through the 10 ohm resistance with S closed.

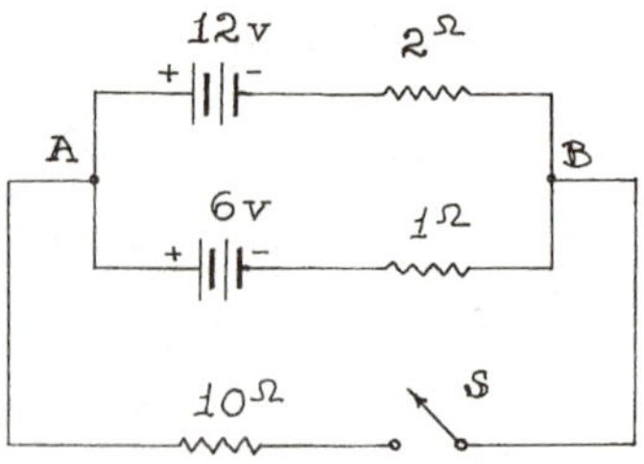

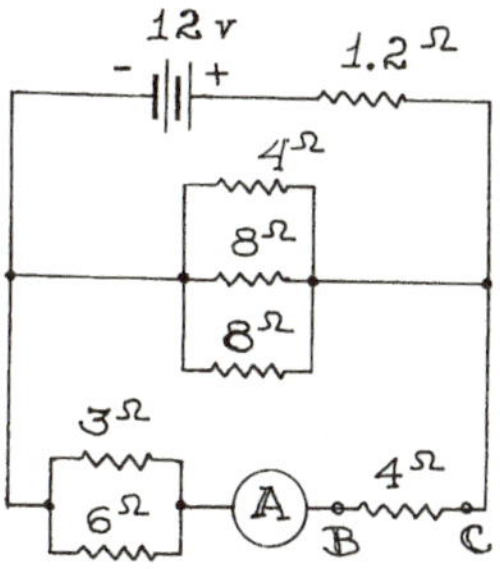

14. (a) Find the equivalent resistance of the accompanying circuit above.
    (b) Calculate the current through A and the voltage drop across BC.
15. A bartender named Thévenin was taking his evening smoke in the basement of his bar. Whilst mulling over the events of the day he leaned up against the gas main and received a sizeable electric shock. This bartender decided to investigate the properties of the electrical phenomenon of which he was now so well aware. After borrowing a multimeter from the pinball parlor next door, he measured the open circuit voltage between the gas main and an adjacent water pipe. This open circuit voltage was 100 volts DC. The bartender then visualized the possibility of heating a portion of his basement for free. To investigate further he measured the short circuit current between the gas main and the water pipe and found it to be 5 amperes DC.
    (a) What should the resistance of the electric heater be for maximum heating from this source?
    (b) What is the maximum heat output from the heater if the conversion factor is 4.18 joules/calorie?
16. In the corresponding figure,

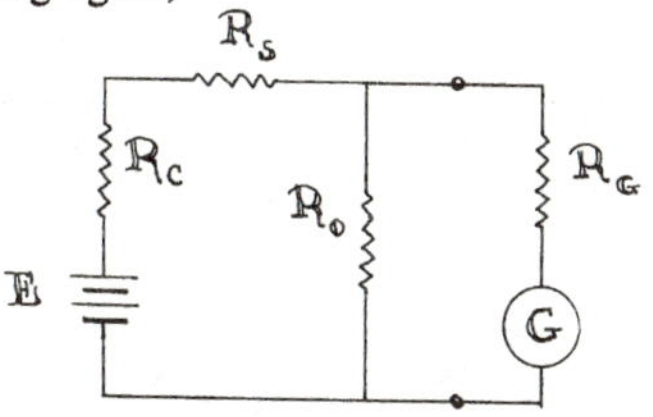

    (a) Calculate the current through G by calculating the equivalent resistance of the circuit.
    (b) Calculate the current through G using Thévenin's theorem.
17. Prove that maximum power is dissipated in the load resistor of the circuit shown when $R_L = R_E$. R_E and E are understood to be fixed.

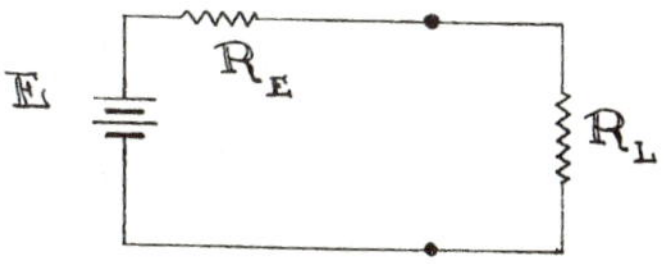

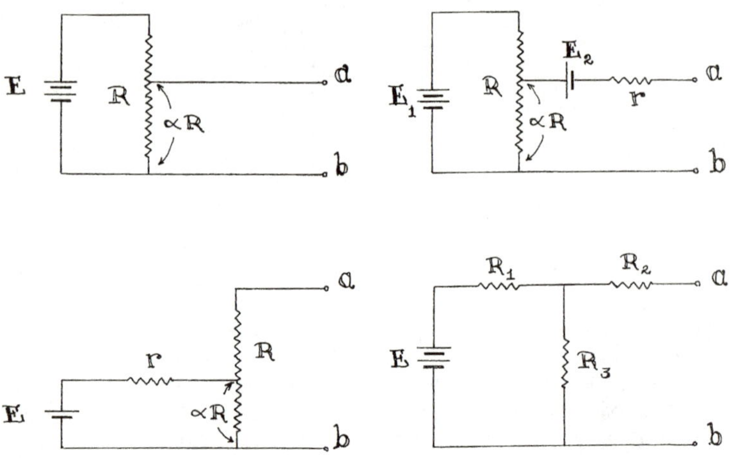

18. Replace the circuits above to the left of the terminals a and b by an equivalent source of emf in series with a resistance, using Thévenin's theorem. In all cases, $0 < \alpha < 1$.

19. In the accompanying circuit at the right, all resistances are 1 ohm.
   (a) Calculate the current flowing in resistor x.
   (b) Calculate the current flowing in resistor y.

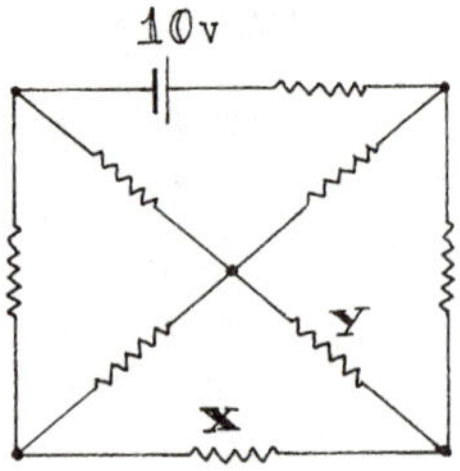

20. Calculate the galvanometer current in the Wheatstone bridge shown below, using Thévenin's theorem and a delta-wye transformation.

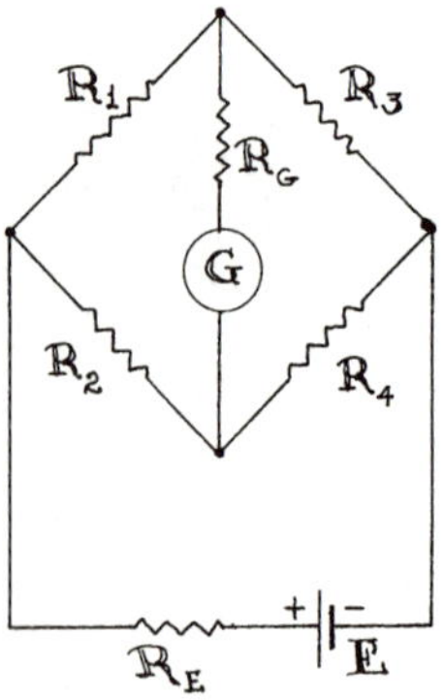

21. If the bridge shown is out of balance by 1 ohm, how much current flows in the galvanometer?

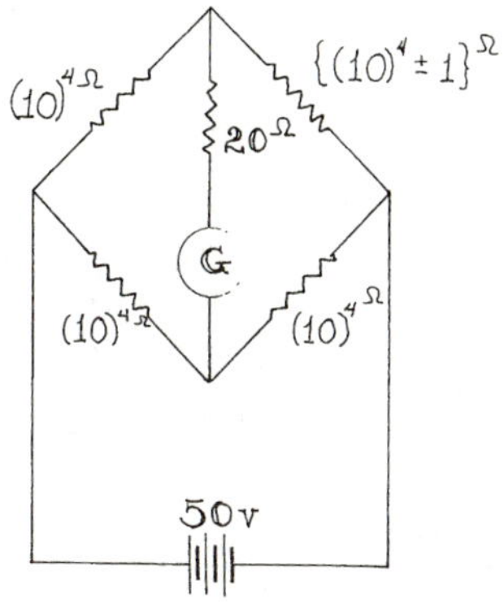

22. Find the resistance between the terminals a and b shown in the diagram.

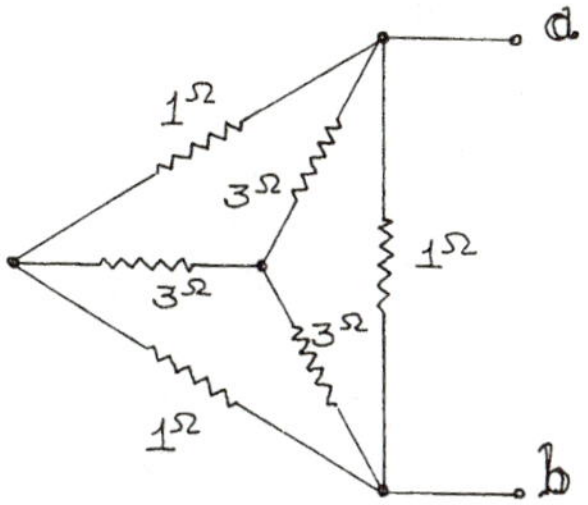

CHAPTER VI

1. A particle of mass m and charge q moves in a region where $\mathbf{B} = B_0\,\mathbf{k}$ and $\mathscr{E} = \mathscr{E}_0\,\mathbf{j}$. At $t = 0$, $\mathbf{v} = v_0\,\mathbf{i}$. Under what conditions will the velocity remain constant?
2. A beam of electrons reaches a velocity **v** after falling through a potential difference V. The final beam is cylindrical and has a uniform volume charge density ρ. The cylindrical volume has a radius R. Show that the total electrostatic and electromagnetic radial acceleration of an electron on the surface ($r = R$), in a region free of external fields, is given by

$$a = \frac{I}{2\pi\epsilon_0}\left[\frac{e}{2mV}\right]^{1/2}\frac{1}{R}\left(1 - \frac{v^2}{c^2}\right)$$

where $$I = \pi R^2 \rho v.$$

3. At a time $t = 0$, two electrons are moving parallel to the z axis with a constant and equal velocity of 0.99c (that is, $\beta = 0.99$). They are separated by a distance of 10^{-10} m. Considering both the electric and magnetic

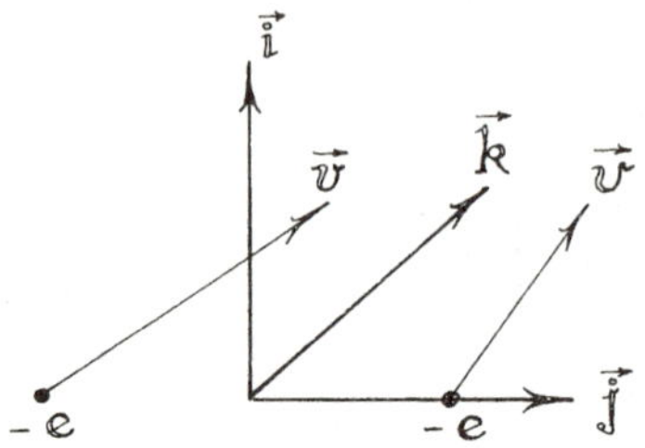

interaction, compute the force between the electrons. Assume the position vector between the electrons to be perpendicular to their velocities.

4. An electron moves in a region containing crossed electric and magnetic fields. The magnetic field is constant $\mathbf{B} = -B_0\,\mathbf{k}$, and the electric field is constant $\mathscr{E} = -\mathscr{E}_0\,\mathbf{j}$. The electron starts from rest at the origin at $t = 0$.
(a) Show that the parametric equations of motion are that of a cycloid,

$$x = \frac{\mathscr{E}_0}{\omega_c B_0}\{\omega_c t - \sin\omega_c t\}$$

and

$$y = \frac{\mathscr{E}_0}{\omega_c B_0}\{1 - \cos\omega_c t\}.$$

Define ω_c.
(b) What effect does an initial x component of velocity have upon the orbit (that is, $\mathbf{v}_0 = v_0\,\mathbf{i}$)? Sketch the orbits for a and b.

5. An alpha particle of velocity 10^6 m/sec moves in an orbit perpendicular to a constant magnetic field of 1 weber/m^2.
(a) Find the radius of the orbit and the time required for one revolution.
(b) How fast must an electron be moving to trace out the same circular orbit?

6. A region containing crossed electric and magnetic fields is established. The magnetic field is constant in space and time, $B = B_0\,\mathbf{k}$. The electric field vector is parallel to the xy plane,

$$\mathscr{E}(t) = -\{\cos\omega t\,\mathbf{i} + \sin\omega t\,\mathbf{j}\}.$$

An electron starts from rest at the origin at $t = 0$. $\mathscr{E}_0 = 10^5$ volts/m, and ω_c is the cyclotron frequency.
(a) How long will it take the electron to reach an energy of 10^3 electron volts?
(b) At this energy what is the location of the electron?
(c) How many revolutions about the origin has it made?

7. A cylindrical beam of electrons (of charge $q = -e$ and mass m_0) has an initial velocity of $\mathbf{v}_0 = v_0\,\mathbf{k}$. Initially the volume charge density $\rho = -ne$

is constant, n being the number of electrons per unit volume. Let R be the radius of the cylinder. Consider the fields acting upon an electron at a distance r from the **k** axis.

(a) Show that the electric field acting upon the electron at any time t is given by

$$\mathscr{E} = -\frac{e}{\epsilon_0}\frac{\mathbf{r}}{r^2}\int_0^r n(r')r'\,dr'$$

where n(r′) is the volume distribution of electrons at a time t, r being the perpendicular distance from the cylinder axis.

(b) Show that the magnetic field at **r** is given by

$$\mathbf{B} = -\mu_0 e\,\mathbf{v} \times \frac{\mathbf{r}}{r^2}\int n(r')r'\,dr'.$$

(c) Show that the initial acceleration (at t = 0) of an electron at r < R is

$$\frac{d^2\mathbf{r}}{dt^2} = \frac{e^2 n}{2m_0\epsilon_0}\{1 - \beta^2\}\,\mathbf{r}$$

where $\beta = v/c, \quad 1/c^2 = \epsilon_0\mu_0,$

and $\mathbf{r} = x\,\mathbf{i} + y\,\mathbf{j}.$

(d) Show that the position of an electron at a time δt (shortly after t = 0) is to first order

$$r = r_0 \cosh(\alpha\,\delta t)$$

where r_0 is the position at t = 0 and

$$\alpha = \left\{\frac{e^2 n_0}{2m_0\epsilon_0}(1 - \beta^2)\right\}^{1/2}.$$

(e) Will the position vector **r** increase in magnitude more rapidly than cosh αt or less rapidly, as the beam spreads out?

8. A 60-in. diameter cyclotron accelerates deuterons to 16 million electron volts. What is the magnitude of the magnetic induction field necessary to operate such an instrument? What oscillator frequency is necessary?

9. Electrons are emitted from one plate of a parallel plate condenser with

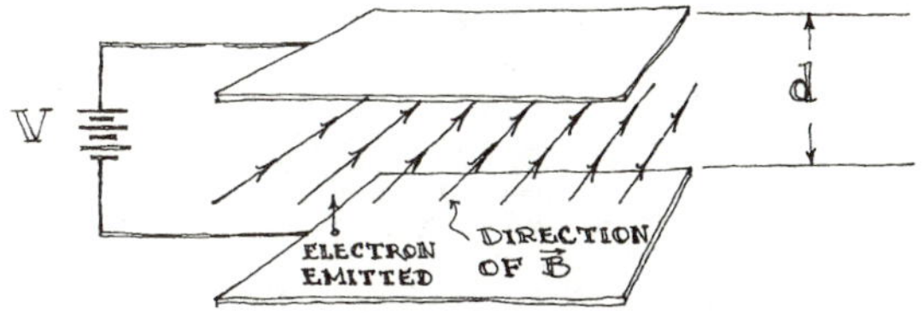

essentially zero velocity. If the plates, separated by a distance d, are parallel to a constant magnetic field B and the voltage between the plates is V, compute the orbit of the electron.

10. The potential difference between the axis and the wall of a cylindrical conducting shell is 10^3 volts. A constant, axially directed magnetic induction

field B is superposed. If the magnitude of the magnetic field is 0.1 weber/m^2, find the orbit of an electron emitted radially from the axis with essentially zero velocity. Plot this orbit.

11. A plane circular wire loop of radius R carries a current I. The plane of the loop is perpendicular to a constant magnetic field $\mathbf{B} = B_0\,\mathbf{k}$. What is the total force on half the loop? From this obtain the tension in the wire.

12. A current loop is formed by a semicircular conducting wire and a straight section of wire. The current carried in the loop is 10 amperes.

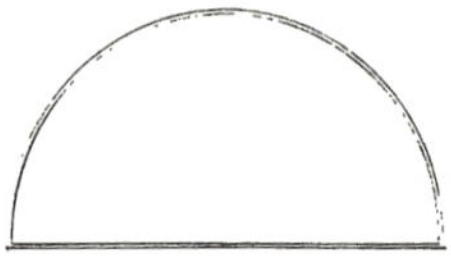

(a) Find the magnetic moment.
(b) The current loop is placed in a constant external field $|\mathbf{B}| = 0.01$ weber/m^2. What is the torque on the loop when the normal to the loop makes an angle of 30° to the magnetic induction vector?
(c) Prove that when the loop normal **n** is parallel to **B**, the loop is in equilibrium.

13. The coil of a d'Arsonval galvanometer has 100 turns and encloses an area

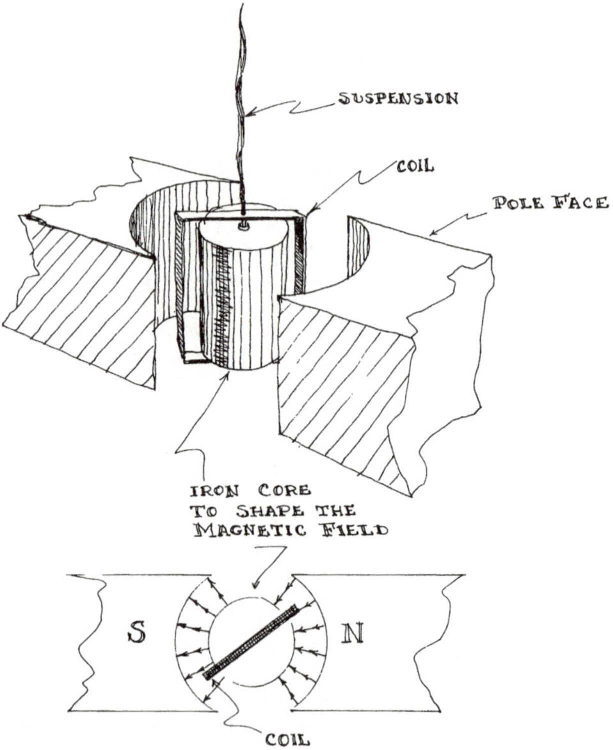

of 5 cm^2. The magnetic induction in the region of the coil is everywhere 0.10 weber/m^2. The torsional constant of the suspension is 10 newton-m/degree. If the field is always parallel to the face of the coil for reasonable rotations, find the angular deflection per microampere in the coil.

14. A magnetic moment $\mathbf{M} = M_0\,\mathbf{k}$ resides at the center of a circular current loop. The circle is in the xy plane; the radius is R and the current is I.
(a) If the magnetic moment is moved a distance z along the z axis, compute the work done.
(b) If the magnetic moment is displaced a small distance along the z axis and then released, compute its period of linear oscillation about the xy plane.

15. The following five points are connected by straight wire segments: (0,0,0), (0.1,0,0), (0,0.1,0), (0,0.1,0.1), and (0,0,0.1). A current I flows through the

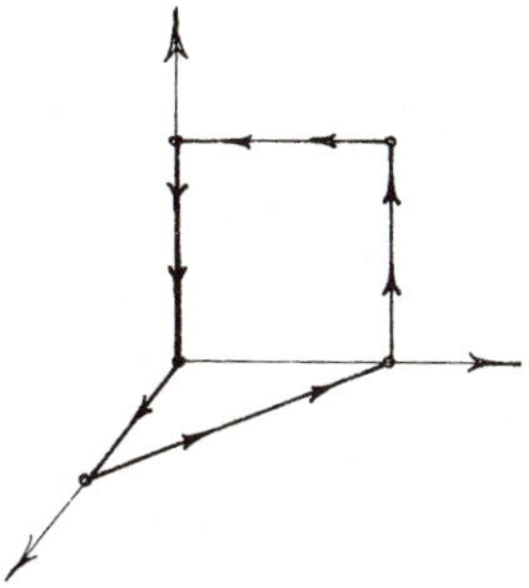

system in the direction described by the ordering of the points. The system is in a region containing a constant magnetic field $\mathbf{B} = 0.2\,\mathbf{k}$ weber/m^2.
(a) Compute the potential energy of this loop.
(b) Compute the torque on this loop about the origin.

16. A circular loop of wire of radius R, lying in the xy plane and centered at the origin, carries a current I. A magnetic field varies in the plane of the loop such that the z component goes as $(a/R)B_0e^{-z^2/\alpha^2}$, where a is the cylindrical variable

$$a = \sqrt{x^2 + y^2},$$

and B_0 is a constant. Compute exactly the energy stored in the loop in the presence of the external field.

17. A square loop of wire of side R, lying in the xy plane and centered at the origin with the sides parallel to the x and y axes, carries a current I. A magnetic field present in the region of the loop has the form

$$\frac{\mu_0 I}{2\pi}\,\frac{\mathbf{i} \times (\mathbf{r} - 2R\,\mathbf{j})}{|\mathbf{r} - 2R\,\mathbf{j}|^2}\,.$$

Using the exact relation for the potential energy, compute the stored energy.

18. A magnetic dipole $\mathbf{M} = -M_0\,\mathbf{k}$ is moving with a velocity $\mathbf{v} = v_0\,\mathbf{k}$ at $t = 0$, and its position at $t = 0$ is $x = 0$, $y = 0$, $z = -\alpha$. A magnetic field $\mathbf{B}$ has a nonzero value in the region $-\alpha^{-1} \leqslant z \leqslant \alpha^{-1}$, and the shape of the axial field is given by

$$\mathbf{B}(0,0,z) = \frac{B_0\,\mathbf{k}}{1 + \alpha^2 z^2}.$$

Outside this region the field is zero. The mass of the dipole is m, and $v_0 = \sqrt{MB/2m}$.
(a) How far into the field will the dipole penetrate?
(b) What is its velocity (magnitude and direction) as $t \to \infty$?

19. What is the cyclotron frequency of an electron in the earth's magnetic field? (Look up the field at your location.)

20. A 1-meter conducting rod is free to rotate about one end. If a constant field of 0.05 weber/m^2 is present parallel to the axis of rotation, find the value of the current passing through the rod which will raise the rod to a horizontal position. Assume that the mass per unit length of the rod is 0.05 kg/m.

21. Carry out the following demonstration that it is impossible to measure the intrinsic magnetic moment of the electron ($e\hbar/2m$) by a Stern-Gerlach experiment. It will be shown that in such an experiment the Lorentz force deflection is much greater than the deflection due to the magnetic moment.

Assume that an electron given a velocity of 10^6 m/sec in the $\mathbf{j}$ direction passes into an inhomogeneous magnetic field. The geometry of the magnetic field is such that there is symmetry relative to the yz plane and the moment deflections are in the z direction. The electron enters the magnetic field at $t = 0$ at the point (0,0,0) with its magnetic moment in the $-\mathbf{k}$ direction, $\mathbf{M} = -M_e\,\mathbf{k}$ ($M_e = e\hbar/2m_0$). The magnetic field is given by

$$\mathbf{B} = b_1(x,z)\,\mathbf{i} + [b_2(x,z) + B_0]\,\mathbf{k}.$$

There is no x or y component of the field at $x = 0$; thus $b_1(0,0) = 0$, and $(\partial b_2/\partial z)_{x=0}$ = a constant = ζ_0.
(a) The electron will be deflected initially in the $-x$ direction (effect of translational motion). Show this. Then show that after the electron is deflected there is more than one force component.
(b) Using Maxwell's second equation, show that the field component in the $\pm x$ direction is given by $b_1\,\Delta x = -\zeta_0\,\Delta x$ for small deviations from $x = 0$.
(c) Show that the translational force on the magnetic moment is $-M_e\zeta_0\,\mathbf{k}$.
(d) Show that after the electron has been deflected from the $\mathbf{j}$ axis a distance $\Delta x = -\delta$, the total z component of the translational force is given by

$$F_z = \{-M_e + e v_0 \delta\}\zeta_0$$

where $\mathbf{v}_0 = v_0\,\mathbf{j}$.
(e) How far has the electron traveled in the $-x$ direction when $F_z = 0$? How far has the electron traveled along the y axis when $F_z = 0$?
(f) Does the interaction between $\mathbf{v}_0$ and $\mathbf{B}$ in the $\mathbf{k}$ direction become larger than $M_e\zeta_0$ as the motion continues beyond the point specified by $F_z = 0$?

Hence show that it is not possible to measure the magnetic moment of the free electron by such an experiment.

22. Assume that the electron is a homogeneously charged sphere of radius 10^{-13} cm. What would have to be the angular velocity of this sphere (rotating about a diameter) to account for the electron spin magnetic moment? For this angular velocity, find the velocity of a point on the equatorial circle of this rotating sphere.
23. Find the deflection of a singly charged silver ion after a traversal in the plane of symmetry of a 1-meter long inhomogeneous magnetic field. In the symmetry plane the field varies as $B_0[1 + (z/L)]$ where $B_0 = 1$ weber/m^2 and $L = 10^{-2}$ m. Assuming that the ion moves with a velocity of 10^3 m/sec, find the approximate lateral deflection in one meter. Initially the beam moves along the y axis.

CHAPTER VII

1. Two long straight wires lie parallel to the z axis. One wire carries a current of 10 amperes in the +z direction; it lies on the y axis at $5(10)^{-3}$ **j** m. The other wire carries a current of 5 amperes in the $-$z direction; its position

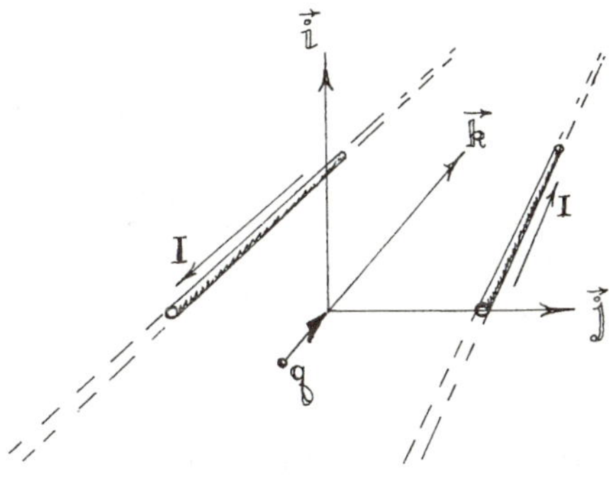

on the y axis is given by $-5(10)^{-3}$ **j** m. A proton ($q = +1.6(10)^{-19}$ coulomb) is ejected along the z axis with an initial velocity of 10^8 cm/sec. What is the magnitude and direction of the initial force on the proton?
2. Two long straight wires are parallel to the z axis. One wire is at $x = 0$, $y = +l$, with current I in the $+$**k** direction. The other wire is at $x = 0$, $y = -l$ and its current is in the $-$**k** direction. Find the force on a magnetic moment **M** $=$ M **i** lying in the xz plane.
3. A flexible copper solenoid forms a helix of radius R and a total of N turns, and an initial length x. The helix is attached to a support and to a mass

m which rests on a table below. What value of current I is required in order to lift the mass from the table, in terms of N, x, R, m and g?

4. Four long straight wires are parallel to the z axis. The current direction and position of each wire are as follows: $-\mathbf{k}$, $(0,-1,z)$; $-\mathbf{k}$, $(1,0,z)$; $+\mathbf{k}$, $(0,1,z)$; $+\mathbf{k}$, $(-1,0,z)$. Compute the field **B** at (0,0,0).
5. Two long straight wires are parallel to the z axis and carry currents I in the $-z$ direction. The position of one wire is (0,1,0) meters and that of the other is (0,−1,0) meters. A particle of mass m and charge q is fired in a direction making an angle α to the axis in the yz plane. The initial particle speed is v_0. Assuming α a small angle, find the frequency of the oscillation the particle makes about the z axis.
6. A toroidal coil of N = 500 turns carrying a current of 1 ampere has a square cross section 6 cm on an edge. The inner radius of the toroid is 6 cm and the outer radius is 12 cm.
   (a) What is the value of the field inside of the coil? Use the circuital law.
   (b) What is the approximate value of the field at the center of the toroid?
7. A coil of N turns is wound upon the surface of a cone as shown in the diagram. What is the field at the center of the base of the cone?

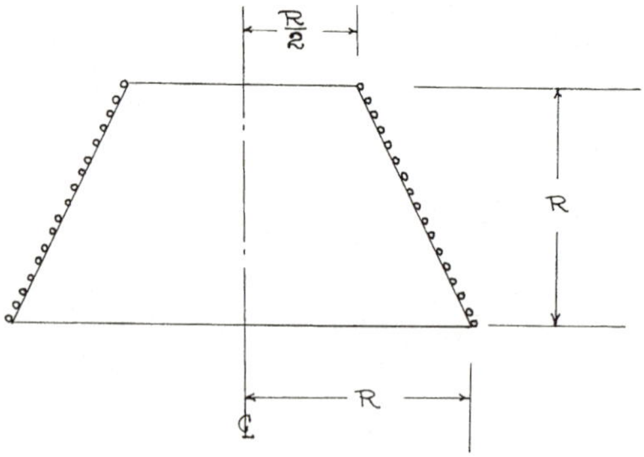

8. Find the direction and magnitude of **B** at a point P in the vicinity of a current-carrying strip of infinitesimal thickness and width b. The current is in the $+\mathbf{k}$ direction, and the strip extends from $-\infty$ to $+\infty$ along the z axis.
9. Two circular plane parallel coils coaxial on the z axis each having N turns and carrying a current I in the same direction are separated by a distance R. The effective radius of each coil is R.
   (a) Obtain an expression for the axial field between the coils.
   (b) The origin is midway between the loops. Using Maxwell's second equation, find the radial component of the field at a distance r from the axis in the xy plane.
   (c) At what distance from the z axis is the radial component of the field 10% of the axial field?

10. A solenoid of radius R is L meters long and has N turns per meter. A current I is passed through the coils of the solenoid and the solenoid is placed in a constant external magnetic field **B**.
(a) Compute the torque on the solenoid about its center if the solenoid axis makes an angle α with the direction of the magnetic field.
(b) Consider the solenoid as a bar magnet of length L. If the torque on a bar magnet can be expressed as mLB sin α where m is the strength of the so-called magnetic pole, what is the equivalent pole strength of the solenoid?
11. A long, solid cylindrical conductor of radius b carries a constant density current $\mathbf{J} = J_0 \mathbf{k}$. A cylindrical hole of radius a with axis parallel to the cylinder axis is cut in the conductor. c is the displacement of the axis of the hole from the axis of the conductor, and $c + a < b$. Find the general expression for the field outside and inside of the conductor.
12. A toroid of square cross section carries 1000 turns of wire. The inner radius of the toroid is 5 cm and the outer radius is 10 cm. A long straight wire carrying 10 amperes passes along the axis of rotational symmetry of the toroid. If the current in the toroidal windings is 10 milliamperes, compute the energy in joules stored in the mutual interaction of the toroid and the wire.
13. A cylindrical straight wire of radius R carries a constant current density parallel to its axis. Prove that curl $\mathbf{B} = \mu_0 \mathbf{J}$ on the interior of the wire. Use the circuital law to develop **B** in the interior.
14. A hollow cylinder of outer radius b and inner radius a carries a constant current density $\mathbf{J} = J_0 \mathbf{k}$. Derive the expressions for the magnetic induction field in the three regions

$$0 \leqslant r \leqslant a, \quad a \leqslant r \leqslant b, \quad \text{and} \quad b \leqslant r.$$

15. A coaxial cable is formed of a solid cylindrical conductor of radius $(a - \delta)$ and an outer cylindrical sheath of outer radius b and inner radius $(a + \delta)$. The two conductors are insulated from one another. A total current I flows in each conductor, in opposite directions in each. The current is uniformly distributed in each, i.e., $\mathbf{J}_1 = -\mathbf{J}_2$, and $\mathbf{J}_1$ and $\mathbf{J}_2$ are constant.
(a) Compute the current densities in each conductor.
(b) Compute the magnetic induction field for the regions $r > b$, $b > r > a + \delta$, $a + \delta < r < a - \delta$ and $a - \delta > r > 0$.
16. A spherical volume of radius R contains a constant volume density of charge ρ. The sphere rotates about a diameter with a constant angular velocity $\boldsymbol{\omega}$.
(a) Show that the magnetic induction field **B** at the center is

$$\mathbf{B}_0 = \tfrac{1}{3}\mu_0 \rho R^2 \boldsymbol{\omega}.$$

(b) What is the value of **A**, the vector potential at the center?
17. A long, solid cylindrical conductor of radius R carries a constant current density $\mathbf{J} = J_0 \mathbf{k}$ and a total current I. The axis of a cylindrical cavity of radius R/2 is located a distance R/2 from the axis of the conductor. A long thin straight wire carrying a current I in the $-\mathbf{k}$ direction lies in the plane of the conductor axis and cavity axis and is a distance 3R/2 from the

center of the conductor. What is the direction and magnitude of the force per unit length on the wire?

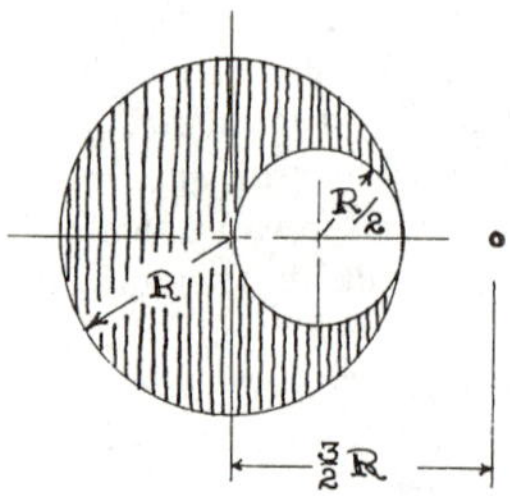

18. A cylindrical cavity of radius a is drilled into a long cylindrical conductor of radius R ($a < R$) with the axis parallel to the axis of the conductor. The distance between the axes is c. If a total current I is carried by the conductor, show that the energy *per unit length* stored in the cavity is given by

$$U_{\text{u.l.}} = \frac{\mu_0 I^2 c^2 a^2}{8\pi(R^2 - a^2)^2}.$$

19. A long straight wire lying along the z axis carries a current I in the positive z direction. A magnetic moment $\mathbf{M} = M_0 \mathbf{j}$ is placed on the x axis a distance x_p from the wire.
(a) What is the potential energy of **M** in the field of the wire?
(b) Show that the translational force on **M** in the position described is given by

$$\mathbf{F} = -\mu_0 \frac{MI\,\mathbf{i}}{2\pi x_p^{\,2}}.$$

20. Through the center of a circular wire loop carrying a current I passes a long straight wire carrying an equal current I. The smallest angle the plane of the loop makes with the wire is 60°. Is there a net force on the wire? Explain your answer.
21. Consider the spherical section shown. The volume contains a constant

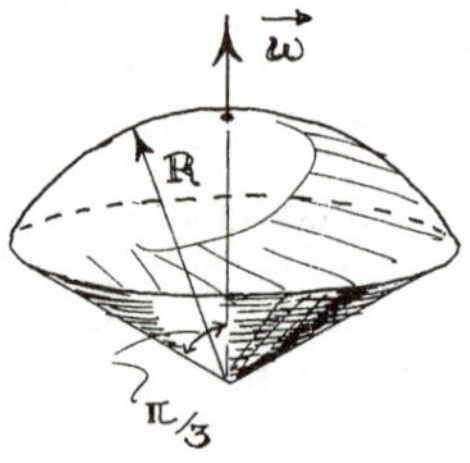

density of charge ρ which spins about the **k** axis with a constant angular velocity $\boldsymbol{\omega}$. Show that the magnetic induction field at the origin is given by

$$\mathbf{B}_0 = \tfrac{5}{96}\mu_0\rho R^2\boldsymbol{\omega}.$$

22. Find the magnetic induction field inside of a spherical shell of surface charge density σ and radius R which spins about a diameter with a constant angular velocity $\boldsymbol{\omega}$. Using the magnetic vector potential which provides this field, show that the total magnetic energy stored in the spinning sphere is

$$U = \frac{\pi^2}{8}\mu_0\sigma^2\omega^2R^5.$$

23. A disk of charge spins about its symmetry axis with a constant angular velocity $\boldsymbol{\omega}$. The surface charge density σ is constant and the radius of the disk is R. Find the axial field at any point on the symmetry axis.
24. A magnetic moment $\mathbf{M}_1 = M_0\,\mathbf{k}$ is fixed at $r_1 = a\,\mathbf{i}$, and $\mathbf{M}_2 = -M_0\,\mathbf{k}$ is fixed at $r_2 = -a\,\mathbf{i}$. Compute the far field vector magnetic potential.
25. A hollow sphere of radius R has three axially symmetric conducting regions on the surface, as shown in the figure. Positive charge of constant surface

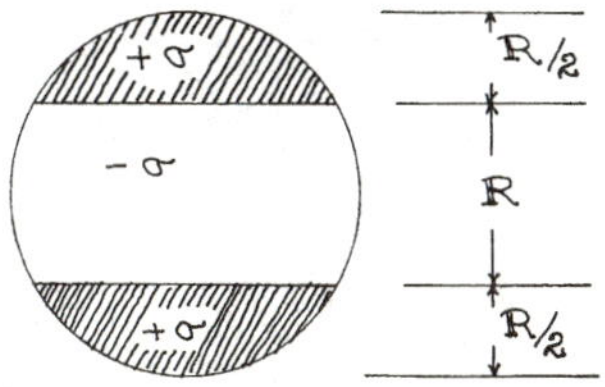

density $+\sigma$ is placed on the upper and lower cap. Negative charge of constant surface density $-\sigma$ is placed on the surface in the midregion. The sphere spins about the axis of symmetry with a constant angular velocity $\boldsymbol{\omega}$. Compute the magnetic induction field at the center O.

CHAPTER VIII

1. A set of parallel conducting rails is connected to a battery as shown on the next page. A constant magnetic field **B** is directed perpendicular to the plane of the rails. A conducting cross bar is constrained by an external force $\mathbf{F}_{ex}$ to slide at a constant velocity $\mathbf{v}_0$.
(a) If the current is I, calculate the rate at which the external force does work on the system.
(b) Compare this with the rate at which magnetic flux is swept out.

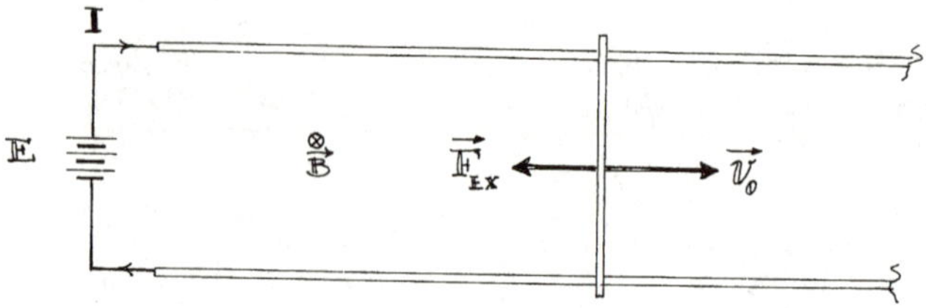

2. A set of parallel conducting rails is connected to a battery (as in Problem 1) of emf E. Each rail has a resistance of r ohms per meter. A constant magnetic field **B** is directed perpendicular to the plane of the rails. If a resistanceless crossbar starts from rest at x_0 at $t = 0$, the total resistance at x_0 being $2r_0$, find
   (a) The force on the crossbar after any time t,
   (b) Its velocity after a time t,
   (c) Its position as a function of time.
3. A circular conducting loop of wire in a magnetic field **B** rotates about a diameter with a constant angular velocity **ω**. If **ω** and **B** are parallel,

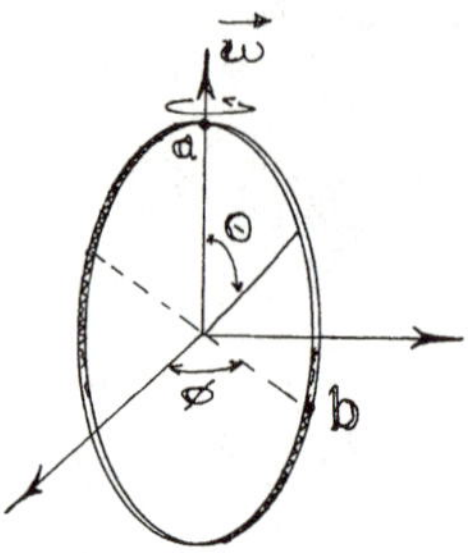

show that the emf developed between points a ($\theta = 0$) and b ($\theta = \pi/2$) is given by

$$E = \omega BR^2/2$$

where R is the radius of the loop.
4. Two connected square loops of wire lie in the same plane as a long straight

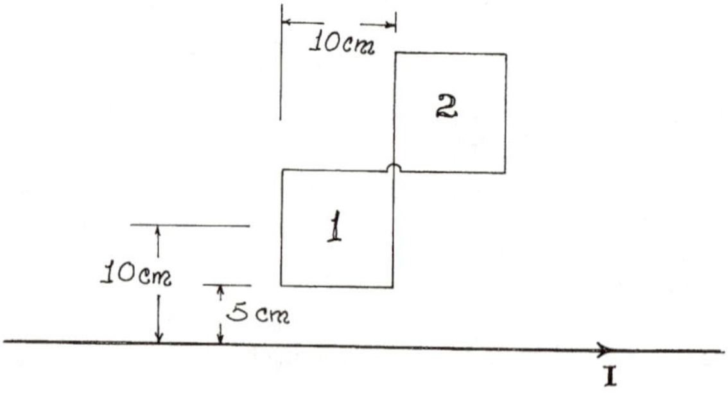

wire carrying a current I. The square loops are 10 cm on a side and each has two sides parallel to the wire. The center of loop 1 is 10 cm from the wire, while the center of loop 2 is 20 cm from the wire. Assuming $I = I_0 \sin \omega t$, compute the total emf induced in the loops.

5. A long straight wire carrying 25 amperes lies upon a table of nonmagnetic material. A square loop of wire is also placed upon the table. The loop is 6 cm on a side with two sides parallel to the straight wire, and its center is 10 cm from the straight wire. If the square carries a current of 4 amperes, what is the resulting magnetic force on it?
6. A solenoid of length L and radius a is made up of N circular loops carrying a current I. The solenoid is placed with its axis along the symmetry axis of a large circular current loop I of radius R ($R \gg a$).

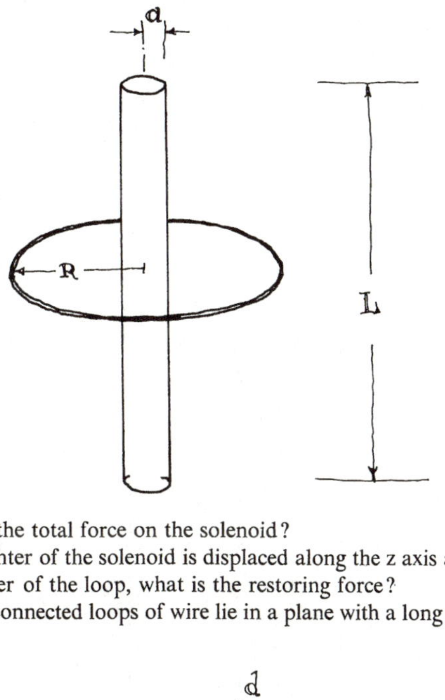

(a) What is the total force on the solenoid?
(b) If the center of the solenoid is displaced along the z axis a distance δ **k** from the center of the loop, what is the restoring force?

7. Two square, connected loops of wire lie in a plane with a long straight wire.

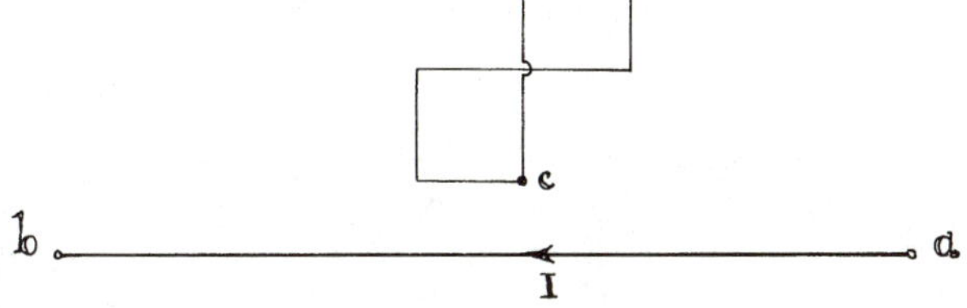

Current increases in the long straight wire as

$$I(t) = I_0(1 - e^{-\alpha t});$$

the direction of positive current is from a to b. If the sides of the two squares are equal and if the geometry of the loop is as shown, does the induced current flow from c to d or from d to c? Explain.

8. A long wire is connected to a source of emf as shown below at the left. A single square loop of wire of resistance R is located in the plane of the long wire a distance b from it.
(a) Compute the mutual inductance between the long wire and the loop.
(b) At t = 0 the switch S is closed. Neglecting the self-inductance of the loop and the connections of the long wire circuit, write the two differential equations which specify the voltage conditions in the two circuits.
(c) By differentiation obtain the equation for I_1, and show that the time constant for the current I_1 is

$$(\mu_0 b/2\pi R) \log 2.$$

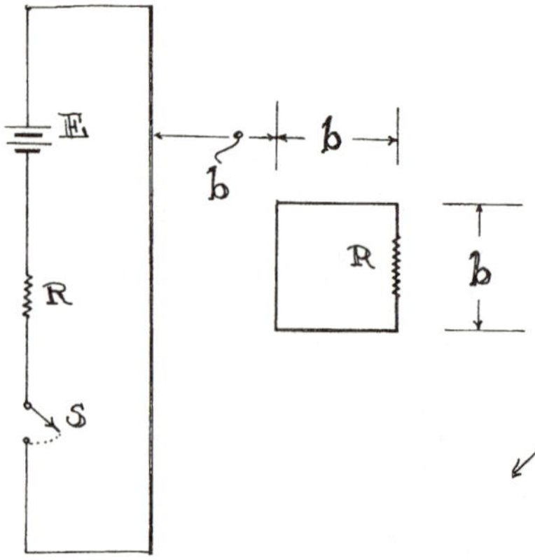

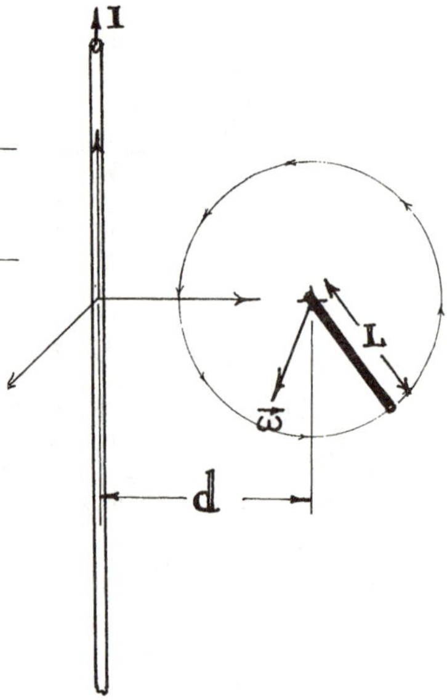

9. A long straight wire carries a current of I amperes above right. A conducting straight rod of length L is rotated about one of its ends with an angular velocity ω. The long straight wire is in the plane of rotation of the rod and the axis of rotation is d meters from the wire ($d > L$).

(a) Find the emf developed between the ends of the rod.
(b) What is the average emf developed over one cycle?

10. A long straight wire lies on the central axis of a toroidal coil having N turns and rectangular cross section. Using the dimensions shown, compute the mutual inductance between the coil and the wire.

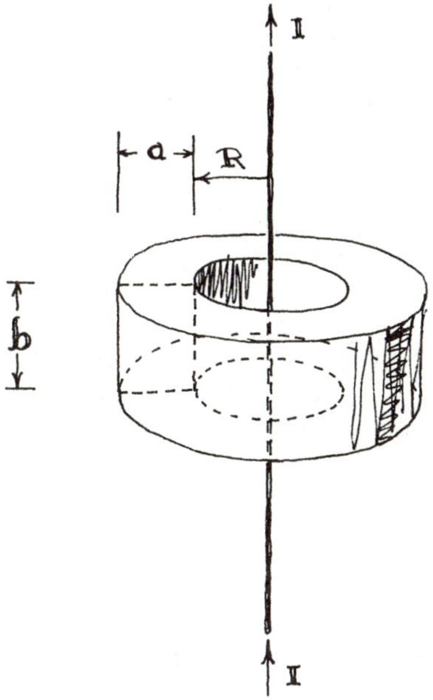

11. A long straight wire lies on the central axis of the toroidal coil of Problem 10. The toroid has 1000 turns of wire and a total resistance of 10 ohms. If the long wire carries an alternating current of 10 sin 377t amperes, what is the average power dissipated in the toroid?
12. A slender metallic rod of length l rotates about a perpendicular axis through one of its ends, with an angular velocity $\boldsymbol{\omega}$. If a uniform magnetic field $\mathbf{B}$ is parallel to the vector $\boldsymbol{\omega}$, find the emf developed across the length of the rod.
13. A rectangular coil of N turns and area A rotates with an angular velocity $\boldsymbol{\omega}$ about an axis which is in the plane of the coil and perpendicular to a magnetic field $\mathbf{B}$. The axis of rotation bisects two opposite sides. The emf developed in the coil is delivered to a load resistance.

    (a) Show that the torque necessary to turn the coil is

$$\tau = -(\mathrm{IBAN} \sin \omega t)\frac{\boldsymbol{\omega}}{\omega}$$

    (b) Hence show that the power developed in the load is equal to the power delivered in turning the coil (assume the coil resistance to be zero).
14. Consider two concentric coplanar coils having N_1 and N_2 turns respectively,

and radii R_1 and R_2. If $R_1 \gg R_2$, so that the field of coil 1 is approximately uniform over coil 2, show that the mutual inductance is

$$L_{12} \simeq \mu_0 \pi R_2^2/2R_1.$$

15. Prove that if two coils of self-inductances L_{11} and L_{22} with orientation resulting in a mutual inductance L_{12} are connected in series without changing their geometry, the measured inductance of the system is $L_{11} + L_{12} \pm 2L_{12}$.
16. A solid copper disk is constrained to rotate about its axis of symmetry. A current of 10 amperes is admitted at a point on the periphery and removed at the center along the axis of symmetry. The disk has a radius of 10 cm

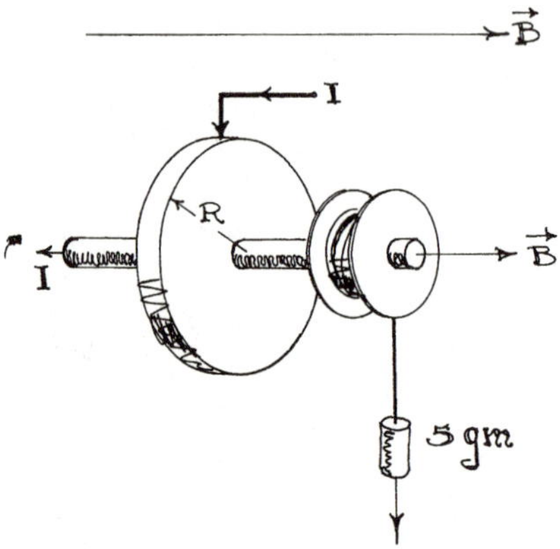

and is situated in a constant magnetic field B = 0.1 weber/m² directed along the axis. A pulley having a radius of 5 cm is connected to the disk by an axle. A mass of 5 g is hung over the pulley in a clockwise fashion. Does the 5 g mass move up or down? Explain your answer.
17. A copper disk 10 cm in radius rotates inside a long solenoid of radius 11 cm,

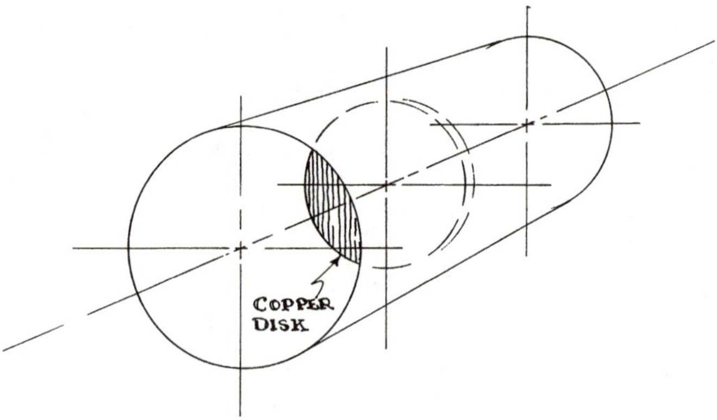

the axis of the disk lying along the axis of the solenoid. The solenoid is 1 meter long, has 1000 turns, and carries a current of 8 amperes per turn. If the disk rotates at 1200 rpm, calculate the induced emf between the axis and the rim of the disk.

18. Two long straight parallel wires of radius δ form a transmission line. If the wires are $2l$ meters apart, compute the self-inductance per unit length of the system.
19. A spherical conducting shell of radius R rotates about a diameter in a constant magnetic field **B** with a constant angular velocity $\boldsymbol{\omega}$ in the direction of **B**. Find the emf between the point a on the axis and a point b on the equatorial circle.
20. A solid copper disk is free to rotate about its axis of symmetry. By means of brushes it is connected to a source of constant emf as shown. A

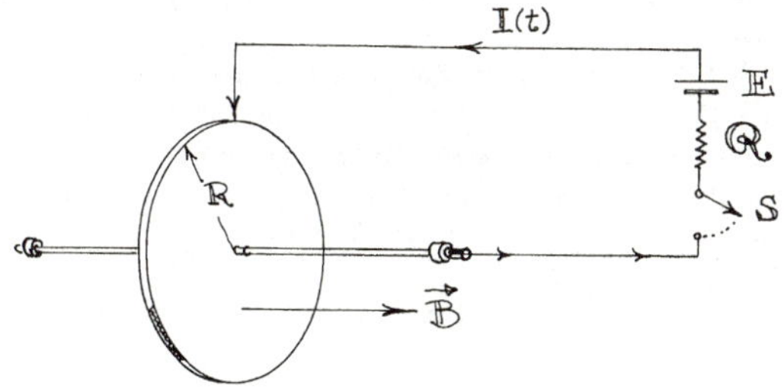

constant, axially directed magnetic field is superposed. At t = 0 the switch S is closed and the wheel rotates.

(a) Show that the magnitude of the torque at any time t is

$$|\tau| = I(t)BR^2/2.$$

(b) If $\mathscr{J}$ is the moment of inertia of the wheel about the axis, show that the equation of motion is

$$\left(\mathscr{R}\frac{2\mathscr{J}}{BR^2}\right)\frac{d\omega}{dt} + \left(\frac{BR^2}{2}\right)\omega = E,$$

where $\mathscr{R}$ = the resistance of the total circuit. Solve for $\omega(t)$.

(c) If $\omega = 0$ at $t = 0$, show that ω approaches a limiting value of $2E/BR^2$.

21. An electron at rest is one meter from the axis of an infinitely long solenoid at time t = 0. The solenoid has a radius 10^{-1} m and carries 10^4 turns/m. If the current varies in time as

$$I(t) = I_0 \sin \omega t,$$

where $I_0 = 1$ ampere and $\omega = 10^6$ rad/sec. In the interval near t = 0, find the force exerted upon the electron at t = 0. (*Hint:* Although the magnetic field is zero, the vector potential is not, and $\mathscr{E} = -\partial\mathbf{A}/\partial t$.)

22. An electron is injected into a region containing a homogeneous, unidirectional magnetic field **B**(t). The magnetic field varies sinesoidally. Assuming the

injection velocity of the electron to be essentially zero, discuss characteristics of the magnetic field **B**(t) which would produce a stable orbit for the electron.

CHAPTER IX

1. A permanently magnetized cylinder of 10 cm radius is known to have a homogeneous, isotropically distributed magnetic moment $\mathbf{m}_v = (10^7/4\pi)\,\mathbf{k}$ amp/m. A spherical hole of radius 1 cm is cut at the center of the cylinder. Find the ratio of the magnitude of the magnetic induction field at the center of the hole to the magnitude of the axial **B** field in the cylinder proper.
2. An iron cylinder of radius R and length l has a relative permeability μ. The iron has a permanent magnetization of $\mathbf{m}_v$, where $\mathbf{m}_v = m_v\,\mathbf{k}$ ($\mathbf{k}$ = symmetry axis).
   (a) What is the magnetic induction field **B** at the center of the cylinder?
   (b) What is the field at either end, on the symmetry axis?
3. A toroidal coil is wound on an iron core. The core has a relative permeability of 800. If the winding has a resistance of 2 ohms, compute the time constant of the coil. The time constant is given by the inductance times the resistance, LR.
4. A long straight copper wire 1 cm in diameter is surrounded coaxially by a

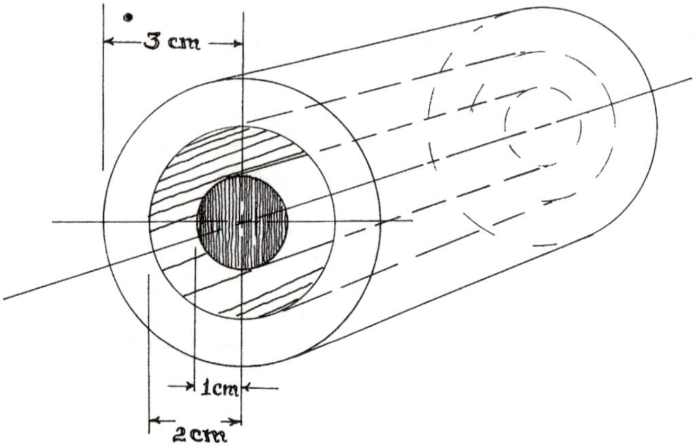

long hollow iron cylinder of $\mu = 1000$, inner radius 2 cm, and outer radius 3 cm. The wire carries a steady current of 20 amperes.
(a) Compute the total flux Φ per meter in the iron.
(b) The amperian currents induced on the iron cylinder surfaces are parallel or antiparallel to the current in the wire, and uniformly distributed over the

surfaces. Find the magnitudes and directions of the amperian currents on the outer and inner surfaces of the cylinder.

(c) Compute the intensity of magnetization $\mathbf{m}_v$ at a point 2.5 cm from the axis of the wire.

(d) Prove that the amperian current density inside the iron is zero.

(e) Show that the magnetic field outside the iron is the same as if the iron were absent.

5. The diamagnetic susceptibility of a monatomic gas is $\chi_m = -10^{-8}$. If the number of electrons per atom is 10 and the average radius of the orbits is 10^{-10} m, find the number of atoms per cubic meter.
6. A hollow glass toroid of rectangular cross section is wrapped with 1000 turns of wire, and a steady current of 10 amperes is maintained in the windings. The core (hollow section) is filled with a paramagnetic gas at S.T.P. At this temperature the paramagnetic susceptibility is 10^{-5}. Very slowly at constant volume the temperature is lowered to -100°C. Discuss the change in the energy stored in this system.
7. The hysteresis of a magnetic material is specified as

$$B(NI) = B\{1 - e^{-\alpha N|I|}\}$$

where $\alpha = 10^{-3}$ ampere turns per meter and $N =$ the turns per meter.

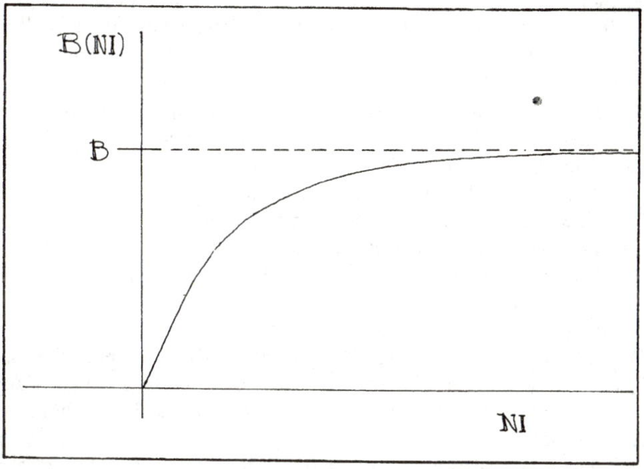

(a) Find the effective permeability for very weak fields.

(b) A toroidal core is formed of this material and a radial gap subtending $\pi/12$ radians at the center is cut in the toroid. The cross section is circular and of radius 0.01 m while the radius of the toroid is 0.25 m. 1000 turns of wire are wound about the core and a current of 1 ampere is passed through the winding. Find the magnetic moment per unit volume in the core.

8. Consider again the core, gap, and primary windings described in Problem 7.

Assume that now 10 turns of secondary winding are wrapped about the core. If the current in the primary increases linearly with time, what is the induced emf in the secondary?

9. A soft iron hoop has a circular cross section πr^2 and a mean radius R ($r \ll R$). A cylindrical bar of radius r lies along a diameter, joining the hoop at each end. Ten turns of wire are wound about the cylinder and 10 turns are wound in the same sense about a portion of one of halves of the hoop. Now suppose that $R \gg r$ and a gap of length $d \ll R$ is cut into the hoop. The relative permeability of the iron is μ.
(a) If $R = 0.5$ m, $r = 0.01$ m, $d = 0.001$ m, and $\mu = 10^3$, compute the field in the various sections of the iron when the current passing through each set of coils is 10 amperes.
(b) Does it make a difference whether you have taken the gap on arm of the hoop or on the crossbar? Why?
10. A permanently magnetized, solid sphere of radius R_1 contains a uniform magnetic moment per unit volume $\mathbf{m}_v = m_v\,\mathbf{k}$. The sphere is surrounded by a concentric spherical shell of magnetic material, say soft iron, and the outer radius of the shell is R_2. Find the **B** and **A** fields in all regions. Assume that the soft iron is homogeneous and isotropic, having a relative permeability μ_2.
11. A cylindrical Alnico magnet 2 cm in diameter is permanently and uniformly magnetized in the axial direction up to a field of 15 webers/m^2. The cylinder is 10 cm long. Without destroying the magnetization, the cylinder is cut in half. If the two halves lie on the same axis with the faces of the cut separated by 10^{-3} m, what is the force between them?
12. A cylinder of magnetic material is permanently magnetized to a uniform axial magnetization inside the cylinder. If the internal field is 0.5 weber/m^2, the radius 0.02 m, and the length 0.5 m, compute
(a) The $\mathbf{m}_v$ and **H** vectors inside the cylinder,
(b) The pole strength σ_m at the ends of the cylinder,
(c) The magnetic moment of the entire cylinder,
(d) The torque exerted when the cylinder is placed in a uniform external field of 10^{-2} weber/m^2.
13. A cylinder of magnetic material of relative permeability μ carries a total DC current I directed along the symmetry axis. Compute the magnetic field at every point on a cross section and the current density as a function of the distance from the axis. Assume that the medium is homogeneous and isotropic.
14. A long cylindrical solenoid of N turns, length l, and radius R carries a current I. A cylindrical iron core of radius a is inserted concentric with the solenoid.
(a) Neglecting end effects, find the self-inductance.
(b) Assume that a secondary circular coil is wound about the central portion of the solenoid. What is the mutual inductance?
15. (a) If in part (a) of Problem 14 the cylindrical core is withdrawn until a length x remains inside the solenoid, compute the self-inductance.
(b) Assuming that a steady current I flows in the coils of the solenoid,

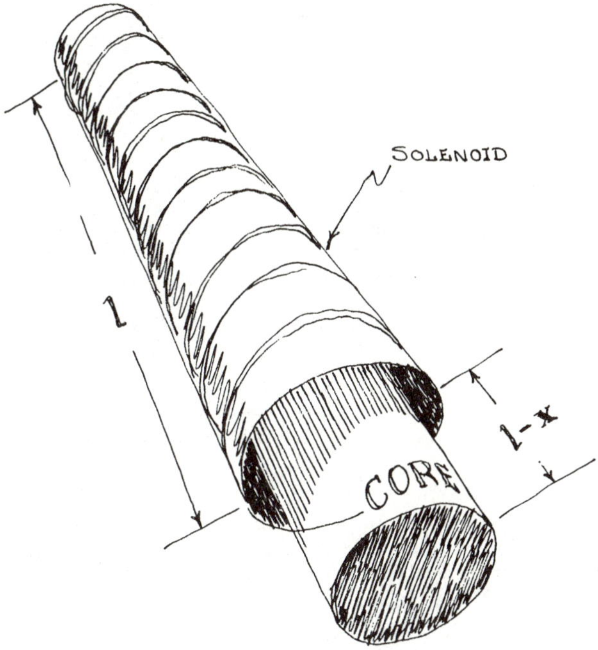

compute the approximate restoring force on the iron core. To do this, assume that $a \ll R$ and that the energy change in moving the core from x to $x + dx$ is given entirely by the change in the volume containing polarized material.

16. (a) A spherical volume of radius R contains an isotropic magnetic medium uniformly magnetized. Compute the amperian surface currents if the magnetic moment per unit volume is $m_v \mathbf{k}$.
    (b) What is the energy stored in such a system?
17. A toroid having an interior radius of 10^{-1} meter and a square cross section $5(10)^{-2}$ meter on a side is wound with 1000 turns of wire. A gap is cut in the toroid, and the faces of the gap subtend an angle of 0.1 radian relative to the axis of symmetry. Compare the **B** and **H** fields of this system when the toroid is made of aluminum and when the toroid is made of graphite. The temperature is 20°C, and the susceptibilities are $2.2(10)^{-5}$ for aluminum and -10^{-4} for graphite.
18. The table on page 486 gives an example set of values for B vs. H in ampere turns/meter for steel. The return from $H = -180$ to $H = +200$ is accomplished by using the values of the corresponding positive H values with the B signs reversed. For example, on the return for $H = -180$, $B = -1.00$; for $H = -160$, $B = -0.992$; and for $H = 0$, $B = -0.492$. Make a graph of B vs. NI/l. Compute m_v and graph this as a function of B; also plot m_v vs. H.

Ampere turns/m	B, weber/m²	Ampere turns/m	B, weber/m²
180	1.000	−20	0.120
160	0.992	−40	−0.332
140	0.981	−60	−0.565
120	0.961	−80	−0.710
100	0.935	−100	−0.810
80	0.895	−120	−0.880
60	0.845	−140	−0.940
40	0.771	−160	−0.981
20	0.672	−180	−0.992
0	0.450	−200	−1.000

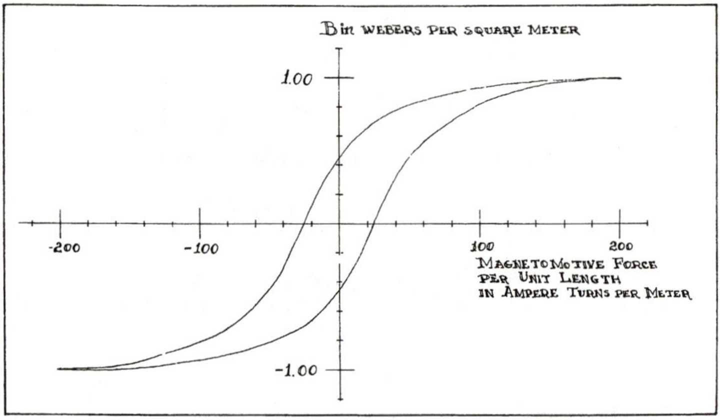

19. (a) Consider a hysteresis loop such as the one in the previous problem. Show that the work done against an induced emf is

$$\delta W = NI\,\delta\Phi.$$

Because $\oint \mathbf{H}\cdot d\mathbf{r} = NI$, show that

$$\delta W = \iiint_{\text{All Space}} \mathbf{H}\cdot(\delta\mathbf{B})\,d\tau.$$

Hence show that the energy loss per unit volume for a displacement along the hysteresis loop is

$$u_{a\to b} = \int_a^b \mathbf{H}\cdot d\mathbf{B}.$$

(b) Now demonstrate that the loss per unit volume in one cycle is

$$u = \mu_0 \oint \mathbf{H} \cdot d\mathbf{m}_v.$$

Here we have assumed a toroid made of one homogeneous material with no gaps; then $\mathbf{H}$ is a measure of the ampere turns per unit length.

20. Assume that the toroid of Problem 17 is made of the steel characterized by the hysteresis data of Problem 18. Compute the hysteresis energy loss for one complete current cycle.

21. A cylindrical Alnico magnet is permanently magnetized to provide a constant internal field of 0.5 weber/m^2. The radius of the magnet is $2(10)^{-2}$ m and the height is 10^{-1} m. A circular coil of radius $3(10)^{-2}$ m is stationary and positioned in the median plane of the cylindrical magnet with the cylinder lying along its symmetry axis. At $t = 0$ the cylinder is moved relative to the coil along the z axis such that the field linkage between magnet and coil goes as

$$\Phi(t) = \Phi_0, \qquad 0 \leqslant t \leqslant 10^{-1}\ \text{sec}$$
$$\Phi(t) = \Phi_0(10)^2(-t + 1.1(10)^{-1}), \quad 10^{-1} \leqslant t \leqslant 1.1(10)^{-1}\ \text{sec}$$
$$\Phi(t) = 0, \qquad 1.1(10)^{-1} \leqslant t.$$

Assume that the leads of the coil are connected to a resistance of 10 ohms. Compute the current in the coil plus resistance circuit.

22. Steinmetz found an empirical formula which roughly describes the energy loss per cycle in a hysteresis loop:

$$u_{cycle} = \eta(B_{max})^{1.6}$$

where B_{max} is the maximum flux density, and η is the Steinmetz coefficient.
(a) Find η for the hysteresis cycle described in Problem 18 where $B_{max} =$ 1 weber/m^2.
(b) Develop the relation giving the hysteresis power loss per unit volume when the exciting coils for this medium are subjected to a sinesoidal emf which provides a peak field of 1 weber/m^2. Compute the loss for frequencies of 60 cycles/sec and 1000 cycles/sec.

23. A primary of N_1 turns and a secondary of N_2 turns are wound upon a

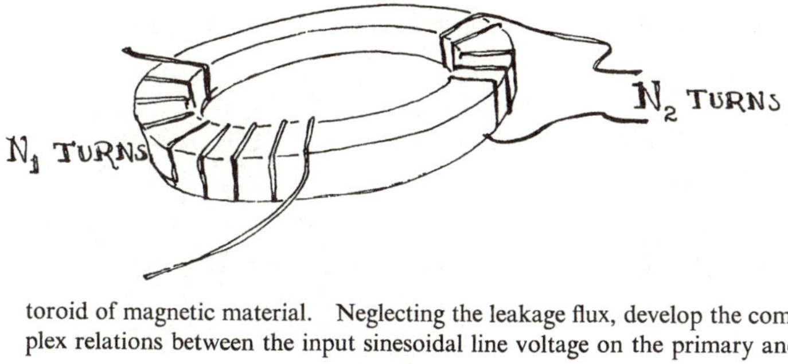

toroid of magnetic material. Neglecting the leakage flux, develop the complex relations between the input sinesoidal line voltage on the primary and

(a) The emf induced in the primary coil,
(b) The emf induced in the secondary coil,
(c) The current in the secondary when a load resistance R is connected across the terminals.
(d) Find the relation between the primary and secondary currents.

CHAPTER X

1. A 100 megacycle signal is delivered by an aluminum cable 5 cm in diameter. The conductivity is $(10)^8/2.83$ mho/m.
   (a) If the paramagnetic susceptibility is $2.3(10)^{-5}$, what is the penetration depth? The current density goes as $e^{-\alpha z}$, the penetration depth being defined as $1/\alpha$.
   (b) Does the penetration depth increase or decrease with χ_m? Explain.
   (c) Will α^{-1} increase or decrease as the temperature is increased?
2. (a) Consider an absorbing dielectric medium containing N_v electrons per unit volume. Given that an electron of mass m, when displaced a distance x from equilibrium with respect to an atom or molecule, is acted upon by a linear restoring force $-\omega_0^2 mx$, show that an electromagnetic wave $\mathscr{E} = \mathscr{E}_0 e^{i\omega t}\,\mathbf{i}$ will produce a maximum deflection of the electron

$$x_{max} = \frac{e\mathscr{E}_0/m}{\omega_0^2 - \omega^2}.$$

   (*Hint:* Solve the equation of motion $m\ddot{x} + \omega_0^2 mx = e\mathscr{E}_0 e^{i\omega t}$.)
   (b) Since an electron radiates when it accelerates, there is radiation damping. If the total radiation per second is $I(t) = \mu_0 e^2 a^2/6\pi c$ where a is the acceleration, we can assume a radiation damping force

$$F_d = -\gamma \frac{dx}{dt},$$

   and, near resonance, $$\gamma = \frac{\mu_0 \omega_0^2 e^2}{6\pi c}.$$

   Show that the maximum amplitude of oscillation of the electron is now

$$x_{max} = \frac{e\mathscr{E}_0/m}{\omega_0^2 - \omega^2 + (i\omega\gamma/m)}.$$

   (c) Show that the dipole moment per unit volume in terms of x_{max} is

$$p_v = \frac{N_v e^2 \mathscr{E}_0/m}{\omega_0^2 - \omega^2 + (i\omega\gamma/m)}.$$

   Also show that the dielectric constant for the wave is given by

$$\epsilon = 1 + \frac{N_v e^2/\epsilon_0 m}{\omega_0^2 - \omega^2 + (i\omega\gamma/m)}.$$

3\. (a) Apply Maxwell's equations to this dielectric and

$$\left\{\nabla^2 - \epsilon_0\epsilon\mu_0 \frac{\partial^2}{\partial t^2}\right\}\mathscr{E} = 0.$$

ϵ is now a complex number $\epsilon = (n - i\alpha)^2$.
Show that a solution is

$$\mathscr{E}(x,t) = \mathscr{E}_0 e^{-\alpha x/c}\, e^{i\omega(t-(nx)/c)}\, \mathbf{i},$$

where $n = c/v$ is the index of refraction and α is the absorption coefficient.
(b) Separate the real and complex parts of ϵ in part (c) of Problem 2 and show that when the frequency-dependent component is small,

$$n \simeq 1 + \frac{1}{2}\frac{(N_v e^2/m\epsilon_0)(\omega_0^2 - \omega^2)}{(\omega_0^2 - \omega^2)^2 + (\omega^2\gamma^2/m^2)},$$

and

$$\alpha \simeq \frac{1}{2}\frac{(N_v e^2/m\epsilon_0)(\omega\gamma/m)}{(\omega_0^2 - \omega^2)^2 + (\omega^2\gamma^2/m^2)}.$$

(c) Make a rough plot of n and α vs. ω. The region to the left of ω_0 is called the region of normal dispersion and shows for many dielectrics an index of refraction for visible frequencies increasing slowly with frequency. The region near ω_0 is called the region of anomalous dispersion, showing a rapid variation of n with ω accompanied by a high absorption of the radiation.

4\. A parallel plate condenser has circular plates of area 10^{-2} m^2. The plate separation is 10^{-2} m, and the gap region is filled with a dielectric $\epsilon = 3$. Connected in series with the condenser are an open switch, a 10 ohm resistor, and a 1000 volt battery. At $t = 0$ the switch is closed. Neglecting edge effects, compute
(a) The electric and magnetic field vectors as a function of time,
(b) The Poynting vector and the total energy stored at any time t,
(c) The energy stored in the polarization of the dielectric as a function of time,
(d) The dielectric displacement current.

5\. Consider again the parallel plate condenser of Problem 4, but with a vacuum gap. The gap width is d and the radius of the plates is R. A rectangular loop of wire, of width d and length R, is constructed between the plates with the plane of the loop perpendicular to the condenser plates. If the loop contains 10 turns of wire and the plane of the loop extends radially from the center to the edge, compute the induced emf in the loop as a function of time after the switch in the external circuit is closed.

6\. In Problem 4 compute the vector magnetic potential and electric potential as a function of time. Discuss the uniqueness of these potentials. Show a gauge transformation.

7\. The electric field and current density are parallel to the axis of a long cylindrical wire of conductivity β. By integrating the Poynting vector over the entrance face, show that the power dissipated in the length l is RI^2.

8\. Optically active dielectrics are characterized by a dielectric tensor which has a small imaginary component. To represent such a medium, relate **D** and $\mathscr{E}$

by a dielectric tensor $\mathfrak{E}''$,

$$\mathbf{D} = \epsilon_0 \mathfrak{E}'' \mathscr{E},$$

where $\quad \mathfrak{E}'' = \mathfrak{E} + i\gamma \quad$ or $\quad \epsilon''_{mn} = \epsilon_{mn} + i\gamma_{mn}.$

The first term $\mathfrak{E}$ is the ordinary dielectric tensor. The additional terms γ_{jk} do *not* have the ohmic dissipative character they would have for a metal. Therefore, the γ_{jk} cannot contribute to the energy density as computed from the Maxwell equations.

(a) Show that if the γ_{jk} are the components of an antisymmetric tensor, i.e., if $\gamma_{mn} = -\gamma_{nm}$, and $\gamma_{mm} = 0$, then, γ does not contribute to the energy density u.

(b) Show that this antisymmetric tensor can be replaced by a vector product operation

$$\mathbf{D} = \epsilon_0 \mathfrak{E} \cdot \mathscr{E} - \epsilon_0 i \mathbf{\Gamma} \times \mathscr{E}.$$

Obtain the relation between the components of γ and the components of the rotation vector $\mathbf{\Gamma}$.

CHAPTER XI

1. (a) Find the Poynting vector for electric quadrupole radiation.
   (b) If the angular momentum of the electromagnetic field is given by $\mathbf{r} \times \{(1/c^2)\mathrm{S}\}$, find the total angular momentum carried across the surface of a sphere of radius $\mathbf{r}$.
2. Compute the distribution of charge on the surface of a conducting sphere of radius R which carries surface currents that produce a magnetic dipole type of radiation field.
3. (a) Derive the electric and magnetic components of an electric quadrupole radiation field.
   (b) Compute the associated charge distribution on the surface of a sphere of radius R.
4. Consider two long straight parallel wires each having a radius δ. The set of parallel wires is driven by a generator at a location far removed from the point of examination, at a location $x \to -N$, assuming that the wires are parallel to the x axis. There is some appropriate termination connecting the two wires at the opposite end, at $x \to +N$.

   We assume a general case in which the transmission line has a parallel capacitance per unit length C′, a series inductance per unit length L′, a series resistance per unit length R′, and a parallel shunting resistance per unit length $\dfrac{1}{G}$. These are parameters per unit length of two parallel conductors.

   A section of the line between x and $x + \Delta x$ is now considered. The parallel parameters $C'\,\Delta x$ and $G\,\Delta x$ are concentrated at the center of the section. The series parameters are split in half and distributed symmetrically relative to the midpoint, as shown in the diagram.

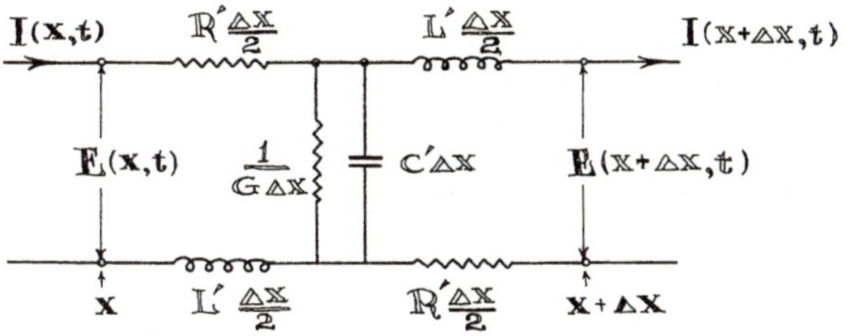

(a) Show that the loop voltage equation for the section Δx gives

$$R'I + L'\frac{\partial I}{\partial t} = -\frac{\partial E}{\partial x}. \tag{1}$$

(*Hint:* Start with

$$E(x,t) = R'\frac{\Delta x}{2} I(x,t) + L'\frac{\Delta x}{2}\frac{\partial I}{\partial t}(x,t) + E(x+\Delta x,t)$$

$$+ R'\frac{\Delta x}{2} I(x+\Delta x,t) + L'\frac{\Delta x}{2}\frac{\partial I}{\partial t}(x+\Delta x,t).$$

Rearrange; divide by Δx; expand $I(x + \Delta x,t)$ in a power series in Δx; and take the limit as $\Delta x \to 0$.)

(b) Show that the conservation of current at the connection to the parallel branch gives

$$GE + C'\frac{\partial E}{\partial t} = -\frac{\partial I}{\partial x}. \tag{2}$$

(*Hint:* The current through the parallel branch is $EG\,\Delta x + C'\,\Delta x\,\frac{\partial E}{\partial t}$. Show this.)

(c) By differentiation of equations (1) and (2) show that

$$\frac{\partial^2 E}{\partial x^2} = L'C'\frac{\partial^2 E}{\partial t^2} + \{R'C' + L'G\}\frac{\partial E}{\partial t} + R'GE,$$

and $$\frac{\partial^2 I}{\partial x^2} = L'C'\frac{\partial^2 I}{\partial t^2} + \{R'C' + L'G\}\frac{\partial I}{\partial t} + R'GI.$$

(d) The equations for a lossless line are obtained when $R' = G = 0$. Using a radius δ for the wires and a separation of $2l$ meters, obtain the expression for the velocity of propagation in terms of δ, l, ϵ_0 and μ_0. Discuss the solutions $E(x,t)$ and $I(x,t)$.

(e) Under the condition that $\frac{R'}{L'} = \frac{G}{C'}$ the system is called a distortionless line.

Show that

$$E(x,t) = e^{-act}v(x,t)$$

is a solution to the equation for the distortionless line. Find values for a which give the defining equation for v as

$$\frac{\partial^2 v}{\partial x^2} = \frac{1}{c^2}\frac{\partial^2 v}{\partial t^2}.$$

(f) The leakage-free, noninductive cable results when $G = 0$ and $L' = 0$. Show that under these conditions

$$\frac{\partial^2 E}{\partial x^2} = R'C'\frac{\partial E}{\partial t}.$$

Explain why such a cable must be artificially loaαed with series inductances when used for long-distance transmission.

5. A long solenoid of length L and radius a has a total of N turns. The solenoid is connected to a battery E and a switch S. Assume that the total resistance of the N turns plus the external circuit is R. At $t = 0$ the switch S is closed, connecting the battery E to the solenoid.
(a) Find the magnitude of the magnetic induction field **B** in the solenoid as a function of time.
(b) Find the electric field $\mathscr{E}$ generated by **B** inside the solenoid. Show that

$$\oint \mathscr{E} \cdot d\mathbf{r},$$

evaluated at the inner surface of the solenoid ($r = a$) gives the back emf.
(c) Show that the total energy transported during the transient period is $\frac{1}{2}LI^2$ where I is the final steady state current E/R; that is, compute

$$\frac{\partial U}{\partial t} = \iint_{\substack{\text{Inner}\\ \text{Surface}}} \mathbf{S} \cdot \mathbf{n}\, dS$$

where **S** is the Poynting vector, and show that

$$U = \int_0^\infty \frac{\partial U}{\partial t}\, dt = \tfrac{1}{2}LI^2.$$

6. A plane electromagnetic wave in free space is given by

$$\mathscr{E}(\mathbf{r},t) = \frac{\mathscr{E}_0}{\sqrt{2}}(\boldsymbol{\epsilon}_1 + \boldsymbol{\epsilon}_2)e^{ikz}e^{-i\omega t}$$

where $\mathscr{E}_0 = 10^{-5}$ volt/m, $\omega = 10^7$ rad/sec

and $\boldsymbol{\epsilon}_j$ is a unit base vector.
(a) Calculate the amplitude of the magnetic field vector.
(b) Compute the Poynting vector for this wave.

7. (a) Consider an electromagnetic wave of $\omega = 10^{10}$ rad/sec incident from a vacuum upon a conducting plane. If the relative permeability μ_2 of the conductor is 1 and the conductivity β_2 is $6.5(10)^7$ mhos/m, find the reflection coefficient for normal incidence.
(b) Now let this wave be incident on the plane conductor at an angle of 45°.

Find the coefficient of reflection for polarizations parallel to and perpendicular to the plane of incidence.

(c) What is the coefficient of reflection for a randomly polarized wave incident at $\theta = 45°$?

8. Consider a plane polarized wave incident upon a plane boundary between two homogeneous isotropic dielectrics ϵ_1 and ϵ_2. Show that the ratio of the reflected to incident amplitudes, when $k < k'$, is

$$\frac{F_1''}{F_1} = \frac{\tan\theta' - \tan\theta}{\tan\theta' + \tan\theta}, \qquad \frac{F_\perp''}{F_\perp} = -\frac{\tan(\theta - \theta')}{\tan(\theta + \theta')}.$$

(a) Using Snell's law, show that $F_\perp''$ is zero when the angle of incidence θ_P is given by

$$\tan\theta_P = k'/k.$$

This is called Brewster's law, and θ_P is the angle of incidence which produces a plane polarized reflected wave for which the polarization is perpendicular to the plane of incidence. Notice that there is also no reflected $F_\perp''$ in the trivial case $\epsilon_1 = \epsilon_2$.

(b) Demonstrate that when the incident wave is randomly polarized, the reflected wave for an incident angle θ_P is plane polarized.

9. In the case of reflection of an electromagnetic wave at a plane boundary between two homogeneous and isotropic dielectrics ϵ_1 and ϵ_2 where $\epsilon_1 > \epsilon_2$, total reflection can occur.

(a) Find the incident angle θ_0 corresponding to the onset of total reflection. Use Snell's law.

(b) When the angle of incidence is greater than θ_0, the conservation of phase at the boundary still holds, and therefore Snell's law holds. Under these conditions the angle of refraction θ' is imaginary. Show that when

$$\sin\theta' = \left(\frac{k}{k'}\right)\sin\theta > 1,$$

$$\cos\theta' = \pm i\sqrt{\left(\frac{k^2}{k'^2}\right)\sin^2\theta - 1}.$$

(c) Show that only the positive sign for $\cos\theta'$ gives a well-behaved exponentially damped function in the region $0 \leqslant z < \infty$. (*Hint:* Examine e^{ikr} in the region below the boundary.)

(d) Show that under conditions of total reflection,

$$F_1'' = e^{-i\delta_1}F_1 \quad \text{and} \quad F_\perp'' = e^{-i\delta_\perp}F_\perp,$$

where the δ's are defined by

$$\tan\frac{\delta_1}{2} = \frac{[\sin^2\theta - (k'^2/k^2)]^{1/2}}{\cos\theta}$$

and

$$\tan\frac{\delta_\perp}{2} = \left(\frac{k^2}{k'^2}\right)\tan\frac{\delta_1}{2}.$$

The δ's are defined from the general relations between F, F′, and F″.

(e) The relative phase shift of the two amplitudes F_1'' and $F_\perp''$ of the reflected wave is δ, where

$$\delta = \delta_1 - \delta_\perp.$$

Show that in all cases in which $\delta > 0$, the reflected wave is elliptically polarized.

(f) For the two cases in which $\tan(\delta/2)$ is $\frac{1}{2}$ and 1, plot the magnitude and direction of the electric vector $\mathbf{F}''$ over one period. What are the lengths of the semimajor and semiminor axes in these two cases? What ratios ϵ_2/ϵ_1 produce these two phase shifts?

10. An electromagnetic wave incident upon a conducting surface induces currents which interact with the fields to produce a net pressure upon the surface. Consider a plane monochromatic wave normally incident upon a plane conducting surface; the wave moves along the negative z axis, and the normal to the plane is $\boldsymbol{\epsilon}_3$. Assume that the surface is a perfect conductor.
(a) Show that the combination of the incident with the reflected wave forms a standing wave.
(b) In part (a) one finds that **H** is a maximum at the boundary. Compute the induced surface current.
(c) Using the Lorentz force on an infinitesimal element of area, show that the radiation pressure P is given by

$$\mathbf{P} = -\epsilon\, |\mathscr{E}_0|^2\, \boldsymbol{\epsilon}_3$$

where $\mathscr{E}_0$ is the amplitude of the incident electric wave.

11. (a) Show that the time average of the energy density for a plane wave moving along $-\boldsymbol{\epsilon}_3$ is

$$\langle u \rangle = \tfrac{1}{4}\mathscr{R}\,\{\mathscr{E}^* \cdot \mathbf{D} + \mathbf{B}^* \cdot \mathbf{H}\},$$

and the time average of the Poynting vector is

$$\langle \mathbf{S} \rangle = -\frac{c^2}{\sqrt{\mu\epsilon}}\, \langle u \rangle\, \boldsymbol{\epsilon}_3.$$

(b) Show that when this radiation is normally incident on the perfect conducting plane of Problem 10, the radiation pressure is

$$\mathbf{P} = -2\langle u \rangle\, \boldsymbol{\epsilon}_3 = \frac{2\sqrt{\mu\epsilon}}{c^2}\, \langle \mathbf{S} \rangle$$

12. Radiation pressure can be interpreted in terms of an impulse set up by a change in the momentum of an impinging field. Using the analogy of fluid flow, the momentum in a cylinder of cross-sectional area dA and length (c dt) is

$$dp = \frac{dp}{d\tau}\,(c\,dt)\,dA$$

where $dp/d\tau$ is the momentum density (momentum per unit volume).
(a) Show that the pressure is given by

$$\mathbf{P} = \frac{d\mathbf{F}}{dA} = 2c\,\frac{d\mathbf{p}}{d\tau},$$

and from this (and Problem 11) show that

$$\frac{dp}{d\tau} = \frac{1}{2c^2} \mathscr{R}\langle \mathbf{S} \rangle.$$

13. A circular cylinder 1 meter in diameter and 100 meters long is conducting over half its surface area, that is, over a semicylindrical portion 100 meters long. If the conducting area radiates uniformly 10 kilowatts at 10 megacycles, compute the total force on the cylinder.
14. A 1000 megacycle transmitter sends out pulses 1 μ sec long at a power level of one megawatt. Compute the momentum of one pulse. Compare this momentum to that of an electron traveling with an energy of 1 Mev.

CHAPTER XII

1. The circuit shown in the diagram pertains to the conditions existing before t = 0. The condenser initially carries a voltage equal to that of the battery. The switch S is closed at t = 0.

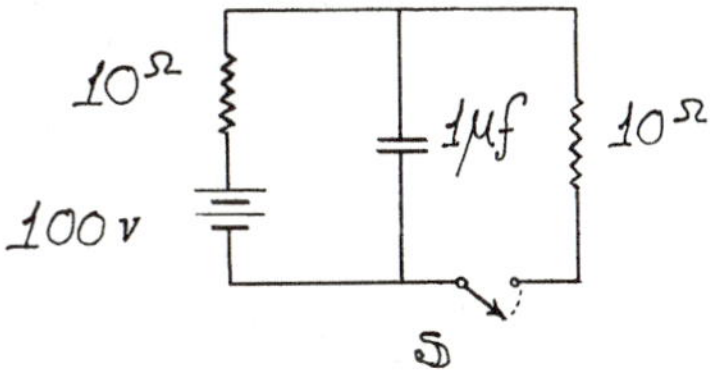

(a) What is the voltage on the condenser a long time after S has been closed?
(b) What is the voltage across the condenser 10^{-5} sec after S has been closed?

2. The circuit shown has been connected for a considerable length of time. If the switch S is closed when t = 0, find the currents and charges during the transient periods for both circuits. What are the time constants?

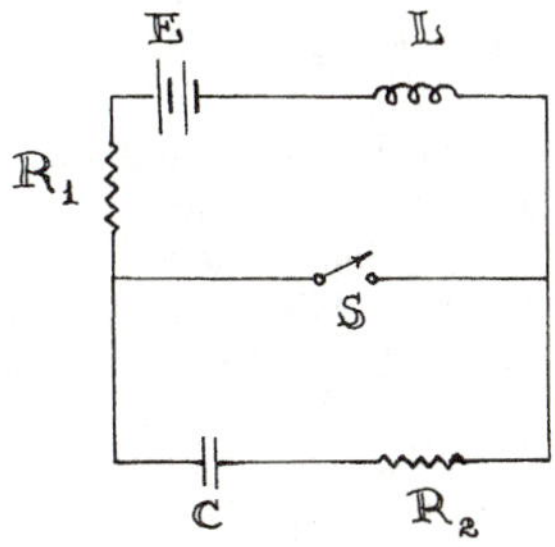

3. Find the total energy dissipated in the resistor during the overdamped discharge in an undriven RLC series circuit if there is no electromotance and the initial charge is Q_0.
4. Show that if a condenser C discharges through a critically damped LRC series circuit, the current is a maximum 2L/R sec after shorting. What is the maximum transient power delivered to the resistor?
5. A battery of potential V is connected to a series circuit made up of an uncharged condenser C and inductance L with resistance R, at a time t = 0. Find the current as a function of time. Show that the maximum potential difference that appears across the condenser is $V(1 + e^{-\delta/2})$, where δ is the logarithmic decrement of the circuit.
6. In the AC current shown,

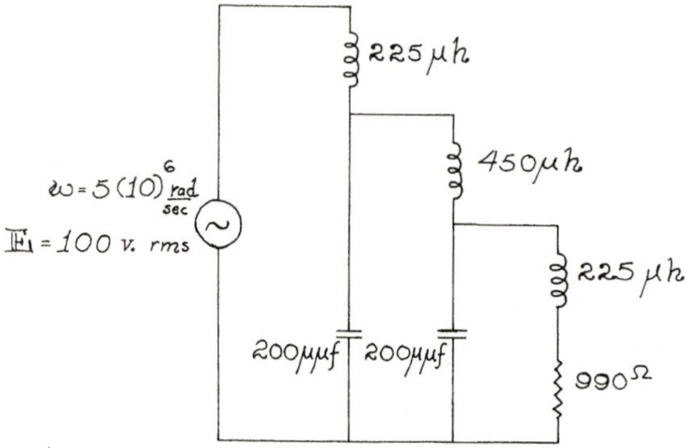

(a) What current is drawn from the generator?
(b) What power is delivered to the circuit?

7. A rectangular galvanometer coil 1 cm wide and 2 cm high has 1000 turns. For any angular orientation of the coil the external magnetic field is parallel to the plane of the coil. The field is essentially constant and has a magnitude of 0.2 weber/m² for reasonable deflections. The suspension has a torque constant of $4(10)^{-10}$ newton-m/rad. The galvanometer coil has a damping constant of R newton-m-sec and a moment of inertia of $\frac{1}{4}(10)^{-6}$ kg-m². At a time t = 0 a current of 1 microampere is suddenly passed through the coil and it continues at this level.
(a) If at t = 0, $\theta = 0$, what is the final deflection of the galvanometer in radians?
(b) What should be the value of R for critical damping?
8. A capacitance C and an inductance L with internal resistance R are connected in parallel. Find the angular frequency for which the impedance of the combination is a maximum. What is the impedance at this frequency?
9. Show that the condition for the resonance of a parallel LC circuit with variable inductance is

$$L^2 = \frac{Q^2}{\omega^2 C(1 + Q^2)},$$

assuming the resistance of the coil to increase with the inductance in such a way that the Q remains constant. Show that the condition for maximum impedance under these conditions is $L = 1/\omega^2 C$. Find the impedance presented by the circuit under these two conditions.

10. A 60-cycle AC series circuit contains a resistance of 2 ohms and an inductance of 10 mh. What is the power factor? What capacitance placed in the circuit will make the power factor unity? Find the ratio of the currents for the two conditions.
11. A 10 μh inductance is connected in series with a capacity of 100 $\mu\mu$f. Across this combination in parallel is another inductance of 10 μh. What are the frequencies of maximum and minimum impedance?
12. Three impedances Z_1, Z_2, and Z_3 are connected in parallel across a 60-cycle emf of 40 volts rms. $Z_1 = 10$, $Z_2 = 20 + i20$, $Z_3 = 3 - i40$ ohms.
    (a) Find the admittance, conductance, and susceptance of each branch.
    (b) Find the resultant conductance and susceptance of the three parallel branches.
    (c) Find the current in each branch, and the resultant current.
    (d) Draw the vector diagram of the circuit.
13. In the circuit shown, calculate the current in a load impedance Z_L =

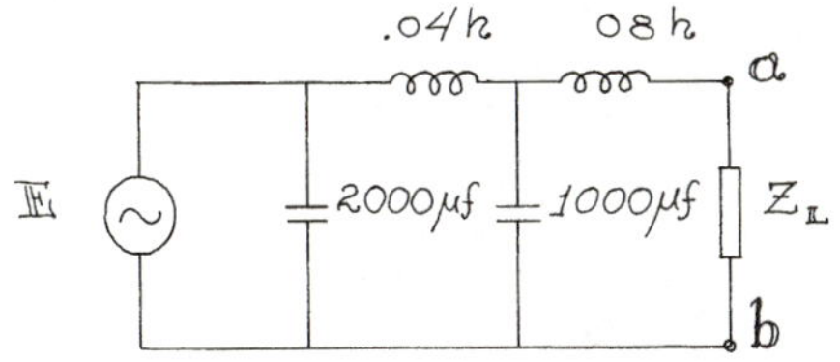

$10 - i7.5$ ohms connected across the terminals a, b. Assume the generator frequency to be 60 cycles.

14. Find the transient and steady-state currents in a series LRC circuit which is energized at time $t = 0$ with an alternating voltage $E_m \sin(\omega t + \theta)$. Assume the initial charge on the capacitor to be Q_0.
15. In the circuit shown, the rms terminal voltage of the AC generator is 100

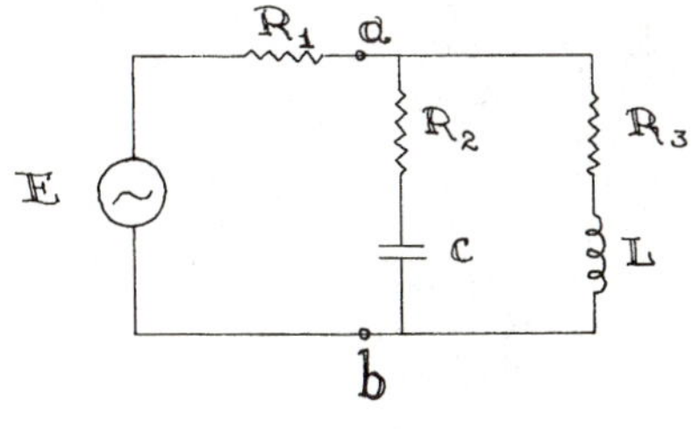

volts, f = 60 cycles/sec, C = 20 μf, L = 0.25 h, and $R_1 = R_2 = R_3 = 10$ ohms.

(a) Find the equivalent impedance between points a and b.
(b) Find the current in each resistor.
(c) Construct the vector diagram of the circuit.

16. Find the natural frequencies, time constants, and transient current and

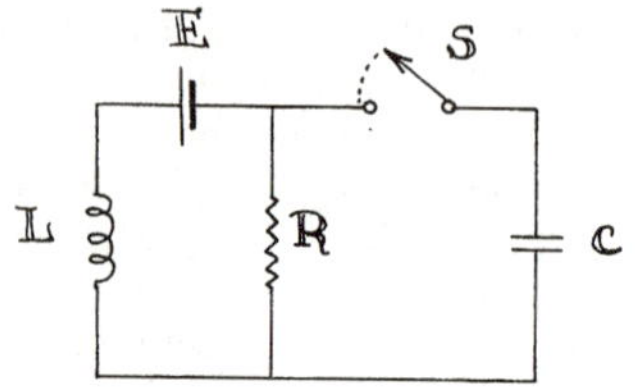

charge in the circuit shown. Assume the switch S to be closed at t = 0.

17. Show that if RC $\ll$ 1, then

$$e_0 = RC\frac{de_i}{dt}$$

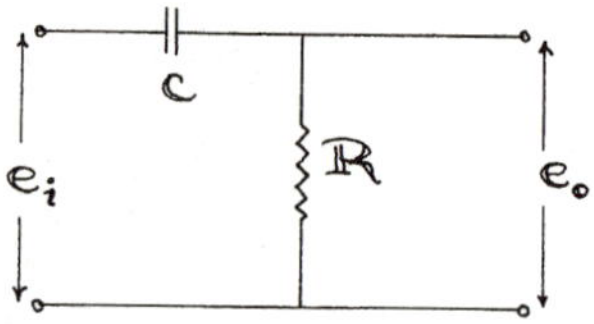

where e_i is the input signal in the circuit, and e_0 is the output signal.

18. Consider the circuit shown.

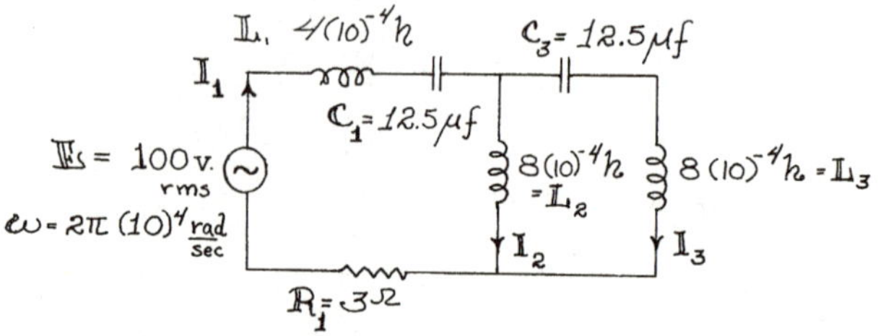

(a) Compute the current in L_1.
(b) Compute the current in L_3.
(c) Is the current in L_2 leading or lagging E?
(d) Compute the phase angles between E and I_1, E and I_2, and E and I_3.

Index

A CATALOG OF SELECTED

DOVER BOOKS

IN SCIENCE AND MATHEMATICS

A CATALOG OF SELECTED

DOVER BOOKS

IN SCIENCE AND MATHEMATICS

Astronomy

BURNHAM'S CELESTIAL HANDBOOK, Robert Burnham, Jr. Thorough guide to the stars beyond our solar system. Exhaustive treatment. Alphabetical by constellation: Andromeda to Cetus in Vol. 1; Chamaeleon to Orion in Vol. 2; and Pavo to Vulpecula in Vol. 3. Hundreds of illustrations. Index in Vol. 3. 2,000pp. 6⅛ x 9¼.
23567-X, 23568-8, 23673-0 Three-vol. set

THE EXTRATERRESTRIAL LIFE DEBATE, 1750–1900, Michael J. Crowe. First detailed, scholarly study in English of the many ideas that developed from 1750 to 1900 regarding the existence of intelligent extraterrestrial life. Examines ideas of Kant, Herschel, Voltaire, Percival Lowell, many other scientists and thinkers. 16 illustrations. 704pp. 5⅜ x 8½. 40675-X

A HISTORY OF ASTRONOMY, A. Pannekoek. Well-balanced, carefully reasoned study covers such topics as Ptolemaic theory, work of Copernicus, Kepler, Newton, Eddington's work on stars, much more. Illustrated. References. 521pp. 5⅜ x 8½.
65994-1

AMATEUR ASTRONOMER'S HANDBOOK, J. B. Sidgwick. Timeless, comprehensive coverage of telescopes, mirrors, lenses, mountings, telescope drives, micrometers, spectroscopes, more. 189 illustrations. 576pp. 5⅝ x 8¼. (Available in U.S. only.)
24034-7

STARS AND RELATIVITY, Ya. B. Zel'dovich and I. D. Novikov. Vol. 1 of *Relativistic Astrophysics* by famed Russian scientists. General relativity, properties of matter under astrophysical conditions, stars, and stellar systems. Deep physical insights, clear presentation. 1971 edition. References. 544pp. 5⅝ x 8¼. 69424-0

Chemistry

CHEMICAL MAGIC, Leonard A. Ford. Second Edition, Revised by E. Winston Grundmeier. Over 100 unusual stunts demonstrating cold fire, dust explosions, much more. Text explains scientific principles and stresses safety precautions. 128pp. 5⅜ x 8½. 67628-5

THE DEVELOPMENT OF MODERN CHEMISTRY, Aaron J. Ihde. Authoritative history of chemistry from ancient Greek theory to 20th-century innovation. Covers major chemists and their discoveries. 209 illustrations. 14 tables. Bibliographies. Indices. Appendices. 851pp. 5⅜ x 8½. 64235-6

CATALYSIS IN CHEMISTRY AND ENZYMOLOGY, William P. Jencks. Exceptionally clear coverage of mechanisms for catalysis, forces in aqueous solution, carbonyl- and acyl-group reactions, practical kinetics, more. 864pp. 5⅜ x 8½.
65460-5

THE HISTORICAL BACKGROUND OF CHEMISTRY, Henry M. Leicester. Evolution of ideas, not individual biography. Concentrates on formulation of a coherent set of chemical laws. 260pp. 5⅜ x 8½. 61053-5

A SHORT HISTORY OF CHEMISTRY, J. R. Partington. Classic exposition explores origins of chemistry, alchemy, early medical chemistry, nature of atmosphere, theory of valency, laws and structure of atomic theory, much more. 428pp. 5⅜ x 8½. (Available in U.S. only.) 65977-1

GENERAL CHEMISTRY, Linus Pauling. Revised 3rd edition of classic first-year text by Nobel laureate. Atomic and molecular structure, quantum mechanics, statistical mechanics, thermodynamics correlated with descriptive chemistry. Problems. 992pp. 5⅜ x 8½. 65622-5

Engineering

DE RE METALLICA, Georgius Agricola. The famous Hoover translation of greatest treatise on technological chemistry, engineering, geology, mining of early modern times (1556). All 289 original woodcuts. 638pp. 6¾ x 11. 60006-8

FUNDAMENTALS OF ASTRODYNAMICS, Roger Bate et al. Modern approach developed by U.S. Air Force Academy. Designed as a first course. Problems, exercises. Numerous illustrations. 455pp. 5⅜ x 8½. 60061-0

DYNAMICS OF FLUIDS IN POROUS MEDIA, Jacob Bear. For advanced students of ground water hydrology, soil mechanics and physics, drainage and irrigation engineering and more. 335 illustrations. Exercises, with answers. 784pp. 6⅛ x 9¼. 65675-6

ANALYTICAL MECHANICS OF GEARS, Earle Buckingham. Indispensable reference for modern gear manufacture covers conjugate gear-tooth action, gear-tooth profiles of various gears, many other topics. 263 figures. 102 tables. 546pp. 5⅜ x 8½. 65712-4

MECHANICS, J. P. Den Hartog. A classic introductory text or refresher. Hundreds of applications and design problems illuminate fundamentals of trusses, loaded beams and cables, etc. 334 answered problems. 462pp. 5⅜ x 8½. 60754-2

MECHANICAL VIBRATIONS, J. P. Den Hartog. Classic textbook offers lucid explanations and illustrative models, applying theories of vibrations to a variety of practical industrial engineering problems. Numerous figures. 233 problems, solutions. Appendix. Index. Preface. 436pp. 5⅜ x 8½. 64785-4

STRENGTH OF MATERIALS, J. P. Den Hartog. Full, clear treatment of basic material (tension, torsion, bending, etc.) plus advanced material on engineering methods, applications. 350 answered problems. 323pp. 5⅜ x 8½. 60755-0

A HISTORY OF MECHANICS, René Dugas. Monumental study of mechanical principles from antiquity to quantum mechanics. Contributions of ancient Greeks, Galileo, Leonardo, Kepler, Lagrange, many others. 671pp. 5⅜ x 8½. 65632-2

Physics

OPTICAL RESONANCE AND TWO-LEVEL ATOMS, L. Allen and J. H. Eberly. Clear, comprehensive introduction to basic principles behind all quantum optical resonance phenomena. 53 illustrations. Preface. Index. 256pp. 5⅜ x 8½. 65533-4

ULTRASONIC ABSORPTION: An Introduction to the Theory of Sound Absorption and Dispersion in Gases, Liquids and Solids, A. B. Bhatia. Standard reference in the field provides a clear, systematically organized introductory review of fundamental concepts for advanced graduate students, research workers. Numerous diagrams. Bibliography. 440pp. 5⅜ x 8½. 64917-2

QUANTUM THEORY, David Bohm. This advanced undergraduate-level text presents the quantum theory in terms of qualitative and imaginative concepts, followed by specific applications worked out in mathematical detail. Preface. Index. 655pp. 5⅜ x 8½. 65969-0

ATOMIC PHYSICS (8th edition), Max Born. Nobel laureate's lucid treatment of kinetic theory of gases, elementary particles, nuclear atom, wave-corpuscles, atomic structure and spectral lines, much more. Over 40 appendices, bibliography. 495pp. 5⅜ x 8½. 65984-4

AN INTRODUCTION TO HAMILTONIAN OPTICS, H. A. Buchdahl. Detailed account of the Hamiltonian treatment of aberration theory in geometrical optics. Many classes of optical systems defined in terms of the symmetries they possess. Problems with detailed solutions. 1970 edition. xv + 360pp. 5⅜ x 8½. 67597-1

THIRTY YEARS THAT SHOOK PHYSICS: The Story of Quantum Theory, George Gamow. Lucid, accessible introduction to influential theory of energy and matter. Careful explanations of Dirac's anti-particles, Bohr's model of the atom, much more. 12 plates. Numerous drawings. 240pp. 5⅜ x 8½. 24895-X

ELECTRONIC STRUCTURE AND THE PROPERTIES OF SOLIDS: The Physics of the Chemical Bond, Walter A. Harrison. Innovative text offers basic understanding of the electronic structure of covalent and ionic solids, simple metals, transition metals and their compounds. Problems. 1980 edition. 582pp. 6⅛ x 9¼. 66021-4

HYDRODYNAMIC AND HYDROMAGNETIC STABILITY, S. Chandrasekhar. Lucid examination of the Rayleigh-Benard problem; clear coverage of the theory of instabilities causing convection. 704pp. 5⅜ x 8¼. 64071-X

INVESTIGATIONS ON THE THEORY OF THE BROWNIAN MOVEMENT, Albert Einstein. Five papers (1905–8) investigating dynamics of Brownian motion and evolving elementary theory. Notes by R. Fürth. 122pp. 5⅜ x 8½. 60304-0

THE PHYSICS OF WAVES, William C. Elmore and Mark A. Heald. Unique overview of classical wave theory. Acoustics, optics, electromagnetic radiation, more. Ideal as classroom text or for self-study. Problems. 477pp. 5⅜ x 8½. 64926-1

METHODS OF THERMODYNAMICS, Howard Reiss. Outstanding text focuses on physical technique of thermodynamics, typical problem areas of understanding, and significance and use of thermodynamic potential. 1965 edition. 238pp. 5⅜ x 8½. 69445-3

TENSOR ANALYSIS FOR PHYSICISTS, J. A. Schouten. Concise exposition of the mathematical basis of tensor analysis, integrated with well-chosen physical examples of the theory. Exercises. Index. Bibliography. 289pp. 5⅜ x 8½. 65582-2

RELATIVITY IN ILLUSTRATIONS, Jacob T. Schwartz. Clear nontechnical treatment makes relativity more accessible than ever before. Over 60 drawings illustrate concepts more clearly than text alone. Only high school geometry needed. Bibliography. 128pp. 6⅛ x 9¼. 25965-X

THE ELECTROMAGNETIC FIELD, Albert Shadowitz. Comprehensive undergraduate text covers basics of electric and magnetic fields, builds up to electromagnetic theory. Also related topics, including relativity. Over 900 problems. 768pp. 5⅝ x 8¼. 65660-8

GREAT EXPERIMENTS IN PHYSICS: Firsthand Accounts from Galileo to Einstein, edited by Morris H. Shamos. 25 crucial discoveries: Newton's laws of motion, Chadwick's study of the neutron, Hertz on electromagnetic waves, more. Original accounts clearly annotated. 370pp. 5⅜ x 8½. 25346-5

RELATIVITY, THERMODYNAMICS AND COSMOLOGY, Richard C. Tolman. Landmark study extends thermodynamics to special, general relativity; also applications of relativistic mechanics, thermodynamics to cosmological models. 501pp. 5⅜ x 8½. 65383-8

LIGHT SCATTERING BY SMALL PARTICLES, H. C. van de Hulst. Comprehensive treatment including full range of useful approximation methods for researchers in chemistry, meteorology and astronomy. 44 illustrations. 470pp. 5⅜ x 8½. 64228-3

STATISTICAL PHYSICS, Gregory H. Wannier. Classic text combines thermodynamics, statistical mechanics and kinetic theory in one unified presentation of thermal physics. Problems with solutions. Bibliography. 532pp. 5⅜ x 8½. 65401-X